View from just east of the summit of Sgorr Dhonuill, looking north. In the foreground are prominent cliffs of resistant porphyritic granite, occupying the centre of the Ballachulish Igneous Complex. Below the cliffs are the forested slopes of Gleann a' Chaolais, underlain by diorite of the igneous complex. Across Loch Leven, spanned by the Ballachulish Bridge, lie North Ballachulish and the hills of Nether Lorn, underlain by Dalradian metasediments. The small tidal island forming a promontory on the west side of the small anchorage east of North Ballachulish contains incipiently cordierite-spotted phyllites of the Leven Schist, and marks the outer margin of the thermal aureole in this area

G. Voll, J. Töpel, D.R.M. Pattison, F. Seifert (Eds.)

Equilibrium and Kinetics in Contact Metamorphism

The Ballachulish Igneous Complex and Its Aureole

With 206 Figures and 37 Tables

Springer-Verlag Berlin Heidelberg GmbH

Professor Dr. Gerhard Voll
Mineralogisch-Petrographisches
Institut der Universität
Zülpicher Strasse 49
W-5000 Köln, FRG

Dr. Jutta Töpel
Bayerisches Geoinstitut
Universität Bayreuth
Postfach 10 12 51
W-8550 Bayreuth, FRG

Professor Dr. David R.M. Pattison
The University of Calgary
Department of Geology and
Geophysics
2500 University Drive N.W.
Calgary, Alberta T2N 1N4
Canada

Professor Dr. Friedrich Seifert
Bayerisches Geoinstitut
Universität Bayreuth
Postfach 10 12 51
W-8550 Bayreuth, FRG

ISBN 978-3-642-76147-8 ISBN 978-3-642-76145-4 (eBook)
DOI 10.1007/978-3-642-76145-4

Library of Congress Cataloging-in-Publication Data

Equilibrium and kinetics in contact metamorphism: the Ballachulish Igneous Complex and its aureole/G. Voll... [et al.], eds. p. cm. Includes bibliographical references and index.
ISBN 978-3-642-76147-8
1. Rocks, Metamorphic – Scotland – Highlands. 2. Intrusions (Geology) – Scotland – Highlands. 3. Petrology – Scotland – Highlands. I. Voll, G. (Gerhard), 1927– . II. Title: Ballachulish Igneous Complex and its aureole. QE475.A2E68 1991 552'.4'094115 – dc20 91-12482

Softcover reprint of the hardcover 1st edition 1991

Typesetting: International Typesetters Inc., Manila, Philippines

32/3145-543210 – Printed on acid-free paper

Preface

With the new global tectonics approach in the Earth Sciences, the quantitative aspects of the dynamics of rock-forming processes came into focus: geologists are no longer satisfied knowing the pressure-temperature conditions of the formation of a metamorphic rock or of the emplacement of a magmatic body, but instead would like to learn the time history of these rocks as well, i.e., derive the temperature-pressure-time path and relate it to a tectonic process. To achieve this goal, a knowledge of both pressure-temperature-dependent equilibria and the time scales at which these equilibria may be attained are essential. However, the latter kinetic information is much more difficult to retrieve than that on equilibria: whereas equilibria are controlled by state variables, and proper laboratory experiments may be directly applied to equilibrium natural assemblages, kinetics also depends on factors other than state variables, such as grain size, dislocation density, and especially time (rate of heating, duration of annealing, rate of cooling). Extrapolation of kinetic data obtained at high temperatures on laboratory time scales to more realistic lower temperatures and geological time scales are dangerous because, for example, of possible changes from an intrinsically controlled defect regime to an extrinsic one as temperature is lowered, or from an interface-controlled to a diffusion-controlled reaction mechanism.

Progress in our understanding of the kinetics of rock-forming processes can, therefore, only be expected by a careful comparison of laboratory and field data. With the time history of the evolution of a natural rock being in general poorly known and the factors controlling the kinetics difficult to reconstruct, it is essential to use as many data and techniques as possible, but at the same time keep the number of complicating factors low.

In this book, such a multilateral approach is attempted for *one* contact aureole, that of the Ballachulish Igneous Complex in the Scottish Highlands. It has been selected because of the large variety of different rock types contained and the relatively simple cylindrical geometry. Also, because of its size, the time scale of contact metamorphism in this aureole (in contrast to that of regional metamorphism) is closer to that of laboratory studies than in regional metamorphism, and a chance might exist for a linkage of results. It is probably not an exaggeration to say that this is the most comprehensive study of a single igneous complex-contact aureole system yet published.

The work on the Ballachulish igneous body and its aureole presented here started independently in two places: at the Grant Institute of Geology of Edinburgh University, and in a priority program run by the Deutsche Forschungsgemeinschaft, entitled "Kinetics of rock- and mineral-forming processes" in which a number of groups in geophysics, petrology, mineralogy and geochemistry from different German universities participated. It was only by a combination of activities that the complex problem could be tackled.

The editors and authors would like to express their gratitude for support to the Deutsche Forschungsgemeinschaft, Bonn. The following colleagues provided in depth reviews of the manuscripts, helpful discussions or supplementary data: J.R. Ashworth (Birmingham, UK), M.H.P. Bott (Durham, UK), T. Chacko (Edmonton, Alberta, Canada), G. Droop (Manchester, UK), J. Farver (Providence, R.I., USA), C. France-Lanard (Nancy, France), E. Ghent (Calgary, Alberta, Canada), M. Gottschalk (Tübingen, FRG), C.M. Graham (Edinburgh, UK), J. Hoefs (Göttingen, FRG), H. Jaffe (Amherst, Mass., USA), W. Johannes (Hannover, FRG), P. Metz (Tübingen, FRG), W.F. Müller (Darmstadt, FRG), W.S. Pitcher (Liverpool, UK), A. Putnis (Cambridge, UK), P. Ribbe (Blacksburg, Va., USA), J. Rice (Eugene, Or., USA), D.C. Rubie (Bayreuth, FRG), D. Rumble III (Washington, D.C., USA), L. Rybach (Zürich, Switzerland), S. Sheppard (Nancy, France), W.E. Stephens (St. Andrews, Scotland), A.B. Thompson (Zürich, Switzerland), V. Trommsdorff (Zürich, Switzerland), R. Uhrenbacher (Kiel, FRG), J.W. Valley (Edinburgh, UK), G. Werding (Bochum, FRG), W. Wünch (München, FRG), R.A. Yund (Providence, R.I., USA). Technical assistance was provided by T. Baller (Bochum, FRG), P. Dalgetty (Edinburgh, UK), T. Fockenberg (Bochum, FRG), B. Kennedy, E. Löbl (Bayreuth, FRG). To all these persons and institutions our sincere thanks are due.

August 1991 The Editors

Contents

Part V Concluding Discussion

Contributors

Blümel, P.
Institut für Mineralogie, Technische Hochschule Darmstadt,
Schnittspahnstr. 9, W-6100 Darmstadt, FRG

Buntebarth, G.
Institut für Geophysik, TU Clausthal, Arnold-Sommerfeld-Str. 1,
W-3392 Clausthal-Zellerfeld, FRG

Feuer, H.
Institut für Kristallographie der Universität, Senckenberganlage 30,
W-6000 Frankfurt/Main 11, FRG

Fuess, H.
Fachgebiet Strukturforschung, FB Materialwissenschaften,
TH Darmstadt, Petersenstr. 20, W-6100 Darmstadt, FRG

Harmon, R.S.
Dept. of Geol. Sci., Southern Methodist University,
Dallas TX, 75275, USA
present address:
NERC-Scientific Services, Isotope Geology Centre, 64 Grays Inn Road,
London WX 1X 8GN, UK

Harte, B.
Grant Institute of Geology, University of Edinburgh,
West Mains Road, Edinburgh EH9 3JW, UK

Heuss-Aßbichler, S.
Institut für Mineralogie und Petrographie der Universität,
Theresienstr. 41, W-8000 München 2, FRG

Hoernes, S.
Mineralogisch-Petrologisches Institut der Universität,
Poppelsdorfer Schloß, W-5300 Bonn 1, FRG

Krause, C.
Institut für nukleare Entsorgungstechnik,
Kernforschungszentrum Karlsruhe, W-7500 Karlsruhe, FRG

Kroll, H.
Institut für Mineralogie der Universität, Corrensstr. 24,
W-4400 Münster/W., FRG

Linklater, C.M.
Grant Institute of Geology, University of Edinburgh, West Mains Road,
Edinburgh EH9 3JW, UK

MacLeod-Kinsel, S.
Dept. of Geol. Sci., Southern Methodist University,
Dallas TX, 75275, USA

Maresch, W.
Institut für Mineralogie der Universität, Corrensstr. 24,
W-4400 Münster/W., FRG

Masch, L.
Institut für Mineralogie und Petrographie der Universität,
Theresienstr. 41, W-8000 München 2, FRG

Meissner, R.
Institut für Geophysik, Universität Kiel, Olshausenstr. 40–60,
W-2300 Kiel FRG

Pattison, R.D.M.
The University of Calgary, Department of Geology and Geophysics,
2500 University Drive N.W., Calgary, Alberta T2N 1N4, Canada

Rabbel, W.
Institut für Geophysik, Universität Kiel, Olshausenstr. 40–60,
W-2300 Kiel, FRG

Schreyer, W.
Institut für Mineralogie, Ruhr- Universität Bochum,
W-4630 Bochum, FRG

Schröpfer, L.
Institut für Kristallographie der Universität, Senckenberganlage 30,
W-6000 Frankfurt/Main 11, FRG

Strong, D.F.
Dept. of Earth Sci., Memorial University, St. Johns, New Foundland, Canada

Troll, G.
Institut für Mineralogie und Petrographie der Universität, Theresienstr. 41, W-8000 München 2, FRG

Voll, G.
Mineralogisch-Petrographisches Institut der Universität, Zülpicher Str. 49, W-5000 Köln 1, FRG

Weiss, S.
Institut für Mineralogie und Petrographie der Universität, Theresienstr. 41, W-8000 München 2, FRG

Part I
Introduction

1 The Setting of the Ballachulish Intrusive Igneous Complex in the Scottish Highlands

B. Harte and G. Voll

The Ballachulish igneous complex is one of a series of relatively late granitoid complexes in the Caledonian orogenic zone and intrudes metasediments of the Dalradian Supergroup in the Scottish Highlands (Fig. 1.1). For those who are unfamiliar with the Scottish Highlands, the purpose of this chapter is to give some background to the nature of the Caledonian orogeny in Scotland, the Dalradian Supergroup and the late Caledonian granitoids. The aim, in conjunction with the following chapter on local regional geology, is to provide the reader with some background on crustal conditions and tectonic relations of intrusion, and on the broad history of development of the aureole rocks prior to contact metamorphism. Clearly these features are relevant to models of pressure-temperature-time relationships in the intrusive complex and its aureole, and to examining kinetic controls on the crystallization of the magmatic and metamorphic rocks.

1.1 Major Geological Provinces and Lineaments in Scotland

The margins of the Caledonian orogenic zone and the distribution of late Caledonian granites both encompass the major part of Scotland, and it is therefore useful to briefly outline major aspects of Scottish geology. The map and cross-section of Scotland, Figs. 1.1 and 1.2, show the major tectonostratigraphic provinces and their bounding lineaments in this geologically diverse country. Each of these features is briefly characterized below.

Northwest Foreland Province. This consists of a metamorphic basement (Lewisian) with a little-metamorphosed cover of both Proterozoic (Torridonian) and Cambro-Ordovician sediments. The Lewisian basement complex has undergone upper amphibolite and granulite facies metamorphism (Scourian event) in the Archaean, and amphibolite facies metamorphism probably at several times in the Proterozoic. The northwestern province is generally accepted as having formed part of the Laurentian continental plate, and is clearly 'foreland' to the Moine Schists of the Northern Highlands province as shown by the relationships along the Moine thrust zone.

G. Voll, J. Töpel, D.R.M. Pattison, F. Seifert
Equilibrium and Kinetics
in Contact Metamorphism
©Springer-Verlag Berlin Heidelberg 1991

Moine Thrust Zone. Along this zone the Moine Series metasediments are thrust over the Lewisian, Torridonian and Cambro-Ordovician of the Northwest Foreland. Imbrication of the foreland implies substantial (> 50 km) movement to the WNW, and the age of this thrusting is ca. 430 Ma.

Northern Highland Province. It consists dominantly of greenschist and amphibolite facies metasedimentary rocks (largely psammites) of the Moine Series. Regional metamorphism occurred most recently in the Ordovician-Silurian, but there is also evidence of Precambrian metamorphism at ca. 1025 Ma (Grenvillian) and 750 Ma (Morarian).

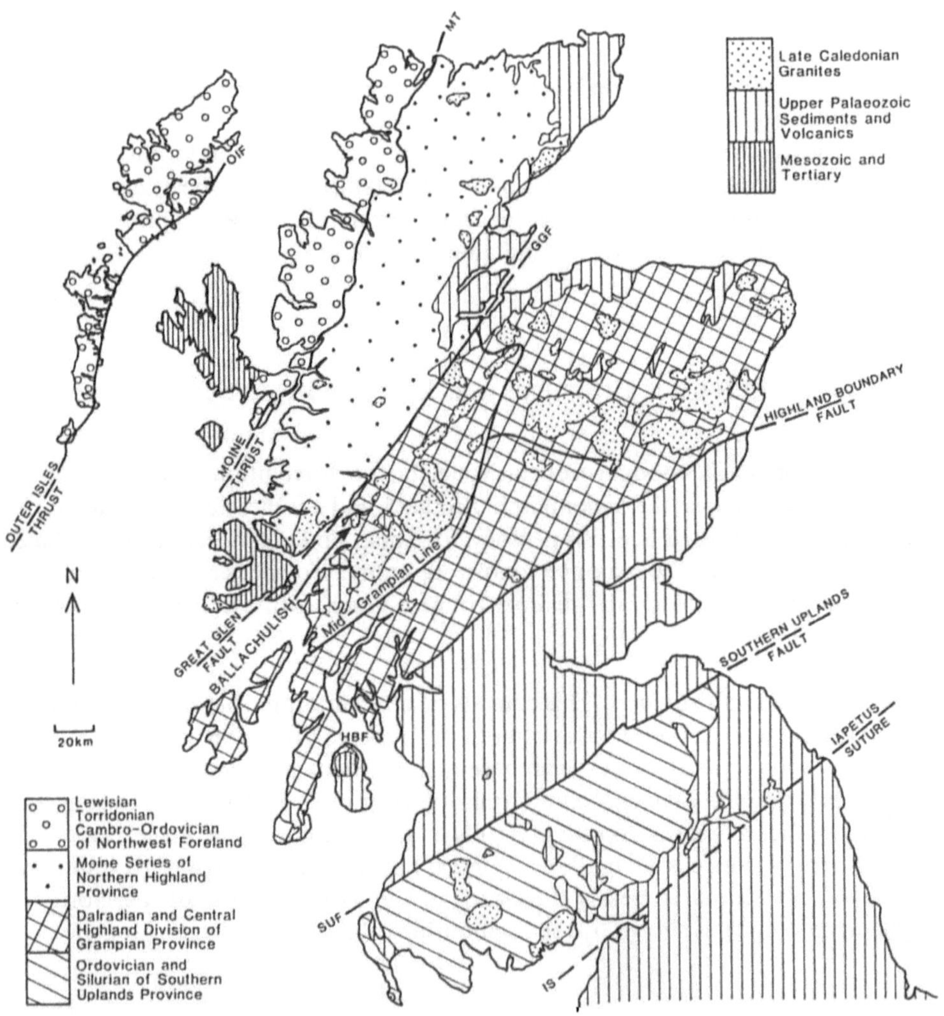

Fig. 1.1. Major tectonostratigraphic provinces and their bounding lineaments in Scotland, also showing the distribution of late Caledonian granites

Great Glen Fault. This was identified as a strike-slip fault many years ago but there has been considerable debate over the timing, direction and extent of lateral translation. Despite similarities between some of the metamorphic rocks on either side, it may separate distinct terranes which came together in the Lower Palaeozoic.

Grampian Highland Province. It consists largely of greenschist to amphibolite facies metamorphic rocks. Unlike the Northern Highland province, it is dominated by the Dalradian Supergroup of varied metasedimentary and meta-igneous rocks, which are late Precambrian in age, and were regionally deformed and metamorphosed in the Grampian 'orogeny' prior to the mid-Ordovician. The lower part of the established Dalradian succession, psammites of the Grampian Group, structurally overlies a series of often gneissose rocks (Central Highland Division).

Highland Boundary Fault Zone. A complex fault zone, which was probably active throughout the Lower Palaeozoic and into the Devonian. It forms a terrane boundary

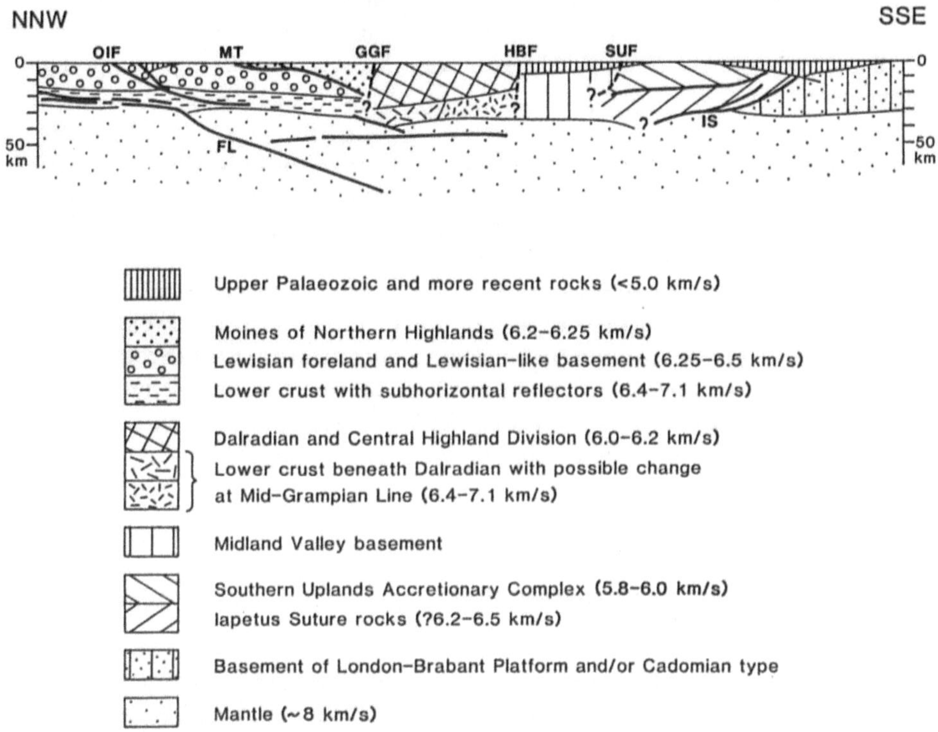

Fig. 1.2. Vertical cross-section from NNW to SSE across Scotland, showing tentative boundaries extrapolated to depth of the tectonostratigraphic provinces and lineaments of Fig. 1.1. Based predominantly on seismic data from the LISPB profile (Bamford et al. 1977; Bamford 1979) and BIRPS data (e.g., Hall et al. 1984; Freeman et al. 1988; Snyder and Flack 1990 and Snyder, pers. com.) in conjunction with geological data (see text). *OIF* Outer Isles Fault; *MT* Moine Thrust; *FL* Flannan Line; *GGF* Great Glen Fault; *HBF* Highland Boundary Fault zone; *SUF* Southern Uplands Fault; *IS* Iapetus Suture

to the Dalradian rocks of the Grampian Highland province. Emplaced along the fault zone are slivers of the Highland Border Complex which consist of Ordovician shales, cherts, limestones and arenites, as well as older pillow lavas and serpentinites. These lithologies are in a state of low-grade metamorphism and are thought to represent remnants of a marginal basin or basins.

Midland Valley Province. The surface outcrop consists largely of Devonian and Carboniferous sediments and associated igneous rocks. As a consequence there is much uncertainty about the relations of this block during the Lower Palaeozoic with respect to the Grampian Highland (Dalradian) province to the north and the Southern Uplands (Ordovician-Silurian) province to the south. Inliers of Ordovician-Silurian rocks are seen locally and there is clear evidence that the Midland Valley province has a continental crustal basement. In the SW of the province, both the Girvan complex of Llanvirn-Ashgill sedimentary rocks, and the Ballantrae ophiolite complex appear to be thrust onto the Midland Valley block from the southeast. These complexes probably represent distinct terranes to that of the main Midland Valley block.

Southern Uplands Fault. This is the lineament along which the Southern Uplands imbricated sequence of Ordovician and Silurian rocks is emplaced against the Midland Valley province; probably with both thrust and strike-slip components of movement.

Southern Uplands Province. It consists largely of an imbricate stack of weakly metamorphosed greywackes and shales of Ordovician and Silurian age, whose formation is usually associated with the closure of the Lower Palaeozoic Iapetus Ocean (relatively situated to the south of Scotland). The rocks are often considered to represent a fore-arc accretionary prism associated with subduction to the NW, but there are other views concerning their disposition and that of the associated volcanic arc.

Craig (1991) provides a comprehensive account of all aspects of Scottish geology. A restricted set of reviews, covering much of the material in this introduction is provided by: Fettes et al. (1984), Stephens and Halliday (1984), Bluck (1985), McKerrow (1987), Harte (1988), Soper (1988), Thirlwell (1988b).

1.2 The Caledonian Orogenic Belt

Traditionally, the Caledonian orogenic belt as been regarded as a major period of regional deformation and metamorphism of largely Lower Palaeozoic age, affecting sediments of a variety of depositional ages ranging from Proterozoic (Moine Series) to Silurian (in the Southern Uplands).

Extensive radiometric dating studies over the past 30 years have modified this picture substantially in showing several episodes of deformation and metamorphism in the Moine Series, whilst Dalradian depositional, deformation and metamorphic ages have been pushed into or towards the Precambrian. Thus, while events recorded at the margins of the orogenic belt (namely the Moine Thrust and the Southern Upland imbricate prism) have remained Ordovician-Silurian in age, the deformation and regional metamorphic events in the interior of the Scottish Highlands have been recognized to be much more complex, and certainly *not only* of Ordovician-Silurian age.

A brief outline of the present views of Moine and Dalradian history and the broader relationships of the late Caledonian granites is given below. Much of the radiometric dating evidence is summarized by Harte (1988) and Table 1.1 is largely taken from that paper and updated.

In the Moine Schists of the southern part of the Northern Highland province, a history of repeated metamorphism and deformational events is recognized (Table 1.1). Thus, major events are thought to have occurred at ca. 1025 Ma ('Grenvillian'), as well as ca. 465–425 Ma ('Caledonian'); there is possibly another significant event at ca. 750 Ma ('Morarian') dated mainly from deformed pegmatites.

The development of the Moine Thrust zone (Elliot and Johnson 1980), bringing the Northern Highland Moines into their present position on the Northwest Foreland, occurred at about 430 Ma and generated widespread mica K-Ar ages with similar ages to the thrusting (Johnson et al. 1985). The major Lower Palaeozoic events of high-temperature metamorphism and deformation are dated largely at 465–430 Ma in the southern Moines (Table 1.1), and appear to be partly associated with a major ductile thrust (Sgurr Beag Slide) along which slices of Lewisian basement appear (Tanner 1970; Powell et al. 1981; Roberts et al. 1984). In several models, movement on internal Moine slides, like the Sgurr Beag Slide, is linked with the Moine Thrust movements, and may be further extended to the Outer Isles Fault and other subparallel features (Fig. 1.2) that show on seismic reflection profiles (Butler and Coward 1984; Snyder and Flack 1990; but see also Soper 1988).

For many years the Moine metasedimentary series, which includes extensive psammitic rocks, was considered to extend south of the Great Glen Fault into the Grampian Highland province. However, widespread psammites in the central Grampian Highlands, lying essentially between the base of the Dalradian Appin Group (line DB on Fig. 1.3) and the area of Central Highland Division rocks, are now classified as Grampian Group and considered to form the lowermost group of the Dalradian Supergroup (Harris et al. 1978). The Grampian Group has been largely described (Piasecki and van Breemen 1983) as resting upon the gneisses forming the Central Highland Division (Fig. 1.3) with a tectonic contact (the Grampian Slide). Piasecki and van Breemen (1983) date pegmatites in these contact zones at 700–750 Ma (Table 1.1). To what extent the Central Highland Division has a separate history from the Dalradian has been widely debated (see Harte 1988, for a review; and Lindsay et al. 1989, for a recent contribution).

Until 1989, the Dalradian Supergroup was considered to have been deposited across the Precambrian-Cambrian boundary, with major nappe-forming events in the Cambrian, and major metamorphism at ca. 500 Ma and major uplift at ca. 450 Ma (e.g., Fettes et al. 1984; Harte 1988). However, Rogers et al. (1989) have redated the Ben Vuirich granite in the central Grampian Highlands (BV on Fig. 1.3) and obtained a concordant U/Pb age of ca. 590 Ma. Since this key intrusion appears to have been emplaced between the second (D2) and third (D3) phases of Dalradian deformation, the new date implies that Dalradian sedimentation is essentially Precambrian, and the

Table 1.1. Comparative ages of events in parts of the Northern and Grampian Highland provinces

	Grampian Highlands Province: south-central and eastern area[a]	Grampian Highlands Province: north-central area	Northern Highlands Province: southern area	
			ca. 1000-1050 Ma Grenvile orogeny	
		ca. 750 Ma Grampian slide and pegmatites	ca. 800-750 Ma Morarian pegmatites	
600	D1 and D2 nappe structures Ben Vuirich granite	Nappe structures		600
550			Carn Chuinneag granite	550
500	D3 folding, peak metam Newer Gabbro intrusions. Granitic intrusions	Deformation and metamorphism?		500
450	K/Ar mica cooling ages and uplift with D4 structures and Highland Border Downbend formation (460–440 Ma)	Deformation and metamorphism (G. Kyllachy granite)	Deformation and metamorphism (Glen Dessary syenite)	450
	late Caledonian granitoid intrusions	K/Ar mica cooling ages, late Caledonian granitoid intrusions	Moine Thrust K/Ar mica cooling ages late Caledonian granitoid intrusions	
400				400

Events are placed in their correct relative position with respect to ages given at right and left margin (note scale is not uniform).
[a]D prefixes refer to deformation phases. Where listed events in 600–400 Ma. period are accompanied by a vertical line then radiometric age evidence exists. References giving sources of all data: van Breemen et al. 1979; van Breemen and Piasecki 1983; Brook et al. 1976; Dempster 1985; Dewey and Pankhurst 1970; Harte et al. 1984; Harte 1988; Pankhurst 1982; Piasecki and van Breemen 1983; Rogers et al. 1989.

early major nappe-forming event either Precambrian or lowermost Cambrian (see also Halliday et al. 1989). Major (highest grade) Dalradian metamorphism may still be dominantly ca. 500 Ma, and major uplift still at ca. 450 Ma (Table 1.1).

The Northern and Grampian Highland provinces show little evidence of any common history prior to about 460 Ma (mid-Ordovician). In particular, the Grampian orogenic events of major deformation and metamorphism find no expression in the Northern Province (Powell et al. 1988; Harte 1988). Events of similar ages are seen at ca. 450 Ma, when there is evidence of major deformation and regional metamorphic events in the Moines of the Northern Highland Province, some tectonothermal events in the northcentral part of the Grampian Highland province, and major uplift (associated with the Highland Border Downbend, see Sect. 1.3) in the southern part of the Grampian Highland province (Table 1.1). However, these events are different in character and the thrust-slide style of Caledonian tectonics in the Northern Highlands has no obvious counterpart in the Grampian Highlands. The major 465–425 Ma events of the Northern Highlands and Moine Thrust zone have been linked with the closure of an ocean between Laurentia and Baltica and the Scandinavian Scandian orogeny (but see Soper 1988).

The lack of evidence of any common orogenic history prior to 460 Ma for the Northern and Grampian Highland provinces, in conjunction with probable major differences in basement rocks, geophysical characteristics, and their overlying stratigraphic successions, led Harte (1988) to suggest that the Grampian Highlands were an exotic terrane with respect to the Northern Highlands and Northwest Foreland provinces. This exotic terrane could have been brought near to its present relative position in the Ordovician by movement on the Great Glen Fault (along which several sets of movement may have occurred). Whilst accepting an allochthonous relationship for the Grampian Highlands, Soper (1988), does not consider it exotic, because he believes the Dalradian and Moine sediments to have been deposited at the margin of the Laurentian craton (represented by the Northwest Foreland province in Scotland) and to have been stacked back on to the craton by subsequent tectonic movements.

Clear evidence of a close spatial association between the Northern and Grampian Highland provinces is seen at 430–400 Ma by the presence in both provinces of: (1) wrench faults of usually small sinistral displacement, which appear related to the Great Glen Fault; and (2) late Caledonian granites (Sect. 1.4). Indeed, Watson (1984) has emphasized that these two sets of features also show an association with one another across the whole of the Scottish Highlands. Thus late granite bodies are emplaced along wrench faults and also fractured and offset by them (a feature seen at Ballachulish, see Weiss and Troll, Chap. 4, this Vol.).

Following Soper (1988) and McKerrow (1988), the wrench faulting may be linked with major displacement along the Highland Boundary Fault, and all may be related to the oblique collision of a crustal block (the Cadomian block of Gondwanaland affinity) with Scotland along the approximate line of the Iapetus Suture (Figs. 1.1, 1.2). These events are Silurian and early Devonian in age (Thirlwell 1988b) and would have played a major role in bringing together the array of the Scottish geological provinces seen in Figs. 1.1 and 1.2.

Thus the Scottish Caledonian 'orogeny' is most simply viewed as the series of events by which various terranes, from the Northern Highland province southwards, were assembled, during late Ordovician, Silurian and early Devonian time, to form the present major features of Scottish geology. Much of this assembly took place along fault zones (thrust and strike-slip), and is partly overlapped in time by the emplacement of the late Caledonian granites, to which the Ballachulish intrusive complex belongs.

1.3 The Dalradian Supergroup

The Dalradian Supergroup of metamorphosed dominantly sedimentary rocks is believed to have been largely deposited in late Proterozoic time (Halliday et al. 1989; Rogers et al. 1989). The original sediments were very varied in nature, thus leading to the recognition of a well-defined stratigraphic succession, which has in turn allowed the elucidation of complex major structures. Detailed structural and metamorphic histories have been determined throughout the Grampian Highlands province.

1.3.1 Lithology

The supergroup is divided into four lithostratigraphical groups (Harris and Pitcher 1975; Harris et al. 1978), as itemized below. The outcrops of these groups and their subgroups dominantly run in relatively long and narrow units with a NE-SW (Caledonian) trend, and in the central part of the Highlands, they show a broadly regular disposition with the oldest (Grampian Group) in the NW (northwest of line DB in Fig. 1.3) and the youngest (Southern Highland Group in the SE. The thicknesses recorded below for each group are probably maxima, and the estimated cumulate thickness of about 25 km is nowhere exposed in a continuous sedimentary sequence. As detailed below, the sedimentary sequence changes from that of a stable shelf environment, in the lower part, to an unstable one in which fault-bounded basins developed as the crust became thinned and locally ruptured (Anderton 1985).

Southern Highland Group (3–4 km). This is a sequence of metamorphosed muds, siltstones, greywackes and grits (quartzose greywackes). They are indicative of turbiditic deposition in proximal fans, with extreme lateral facies changes making it impossible to trace formations for more than short distances (Harris and Fettes 1972). Volcaniclastic sediments and lavas occur within this group and the immediately underlying Argyll Group.

Argyll Group (12 km). An exceptionally varied sequence; it includes rocks of metamorphosed clastic wedges, limestones, mudstones and quartzites. Lateral facies variations are common, and basins of deposition controlled by NW-SE and NE-SW faults have been recognized (Litherland 1982; Anderton 1985). In broad terms, this group represents the transition from relatively stable shelf conditions shown by the underlying Appin Group to unstable tectonic conditions.

Appin Group (6–7 km). A metamorphosed sequence of generally mature sandstones, limestones and mudstones, this group has extensive continuity of formations along the strike. This group indicates basin deepening with respect to the underlying Grampian Group, but nonetheless shows relative tectonic stability during deposition.

Grampian Group (1–2 km). dominated by metamorphosed quartzose and feldspathic sandstones with siltstones and mudstones, characteristic of intertidal environments (Hickman 1975).

1.3.2 Structure

Large, partly recumbent, fold structures whose limbs often appeared to be modified by low-angle faults or slides were recognized during early mapping in the Ballachulish region (Bailey 1910). The detailed complexity of local structural sequences within the Dalradian was also recognized at an early stage (Clough 1897).

The earlier or primary phases of deformation have been resolved into at least two extensively recorded events (D_1-D_2 of: Harte and Johnson 1969; Harris et al. 1976; Roberts and Treagus 1979; Thomas 1979; Harte et al. 1984). The major, tight to isoclinal, recumbent and nappe-like, Dalradian fold structures developed during these primary deformation phases. Refolding of the primary structures is particularly obvious in some episodes of later secondary deformation (referred to as D_4 by Harte and Johnson 1969; Harris et al. 1976; Harte et al. 1984; but described as D_3 by Roberts and Treagus 1979). The D_3 deformation phase of Harte and Johnson, Harris et al., and Harte et al. (op. cit.) is often roughly congruent with the primary deformation structures, but now appears to be considerably younger in age (see Table 1.1 and Rogers et al. 1989). The extensive zone of secondary (largely post-metamorphic and late-stage) D_4 deformation associated with the Highland Border Downbend (see below) is associated with uplift (Harte et al. 1984; Dempster 1985) and estimated to be ca. 450 Ma in age (Table 1.1).

The primary structure of the Southwest Highlands (see review and synthesis by Roberts 1974) is roughly that of an asymmetrical and involuted mushroom structure. There is a major SE-facing antiform (the Ardrishaig Antiform, which represents the Tay Nappe near its "root" zone) occurring to the SE of a synformal structure (the Loch Awe Syncline) with a relatively steeply dipping axial plane, and to the NW of this synform is a NW-facing antiform (Islay Anticline). Thus the Loch Awe Syncline marks a zone of divergence for NW- and SE-facing structures (Figs. 1.3, 1.4). The Tay Nappe (Shackleton 1958) is refolded by roughly monoformal secondary folds, which generate the Knapdale Steep Belt and the Highland Border Steep Belt (Figs. 1.3, 1.4). These antiformal secondary folds create a composite box-like structure (the Cowal Antiform), whose gently dipping portion continues into the Central Highlands as the Tay Flat Belt. The Highland Border Steep Belt may be traced along the Highland Border, immediately north of the Highland Boundary Fault, across the whole of Scotland, and the secondary antiformal structure associated with it is referred to as the Highland Border Downbend (Figs. 1.3, 1.4).

Tracing the primary structures of the Southwest Highlands into the Central Highlands has proved difficult because of a deepening of structural level and the intervention of large late Caledonian granite complexes (Fig. 1.1). The upward and NW-facing structure of the Islay Anticline finds continuity in the more complex NW-facing structures of the Ballachulish-Appin area, which are refolded by the

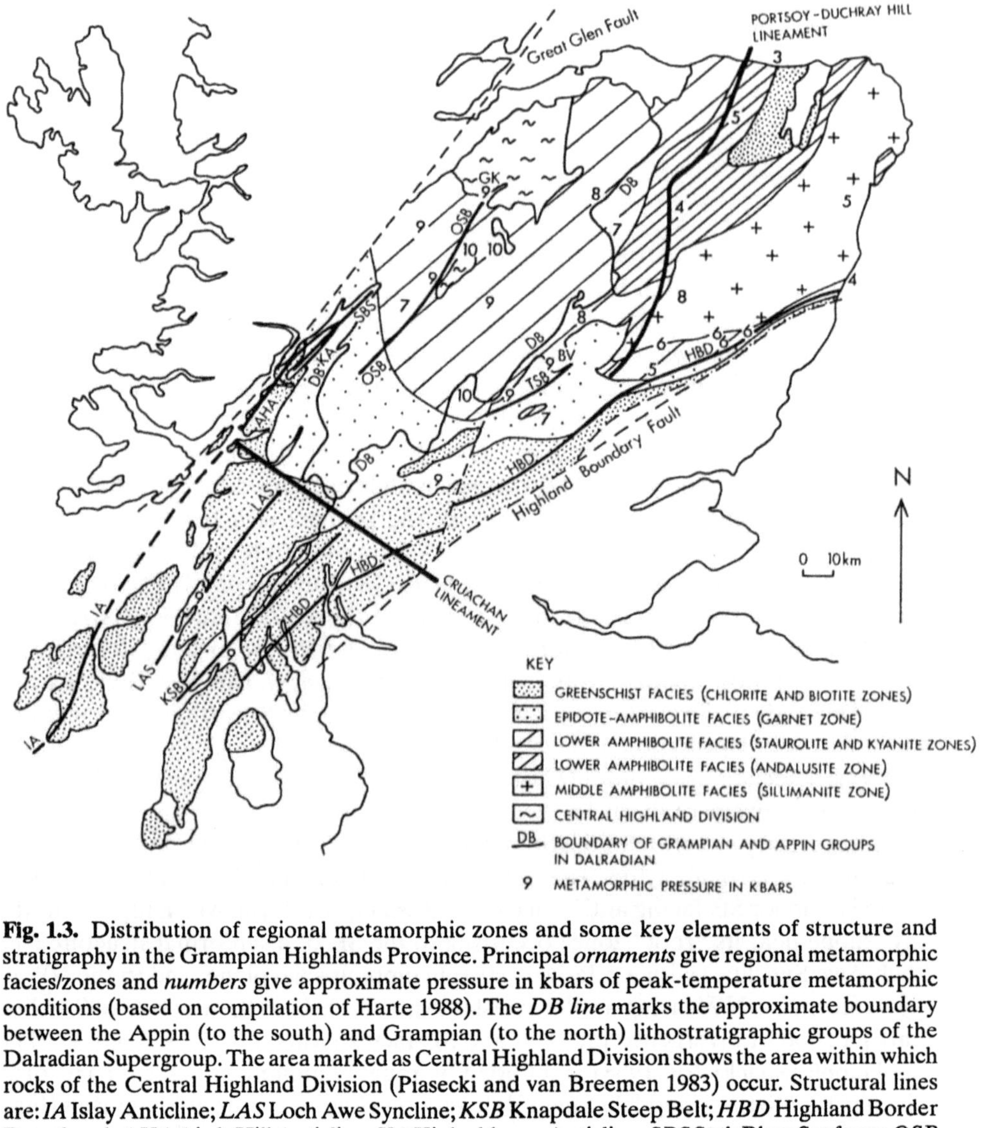

Fig. 1.3. Distribution of regional metamorphic zones and some key elements of structure and stratigraphy in the Grampian Highlands Province. Principal *ornaments* give regional metamorphic facies/zones and *numbers* give approximate pressure in kbars of peak-temperature metamorphic conditions (based on compilation of Harte 1988). The *DB line* marks the approximate boundary between the Appin (to the south) and Grampian (to the north) lithostratigraphic groups of the Dalradian Supergroup. The area marked as Central Highland Division shows the area within which rocks of the Central Highland Division (Piasecki and van Breemen 1983) occur. Structural lines are: *IA* Islay Anticline; *LAS* Loch Awe Syncline; *KSB* Knapdale Steep Belt; *HBD* Highland Border Downbend; *AHA* Airds Hill Anticline; *KA* Kinlochleven Anticline; *SBS* Stob Bhan Synform; *OSB* Ossian Steep Belt; *TSB* Tummel Steep Belt. The two 'lineaments' mark zones of discontinuity in metamorphic and other features (see text); the Portsoy-Duchray Hill Lineament extends from the position marked to the western edge of the low-pressure (Buchan) amphibolite facies. *BV* marks the location of the Ben Vuirich older granite whose age constrains many aspects of timing of Dalradian events (see text)

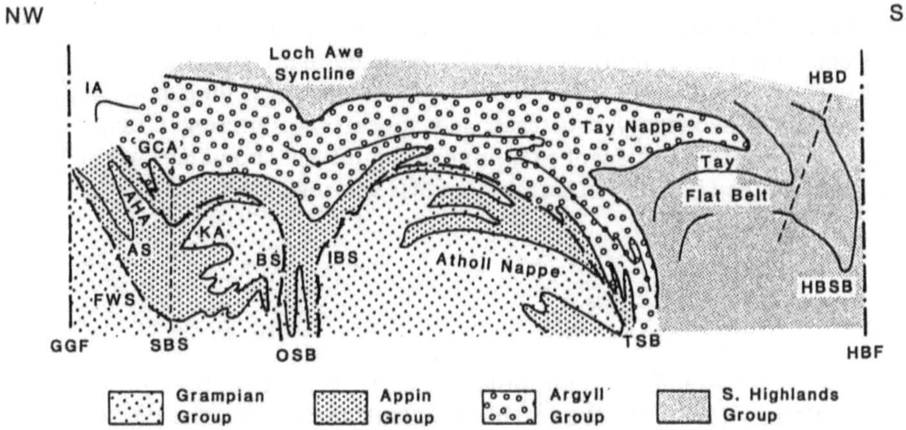

Fig. 1.4. Perspective cross-section from NW to SE of the Grampian Highlands Province, showing major tectonic structures in the Dalradian Supergroup; lower structural levels exposed in Central Highlands, higher structural levels exposed in Southwest and Southern Highlands (after Thomas 1979). *IA* Islay Anticline; *AHA* Airds Hill Anticline; *KA* Kinlochlevel Anticline; *AS* Appin Syncline; *GCA* Glen Creran Anticline; *SBS* Stob Bhan Synform; *OSB* Ossian Steep Belt; *TSB* Tummel Steep Belt; *HBSB* Highland Border Steep Belt; *HBD* Highland Border Downbend; *FWS* Fort William Slide; *BS* Ballachulish Slide; *IBS* Italy Boundary Slide; *HBF* Highland Boundary Fault zone; *GGF* Great Glen Fault

secondary major Stob Bhan Synform (as illustrated in Figs. 1.3, 1.4 largely after Roberts and Treagus 1977; but see Voll 1964; Hickman 1978, and Litherland 1982, for other viewpoints). The Tay Nappe continues as a fold complex (Mendum and Fettes 1985) into the Central Highlands, and is the dominant major primary major structure to the southeast of the Tummel Steep Belt (Figs. 1.3, 1.4). For some time the Tummel Steep Belt was seen as the approximate line of continuity of the Loch Awe Syncline, but it is now considered to be largely a later structure (Bradbury et al. 1979). Roberts and Treagus (1977, 1979) correlated the Loch Awe Syncline with the Ballachulish Syncline of the Ballachulish-Loch Leven area, though this was not accepted by Hickman (1978). Thomas (1979) has argued that the Loch Awe Syncline is represented in the Central Highlands by the Ossian Steep Belt (Fig. 1.3). This is the interpretation shown in Fig. 1.4, where the Ossian Steep Belt, like the higher-level Loch Awe Syncline, is again seen as a zone of divergence separating the SE-facing Atholl and Tay Nappes from the NW-facing structures of the Ballachulish and Appin areas. The interpretations of the nature of slides, like the Iltay Boundary Slide and the Ballachulish Slide in Fig. 1.4, have varied from Bailey's view of low-angle tectonic dislocations developed during primary folding (see also Roberts and Treagus 1977), to changes in stratigraphy due to sedimentary facies variation (Voll 1964; Hickman 1978), and extensional faults developed during sedimentation (Soper and Anderton 1984).

There has been extensive debate concerning the correct overall interpretation of the major primary structures, with much depending on whether or not the Loch Awe Syncline and Ossian Steep Belt should be regarded as a root zone, from which fountains of nappes of opposed facing arise on either side (e.g., Roberts and Treagus

1979; Thomas 1979; Shackleton 1979; Litherland 1982). The extent to which some of the steep belts represent primary and/or secondary major structures has proved difficult to unravel. The Highland Border Downbend is amongst the simpler and most continuous major secondary structures, and appears to represent a major uplift structure, which formed adjacent to a pre-existing tectonic boundary whose latest manifestation is the Highland Boundary Fault (Harte et al. 1984).

1.3.3 Regional Metamorphism

Investigations of regional metamorphism in the Dalradian date back to the classic studies of Barrow (1893, 1912) and Horne (1886), recognizing the occurrence of metamorphic zones in pelitic rocks of the Eastern Highlands. These zones are of both moderate pressure (garnet-kyanite, or Barrovian) and low pressure (cordierite-andalusite, or Buchan) types. Subsequent mapping recognized a broadly Barrovian sequence involving garnet and kyanite throughout most of the Grampian Highlands (Fig. 1.3); whilst in more recent work details of mineral chemistry, reactions, facies series and mineral growth histories have been described and augmented by P-T estimates (see reviews by Atherton 1977; Harte 1988).

Given that the zonal boundaries in Fig. 1.3 largely provide a guide to temperature variations, a selection of pressure estimates from the literature has been added to Fig. 1.3. The lack of correlation of temperature with pressure and structural depth is evident, even with attention restricted to the broadly Barrovian area. The pattern of metamorphism is thus more complicated than can be explained with a simple depth model, such as that advanced by Kennedy (1948), on the basis of the deceptively simple pattern of metamorphic zones. Sophisticated thermal models (Richardson and Powell 1976; Dempster 1985) have added much to an understanding of the time dimension in regional metamorphism, but it is evident that sets of such models are necessary to take account of the three-dimensional spatial variation in factors such as heat production, and heat transport by melts (e.g., Harte and Hudson 1979) or other fluids. Harte (1988) has suggested that domains or provinces of regional metamorphism may be recognized across the Dalradian outcrop on the basis of P-T-time relations, and that the boundaries of these domains often show some coincidence with tectonostratigraphic discontinuities, thereby suggesting the involvement of a long-standing basement control (see also Fettes et al. 1986). The most prominent of these boundaries are those marked by the Cruachan Lineament (Graham 1986) and the Portsoy-Duchray Hill Lineament in Fig. 1.3. Harte and Hudson (1979) and Harte and Dempster (1987) have emphasized the likely importance of lateral heat flow at the tectono-stratigraphic boundaries to metamorphic domains in controlling the distribution of metamorphic zones; this is probably of considerable significance in the original Barrovian area of the Southeast Highlands. The causative links between tectonic thickening of the crust and consequent enhanced heat production leading to regional metamorphism (Richardson and Powell 1976; Dempster 1985), have become less clear with the new dating of the Ben Vuirich granite (see Sect. 1.2). Thus the early primary deformation with which thickening may be associated now appears to have occurred prior to 590 Ma (Rogers et al. 1989), whilst the major metamorphism is possibly distinctly later at around 500 Ma (Table 1.1).

1.4 The Late Caledonian Granites

Although usually referred to as granites, these late Caledonian intrusions are commonly composite and often include a range of rock types from granites to diorites, with even ultramafic rock types present in some cases. They are referred to as 'late' because in all provinces they were intruded subsequent to regional metamorphism (though some of this metamorphism may now be described as pre-Caledonoid, Sect. 1.2). These same age relations have also led to them being referred to as Newer Granites, in distinction to the Older Granites (e.g., Ben Vuirich) which were affected by regional metamorphism and deformation.

Figure 1.1 shows that the late Caledonian 'granites' are found extensively across the Northern Highland, Grampian Highland and Southern Upland provinces. In Argyll and adjacent areas (within 60 km of Ballachulish) there are extensive volcanic sequences of lower Old Red Sandstone age, which, in some cases (Glen Coe, Ben Nevis), are intimately associated with the intrusive granitoid rocks. In the Midland Valley province, there are extensive volcanic sequences of lower Old Red Sandstone age, even though there are few late Caledonian granitoid intrusions in this province. Although often loosely regarded as Devonian in age, the base of the continental facies sediments of the Old Red Sandstone locally extends into the Wenlock, and the Siluro-Devonian boundary is placed at ca. 412 Ma (Thirlwell 1988a,b). Extensive age determinations (reviewed by Thirlwell 1988b) for the late Caledonian magmatism (both volcanic and plutonic) yield approximate ages of: 430–408 Ma for the Scottish Highlands; 415–410 Ma for the Midland Valley; with the Southern Upland province giving 414–407 Ma for the northern part and 398–391 Ma for the southern part.

In general terms, the granitoid rocks are mainly metaluminous with high K and calc-alkalic affinities (Stephens and Halliday 1984). Thirlwell (1988a,b) has emphasized that both the volcanic and plutonic rocks show geochemical characteristics and spatial variations typical of subduction-related rocks; with Nb, P and Ti depletion relative to REE, and some enrichment in Ba, K and Sr. Application of the classification of White and Chappell (1983) to Scotland has been disputed (Plant et al. 1983; Stephens 1988), but Stephens and Halliday classify the great majority of the late Caledonian granitoids as I-type, with some representation of A-type in the Cairngorm suite (see below), and S-type features restricted to plutons in the south of the Southern Upland province.

Regional variations in the geochemistry of both the plutonic and volcanic rocks have been documented (for reviews and references see: Stephens and Halliday 1984; Stephens 1988; Thirlwell 1988a,b). A major distinction is recognized between the plutonics and volcanics of the Argyll suite (including Ballachulish) and those of the Southern Highlands and Midland Valley, with the boundary between these areas placed at the Mid-Grampian Line (Fig. 1.1). The Argyll suite is particularly characterized by high Ba and Sr, with low Nd isotope ratios and old radiogenic Pb in zircons. To the NE in the Grampian Highland province (Fig. 1.1), another suite of plutonic complexes (the Cairngorm suite) shows differences to both the Argyll and South of Scotland suites and consists more extensively of granites without the extensive diorites and granodiorites of the other two areas. In the Southern Upland province, the granitoid complexes of the north show geochemical similarities to those of the Southern Highlands and Midland Valley (south of Scotland suite), but in the south of the Southern Uplands province, the granitoids show differences in geochemistry, as well as being distinctly younger.

On the basis of trace element and isotope compositions, the importance of crustal and mantle sources have been noted for the plutonic and volcanic rocks, with a distinct change in these sources between the Argyll area and the Southern Highland/Midland Valley area (Stephens and Halliday 1984; Thirlwell 1988a,b; and references therein). Stephens and Halliday (1984), following Halliday (1984), suggest a change in both crustal basement and mantle lithosphere across the Mid-Grampian Line (Figs. 1.1, 1.2). Chappell and Stephens (1988) advocate an origin for the extensive I-type granites which principally involves the partial melting in the crust of older igneous rocks, which themselves have been derived from the mantle.

A number of causes and tectonic relationships for the late Caledonian magmatism have been proposed, depending principally upon the relative dates of the igneous activity and the closure of the Iapetus Ocean with the associated termination of subduction. Watson (1984) considered that much of the magmatism post-dated closure of the Iapetus Ocean during the Silurian at ca. 420 Ma, and she related the magmatism to movement on block faults, which induced melting in a mantle previously metasomatized during earlier subduction events. Following this, Zhou (1985) suggested comparisons between late Caledonian magmatism and the post-collisional magmatism in parts of the Alpine-Himalayan chain.

However, Thirlwell (1988a,b) emphasizes the geochemical similarities between the rocks of the Scottish Highlands and Midland valley and those of modern volcanic arcs. Excluding the granitoid complexes of the south of the Southern Uplands, Thirlwell (1988b) shows that age determinations allow a correlation of the late Caledonian Scottish magmatism (ages ca. 430–408 Ma) with active subduction at the Iapetus Suture, where closure of the Iapetus Ocean occurred sometime in the interval from 410–396 Ma. In considering the precise position of these late Caledonian granites with respect to the Iapetus Suture, Thirlwell (1988b) notes the potential importance of late Silurian and lower Devonian thrust and strike-slip movements in the Southern Uplands in adjusting the original spatial relationships. Soper (1988) emphasizes the evidence that positional relationships throughout the Scottish Caledonides cannot be read from a present-day map, because they will have been modified by the continuation into the lower Devonian of strike-slip movements along the major fault and suture zones.

1.5 Summary

The Ballachulish igneous complex is part of an extensive series of granitoid intrusions (late Caledonian granites) found throughout most of Scotland and with an age range of mainly 430 to 410 Ma. An involvement of both lower crustal and mantle magma sources gave rise to the intrusions, which have calc-alkaline affinities. An association with active subduction at the Iapetus Suture in the south of Scotland has been both advocated and denied.

The Ballachulish complex intrudes Dalradian metasediments, which were largely deposited in the Proterozoic and underwent extensive deformation and regional metamorphism a considerable time prior to the intrusion, probably at about 590 and 500 Ma, respectively. The regional deformation and metamorphism events are commonly related to the Grampian 'orogeny', whose relationships to later orogenic events, to both north and south in Scotland, are unknown.

In a restricted and modern sense, the Caledonian orogeny, affecting the whole of Scotland, refers to the assembling together, with considerable thrust and strike-slip fault movement, of the major crustal units which constitute the present major geological provinces of Scotland. This terrane assembly took place largely in the Silurian and early Devonian, closely overlapping the age range of the (late) Caledonian granites.

2 Regional Geology of the Ballachulish Area

D.R.M. Pattison and G. Voll

2.1 Introduction

The purpose of this chapter is to describe the stratigraphy, structure and regional metamorphism of the host rocks to the Ballachulish Igneous Complex. The chapter opens with a description of the geography of the Ballachulish area, and concludes with a discussion of pre-intrusion and post-intrusion uplift in the area.

2.2 Physical Geography

The Ballachulish Igneous Complex is located in Argyllshire on the west coast of Scotland, 20 km south of Fort William and immediately southeast of the junction of Lochs Linnhe and Leven. The area is dominated by Beinn a' Bheithir (see Fig. 2.1), a northward-opening, horseshoe-shaped mountain comprising two main peaks, Sgorr Dhearg (1024 m) and Sgorr Dhonuill (1001 m; see Map 2). To the southeast, across the head of Gleann an Fhiodh, is Sgorr a' Choise (658 m); to the south, across Glen Duror, is Fraochaidh (879 m); and to the west, across Strath Duror, is Ardsheal Peninsula (261 m). The main villages in the area are Ballachulish, by the abandoned slate quarries on the south shore of Loch Leven (see Fig. 2.2), Onich, on the north shore of Loch Leven, Kentallen, northeast of Ardsheal Peninsula, and Duror, in Strath Duror.

Glaciation has moulded the topography of the area. The hills and mountains contain arete ridges, cols, hanging valleys and truncated spurs. The glens are U-shaped valleys. Glacial moraine floors the glens up to an elevation of about 400 m.

Annual rainfall is about 100 in., with the driest period being mid-April to mid-June. Coniferous forest has been planted up to an elevation of 350 m in all of the glens except Gleann an Fhiodh (see Fig. 2.1). The high ground consists of grass, heather and bracken. Outcrop is good in the high ground, but poor in the forested glens.

G. Voll, J. Töpel, D.R.M. Pattison, F. Seifert
Equilibrium and Kinetics
in Contact Metamorphism
©Springer-Verlag Berlin Heidelberg 1991

Fig. 2.1. View looking west down Loch Leven to the Ballachulish area. On the *right-centre* skyline is the rounded shoulder of Creag Ghorm, underlain by porphyritic granite. On the *centre* skyline is Meall a' Chaolais, with a sharp ridge running up towards Sgorr Dhearg, to the *left* out of the picture. This ridge is underlain by Appin Quartzite, Appin Phyllite and Appin Limestone striking into the igneous complex. The prominent white scar running down through the trees below Meall a' Chaolais is the gouge zone of the Ballachulish Fault

2.3 Stratigraphy

The Ballachulish Igneous Complex intrudes Lower and Middle Dalradian metasediments of the Appin and Argyll Groups (Bailey and Maufe 1960; Harris and Pitcher 1975; Litherland 1980). In the study area, the Appin Group comprises the Lochaber, Ballachulish and Blair Atholl Subgroups, whilst the Argyll Group comprises the Creran Succession (see Table 2.1). There has been considerable debate about various aspects of the stratigraphy of the Dalradian (see for example Bailey and Maufe 1960; Voll 1964; Litherland 1970, 1980, 1982; Hickman 1975; Roberts and Treagus 1977; Anderton 1985). This chapter focuses on descriptive aspects of the stratigraphic units in the study area, relying principally on Bailey and Maufe (1960), Litherland (1980) and the authors' observations.

Table 2.1 lists the stratigraphic units in the study area. No fossils have been found in these units. High-precision zircon age dating by Rogers et al. (1989) on the Ben Vuirich Granite, emplaced during regional Caledonian orogenesis that followed Dalradian sedimentation, gives an age of 590 ± 2 Ma; this restricts Lower and Middle Dalradian sedimentation to the late Proterozoic or earlier.

Below, each stratigraphic unit is described briefly, starting from the stratigraphically lowest unit (refer to Map 2).

Fig. 2.2. View looking northeast along Loch Leven from Beinn Bhan. The town of Ballachulish and the abandoned slate quarries are prominent *bottom-centre*, with the village of Glen Coe visible further east along the loch. *Top left* is Ben Nevis, with the Mamore range stretching away to the east

Table 2.1. Stratigraphic units in the Ballachulish area

Group	Subgroup	Formation
Middle Dalradian Argyll Group	Creran Succession	('Banded Leven Schists'[a])
Lower Dalradian Appin Group	Blair Atholl	Cuil Bay Slate[b]
	Ballachulish	Appin Phyllite and Limestone Appin Quartzite Transition Series Ballachulish Slate Ballachulish Limestone
	Lochaber	Leven Schist ('Pelitic Leven Schists'[a]) Glen Coe Quartzite

[a]The terms in parentheses are from Bailey and Maufe (1960).
[b]The Benderloch slide forms a tectonic boundary between the Cuil Bay Slate and Creran Succession.

Glen Coe Quartzite. The Glen Coe Quartzite is a thick, fine-grained, well-bedded quartzite. North of Loch Leven, Roberts (1976) considers the Glen Coe Quartzite to be older Binnein Quartzite.

Leven Schist and Creran Succession. Bailey and Maufe (1960) divided the Leven Schist into two members: the 'pelitic Leven Schists', comprising grey-green phyllites and thinly bedded siltstones, and the 'banded Leven Schists', a sequence of interbedded siltstones, dirty quartzites and variably graphitic slates (Litherland 1980). These two members form the south and southwestern boundary of the igneous complex (see Map 2).

Litherland (1980) revised Bailey and Maufe's (1960) stratigraphy, placing the banded Leben Schists in a new unit, the Middle Dalradian Creran Succession, separated from the pelitic Leven Schists by the Benderloch Slide (see Map 2 and below). Within the Ballachulish area, the Creran Succession comprises two main units: the Creagan Formation, a sequence of semipelitic flags with subordinate striped siltstone, and the upper Salachan Formation, comprising striped siltstones similar to those in the Creagan Formation. The similarity between these units and the pelitic Leven Schists may explain Bailey's original grouping of these units into one formation. Differences between the units become more accented south of the study area near Loch Creran, where Litherland erected his revised stratigraphy (see Fig. 2 of Litherland 1980).

Roberts (1976) considers the pelitic Leven Schists on the northeast flank of the igneous complex to be older Binnein Schists, although Voll (1969) considers the presence of the heavy minerals tourmaline and apatite to be more characteristic of the Leven Schist.

Ballachulish Limestone. The Ballachulish Limestone comprises a lower member of calcareous schist with mm-cm scale dolomitic interbeds, and an upper member of dark, graphitic limestone.

Ballachulish Slate. The Ballachulish Slate is a dark grey to black, graphitic, generally homogeneous slate, with infrequent mm-cm sandy layers. Pyrite (outside the thermal aureole) and pyrrhotite (inside the aureole) are locally abundant.

Transition Series. The Transition Series is a gradational unit between the Appin Quartzite and the Ballachulish Slate. Adjacent to the Appin Quartzite, it comprises 1–10 cm quartzite layers interbedded with thin graphitic phyllite layers. As one goes lower in the unit, the graphitic phyllite layers increase in thickness and abundance, and the lithology becomes dominantly graphitic slate with thin quartzitic interbeds.

Appin Quartzite. Appin Quartzite is a white to grey, feldspathic quartzite, comprising up to 15% alkali feldspar. Cross-bedding and graded bedding is common throughout the unit.

Appin Phyllite and Appin Limestone. Appin Limestone and Appin Phyllite together comprise a single unit, with a higher proportion of limestone occurring towards the lower boundary with the underlying Appin Quartzite. Appin Phyllite ranges from dominantly pelitic lithologies near Onich and in Gleann an Fhiodh, to dominantly semipelitic lithologies near Kentallen. In Coire Guibhsachain, Appin Phyllite comprises pelite and quartzose semipelite interbedded on a scale of 1–10 cm, with conspicuous graded and cross-bedding.

Appin Limestone comprises a basal impure, striped dolomite overlying the Appin Quartzite, and a sequence of dolomite-bearing, pure to impure marble beds interlayered on a scale of 1–10 cm, with pelite, semipelite and calcareous pelite beds in the middle of the phyllite/limestone unit. Locally, dolomitic marble achieves a thickness of several metres, such as in Coire Guibhsachain.

Cuil Bay Slate. The Cuil Bay Slate is a grey to black, variably pyrite-bearing, graphitic slate with rare mm-cm interbeds of graded quartzite.

Creran Succession. See section below on Leven Schist.

2.3.1 Depositional Environment of Metasediments

The Appin and Argyll Groups are interpreted by Anderton (1985) to have been deposited on the northwest margin of a late Precambrian marine gulf. This gulf is thought to have developed over a zone of crustal thinning between the Laurentian and Baltic plates of the Proterozoic supercontinent. The upward-fining sequence of Glen Coe Quartzite to pelitic Leven Schists is thought to represent a gradual deepening of the basin. The Ballachulish Limestone and Ballachulish Slate are deep-water deposits, with the graphitic limestone member and slate representing deposition under anoxic conditions. The coarsening upward sequence of Transition Series to Appin Quartzite indicates filling of the basin, possibly in a prograding delta (Hickman 1975). The Appin Quartzite is interpreted as a tidal shelf deposit, overlain by the shallow water Appin Limestone and Appin Phyllite, the latter of which may represent renewed transgression. Deposition of the Cuil Bay Slate represents a return to deep-water anoxic conditions. Following a period of regression not represented in the Ballachulish area, the Creran Succession ('banded Leven Schists') was deposited in a rapidly deepening, northwest-trending, fault-bounded trough, thought to represent the embryonic Iapetus basin (Litherland 1980; Anderton 1985).

2.4 Structure

2.4.1 Pre-Intrusion Structures

Figure 2.6 shows the main structural features in the study area, after Bailey and Maufe (1960), Roberts (1976), Litherland (1982) and Pattison (1985). Several major northeast-southwest trending folds and slides (ductile faults) define the structural grain of the area. These are, from west to east, the Appin Syncline, Beinn Sgluich Anticline, Glen Stockdale Slide, Airds Hill and Tom Meadhoin Anticlines, Benderloch and Ballachulish Slides, Beinn Donn Syncline, a number of smaller folds of uncertain significance, the Ballachulish Syncline and Sgorr a' Choise Slide. The folds are tight and upright, and generally plunge shallowly northeast or southwest, although the small syncline on the northeast margin of the igneous complex, remapped during this study, plunges steeply to the southwest (Pattison 1985; Weiss 1986). To the west of the Ballachulish and Benderloch Slides, the folds verge northwest, whilst to the east of the slides, the folds verge southeast.

Small-scale deformation features associated with the major folds include tight parasitic folds (see Fig. 2.3), and a penetrative, NE-SW-trending, slaty, axial planar cleavage in pelitic layers. Fracture cleavage is prominent in quartzose layers in the tight folds (see Fig. 2.3). The NE-SW-trending slaty cleavage dominates the fabric of micaceous units throughout the area, which has important implications for heat flow from the igneous complex (Bunteberth, Chap. 18, this Vol.). Parallel to the penetrative cleavage in some pelitic layers is incipient transposed layering and transecting quartz veins. In quartzites, clastic quartz grains are variably flattened and stretched, most especially in the vicinity of slides, where the quartz grains are recrystallized and acquire a strong preferred orientation. Flattening and stretching also affects carbonate grains in marble layers and pyrite grains in the slates. The stretching lineation is generally steeply plunging within the axial planar cleavage, roughly normal to the plunge of the major fold axes.

Southeast of the Sgorr a' Choise Slide, in the Leven Schist, a strong crenulation cleavage with an associated stretching lineation overprints the earlier formed penetrative cleavage and stretching lineation. The latter stretching lineation has approximately the same attitude as the earlier lineation. This later fabric may be associated with large scale folds (e.g., Stob Bhan Synform) to the east of the study area. On the northeast flank of the igneous complex, local reorientation of the steeply dipping, NE-SW-trending bedding and cleavage to an E-W orientation may be associated with the same deformation.

Later deformation episodes are indicated by mm-cm scale refolding of primary layering and penetrative cleavage, and by the development of microscopic- to cm-scale kink folds and crenulations at various orientations (see Fig. 2.4). These small scale features have no expression in any large scale structures in the study area.

2.4.2 Pre-Intrusion Structural Interpretation

The major deformation in the Ballachulish area is associated with early primary folding of the Dalradian during regional Caledonian orogenesis (Litherland 1982). In Perthshire, small-scale structures associated with early folding are crosscut by the 590 ± 2 Ma Ben Vuirich Granite (Rogers et al. 1989), which places a lower limit on the age of early deformation there. This age may also apply to the main deformation in the Ballachulish area, if deformation was not strongly diachronous across the Dalradian.

Fig. 2.3. Tight D_1 minor folds in interbedded quartzite-pelite, Appin Phyllite, Coire Guibhsachain. Note the fanned fracture cleavage in the quartzite, and the thickening of the pelite in the fold hinge. The characteristic axial planar slaty cleavage in the pelite has been obscured by contact metamorphic recrystallization

In the Ballachulish area itself, there is divided opinion about the timing and significance of the major folds and faults and small-scale deformation features. The consensus view, held in general but not in detail, by Bailey and Maufe (1960), Roberts (1976), Treagus (1974), Roberts and Treagus (1977) and Litherland (1982), is that the main NE-SW structures and accompanying penetrative cleavage are D_1 features associated with early nappe development in the southwest of Scotland. Deformation during further nappe evolution resulted in refolding of the earlier NE-SW structures, forming the Stob Bhan Synform to the east of the area and the overprinting crenulation cleavage in the southeast of the area (Bailey and Maufe 1960; Roberts and Treagus 1977). Roberts and Treagus (1977) call this later deformation D_3, following an intervening minor D_2 deformation. In contrast, Hickman (1978) considers that

Fig. 2.4. Late kinking of shiny D_1 cleavage, Leven Schist, southeast of area. This is characteristic of late deformation features in the study area

Roberts and Treagus's (1977) separate D_1 and D_3 structures are of the same generation, categorizing these as D_2, having followed an earlier D_1 isoclinal folding episode associated with nappes not seen in the study area. Based on mapping further to the east, Roberts and Treagus (1977) identified two more deformation episodes, D_4 and D_5, which have their expression in late small-scale features (see above) and in broad E-W structures to the south and east of the study area.

The main point is that in the Ballachulish area, there is one main deformation which has its expression in the major NE-SW trending folds and associated penetrative cleavage. In subsequent chapters, this is referred to as D_1.

2.4.3 Post-Intrusion Structures

Crosscutting the Ballachulish Igneous Complex is the Ballachulish Fault (Pattison 1985; Weiss 1986), which has a sinistral displacement of about 700 m (see Fig. 2.6). The physiographic expression of the fault can be seen in Figs. 2.1 and 2.5. Prominent fault gouge occurs along its trend. A smaller sinistral strike-slip fault crosscuts the igneous complex to the southeast of the Ballachulish Fault, with a displacement of about 100 m (Weiss 1986; see Fig. 2.6); this is probably related to the Ballachulish Fault.

The Ballachulish Fault was not fully recognized until the present study. Aligned 'shatter belts' were noticed in other areas (Bailey and Maufe 1960; Hickman 1975; Roberts 1976), but with the exception of Litherland (1982), these were generally assumed to predate the igneous complex. The Ballachulish Fault and associated shatter belts line up with a wrench fault near Loch Laggan identified from Landsat Imagery by Johnson and Frost (1977). Continuity with this fault gives a total fault length of at least 80 km. Sinistral stratigraphic displacement along the zone of shatter belts both north and south of the Ballachulish area is consistently in the range 600–1000 m, further suggesting that they join up with the Ballachulish Fault (Pattison 1985). The fault makes about a 15° angle with the Great Glen Fault shear zone, which occupies Loch Linnhe immediately west of the study area (see Fig. 2.6). The orientation, sense of movement and amount of displacement suggest that the Ballachulish Fault is a P-type Reidel shear that splays off the main Great Glen shear zone

Fig. 2.5. Prominent gorge (Eas nam Meirleach) formed by the Ballachulish Fault crosscutting the Ballachulish Igneous Complex, Glen Duror, looking northeast. To the *left* of the gorge is light-coloured porphyritic granite, whilst to the *right* is darker, low-weathering hypersthene diorite

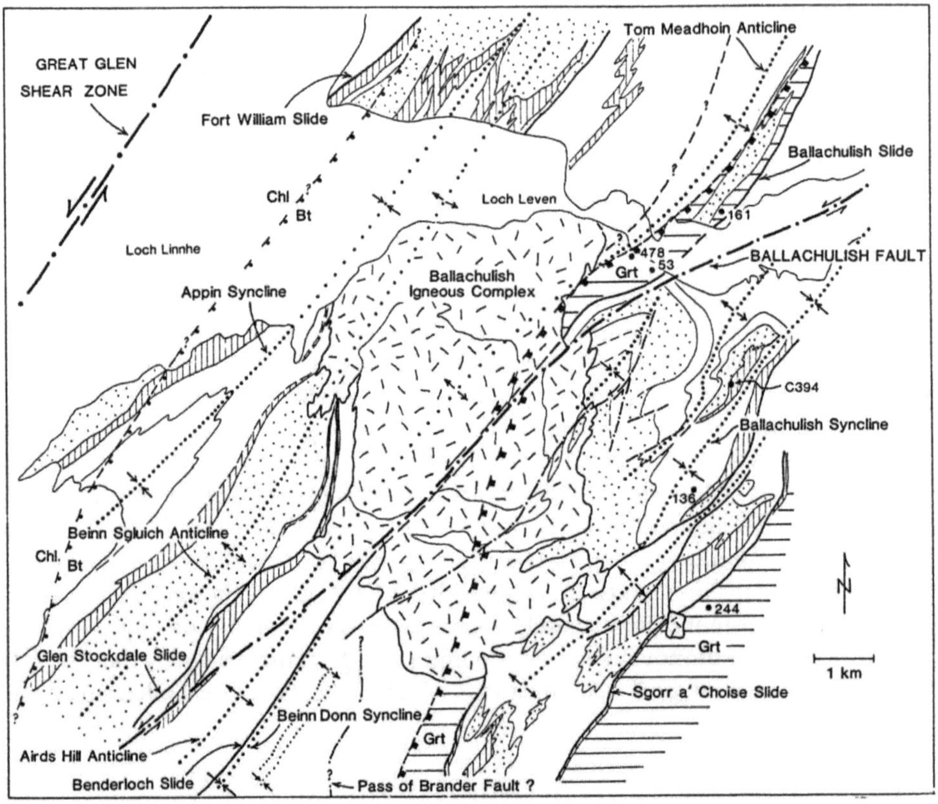

Fig. 2.6. Map of major structural and regional metamorphic features in the Ballachulish area, compiled from Bailey and Maufe (1960), Roberts (1976), Litherland (1982), Pattison (1985) and Weiss (1986). For clarity, only the stratigraphic boundaries (*light lines*), quartzites (*stippled*) and limestones (*vertical line pattern*) have been drawn (for stratigraphy, refer to Map 2). Major D_1 folds (*dotted*) and slides (*medium heavy lines*) have been extrapolated through the Ballachulish Igneous Complex and across Loch Leven. *Ornamented dashed lines* are regional metamorphic isograds (*dots*: biotite isograd; *squares*: garnet isograd), with the ornaments on the high grade side. The *horizontal ruled pattern* refers to garnet-bearing pelites. *Dot-dash lines* represent post-intrusion faults. *Numbers* refer to specimens discussed in text; *numbers with no prefix* are pelitic specimens, whilst C394 is a metacarbonate specimen (see text)

(Johnson 1985 and pers. comm.). Litherland (1982) speculated that the Pass of Brander Fault to the southeast might be the southern extension of the Ballachulish Fault; this seems unlikely, given the straight, well-defined NE-SW trend of the fault south of the igneous complex, in contrast to the speculated arcuate trend of the Pass of Brander Fault (see Fig. 2.6).

It is likely that the Ballachulish Fault was active before and after intrusion of the igneous complex. Unusual striped pelites that appear to have been severely sheared occur in a 10-m-wide zone in the fault zone south of the igneous complex. These have been recrystallized during contact metamorphism to cordierite-bearing hornfels, which suggests that there was movement along the Ballachulish Fault prior to, and possibly during, the intrusion of the igneous complex (Pattison 1985; Weiss 1986).

2.5 Regional Metamorphism

2.5.1 Metamorphism of Pelitic Rocks

Regional metamorphic grade in pelites increases from chlorite grade in the west of the study area to garnet grade in the east and southeast (Bailey 1923; Elles and Tilley 1930), constituting the low-grade portion of a Barrovian metamorphic sequence (Atherton 1977; Fettes 1979; Fettes et al. 1984). This metamorphic sequence is thought to be part of the regional Dalradian metamorphic event, dated by Harper (1967), Pankhurst (1970) and Dempster (1985) at 520–490 Ma.

Mineral abbreviations are listed in Table 8.1 (this Vol.). However, in this chapter, abbreviations refer to natural non-end member compositions rather than to the ideal mineral formulae of Table 8.1. For example, Ms refers to phengitic white mica. Details of mineral chemistry are provided in Pattison (1987 and Chap. 8, this Vol.).

Chlorite Zone. At the west end of the north shore of Loch Leven, phyllites and slates contain the chlorite zone assemblage Ms + Chl + Qtz + Ab + Gr. A number of slate boulders on the shore contain coarse (2–6 mm) chloritoid porphyroblasts, suggesting the nearby presence of a chloritoid zone, but none has been found in outcrop.

Biotite Zone. Biotite first appears sporadically in carbonate-bearing pelites in the Appin Phyllite near Onich, giving rise to the biotite zone. Biotite occurs both as conspicuous 1–4-mm porphyroblasts and as fine tabular grains within the schistosity. The biotite zone, comprising interbedded Ms + Chl + Qtz + Ab and Ms + Chl + Bt + Qtz + Ab pelites, with or without epidote, calcite or dolomite, occupies much of the north shore of Loch Leven and Ardsheal Peninsula.

Biotite formation in dolomite-bearing layers may have been due to reactions of the form:

$$\text{Ms} + \text{Dol} + \text{Qtz} + H_2O = \text{Bt} + \text{Chl} + \text{Cal} + CO_2, \qquad \text{(R1)}$$

or $$\text{Dol} + \text{Kfs} + H_2O = \text{Bt} + \text{Cal} + CO_2. \qquad \text{(R2)}$$

The growth of biotite in carbonate-absent layers may have been due to reactions of the form:

$$\text{Kfs} + \text{Chl} = \text{Bt} + \text{Ms} + \text{Qtz} + H_2O, \qquad \text{(R3)}$$

or $$\text{Ms} + \text{Chl} = \text{Bt} + (\text{Fe,Mg})_{-1}\text{Si}_{-1}\text{Al}_2 + \text{Qtz} + H_2O. \qquad \text{(R4)}$$

Reaction (R4) involves the operation of the Tschermak exchange in coexisting chlorite, muscovite, and biotite (Pattison 1987). Reactions (R1) and (R4) are favoured because K-feldspar has not been observed in any of the assemblages (Pattison 1985). In the vicinity of the garnet isograd, Pattison (1987) demonstrated the progress of reaction (R4) from mineral chemistry variations.

The geographical extent of the biotite zone defined here is much like Bailey's (1923) 'mica conspicuous' zone, rather than the separate Chl and Bt zones of Elles and Tilley (1930), which do not fit in with more recent observations south of the igneous complex (Pattison 1985). Elles and Tilley's Chl and Bt 'zones' are likely the result of differences in bulk composition of the pelitic rocks, rather than representing differences in grade. For similar reasons, the exact location of the biotite-in isograd is very uncertain; it is possible that biotite could be developed in suitable bulk compositions farther to the west than the approximate isograd in Fig. 2.6.

Garnet Zone. Garnet first appears in the assemblage Ms + Chl + Qtz + Grt + Bt in the Leven Schist about 500 m west of the Ballachulish Slide on the north shore of Loch Leven, and about 500 m west of the Ballachulish Slate at the south contact of the igneous complex (see Map 2 and Fig. 2.6). The garnets are small (1–3 mm) and typically altered to chlorite. Southeast of these initial occurrences of garnet in the Leven Schist, there is a broad zone occupied by the Appin Phyllite, Ballachulish Slate and Transition Series in which garnet is essentially absent; a single garnet-bearing specimen (DP136) was found in the Appin Phyllite in Gleann an Fhiodh by Pattison (1985; see Fig. 2.6). Garnet reappears in the Leven Schist in the southeast of the area, where it is more abundant and coarser (3–8 mm) than in the northeasterly occurrences. Southeast of the study area, garnet appears in all pelitic units.

Garnet appears to have formed by the model Fe-Mg continuous reaction:

$$\text{Ms} + \text{Chl} + \text{Qtz} = \text{Grt} + \text{Bt} + H_2O. \qquad \text{(R5)}$$

This reaction commences at lower temperatures in Fe-rich compositions than Mg-rich compositions (e.g., Atherton 1968). Inclusion trails in garnet are generally parallel but sometimes at angles to the D_1 schistosity suggesting that the metamorphic peak was reached during or after the main D_1 deformation.

2.5.1.1 The Bailey Versus Tilley Debate

The cause for the abrupt transition from garnet-bearing Leven Schist to garnet-absent Ballachulish Slate along the shores of Loch Leven was a source of debate between E.B. Bailey and C.E. Tilley in the 1920s and 1930s. Bailey (1923) attributed the break to compositional differences between the pelitic units, laying special emphasis on the 'retarding effect' of graphite on mineral growth in the Ballachulish Slate. Elles and Tilley (1930) considered that the sudden transition was due to post-metamorphic juxtaposition of lower-grade Ballachulish Slate against higher-grade Leven Schist via the Ballachulish Slide, therefore implying that the slide was post-metamorphic rather than syn-metamorphic.

Microprobe analysis of the minerals on either side of the slide shows that minerals in garnetiferous Leven Schist are significantly more Fe-rich than in the immediately adjacent Ballachulish Slate [e.g., Mg/(Mg + Fe) in chlorite, Leven Schist

= 0.37–0.41; in Ballachulish Slate, 0.61; see Fig. 8.22a (this Vol.)]. Because the garnet-forming reaction (R5), Ms + Chl + Qtz = Grt + Bt + H_2O, proceeds at lower temperatures in Fe-rich compositions than Mg-rich compositions, incipient development of garnet proceeded at this location only in the Leven Schist. It is of interest that the presence of graphite, with all other variables being held constant, would tend to lower a(H_2O) and therefore *promote* rather than retard the growth of garnet in the graphitic slates; Pattison (Chap. 16, this Vol.) documents this effect for cordierite-producing reactions in the thermal aureole. With regard to the Bailey-Tilley debate, on balance one would have to rule in favour of Bailey, although probably not for the compositional reasons he gave. Bailey's interpretation of the Ballachulish Slide as a pre- to syn-metamorphic D_1 slide, rather than a post-metamorphic fault, is also consistent with present-day interpretations (e.g., Roberts 1976) and with the textural evidence noted above.

2.5.1.2 Summary of Metamorphism of Pelites

From west to east across the Ballachulish area, there is a smooth increase in grade through chlorite, biotite and garnet zones, with a garnet-absent 'gap' occupied by the Appin Phyllite and Ballachulish Slate whose bulk compositions were too Mg-rich for garnet growth at this grade. Substantiating this compositional argument is the Fe-rich composition of chlorite in the lone garnet-bearing Appin Phyllite sample (DP136; Mg/(Mg + Fe) = 0.40), compared to the more magnesian chlorites in the garnet-absent Appin Phyllite and Ballachulish Slate specimens elsewhere in this zone (0.51–0.72; see Fig. 8.22a, this Vol.). The above metamorphic pattern is in substantial agreement with Bailey's (1923) metamorphic map, and contrasts with the more complicated zonal pattern of Elles and Tilley (1930), Atherton (1977), Fettes (1979) and Fettes et al. (1984).

2.5.2 Metamorphism of Quartzite

Within the Appin Quartzite, recrystallized quartz grains increase in grain size towards the southeast. Clastic K-feldspar grains and their diagenetic adularia rims have been transformed to maximum microcline, consistent with heating during regional metamorphism. Transition to disordered sanidine is only seen further to the east. The oligoclase isograd also lies to the east.

2.5.3 Metamorphism of Carbonates

No regional isograds in carbonates have been delineated. Biotite of probable regional metamorphic origin is developed in many carbonate-bearing assemblages throughout the Ballachulish area (for possible reactions, see Sect. 2.5.1, 'Biotite zone'). Talc has been found in a number of Dol + Qtz-bearing carbonates outside the first development

Table 2.2. Regional garnet-biotite geothermometry

Rock	Garnet[a]				Biotite[b]		Temperatures (°C)[c]			
	X_{Fe}	XMg	X_{Ca}	X_{Mn}	X_{Fe}	X_{Mg}	T	HL	FS	HS
D 53	0.606	0.040	0.200	0.154	0.615	0.385	483	477	435	506
D161	0.650	0.042	0.207	0.095	0.618	0.382	480	473	431	504
D244	0.723	0.033	0.166	0.076	0.745	0.255	533	521	494	556
D478	0.628	0.031	0.228	0.112	0.635	0.365	445	441	390	467

All Fe assumed to be Fe^{2+}.
[a]Garnet: $X_i = X_i/X_{Fe} + X_{Mg} + X_{Mn} + X_{Ca}$.
[b]Biotite: $X_i = X_i/X_{Fe} + X_{Mg}$.
[c]Calibrations: T, Thompson (1976); HL, Holdaway and Lee (1977); FS, Ferry and Spear (1978); HS, Hodges and Spear (1982). The preferred calibration is Hodges and Spear. Garnet compositions are rim compositions. P = 6 kbar. Chemical data from Pattison (1985).

of contact metamorphic cordierite in pelites, but its development concentric to the cordierite isograd (see Map 2) suggests that it is of contact rather than regional metamorphic origin (Masch and Heuss-Aßbichler, Chap. 10, this Vol.).

2.5.4 Pressure-Temperature Conditions of Regional Metamorphism

P-T stabilities of the model biotite- and garnet-forming reactions in pelites have not been experimentally constrained. The petrogenetic grid of Harte and Hudson (1979) suggests garnet-forming reaction (R5) occurs at about 570 °C in the Fe-Mg end-member system, but this is a maximum temperature because Ca and Mn in the natural garnets (see below) shifts the reaction to lower temperatures.

Garnet-biotite geothermometry was applied to four specimens (Pattison 1985); three from near the garnet isograd west of the Ballachulish Slide on Loch Leven, and one from the southeast of the area (see Fig. 2.6). Garnet analyses are from the rims of normally zoned, or 'growth-zoned' (Tracy 1982) crystals, whilst biotite analyses come from fresh, unzoned crystals. The mineral chemical data and derived temperatures are listed in Table 2.2. The calibration of Hodges and Spear (1982) is favoured because this is based on the experimentally reversed calibration of Ferry and Spear (1978) with a correction for Ca and Mn in garnet. Assuming a pressure of 6 kbar (see below), a temperature range of 467–506 °C was recorded for the three northerly specimens. This approximately represents the temperature of the garnet isograd. For the southeast specimen, 556 °C was recorded, consistent with the overall southeasterly increase in grade across the area. Uncertainty in the temperature estimates is estimated as ±50 °C (Hodges and Spear 1982).

In carbonates, a calcite-dolomite thermometry estimate of 500 °C (minimum) was obtained from a Dol + Qtz-bearing sample C394 in the pelitic garnet zone (Masch and Heuss-Aßbichler, Chap. 10, this Vol.; see Fig. 2.6). Although it is possible that 500 °C reflects contact metamorphism (see thermal profiles in Pattison, Chap. 16 and

Buntebarth, Chap. 18, this Vol.), it seems unlikely: talc is absent in this sample, and the contact metamorphic talc isograd, located closer to the igneous complex than C394, occurs in the range 370–480 °C (assuming a pressure of 0.3 GPa; Masch and Heuss-Aßbichler, Chap. 10, this Vol.). Taking 500 °C (minimum) as a regional estimate, it fits in well with the regional garnet-biotite temperature estimates.

The 470–510 °C estimate at the garnet isograd is generally consistent with the lack of evidence for transformation of microcline to sanidine near the garnet isograd in the northeast of the area, according to the temperature estimates of Kroll et al. (Chap. 13, this Vol.). Kroll et al. estimate the transformation temperature as 470–490 °C, based on the location of the transformation relative to a series of isograds of Frey et al. (1980).

Taking a peak metamorphic temperature of about 490 °C at the garnet isograd in the middle of the area and about 550 °C in the southeast of the area, peak metamorphic temperatures in the northwest of the area can be estimated at 400–450 °C, assuming a roughly uniform temperature gradient.

Pressure is much more difficult to estimate, since few geobarometers are applicable to these low-grade assemblages. The Powell and Evans (1983) Chl+Bt+Ms barometer gives pressures ranging from 0.9 to 0.13 GPa at 500 °C and $a(H_2O) = 1$, which seem unreasonably high (Pattison 1985). In Knapdale, 80 km to the south, Graham et al.(1983) estimated pressures of 0.8–0.12 GPa at 510–530 °C in the garnet zone, but these were largely based on the same Chl + Ms + Bt barometer. Pressure estimates in the garnet zone near Roy Bridge, 35 km NE of Ballachulish, using several independent geobarometric methods, range between 0.5 and 0.6 GPa at 530 °C (Richardson and Powell 1976). If one accepts a southwardly increasing pressure gradient from Roy Bridge to Knapdale, then a pressure of 0.6 ± 0.1 GPa seems reasonable for the Ballachulish area (Pattison 1985).

2.6 Small Pre-Intrusion Igneous Bodies

Pre-dating emplacement of the Ballachulish Igneous Complex and post-dating regional metamorphism and deformation, a series of small explosion breccias and mafic-ultramafic pipes were emplaced in the area (see Map 2). These are known as the Appinite Suite (Bowes and Wright 1967). The pipes include (Bailey and Maufe 1960; Bowes and Wright 1967):

1. Olivine + augite + hornblende ± hypersthene ± biotite diorite (Cortlandite)
2. Olivine + augite + hornblende + biotite + plagioclase + K-feldspar diorite (Kentallenite)
3. Augite + biotite + olivine ± hornblende diorite
4. Hornblende + augite + plagioclase ± K-feldspar ± quartz ± biotite diorite (Appinite).

Bowes and Wright (1967) deduced the following sequence of emplacement: (1) explosion breccia pipes; (2) hornblende-rich bodies such as Appinite; (3) augite diorite and Kentallenite bodies; and (4) leucogranite and granodiorite.

The coarse grain size of the bodies, and their association with explosion breccias, suggest that they formed from vapour-rich magmas. The sequence of emplacement suggests an overall decrease in volatile content with time. Wright and Bowes (1979) analyzed the major- and trace-element geochemistry of the suite, and concluded that the bodies represent differentiates of a basic magma of mantle origin, enriched in volatiles.

The Appinite suite is thought to have intruded along the weakened hinges of parasitic folds on the flanks of major folds such as the Appin Syncline (Bowes and Wright 1967). Troll and Weiss (Chap. 3, this Vol.) have shown that to the southeast of the Ballachulish Igneous Complex there is a hybrid transition zone between the monzodiorites and a large Appinite body, suggesting that the Appinite body was emplaced only shortly before the main igneous complex. Thus, the Appinite suite of intrusions may have exploited deeply penetrating extensional faults and fractures which shortly after facilitated the emplacement of the main igneous complex.

2.7 Uplift and Erosion

2.7.1 Pre-Intrusion Uplift

In a detailed study of uplift of the Dalradian in Perthshire and Angus, 80 km E of Ballachulish, Dempster (1985) estimated that at about 410 Ma, garnet and staurolite grade rocks had been uplifted to a depth of 10–15 km and had cooled to about 250 °C. If this is representative of regional uplift of the Dalradian (Dempster 1985), then the ambient temperature of the Dalradian country rocks in the Ballachulish area was probably close to 250 °C at the time and depth of emplacement of the Ballachulish Igneous Complex [412 ± 28 Ma (Troll and Weiss, Chap. 3, this Vol.) and 10 km (Pattison, Chap. 16, this Vol.)]. Ambient temperatures immediately prior to emplacement may have been somewhat higher (e.g., 300 °C), due to higher heat flow accompanying the ascent of the magma.

2.7.2 Post-Intrusion Uplift

A number of lines of evidence suggest that, following emplacement of the Ballachulish Igneous Complex in the Devonian, there was no major tilting during uplift. The Devonian age Glen Coe and Lorne lavas, 10 km to the east and 15 km to the southwest, respectively, are essentially flat lying, as is Devonian Old Red Sandstone further to the south. Droop and Treloar (1981), based on field evidence of Brown (1975), estimated a maximum tilting of 3 km over 35 km along a NW-SE axis in the Glen Etive Complex, 8 km SE of Ballachulish. This translates to a maximum tilting of 0.7 km across the 8-km width of the Ballachulish Igneous Complex and aureole, which is less than the 1-km topographic relief in the area.

2.7.3 Relationship Between the Ballachulish Igneous Complex and the Glen Coe Lavas

The close proximity of the Glen Coe lavas to the Ballachulish Igneous Complex, both of similar age, provides somewhat of a paradox: the igneous complex was emplaced at a depth of about 10 km (Pattison, Chap. 16, this Vol.), whilst the lavas were extruded onto the surface. There are no major faults separating the two that could account for the 10-km difference in level of erosion.

Leven Schists within the Glen Coe cauldron subsidence contain no regional metamorphic garnet, compared to garnet-bearing Leven Schists surrounding the cauldron subsidence (Elles and Tilley 1930). This suggests that the lavas subsided within the caldera. Assuming that the lavas subsided in their present orientation, and noting the diameter of the caldera and height of the surrounding mountains outside the caldera that contain no lavas, Cheeney (pers. comm. 1985) used simple trigonometry to estimate a minimal subsidence of 5000 m. This would reduce the difference in erosion level between the lavas and the Ballachulish Igneous Complex to about 5000 m.

K-Ar age dating of the Lorne Lavas 50 km to the SE, considered to be about the same age as the Glen Coe lavas, gives a mean age of 394 ± 9 Ma (Evans et al. 1971). This is about 18 Ma younger than the 412 ± 28 Ma emplacement age of the Ballachulish Igneous Complex (Weiss and Troll, Chap. 3, this Vol.).

Dempster (1985) estimates rapid rates (0.5 mm/a) of regional uplift in the Dalradian between 410 and 390 Ma, which over 18 Ma could account for up to 9000 m of erosion. In summary, a combination of subsidence of the lavas in the Glen Coe caldera and rapid uplift between the emplacement of the Ballachulish Igneous Complex and the extrusion of the lavas would appear to account for the difference in erosion level between the two centres.

2.8 Summary

The Ballachulish Igneous Complex is emplaced in Lower and Middle Dalradian metasediments comprising quartzites, phyllites, graphitic slates, semipelites and marbles of the Appin and Argyll Groups (see Table 2.1). These are believed to have been deposited in a NE-SW trending gulf in the late Precambrian, with periodic basin deepening and shallowing.

One main deformation, D_1, has affected the rocks in the area, producing a series of NE-SW trending folds and slides and a prominent axial planar slaty cleavage in micaceous units. Later deformation is manifested in a variety of mm-m scale textural and structural features. Regional metamorphism, which probably reached its peak during or after D_1 deformation, increases in grade from west to east through chlorite, biotite and garnet zones. Temperatures at peak regional metamorphism are estimated at 400–450 °C in the west of the study area, 470–510 °C at the garnet isograd in the middle of the area, and about 550 °C in the southeast. Pressure is estimated at about 0.6 GPa.

Subsequent to regional deformation and metamorphism, a series of explosion breccias and small intrusions of mantle-derived mafic-ultramafic magma were emplaced in the Ballachulish area. These immediately preceded the emplacement of the main complex.

Crosscutting the igneous complex and contact aureole is the late NE-SW-trending Ballachulish Fault. Post-intrusion sinistral displacement across the fault is about 700 m. The Ballachulish Fault may have been active before, during and after emplacement of the Ballachulish Igneous Complex.

Pre-intrusion country rocks temperatures at the time and depth of emplacement of the igneous complex (412 ± 28 Ma; 10 km) are estimated to have been 250–300 °C. Following emplacement of the igneous complex, uplift occurred without any major tilting.

Part II
The Intrusive Complex

3 Structure, Petrography and Emplacement of Plutonic Rocks

G. Troll and S. Weiss

3.1 Introduction

The two-phase Ballachulish pluton is composed of a zoned monzodiorite – quartz diorite envelope with distinct flow- and deformation-foliation, surrounding a core of porphyritic granite with hybrid margins. Mesotype members of this suite are constituted of alkaline hypersthene diorites. This provides an unique opportunity to quantify the magmatic crystallization sequence within a relatively "dry" plutonic environment at upper crustal level, in terms of T-X-f_{O_2} evolution (Weiss and Troll, Chap. 4, this Vol.). The present chapter starts with a gross megascopic description of the igneous rocks. Section 3.2 deals with contact relationships, xenoliths and foliation structures, followed by a modal classification of the igneous units (Sect. 3.3.1). A detailed petrography of index minerals, their texture and areal distribution (Sect. 3.3.2) precede the synoptical Sect. 3.4 on magma emplacement and intrusion mechanics.

3.1.1 Field Relationships and Previous Work

The igneous complex of Ballachulish covers an area of only 7.5 × 4.5 km (Fig. 3.1); it is well exposed, with several stream profiles exceeding 600 m in height. The ellipsoid pluton consists of an outer grey diorite, surrounding a younger core of porphyritic pink granite (with hybrid margins). The zoned diorite envelope ranges in composition from monzodiorite to quartz diorite, shows a distinct flow- and deformation-foliation, and contains abundant sedimentary xenoliths at its margins (Walker 1924; Muir 1953a,b). Radiometric data for quartz diorites and granites (Miller and Brown 1965; Harper 1967; P.E. Brown et al. 1968; Hurford 1977; Haslam and Kimbell 1981) indicate a Lower Devonian magmatic event at 412 ± 28 Ma (Fig. 3.14). The plutonic rocks were emplaced at a depth of ca. 10 km within "wet" Dalradian metasediments, producing a 450–800 °C thermal aureole at 0.3 GPa (= 3 kbar) fluid pressure (Pattison, Chap. 16; Masch and Heuss-Aßbichler (Chap. 10, this Vol.).

G. Voll, J. Töpel, D.R.M. Pattison, F. Seifert
Equilibrium and Kinetics
in Contact Metamorphism

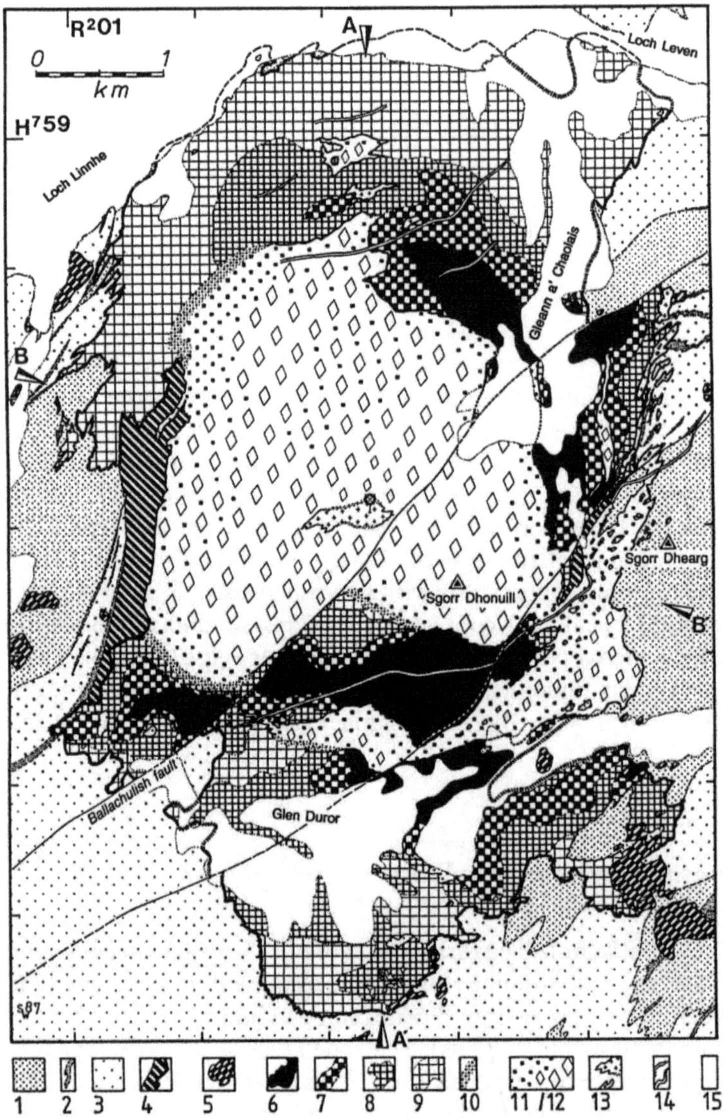

Fig. 3.1. Simplified geological map of Ballachulish igneous complex, based on the 1:10 000 scale field mapping of Weiss (1986). The petrofacial subdivision of the diorite envelope and the central granite refers to Weiss and Troll (1989, Fig. 3). *1,2,3* Quartzites, carbonates and pelites of the Scottish Dalradian (Bailey and Maufe 1960); *4* contact migmatites (Pattison and Harte 1988); *5* appinites (Wright and Bowes 1979); *6* Opx-Cpx monzodiorite ("dry"; Bi+Am 6, Qz 2.5 vol%); *7* Opx-Cpx-Qz monzodiorite ("hydrated"; Bi+Am = 14-26, Qz 3 vol%); *8* Cpx-Am-Qz diorite; *9* Bi-Am-Qz diorite; *10* hybrid transition zone; *11/12* hybrid/porphyritic granite; *13* leucogranite (Haslam and Kimbell 1981); *14* rhyolite dykes; *15* alluvial cover

3.1.2 General Description of Rock Types

Basic Rocks. Basic to intermediate, gas-rich potassic rocks of the Appinite Suite consist of hornblendites, olivine-monzonites and hornblende-rich diorites, forming small stocks and explosion breccia pipes in the periphery of Ballachulish Complex. These mafic bodies, which show strong affinities to mantle-derived magmas (Wright and Bowes 1979), closely predate the main phase of diorite emplacement.

Diorites. These are dark greenish-grey pyroxene diorites which form the inner part of the diorite envelope. They occupy a crescent-shaped area, approaching the country rocks only in SW-SE-E parts of the complex (Fig. 4.1). Hypersthene-rich, medium- to coarse-grained varieties are best exposed in stream profiles on the southern flank of Sgorr Dhonuill (Fig. 3.1). The equigranular diorites are characterized by an adcumulus texture showing andesine+augite+hypersthene+apatite as primocrysts. These phases, which usually do not show optical zoning, are joined by intercumulus opaque oxides and interstitial alkali feldspar (Fig. 3.3a). Besides minor secondary amphibole, black poikilitic Ti-biotite is a conspicuous component of all the orthopyroxene-bearing, quartz-free to quartz-poor monzodiorites. The anhedral biotite flakes show no dimensional orientation to flow- or deformation-structures. Neither areal distribution nor variable volume fraction of biotite seem to be genetically related to the emplacement of the younger granite. To the southeast of Sgorr Dhonuill, where a large granite apophysis cuts the diorites, a very fine-grained marginal facies (Fig. 3.4c) of the hypersthene diorite is exposed, which bordered quartzitic country rocks before the granite intrusion. This "chilled margin" (Cox et al. 1979) indicates a largely

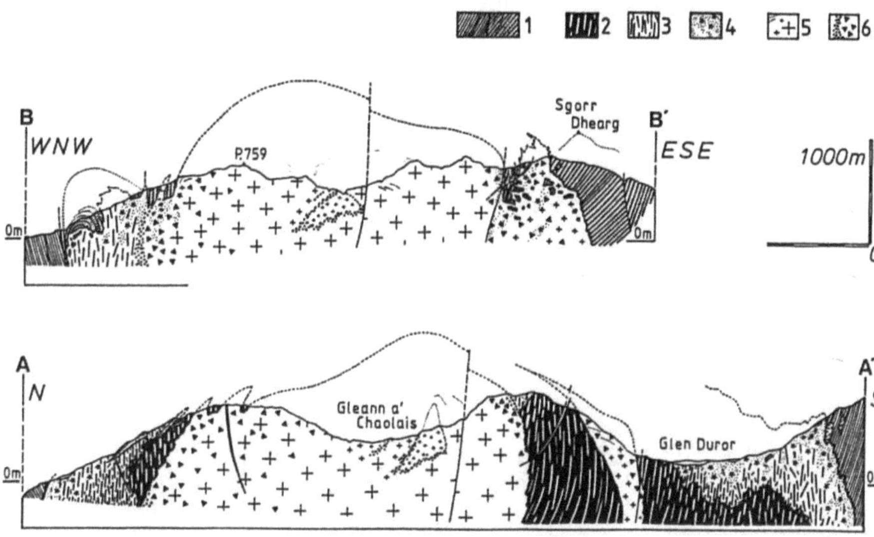

Fig. 3.2. Geological profile sections A-A' and B-B' across the igneous complex (see arrows in Fig. 3.1). *1* Metasediments; *2* monzodiorites (with flow- and deformation-foliation); *3* quartz diorites (with alignment of pelitic xenoliths); *4* fine-grained quartz diorites with xenoliths; *5* granites; *6* transition zone, hybrid granite

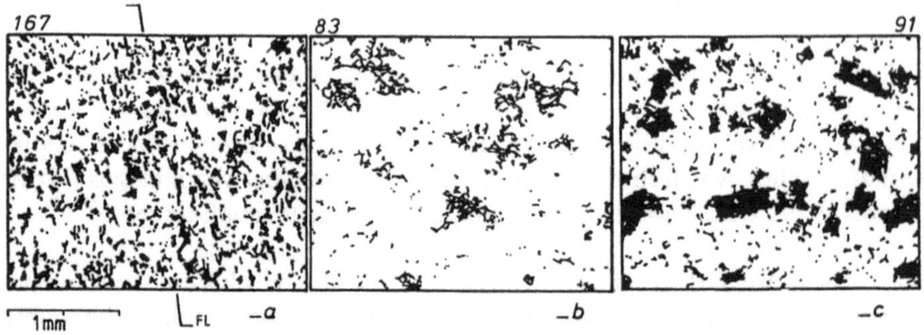

Fig. 3.3. Alkali feldspar textures (*black*) of the type-rocks monzodiorite. **a** with direction of magmatic flow-foliation *FL*; **b** quartz diorite; **c** porphyritic granite. The textures are redrawn from oriented, horizontally cut hand specimens stained for alkali feldspar

liquid state of the initial monzodiorite magma. The textural and grain-size relationships between the "normal" and the "chilled" facies types were studied in some detail (Weiss, Chap. 5, this Vol.).

Quartz diorites. Grey quartz diorites form the marginal northern, northwestern and southern parts of the diorite envelope (Fig. 3.1). The junctions between areas of finer and coarser types of the quartz diorites, as well as the junctions to the monzodiorites, may be rather sharp or merging, a feature already described by Haslam (1968) from the neighbouring Ben Nevis Complex. The grain size is hence very variable, especially in the more leucocratic parts, where xenoliths are abundant (Sect. 3.2.2); a fine-grained marginal facies is mainly confined to the E-SE-SW periphery of the

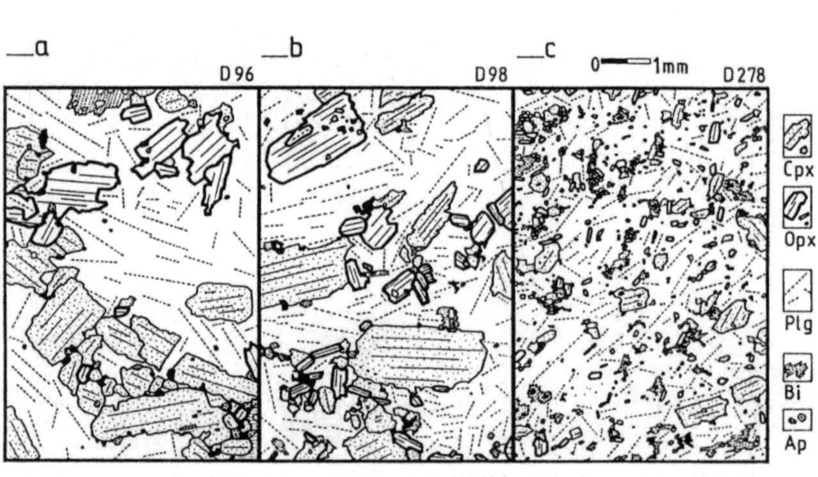

Fig. 3.4. Thin section drawings of three textural types. **a** central facies; **b** normal facies; and **c** 'chilled' marginal facies of Ballachulish monzodiorites. Plg comprises the entire salic matrix, including the flow-foliated fabric of Plg-primocrysts (*dashed*), alkali feldspar and traces of quartz

igneous complex (Fig. 3.8). Here, biotite forms the main mafic phase, whereas stout greenish amphibole (with core relics of augite) joins the paragenesis in the more homogenous parts of the quartz diorites. Poikilitic alkali feldspar reaches a diameter of 0.5–1 cm in this unchilled, medium- to coarse-grained "normal facies" (Fig. 3.3b).

Granites. The main porphyritic granite is fairly homogenous and extends over an area of 8 km^2. Reddish megacrysts of alkali feldspar form euhedral to subhedral tabular crystals, their xenoblastic rims merging with the granodioritic groundmass (Fig. 3.3c). Averaging volume fractions of 5–15% in the Pink Granite (Bailey and Maufe 1960), the perthitic megacrysts reach a length of 0.5–1.5 cm. In the marginal hybrid facies of the main granite, including the broad transitional zone to the quartz diorite envelope (SE of Loch Linnhe, Fig. 3.1), a rudimentary porphyritic texture has developed (Akf $\leq$ 0.5 cm), showing partially resorbed mafic clots of biotite and uralitic amphibole (0.5 cm diam.). A non-porphyritic, fine-grained leucogranite stock occupies the centre of the main granite. This muscovite-bearing aplitic microgranite shows a strong hydrothermal alteration with disseminated Cu-Mo-mineralization (Haslam and Kimbell 1981).

Dykes. Rare rhyolite dykes are 1.5–5-m wide, sometimes displaying a quartz porphyritic texture. They cut the diorites and the porphyritic granite, dipping inwards to the centre of the complex (Figs. 3.1, 3.2). The dykes thus show an affinity to subvolcanic cone-sheet structures, trending subparallel to the earlier granite ring-dykes (Sect. 3.2.1). Microdiorite dykes clearly postdate the emplacement of the igneous complex, crosscutting all magmatic and metasedimentary rocks of the Ballachulish area.

Faults. Possibly related to the Great Glen shear zone is another large lineament, the Ballachulish Fault (Fig. 3.1), with a smaller structure to the southeast, sinistrally displacing the igneous complex and its thermal aureole by ca. 700–800 m (Pattison 1985; Weiss 1986). The deflection of magmatic flow structures in diorites adjoining this fault (Fig. 3.8) suggests early tectonic movements, predating the consolidation of monzodiorites. Likewise, early tectonic shearing is indicated by presence of protomylonitic hornfelsed metapelites in Glen Duror (Pattison 1985; Weiss 1986). Tectonic activity evidently outlasted the emplacement of granites and cone-sheet dykes, since the largest rhyolite dyke shows thinning, displacement and mylonitic shearing within the proximity of both NE-SW fault structures (Weiss 1986, Figs. 10, 11).

3.2 Structural Features of Emplacement and Consolidation of the Plutonic Rocks

3.2.1 The Contact Zones

3.2.1.1 Contacts to the Country Rocks

The contacts to metasediments are generally sharp and discordant, following directions of lithological or structural anisotropy. They comprise bedding planes, the main cleavage s_1, and the set of NNE by SE master and transverse joints, including pre- to syn-intrusive, NE-trending fault zones. Hence the contacts appear mainly steep on outcrop scale, but may nevertheless be controlled by large-scale fold structures trending NE-SW (see Pattison and Voll, Chap. 2, this Vol.).

At the western margin of Ballachulish Complex (near Kentallen), quartz diorites intruded the large quartzite anticline of Beinn Sgluich along E-W transverse joints trending 70–100/70–85° S (see Fig. 3.1 and profile B-B' in Fig. 3.2). The arch structure of the quartzitic units favoured a filling with the intermediate magma, at least in northern core portions of the antiform. This is evident from the course of leucodiorite apophyses in the quartzites, subparallel to current bedding and to narrow semipelite layers (trending 55–60/30–40° N). Likewise, main contacts of quartz diorites and granites to pelitic country rocks on both flanks of the quartzite anticline are stratiform, i.e., they occur as characteristic lit-par-lit injections of leucocratic, near-vertical veins. Southeast of Kentallen, these sheeted contacts are even dipping at 80–85° towards the centre of the igneous complex, due to the westerly vergence of metasedimentary units between the Beinn Sgluich antiform and the Appin syncline. The entire country rock sector between Kentallen and Lagnaha represented a "roof portion" with respect to the quartz diorite magma. This resulted in a relatively high heat flow across the conductive quartzites, with released magmatic fluids first inducing a static partial melting of the adjacent Lagnaha pelite screen and creating a wide zone of contact migmatites (Figs. 3.1, 3.8). Subsequently, the emplacement and "ballooning" of the granite pluton (Sect. 3.4.2) introduced a dynamic stage with plastic deformation and chaotic disruption of the anatectic migmatites (Pattison and Harte 1988).

The southern contact of the igneous complex is best exposed in Coire Dubh, where quartz diorites cut isoclinally folded pelites along transverse joints (120/70° S), and subparallel to bedding and s_1-schistosity (20/75° W). This results in an E-W trending structural contact plane, dipping southwards at 75–80° (see profile A'-A in Figs. 3.1, 3.2). Complex contact relationships characterize the southeastern periphery of the complex, where monzodiorites came into contact with a large mafic stock of the Appinite Suite (Fig. 3.1). Apparently joining the intrusion of appinites within a short time range, the monzodiorite magma was able to assimilate marginal portions of the older hornblende diorites and hornblendites, before these precursors were entirely consolidated. The resulting hybrid transition zones are narrow and contain spherical basic enclaves measuring 0.5–5 m. The entire area forms a mosaic roof zone, where quartzite screens, pelite rafts and basic blocks subsided in fine-grained diorites; this inhomogenous marginal facies frequently displays a schlieren structure.

The eastern contact to quartzitic units at Sgorr Dhearg is vertical or dips steeply towards the east at 80–85°, despite the relatively flat-dipping current bedding in the quartzites. In the northerly adjoining, upturned syncline of Allt Giubsachain, however, which is separated by a fault zone from the quartzites, the intrusion of diorites and granites was stratiform. Semipelites, calc-silicates and carbonates are exposed in this asymmetric, near-vertical syncline. Its western flank is tectonically reduced, with a slightly overturned b-axis plunging southwest at 75–80°. Pre-intrusive strain and shear movements affected several ductile marble layers at the northern margin of the synform, favouring isoclinal folding, tectonic thickening and local disintegration of these incompetent units. On the southwestern flank of the steep synform, granodiorite apophyses and leucogranite veins were injected along lithological boundaries between semipelites and carbonates, into minor fold crests, and along the fan-shaped s_1-schistosity jointing (see Fig. 3.1 and detailed map in Weiss 1986, Fig. 11). Anisotropic lithology and internal structure of the cone-shaped Allt Giubhsachain synform, appearing closed towards northeast, obviously restricted magmatic injection to its SW-flank and to local core portions, hence causing a mainly vertical heat flow.

3.2.1.2 Internal Contacts and Zones of Hybrid Mixing

Sharp, well-defined internal contact planes are only detectable in the middle and eastern portions of the complex, where monzodiorites border the granites. Characteristic exposures are found at an altitude of 550–800 m on the steep slope south of Sgorr Dhonuill, where hypersthene-bearing monzodiorites show subsidences and fracturing at the granite contact, following a set of straight master joints (in 70–90/70–80° S and 140–160/85–90° E). In southwestern and especially in northwestern parts of the igneous complex near Kentallen, however, transitional zones occur in a width of 300–500 m, separating leucocratic quartz diorites and non-porphyritic hybrid granites. This marginal, augite-bearing granite facies forms a tongue-shaped zone (Fig. 4.10a) continuously grading into the more homogenous porphyritic granites, which are free of mafic clots resorbed from the diorites. The contrasting contact relationships between pyroxene-rich monzodiorites and granites in the eastern parts, and between quartz-rich leucodiorites in the western parts of the complex, suggest a temporal and spatial consolidation sequence from "dry" monzodiorites (affected by strain rupturing) to the more fractionated quartz diorites (containing residual hydrous liquid), with the progressively lowered contrast in rheology favouring hybrid magma mixing (Sects. 3.4.1, 3.4.2).

Granite apophyses generally have the form of incomplete ring dykes (Anderson 1937), due to partial subsidences of the diorite envelope along shear fractures. These apophyses are thinning and flattening upwards, always dipping away from the centre of the igneous complex. Their width and dip hence varies with the vertical and lateral distance from the main body of porphyritic granites (as a magmatic centre). These relationships are well illustrated in the northern section of profile A-A' (Fig. 3.2).

Hanging-wall contact planes of the 2–25-m-wide apophyses may be undulating and diffuse, showing partial hybridization against diorite roof portions. This suggests that the diorites of this zone had not reached their solidus when the granites were emplaced, but were sufficiently consolidated to display a brittle behaviour relative to tectonic movements. The largest apophysis of hybrid granites, situated SW of Sgorr Dhearg, attains a width of 200–500 m and is separated from the main porphyritic granites by several diorite screens. The crescent-shaped apophysis is subvertical and widest in its eastern portion (dipping 75–80° E), with progressively reduced thickness towards the southwestern slope of Sgorr Dhonuill. This does nevertheless not result in a significantly reduced outcrop width of the granites, due to the increasingly flat dip (75–50° S) of the apophysis in its western portions (Figs. 3.1, 3.2, profile A'-A). Here, the granite ring dyke branches into flat-dipping, 0.5–2-m-thick aplitic sills (50–15° S), which are petrographically identical with the central leucogranite stock northwest of Sgorr Dhonuill. This fine-grained felsitic microgranite (Haslam and Kimbell 1981) is exposed vertically for more than 300 m southwest of Gleann a' Chaolais, where it shows sharp contacts as well as diffuse transition zones to the surrounding porphyritic granites. The structure of the relatively flat-dipping internal contact zones and the semicircular arrangement of syngenetic aplite sills (N and SW of Sgorr Dhonuill) indicates a possible hood shape for this late leucogranite (Fig. 3.2). Narrow, miarolitic pegmatite sills may crosscut monzodiorites and semipelites (at the E contact), whereas a rare skarn mineralization is developed at the contact with impure calc-silicates. The evolutionary sequence of hybrid granites via leucogranites and aplites to pegmatites, and the tectonic position of late magmatic differentiates (flat-dipping ring-dyke fractures) clearly attributes a pegmatitic-hydrothermal activity to the younger granite magmatism, whereas a similar relationship is not obvious for the diorites.

3.2.2 Metasedimentary Xenoliths and Mafic Enclaves

3.2.2.1 Areal Distribution

Numerous pelite xenoliths are present in the outer, quartz-bearing diorites, where large rafts up to 50 m in length are occasionally preserved (Figs. 3.1, 3.2, 3.5). In the western and northern coastal sectors, leucocratic quartz diorites are heavily contaminated by an unusual number of xenoliths in various stages of resorption (Muir 1953a; Bailey and Maufe 1960). This small-scale fragmentation (Fig. 3.5a) is attributed to piecemeal stoping, acting preferentially in the periphery of Ballachulish Complex. On the other hand, xenoliths are extremely rare in the inner, hypersthene-rich portions of the monzodiorite zone. The porphyritic granites rarely contain xenoliths scattered near their margins, usually in hybrid mixing zones with quartz diorites; on the other hand, numerous monzodiorite blocks and metasedimentary screens became detached from the subsiding diorite envelope and the adjoining country rocks, accompanying the emplacement of the crescent-shaped granite apophysis of the eastern contact (Fig. 3.1). On the western flank of Sgorr Dhearg,

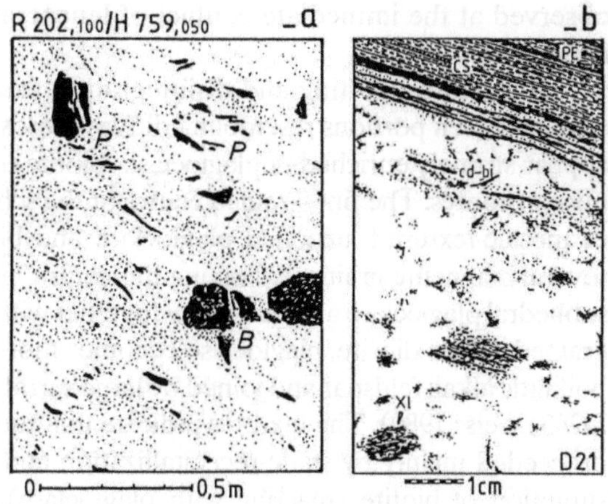

Fig. 3.5 a. Angular pelite xenoliths (P) and elliptic mafic enclaves (B), enclosed in leucocratic quartz diorites; this inhomogenous marginal facies displays fine-grained schlieren portions (*stippled*). The main alignment plane of tabloid xenoliths trends 60/50-80° NW in this subvertical outcrop section (NE-SW road profile near Leitir Mhor/Loch Linnhe) **b** Sheeted contact of leucocratic quartz diorites to interbedded semipelite-calcsilicate metasediments (PE-CS) at the NE margin of the igneous complex. The horizontally cut diorite sample D21 shows fine-grained plagioclase-biotite schlieren (*stippled*), disintegrated cordierite-biotite layers (cd-bi and elliptic micaceous pelite xenoliths (XI, corundum-bearing)

these hybrid granites enclose a huge quartzite screen measuring 750 m in length. It is bordered by several rafts of dolomitic carbonates and calc-silicates, locally interbedded with contact-migmatized pelite layers. Except for some minor tilting, the relative coherency as well as the contact-consistent strike and dip of these *schollen* units suggests a preservation of the primary, pre-intrusive stratigraphy and tectonic structure for metasediments incorporated in the granites. By contrast, the pelitic roof of the southeastern margin of the complex appears largely disintegrated due to block stopping, with dislocated pelite rafts in a size of 25–200 m enclosed by fine-grained diorites (Weiss 1986, Fig. 10).

3.2.2.2 Pelitic Xenoliths

Tabular pelite xenoliths are abundant and evenly distributed in the coastal sectors of the igneous complex. Their size is quite variable, usually 1–15 cm in length; the undeformed angular fragments are usually found in subparallel orientation (Fig. 3.5a). Within the quartz diorites, pelitic xenoliths are generally present as high-grade hornfelses containing cordierite+andalusite+sillimanite±garnet (Qz-present) and corundum+spinel (Qz-absent), whereas a 200–250-m-large raft with garnet+orthopyroxene (Pattison 1985) adjoins the hydrated monzodiorites southeast of Glen Duror (Fig. 3.1). Towards Glen Duror, the diorites continuously grade into amphibole- and biotite-rich varieties; they contain rounded, micaceous pelite xenoliths in a size of usually less than 0.5–2 cm. This advanced resorption stage may also be observed at the immediate contact of leucocratic quartz diorites to pelitic country rocks (Fig. 3.5b).

Progressive melting and disintegration usually converted pelitic xenoliths into dark schlieren portions resembling diffuse mesosomes; they do not contain quartz and appear strongly enriched in plagioclase+biotite+magnetite, relative to the surrounding quartz diorites. The fine-grained centre of the schlieren mesosome is mainly composed of mosaic-textured, unzoned plagioclase and biotite, enclosing fresh cordierite with several magnetite grains. Adjoining this central portion, an inner zone shows abundant subhedral plagioclase and magnetite together with embayed biotite. Towards the coarse-grained quartz diorite, plagioclase becomes more euhedral and zoned, is infiltrated by poikilitic alkali feldspar and joined by large serrate biotite flakes with minor quartz (e.g., D263, Weiss 1986). The textures indicate that the evolution of the schlieren mesosome proceeded mainly by static recrystallization and melting, probably involving the consumption of biotite (reacting with plagioclase). This produced abundant magnetite releasing MgO + H_2O towards the surrounding magma, where the crystallization of magnesio-hornblende was favoured (Sect. 4.2.5, this Vol.).

Since high-grade pelites usually contained the assemblage Qz + Bi + Crd ± Sill + Kf ± Plg ± Ilm prior to migmatization (Pattison and Harte 1988), the restitic appearance of the Qz + Kf-free Crd + Mt + Plg ± Bi mesosome may indicate a selective dehydration melting of biotite (Thompson 1982) in pelitic screens which had been detached from the country rocks, floating and sinking within the hot diorite magma. Even if melting was not very advanced, an effective disaggregation of pelitic screens is possible if a *critical* fraction of anatectic melt is exceeded (ca. 25–35 vol%), changing the rheological properties of the xenoliths from a coherent solid to a viscous "crystal mush" (Van der Molen and Paterson 1979).

Initial xenolith resorption may have involved H_2O-conserving, vapour-absent dehydration melting, such as

Bi + Sill + Qz ± Plg ↔ Crd + Kf + Liquid

(Thompson 1982), since the estimated 1050–1000 °C emplacement-T for the quartz diorite magma is well above the onset of dehydration melting near 720–760 °C at P_{total} = 0.3 GPa, as suggested by Thompson (1982).

Attaining a size of less than 0.3 × 1 cm, fine-grained plagioclase schlieren with abundant magnetite (Fig. 3.6) were occasionally detected in hypersthene-rich monzodiorites (D95a,221a,278). Near these mosaic-textured enclaves, embayed skeletal zircon with anhedral zoning may be preserved as a diagnostic refractory component of largely resorbed metasediments (compare Pidgeon and Aftalion 1978). The centre of the schlieren is rarely composed of cordierite, with several magnetite generations indicating a polyphase, complex reaction sequence (D95a). Some of the anhedral, small-sized orthopyroxene grains which are abundant around the schlieren (Fig. 3.6), might have been derived from the dehydration melting of biotite-rich pelites. The model reaction:

Bt + Qz ± Plg ↔ Opx + Kf + Liquid

requires a minimum temperature of ~800–850 °C at P_{total} = 0.3 GPa (Grant 1986). Although biotite is much more stable in the absence of quartz (Weiss and Troll, Sect.

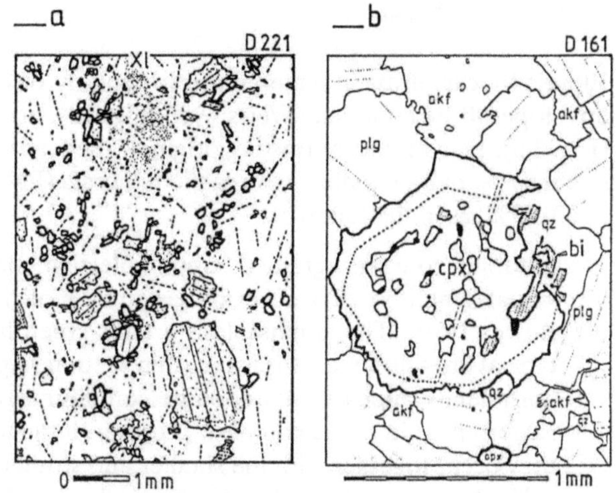

Fig. 3.6 a. Plagioclase-magnetite schlieren portion (XI) in monzodiorite D221. This marginal facies is characterized by clusters of numerous small hypersthene crystals and zoned augite primocrysts (*light stipple*). *Dashed lines* mark plagioclase primocrysts. **b** Twinned zoned augite primocryst (cpx), showing anhedral plagioclase growth inclusions and a late-magmatic rim corrosion. This pyroxene-out reaction, which typifies this hydrated marginal monzodiorite facies (D161), apparently proceeded with alkaline residual liquid to produce biotite and quartz at the expense of augite (*right*)

4.2.3, this Vol.), even a Qz-absent biotite breakdown is possible for small pelite fragments enclosed within the central dry monzodiorites, given the high emplacement temperature of this subzone (near 1100 °C; Weiss and Troll, Sect. 4.3, this Vol.).

3.2.2.3 Quartzitic Xenoliths

Leucocratic quartz diorites at the western and northwestern margin of the igneous complex contain numerous white quartzite xenoliths reaching a size of 10–30 cm. The nearly monomineralic, rounded xenoliths may be rimmed by broad mafic fringes which are rich in pale green augite (narrow inner zone) as well as in greenish-black amphibole with plagioclase (wide outer portion). As pointed out by Muir (1953b), the inner pyroxene corona apparently formed at an early stage, as a result of deficiency of volatiles, with augite first crystallizing from a less contaminated, rather basic quartz diorite magma. Continued contamination with "dry" quartzites and "wet" pelites, as well as further crystal fractionation, caused an enrichment of the magma in $SiO_2 + Al_2O_3 + H_2O$, favouring uralitic replacement of pyroxene, together with magmatic crystallization of plagioclase+amphibole in the outer portion of the corona. The host rock around the corona is modified to a leucocratic, biotite- and sphene-bearing variety which is significantly enriched in quartz (from 11–12 to nearly 20 vol%). These relationships are evident even on a larger scale, since Walker (1924) describes a leucocratic, 10-m-wide contact facies of white granite between quartz diorites and quartzites, 2 km NNE of Kentallen.

3.2.2.4 Mafic Enclaves

Elliptic blocks of mafic material are irregularly distributed throughout the diorite envelope of the complex (Weiss 1986, Fig. 18), usually having a size of 10–40 cm (Fig. 3.5). Due to the contrasting origin of the enclaves, their mineralogy and grain size may be quite variable, with mafic minerals (FM) constituting volume fractions of 30–55%, as in the following:

1. Medium- to coarse-grained, amphibole-rich enclaves (FM = 50–55) represent actual xenoliths, derived from basic rocks of the Appinite Suite (Wright and Bowes 1979). The amphibole is a dark-green hornblende, which may show a typical brownish pleochroism in thin section (e.g., D350 in Weiss 1986). Similar hornblende appears with appinite-contaminated quartz diorites at the south-eastern margin of the complex. It forms elongate crystals up to centimeter size; they occur together with the smaller and stout, blue-greenish amphiboles (with colourless core relics of augite), which elsewhere typify the quartz diorites. Southeast of Glen Duror, advanced hybridization locally modified the margins of large appinite blocks (0.5–5 m in size), thus effacing any sharp contact and producing a close resemblance to the adjoining hydrated monzodiorites (compare Fig. 3.1).
2. Fine-grained, amphibole- and biotite-rich enclaves (FM = 45–50) were detected in the western part of Glen Duror and in the northwestern coastal sector (Leitir Mhor quarries). Petrographically and chemically, (e.g., Ni 100 ppm, D352b in Weiss 1986), the igneous-textured enclaves may resemble appinitic rocks, differing mainly in grain size; normally zoned plagioclase, large zircons and long apatite needles (within poikilitic alkali feldspar) are typical constituents. Such dark microgranular enclaves may indicate a contribution of mantle material during the petrogenesis of granitoid rocks, as pointed out by Didier (1987). A recent Nd-Sr study by Holden et al. (1987) backs this assumption at least for the Strontian and Criffell plutons, strongly suggesting a mantle origin for microdioritic enclaves.
3. Fine-grained, pyroxene-rich enclaves (FM = 30–45) are common in the transitional areas between 'dry' monzodiorites and leucocratic quartz diorites. These angular autoliths represent syngenetic fragments of early crystallized hypersthene diorites that have fallen back into the fractionating magma (compare Tindle and Pearce 1981), undergoing subsequent biotite blastesis and uralitization within the hydrated marginal portions of the monzodiorite subzone (Fig. 3.1), e.g., in Gleann a' Chaolais (D271a in Weiss 1986).

3.2.3 Flow- and Deformation-Foliation in the Diorites

In the quartz diorites, tabular xenoliths (specifically the hornfelses) show a preferred planar orientation which may be attributed to laminar flow in a partly crystallized magma. Fine-grained biotite-rich schlieren, which are thought to have been generated by partial melting of pelitic xenoliths, follow the main xenolith orientation which usually dips at 50–85° towards the margin of the igneous complex.

In the monzodiorites, where xenoliths are scarce, primocrysts of Plg-Cpx-Opx are found in distinctive planar alignment (Fig. 3.40), either in near-vertical orientation or steeply dipping towards the granite pluton at angles of 75 to 85°, giving the overall impression of a steep cone sheet. Figures 3.3a and 3.4 show that the mineral foliation (with interstitial Akf) is marked by elongate plagioclase laths (1.5–7 mm long) and euhedral pyroxenes (1–5 mm long). This adcumulus-type foliation which locally shows pyroxene synneusis (Weiss and Troll 1989, Fig. 11a), is clearly diagnostic for magmatic flow involving megascopic convection currents (Tait et al. 1984; Morse 1986).

A weak deformation foliation superimposes on the primary magmatic foliation, mainly affecting plagioclase in the monzodiorites (Fig. 3.7), and, to a lesser extent, subhedral biotite flakes in augite-rich quartz diorites (Weiss 1986, Fig. 93). A bending of plagioclase is most frequent and intense within pyroxene-rich diorites, where deviations of 2–5° from the c-axis direction of plagioclase are attained (Fig. 3.7). A

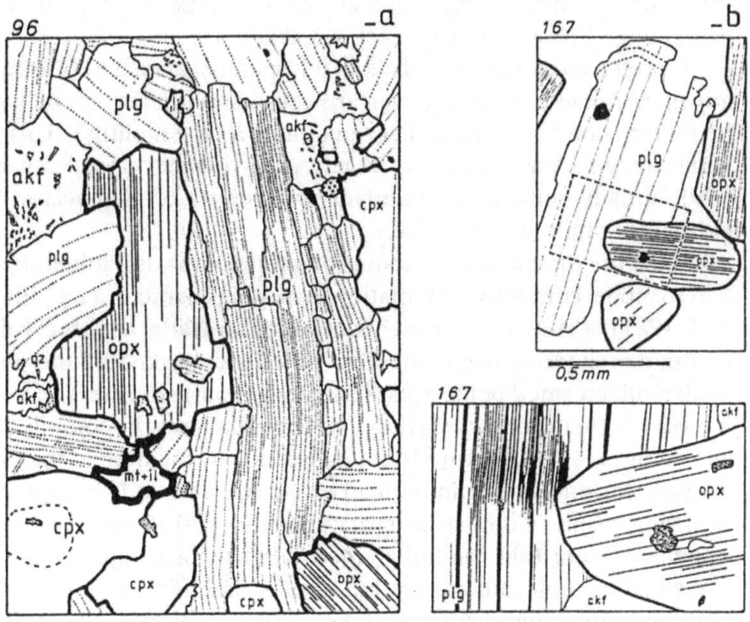

Fig. 3.7 a. Primary adcumulus lamination of Albite-twinned andesine (*stippled*), displaying a distinct secondary deformation-bending and marginal subgrain formation near subophitic hypersthene, within monzodiorite D96 (horizontal section). Symplectitic alkali feldspar (*akf*) and minor quartz (*qz*) occur as interstitial phases to plagioclase (*plg*) and orthopyroxene (*opx*). **b** (*right*): Deformation wedge within growth-twinned plagioclase (*stippled, above*), showing numerous mechanical Albite twins which taper and curve (*black, below*). This deformation-twinning, which involves rotational strain-slip, was induced by a rigid hypersthene crystal (opx, *right*), set up in crosswise orientation to the NNE-SSW adcumulus fabric in monzodiorite D167. Schematic drawing from oriented thin section

marginal subgrain formation may occur (Fig. 3.7a), whereas kink bands or cataclastic features are not detectable optically. In the vicinity of strain-resistant, optically undeformed pyroxenes, the ductile deformation of plagioclase appears most intense. Here, the normal Albite growth twinning is modified by a strong mechanical twinning; it is restricted to small volumes of the grain where fine-scale polysynthetic Albite twins are seen to taper and curve (Fig. 3.7b). The high volume fraction of andesine in monzodiorites (52–65%) hence acts as a stress-supporting framework, preventing the deformation of intercalated rigid pyroxenes (20–28%) by rotational strain-slip and subgrain recrystallization (Tullis 1983).

Due to low strain rate, a ductile flattening of the flow fabric in monzodiorites was apparently not very effective, thus largely preserving this preexisting igneous lamination. Since the (weak) deformation foliation closely traces the (distinctive) magmatic flow foliation, it is diagnostic for a continuous intrusion-deformation sequence (Pitcher 1979), so that it appears quite unrealistic to distinguish between the two textural types both in the field and in thin section, as already stated by Holder (1983). The ductile plagioclase deformation, which is observed within the monzodiorites, does not continue towards the quartz diorites, since these fractionated rocks still contained a considerable amount of residual liquid while the monzodiorites were almost entirely consolidated, at the time of granite emplacement and 'ballooning' (Weiss and Troll, Sect. 4.3, this Vol.).

A preferred orientation of alkali feldspar megacrysts in the granites is detectable only in marginal parts to the western coastal sector, and in the hybrid apophysis at the eastern contact (subparallel to country rock xenoliths). This indicates that the granite magma was emplaced largely as a viscous "crystal mush" diapir (Pitcher 1979), already containing abundant primocrysts of plagioclase, biotite and alkali feldspar (in random orientation).

Since the dimensional orientation of xenoliths and minerals can be effectively used to model convective circulation in magma chambers (e.g., Pitcher 1979; Phillips et al. 1981), a similar approach was made in Ballachulish (Weiss 1986, Fig. 17), tracing directions of magmatic flow which are marked by the planar alignment of tabular, undeformed pelite xenoliths (in the quartz diorites) and of tabloid cumulus primocrysts (in the monzodiorites). It is obvious from the compilation of all structural data in Fig. 3.8, that the trends of magmatic flow foliation in the diorite shell are not everywhere concentric and parallel to the contacts: at first appearance a somewhat surprising result. Comparing structural and petrofacial data (Figs. 3.8, 3.10), it becomes evident that the units "monzodiorite" and "quartz diorite" are not only diversified by their mineralogy, but also by their internal structure. The distinctive, steep concentric flow alignment within the monzodiorite subzone indicates a cylindrical front of cooling and consolidation, subvertically advancing towards the magmatic centre of Ballachulish Complex. On the present level of emplacement, continuing consolidation involved progressive crystal fractionation and country rock assimilation, which specifically changed the rheological properties of the diorite magma in its outer, i.e., marginal portions. The anomalously curved flow trends in NW-N-SE portions of the complex (Fig. 3.8) might well be due to an irregular "blocky" roof structure developing during the emplacement of the outer quartz diorite zone, with large subsiding country rock rafts deflecting the trends of magmatic flow.

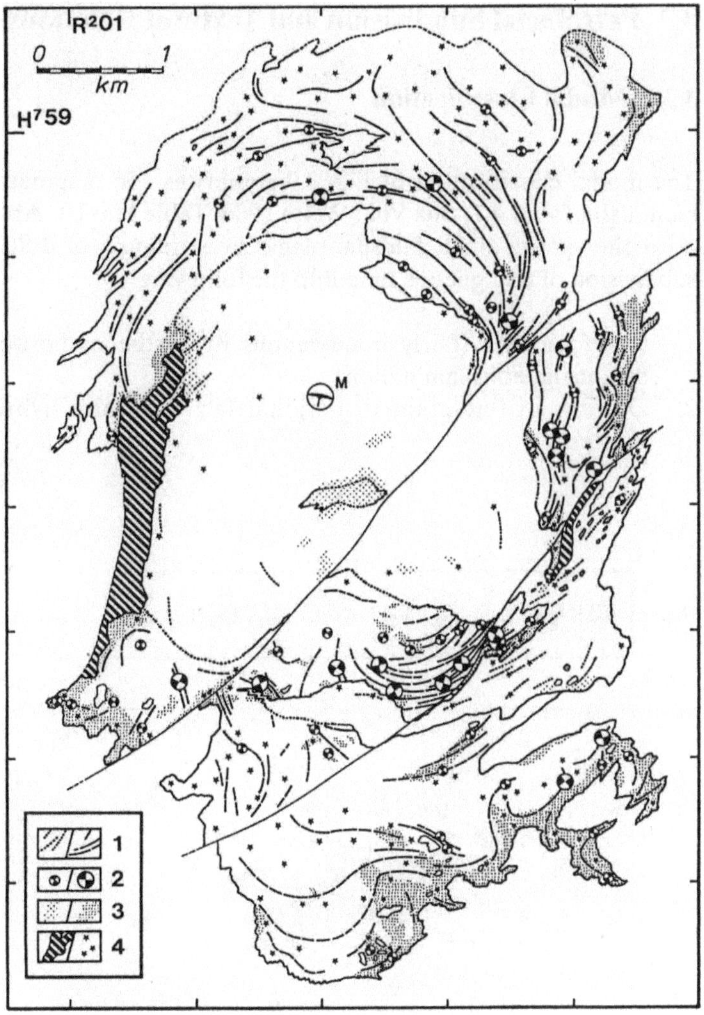

Fig. 3.8. Composite map showing the flow lines of the diorite magma after its emplacement in Ballachulish, as derived from the dimensional orientation of tabloid pelite xenoliths and Plg-Cpx-Opx primocrysts (based on the field data of Weiss 1986, Fig. 17). For lithologies see Fig. 3.1. *1* Trend lines in monzodiorites (*line*), quartz diorites (*long dashes*) and granites (*short dashes*); *2* weak/strong ductile deformation of plagioclase; *3* fine-grained granites/'chilled' and fine-grained marginal facies of the diorites; *4 Schollen* migmatites (Pattison and Harte 1988)/pelitic xenoliths; M magnetic susceptibility measurement (Weiss 1986)

3.3 Petrofacial Subdivision and Textural Relationships

3.3.1 Modal Classification

The modal classification of Fig. 3.9 comprises 156 magmatite samples from Ballachulish (Table 4.1, this Vol.; Weiss 1986, Table IIa+b). Analyses were projected onto the quartz-alkali feldspar-plagioclase triangle of IUGS (1973), allowing a subdivision of the igneous suite into the following:

1. Monzodiorites (fairly homogenous hypersthene diorites, ± quartz-free with distinct igneous lamination)
2. Diorites (± fine-grained marginal facies, locally hybridized with appinitic basites)

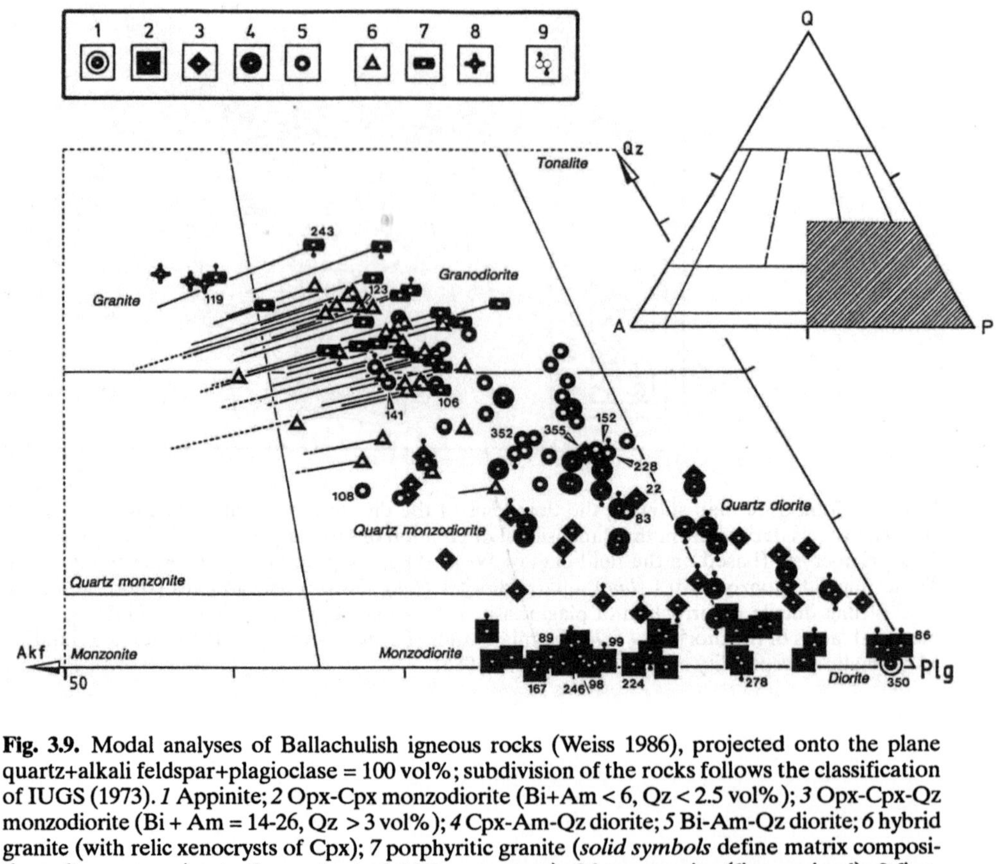

Fig. 3.9. Modal analyses of Ballachulish igneous rocks (Weiss 1986), projected onto the plane quartz+alkali feldspar+plagioclase = 100 vol%; subdivision of the rocks follows the classification of IUGS (1973). *1* Appinite; *2* Opx-Cpx monzodiorite (Bi+Am < 6, Qz < 2.5 vol%); *3* Opx-Cpx-Qz monzodiorite (Bi + Am = 14-26, Qz > 3 vol%); *4* Cpx-Am-Qz diorite; *5* Bi-Am-Qz diorite; *6* hybrid granite (with relic xenocrysts of Cpx); *7* porphyritic granite (*solid symbols* define matrix compositions, *lines* approximate the content of Akf megacrysts); *8* leucogranite (fine-grained); *9* fine-grained facies in general (cf. Figs. 3.1, 3.10 and Table 4.1). Sample numbers refer to Fig. 4.1 and Table 4.1

3. Quartz diorites, quartz monzodiorites and granodiorites (belonging to the inhomogenous, strongly pelite-contaminated suite of outer quartz diorites)
4. Granodiorites, granites (porphyritic normal facies) and quartz monzodiorites (hybrid marginal facies)
5. Monzogranites (fine-grained leucogranites)

Combining the distribution of primary mafic silicates, which show textural evidence of magmatic crystallization from a melt (Weiss and Troll 1989, Fig. 3) with the trends of magmatic lamination for Plg-Cpx-Opx primocrysts and xenoliths (Fig. 3.8), the diorite envelope of Ballachulish Complex may be subdivided into four petrofacial zones (compare Fig. 3.1):

1. Opx-Cpx monzodiorites (Qz < 2.5 vol%; central 'dry' facies contains only 2–6 vol% Bi + Am)
2. Opx-Cpx-Qz monzodiorites (Qz > 3 vol%; 'hydrated' facies of zone 1, with 14–26 vol% Bi+Am)
3. Cpx-Am-Qz diorites (mafic minerals > 20 vol%, Cpx usually ≥ 1–1.5 vol%)
4. Bi-Am-Qz diorites (Cpx ≤ 0.5–1.5 vol%; leucocratic facies with less than 20 vol% mafic minerals)

For a genetic interpretation of these complex lithological relationships, including the introductory description of index minerals in the next section, it appears useful to simplify the modal classification of Fig. 3.9 and to condense the diorite subzones 1 + 2 and 3 + 4 into an inner hypersthene-bearing monzodiorite zone, and an outer quartz diorite zone free of orthopyroxene (combining symbols 2 + 3 and 4 + 5 in Fig. 3.9). Hence, the textural and geothermometrical evaluation of Weiss and Troll (Chap. 4, this Vol.) refers to the three main rock types monzodiorite, quartz diorite and granite (compare Figs. 3.1 and 4.1).

3.3.2 The Index Minerals

Plagioclase. Elongate euhedral plagioclase with very weak or absent zoning constitutes a major fraction (52–61 vol%) in the monzodiorites, reaching 64–65 vol% in the fine-grained marginal facies (D86,278). The andesine is usually affected by a weak ductile deformation bending (Sect. 3.2.3); as within the quartz diorites, corrosion-type boundaries to alkali feldspar may be frequent (Figs. 3.6b, 3.7a). Stout, complexly zoned plagioclase occurs in quartz diorites and in the granites. Core resorptions, oscillations and synneusis effects ('drifting together'; Vance 1969) modify the strong euhedral zoning of plagioclase specifically in the leucocratic quartz diorites, where pelitic xenoliths are abundant. A genetic interpretation of plagioclase zoning is given by Weiss and Troll (Sect. 4.2.2, this Vol.; 1989, Fig. 10).

The Al/Si-ordering state of plagioclase, as deduced from its optical properties (Fig. 3.12a), strongly suggests, despite some scatter resulting from ductile deformation (Fig. 3.7), that low, i.e., Al/Si-ordered, plagioclase is present throughout the igneous suite of Ballachulish, including the chilled diorite sample D278.

Since the sericitization of feldspars, i.e., the partial replacement of plagioclase by secondary muscovite, is mainly controlled by late-stage hydrothermal fluids, the degree of sericitization was estimated in thin section as a measure for the activity of these subsolidus fluids (Fig. 3.10). Generally, the intensity of plagioclase sericitiza-

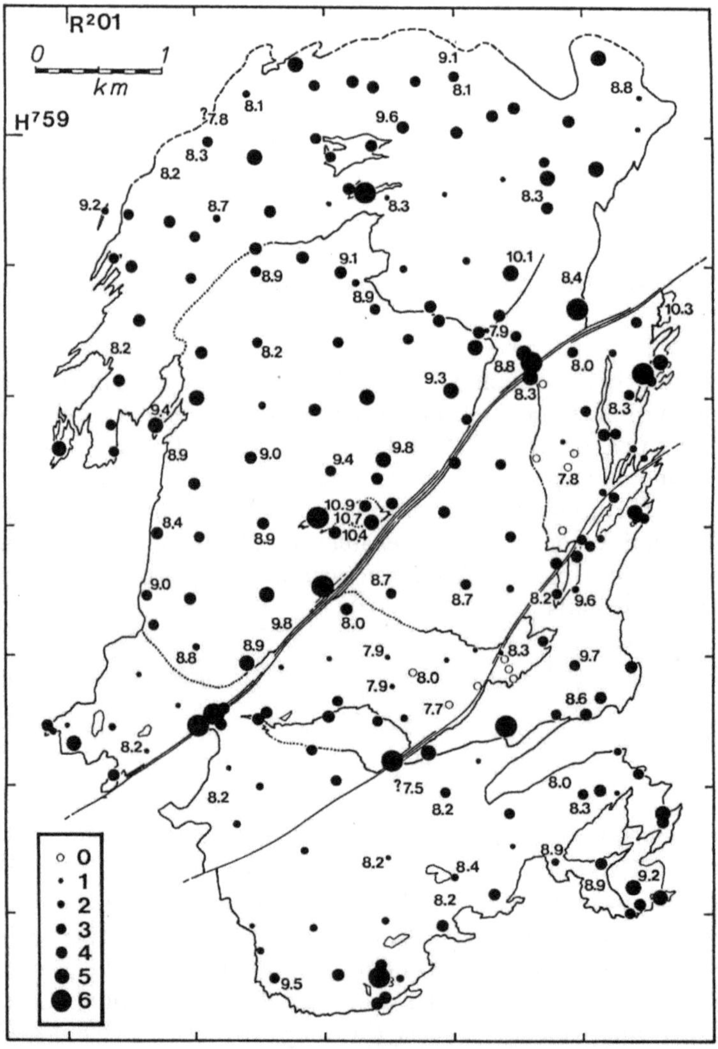

Fig. 3.10. Degree of hydrothermal sericitization of plagioclase within the igneous complex of Ballachulish (Weiss 1986 and unpubl. data). The *circles* give the relative intensity of sericitization estimated in thin section (214 samples), with an empirical scale ranging from *1* (sz. absent) to *6* (sz. complete). *Numbers* refer to the oxygen isotope ratio $^{18/16}O$ [‰] of 66 whole-rock samples from the pluton (Hoernes et al. Chap. 17, this Vol.). For lithologies, see Figs. 3.1, 4.1

tion is positively correlated with the oxygen isotope ratio of the host rock (Hoernes et al. 1985; compare Fig. 3.10); a detailed interpretation, however, has to consider the effects of crystal fractionation and pelite assimilation (Fig. 4.4), which may considerably influence the isotopic composition of the host rocks (Hoernes et al., Sect. 17.5.1, this Vol.). In Ballachulish, the degree of late magmatic hydration (Fig. 4.10), and of post-magmatic hydrothermal alteration (Fig. 17.4), is usually smaller in the diorites than in the granites; it is lowest in the least fractionated, hypersthene-rich monzodiorites. The sericitization is significantly stronger in areas of intense pelite contamination, where it is comparable to the more leucocratic quartz diorites. Other zones of high alteration include the diorite margins and the granite apophyses; the sulphide-bearing central leucogranites and rock portions along the younger faults (see Figs. 3.1, 3.8) appear particularly affected by late fluids.

Alkali feldspar. Alkali feldspar appears in the diorites always as a late interstitial mineral with respect to plagioclase and pyroxenes (±amphibole). Alkali feldspar megacrysts from the granites enclose plagioclase and biotite; they show a distinct mechanical-compositional zoning, which reflects the magmatic zoning of coexisting plagioclase (referring to lower nucleation rates and higher growth rates for Akf; Swanson 1977).

The structural state of alkali feldspar, i.e., the total Al-content of T_1-tetrahedral sites [Σ_{t1}], as deduced from the optic axial angle $2V_x$ (Su et al., 1986), ranges from [0.72–0.80] in Ballachulish monzodiorites over [0.77–0.81] in quartz diorites to [0.79–0.85] in granites (Weiss 1986, Fig. 31). Despite the significant increase in the volume proportion of more Al/Si-ordered alkali feldspar, the overall metastable preservation of orthoclase or intermediate microcline structural states is striking. This is consistent with the observation that a crosshatched twinning, which usually characterizes low microcline (McLaren 1984), was not detectable even in coarse microperthites from the granites. The [Σ_{t1}]-contoured plot of optic axial angle $2V_x$ versus chemical composition in Fig. 3.12b confirms intermediate ordering states in the monzodiorites [$\Sigma_{t1} = 0.75 \pm 0.04$]. In the quartz diorites, [Σ_{t1}] varies considerably (vertical lines in Fig. 3.14b); the effect may be more pronounced in the granites, where a single 1.5-mm grain of Akf had a 47–63° $2V_x$-range (Weiss 1986, Fig. 25a). Such structural inhomogeneities may reflect grain-scale fluctuations in composition ($X_{An,Ab}$) and in fluid activity (Yund 1984). Both parameters, which control Al/Si-ordering (Parsons 1978), preferentially preserved 'stranded' intermediate states for the relatively Or-rich alkali feldspars (Fig. 4.9a) within the 'dry' monzodiorites of Ballachulish.

Microperthitic alkali feldspars were texturally classified according to Smith and Brown (1988). Figure 3.11 shows the areal distribution of perthite types, with textural details reported by Weiss and Troll (1989, Figs. 8–10) as follows:

1. A series of crypto- to microperthites which elsewhere characterize granulites, charnockites and alkali gabbros (Yund et al. 1980; Parsons and Brown 1983), is confined to the hypersthene diorites which are poor in quartz and hydrous phases (Fig. 3.11). This indicates a 'dry', i.e., fluid-absent environment for this pyroxene-rich subfacies (Fig. 4.10a). Here, alkali feldspar textures comprise

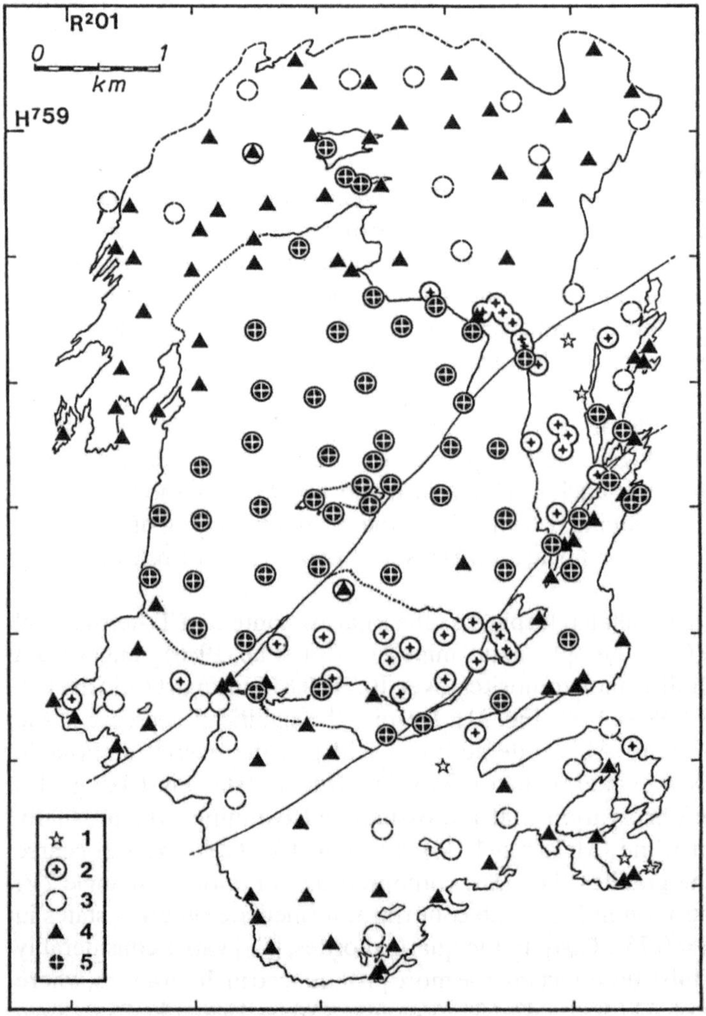

Fig. 3.11. Areal distribution of perthite microtextures in alkali feldspars from Ballachulish igneous complex. For lithologies see Fig. 3.1; textures are shown in Weiss and Troll (1989, Figs. 8-10). *1* No Akf present; *2* film-perthites and symplectite intergrowths; *3* optically homogenous Akf; *4* patch- and flame-perthites; *5* coarse patch- to vein-perthites

optically homogenous cryptoperthites (Weiss 1986, TEM-image of Fig. 34), fine lamellar film perthites and complexly shaped symplectites, including blebs of oligoclase composition.

2. Optically homogenous or patch- to flame-perthitic alkali feldspar characterizes monzodiorites which are Opx-free, the leucocratic quartz diorites and the hybrid granites; coarse patch- to vein-perthites typify the porphyritic granites and the leucogranites. These large, irregularly shaped perthites are referred by Parsons

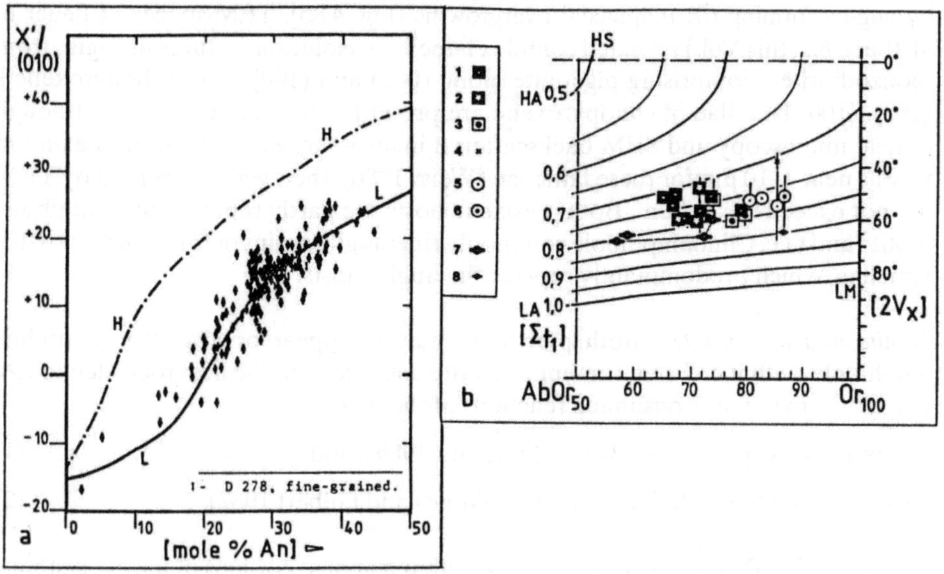

Fig. 3.12. Structural state of plagioclase (**a**) and alkali feldspar (**b**), deduced from optical properties. **a** Plot of extinction angle versus composition and structural state for the a-axis section of plagioclase from Ballachulish diorites and granites. The curves *H* and *L* refer to Al/Si-disordered (high) and -ordered (low) plagioclase (Smith 1974). **b** Al/Si-ordering state of alkali feldspar, estimated from $2V_x$ versus mol% Or (Weiss and Troll 1989, Table II); the diagram is contoured for the Al-content Σt_1 of the T_1 tetrahedral sites in the series high albite-sanidine and low albite-microcline (Su et al. 1986). *Vertical bars* give the $2V_x$-range in single Akf grains. *1,2* = Homogenous and film-perthitic Akf, in monzodiorites; *3* homogenous Akf between symplectites; *4* antiperthites; *5,6* homogenous Akf and matrix between perthites, in quartz diorites; *7,8* megacrysts and groundmass Akf, in the granites

(1978) to deuteric coarsening, favoured by the activity of postmagmatic hydrothermal fluids. This is well documented in the Ballachulish Complex, where deuteric microperthites are commonly associated with plagioclase showing a relatively high degree of hydrothermal sericitization (cf. Figs. 3.10 and 3.11).

Pyroxenes. Turbid subhedral augite, which may show a weak optical zoning, is diagnostic for the diorite suite and may also be metastably preserved in the granites. The optic axial angle of augite increases from monzodiorites ($2V_z$ = 45–59°, core-rim) to quartz diorites ($2V_z$ = 59–62°), while the optic extinction is progressively lowered during re-equilibration and uralitization ($z \wedge c$ = 43–45° to 36–37°). The undeformed, (010)-flattened cumulus phase is joined by clear and unzoned, (100)-flattened hypersthene crystals ($2V_x$ = 51–54°). Whereas inclusions of plagioclase, opaques and apatite are rare in the orthopyroxene (and usually confined to its rims), they are more frequent in the clinopyroxene. Within the marginal facies of monzodiorites, augite cores typically contain numerous plagioclase growth inclusions (Figs. 3.6b, 5.5). Occasionally, Mg-rich hypersthene primocrysts are found (e.g., in D89,99,224), which are corroded and mantled by augite, forming (010)-epitaxial overgrowths (Fig. 4.8b). TEM-studies of Feuer et al. (Sect. 6.2, this Vol.) revealed complex lamellar exsolution textures in augite from monzodiorites, comprising pigeonite along (001) and (100), and orthopyroxene || (100). (100) lamellae of clinopyroxene are present in hypersthene hosts; although optical microscopy and SEM backscattered images suggested apparent maximum widths near 1–10 μm for these lamellae (Weiss 1986), their width observed by TEM did not exceed 0.1–0.4 μm. Both pyroxene hosts are partly replaced by amphibole, biotite and talc. Clinoamphibole commonly rims augite grains or forms narrow (010) lamellae, which predominantly replace the augite hosts.

Biotite and amphiboles. Both pyroxenes may disappear or give way to uralitic amphibole, with increasing content of biotite and quartz in the host rock, due to late magmatic, pyroxene-consuming reactions of the type

3 Opx + K-feldspar + L ↔ Bi + 3 Qz (Grant 1986), and (1)

2 Cpx + Bi + 4 Qz ↔ K-feldspar+Am (Wones and Gilbert 1982). (2)

Hydration Reaction (1) mainly acted within marginal portions of the hypersthene diorites, whereas Reaction (2) is most effective in the quartz diorites, containing poikilitic alkali feldspar and edenitic amphibole (Weiss and Troll 1989, Figs. 10, 11a).

Titaniferous biotite is generally a late phase in the monzodiorites; the dark reddish-brown poikilitic flakes, which coexist with augite and ilmenite, show frequent flexures and kink bands in pyroxene-rich diorites. In the quartz diorites, brown subhedral to euhedral biotite commonly occurs with late magmatic green amphibole and quartz. In the granites, brownish-green euhedral biotite is an early phase, forming growth inclusions together with plagioclase near the core of alkali feldspar megacrysts. Biotite associated with magnetite may be affected by a late-stage hydrothermal alteration in leucocratic quartz diorites, forming disc-shaped prehnite-clinozoisite-sphene-hematite intergrowths below 350–300 °C (Weiss 1986, 3.4.7–4).

Amphiboles exclusively formed at the expense of pyroxenes by topotactic reaction during late- to post-magmatic stages. Three types of amphibole are texturally distinguishable; they were classified by composition into magnesio-hornblende to edenite (Am-I), actinolitic hornblende (Am-II) and actinolite (Am-X; Weiss 1986, 4.4.1 and Table IVd). In monzodiorites, augite is progressively replaced with increasing fractionation stage of the host rock (see D.I. in Table 4.1) by the green magnesio-hornblende to edenite Am-I, whereas hypersthene reacts to colourless, fibrous actinolitic hornblende Am-II. In quartz diorites, augite is mantled or fully replaced by the edenite Am-I; it shares subhedral to euhedral phase boundaries with plagioclase, biotite and quartz, coexisting with magnetite, sphene and alkali feldspar (Weiss 1986, Fig. 59). In quartz diorites D108,152 and in hybrid granite G123, green amphibole encloses colourless core relics and xenocrysts of augite, which showed anomalous optical properties ($2V_z \sim 70°$; $z \wedge c = 25–27°$). Microprobe analysis proved that these high-relief relic grains consist entirely of actinolite Am-X.

Iron-titanium oxides. The volume ratio of ilmenite to magnetite approximates 2:1 in monzodiorites; in the vicinity of hypersthene, ilmenite clusters share grain boundaries with magnetite. In contrast to these two-phase aggregates, primary titanomagnetite and hemoilmenite are mainly found as isolated subhedral grains, separated by silicate phases (Plg and Cpx). Even at 500x magnification (reflected light, oil immersion), the opaque oxides do not show any lamellar exsolutions (e.g., trellis intergrowths of ilmenomagnetite). In leucocratic quartz diorites and in the granites, deuteric alteration and oxidation of the Fe-Ti oxides produced sphene and maghemite.

Zircon. Primary zircon is not present in monzodiorites (containing radioactive apatite), whereas corroded grains with anhedral zoning may occur, originating from resorbed pelitic material (Sect. 3.2.2). Elongate, unzoned zircons crystallized with interstitial quartz and alkali feldspar in the quartz diorites. In the porphyritic granites, stout zircons with strong euhedral zoning are found together with biotite. Pupin showed (1980) that the morphology of zircon may be related to its crystallization temperature. Applying this typology method on zircon from the porphyritic granites of Ballachulish (see Weiss and Troll 1989), an early crystallization near 800–750 (±50) °C in an aluminous monzogranitic magma is suggested. By contrast, zircon from in the quartz diorites was stabilized only at the late magmatic stage within a hyperalkaline residual liquid, probably near 700 ± 50 °C.

3.4 An Emplacement Model for Ballachulish Igneous Complex

3.4.1 The Diorite Pluton

The sketch map of Fig. 3.13a shows an early emplacement stage of the monzodiorite magma, yet already zonally modified by convective fractionation and country rock assimilation towards quartz dioritic compositions, with possible enrichment of a rhyodacitic residual magma in the centre of the complex. Considering the semicircular arrangement of mesotype monzodiorites with their "chilled" eastern margin, and their internal adcumulate texture which has the form of a steeply inclined cylinder (Figs. 3.1, 3.2, 3.8), the first batch of hot and "dry" alkaline diorites was emplaced at a largely liquid state (with ~ 1% dissolved H_2O at ~ 1100 °C; Weiss and Troll, Section 4.3.1, this Vol.). Initially, it constituted only a small elliptical stock measuring ca. 3 × 5 km; this refers to the exposed cross section near sea level (cf. Rabble and Meissner, Chap. 7, this Vol.).

Since the rheological properties of magmas are more dependent on the fraction and size of suspended crystals than on temperature and water content of the melt (Kirkpatrick 1981), the effective viscosity η_e of the monzodiorite magma was calculated for sub-liquidus temperatures, following McBirney and Murase (1984). Mean primocryst diameters of 0.5 mm and 0.8 mm were assumed for the diorites and granites, respectively, by approximating the size for early dynamic growth stages in a melt and thus disregarding static rim zones (for crystal shape and size see Weiss 1986, Figs. 57–60). The used effective diameters of 0.5–0.8 mm were preferred to an average size model (a:b:c-axis) for the fluidally aligned primocrysts, since it is the

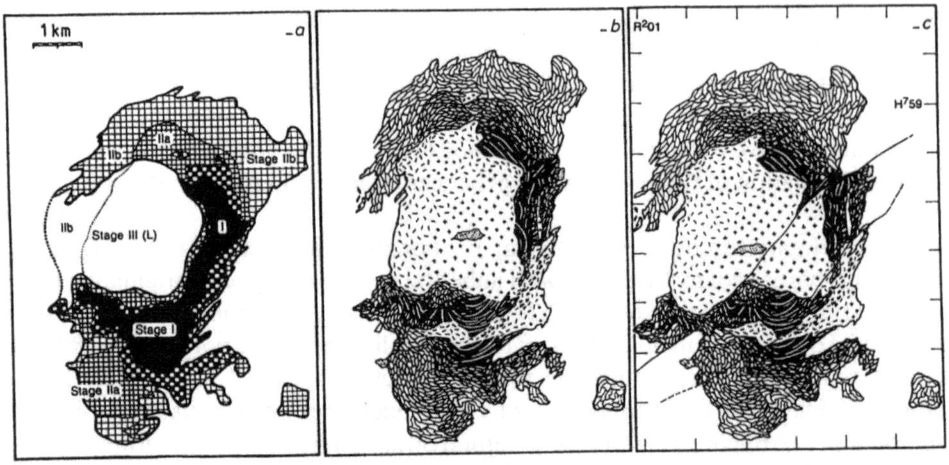

Fig. 3.13. Magmatic evolution of Ballachulish Complex. The sketch map (**a**) illustrates the permissive emplacement of the diorite magma, which was zonally modified by convective fractionation and frame-rock assimilation; the lithologies correspond to Fig. 3.1. Stage **b** accounts for the forceful intrusion of the granites, initially comprising rotational shearing, local subsidences and the propagation of ring dykes within the diorite shell; subsequent in situ expansion ("ballooning") of the central granite pluton induced a weak ductile deformation on mesotype diorites and on contact migmatites. **c** The actual position of the igneous complex shows the effects of postmagmatic sinistral fault displacement

cross section of suspended crystals, which critically controls the flow resistance with respect to laminar convection currents. It should be noted that, given a reasonable thermal gradient and a subcritical effective viscosity ($\eta_e < 10^8$–10^{10} poises; Pitcher 1979), a magma tends to concentrate primocrysts in a "mush zone" (Morse 1986), accumulating the mobile rejected solute in the centre of a pluton, an effect described as convective fractionation (Sparks et al. 1984).

In Ballachulish diorites, convective transport was most effective at initial magma viscosities of 10^3–10^5 poises (for ~ 1100–1050 °C and 0–10 vol% crystals), with Opx-Cpx-Plg primocrysts continuously concentrated, captured and fixed in a subvertical "mush zone" which, at viscosities near 10^8–10^{10} poises containing ~ 30–50 vol% crystals, slowly advanced towards the centre of the igneous complex ("stage I" in Fig. 3.13a). Except for a "chilled" to fine-grained facies confined to the E-SE-SW margins of the intrusion (Fig. 3.8), this convective fractionation produced an even-grained cylindrically shaped thermal envelope of adcumulates (Tait et al. 1984), which still contained ~ 20–25 vol% of trapped alkaline liquid at ca. 950 °C, thus constituting a major source of latent heat even for the late-magmatic T-range (Fig. 4.11). Subsequent to the early stage of monzodiorite emplacement, which involved the centripetal fractional accumulation of a quartz diorite magma in situ, the onset of regional scale tensional movements (Brown 1979) aided the permissive enlargement of the pluton specifically towards its northern and southern margin ("stage IIA" in Fig. 3.13a). Preheated anticlinal roof portions, including the Lagnaha semipelite screen (Pattison and Harte 1988) became highly dissected by block stoping, with numerous small xenoliths drifting and partially dissolving, when the still mobile quartz diorite magma (T ~ 1050–1000 °C, η_e ~ 10^7–10^8 poises for 10–15 vol% crystals) worked its way upwards to the final level of consolidation. This corresponds to the present NE-SW coastal sectors ("stage IIB" in Fig. 3.13a). The presence of complexly zoned plagioclase primocrysts, which show frequent mutual attachment (grain clustering by synneusis), is obviously related to magmatic turbulence during the earlier consolidation stage of these leucodioritic rocks (Vance 1969), which probably contained 5–10 vol% crystals at 1050–1000°C (resulting in η_e of ~ 10^4–10^5 poises).

Marsh has demonstrated (1982, Fig. 6) that thermal and chemical contamination of magmas during country rock incorporation is essentially prevented, if the sinking xenoliths have diameters exceeding 5 m in a basaltic magma (η ~ 10^3 poises) to 50 m in a granitic magma (η ~ 10^7 poises). In Ballachulish there is ample field evidence as well as textural and chemical data, to indicate that the sinking rates of small pelite xenoliths (with effective diameters of less than 5–30 cm) originating from a preheated roof zone, were sufficiently small to favour extensive chemical contamination of the outer quartz diorites (Fig. 4.4). By contrast, larger country rock rafts exceeding a size of 30–50 m were rapidly sinking; they were frozen only within the fine-grained, rapidly solidifying SW-SE marginal pyroxene diorites (Fig. 3.8), which appear less affected by chemical contamination than the leucocratic quartz diorites.

We think that, with advanced fractionation and consolidation of the diorites, a subliquidus residual magma of rhyodacitic composition (D.I. > 70, T ~ 900 °C) accumulated in the centre of the igneous complex. Tectonic fracturing, e.g., related to the Ballachulish lineament (Sect. 3.1.2), may have connected this upper-crustal magma reservoir to a Caledonian central volcano. This is suggested by the neighbouring shallow-level ring complexes of Glencoe and Ben Nevis (Bailey and Maufe 1960), where leucodiorites appear closely associated with andesitic-rhyodacitic lavas, containing biotite-amphibole phenocrysts, occasionally yet with relic augite (Haslam 1968). Even if the postulated release of a rhydacitic residue in Ballachulish was coupled to the subsidence of a central block prior to granite emplacement (Anderson 1937), and even if the country rock block was severely shattered by fault dissection (Pitcher 1979, p. 647) or thermal stressing (Marsh 1982), this large-scale stoping would not leave any detectable thermal or chemical imprint on the rising granite diapir, if the preheated raft units exceeded minimum sizes of only 50–100 m and were hence sinking at a rate of ca. 0.5–1 cm/sec towards bottom levels of the pluton (Marsh 1982).

3.4.2 The Granite Diapir

Textural evidence (Weiss 1986) and mineral stability data (Fig. 4.11) indicate that the granites were emplaced at sub-liquidus temperatures of ~900–850 °C, representing a highly viscous pluton loaded with ca. 30–35 vol% Plg-Akf-Bi primocrysts (with a mean effective diameter of ~0.8 mm). Following McBirney and Murase (1984), a viscosity calculation for this "crystal mush" yields η_e of 10^{11}–10^{13} poises, suggesting a non-Newtonian system with pseudoplastic properties (Pitcher 1979). Nevertheless, syn-intrusive flow and convection may attain a velocity of ~1 cm/s even in highly viscous (10^{12} poise) magmas, as discussed by Rice (1985). Below solid fractions of ca. 50–70 vol%, a gradational transition from Newtonian suspension flow to pseudoplastic creep (granular flow) is observed; the corresponding critical melt fraction, which approximates 30–35 vol%, is less dependent on rock composition than on geometric factors, like grain size and crystal shape (Van der Molen and Paterson 1979). According to the above outlined rheological constraints of granite emplacement, the observed differential overprint of the diorite envelope is essentially controlled by the primary petrofacial inhomogeneity of the zoned diorite envelope (Figs. 3.1, 3.10), specifically in terms of tectonic fracturing and subsequent ductile deformation of mesotype diorites, or of hybridization with leucocratic quartz diorites, as follows:

1. Initially, a "permissive" mode of emplacement involved local subsidences and block dissecting of mesotype inner diorites (Weiss 1986, Fig. 14), while country-rock screens were spalled off from the eastern contact and enclosed by a hybrid ring-dyke apophysis stuffed with small diorite blocks and various metasedimentary xenoliths (Fig. 3.1 and Sect. 3.2.2).
2. When the pseudoplastic granite body was arrested on its present level of emplacement, still being fed with "mushy" magma from depth, this distension diapir expanded in situ (Pitcher 1979), slightly deforming its still ductile monzodiorite envelope and the contact-anatectic country rocks (producing a narrow "chaotic" zone of *schollen* migmatites: Pattison 1985; Pattison and Harte 1988). Modal composition (Table 4.1) and crystallization sequence (Fig. 4.11) indicate that, with the onset of granite "ballooning", the crescent-shaped zone of hypersthene-rich monzodiorites (e.g., D89,98,224; see Figs. 4.1, 5.1) was still at a temperature near 950–900 °C and contained ca. 15–10 vol% alkaline pore liquid. Even the transitional, pyroxene-bearing quartz diorites (Cpx > 1.5–2.5 vol%, e.g., D22,355) had liquid fractions of *less* than 20–25 vol%, and are consequently showing deformed plagioclase and biotite, like the mesotype monzodiorites. The leucocratic quartz diorites (e.g., D352,108), by contrast, still exceeded a rheologically critical fraction of 25–35 vol% residual melt at T ~750–800 °C, and remained hence unaffected by mineral deformations.
3. Zones of rheological unconformity developed specifically near the western granite margin where the rising granite diapir came into contact with leucocratic portions of the quartz diorite envelope, which still had more than 30–40 vol% intercumulus melt and were consequently displaying suspension-like properties. The resulting shear flow ('forced convection') was presumably intensified by the presence of rigid xenoliths and Cpx-Plg-Akf primocrysts (compare Kouchi and Sunagawa 1985, Figs. 1,5c), which considerably increased the efficiency of hybrid mixing of leucodiorites and granites, hence generating broad transitional zones of granodioritic bulk composition (Sect. 3.2.1) and a marginal granite facies bearing augite xenocrysts (Cpx > 0.5 vol%, Fig. 4.10a).

3.4.3 A Conceptual P-T-t Path for the Ballachulish Complex

Figure 3.14 shows two possible temperature-time paths for metasediments and magmatic rocks of the Ballachulish area, deduced from thermobarometric data and isotopic ages (compiled in Weiss 1986).

Accompanying thermal relaxation which followed the Caledonian tectogenesis, crustal units of the SW Grampian Highlands underwent a regional metamorphic thermal maximum at 520–490 Ma (Dempster 1985). For this Lower Ordovician event, Dalradian metasediments of the Ballachulish area record a NW-SE regional temperature gradient of 475–525 ± 50 °C at a pressure of 0.6 ± 0.15 GPa (Pattison and Voll, Chap. 2, this Vol.). The data are consistent with P-T results of ~0.7 GPa/530 °C for the Perthshire transect, ca. 80 km ENE of Ballachulish (Dempster 1985). This

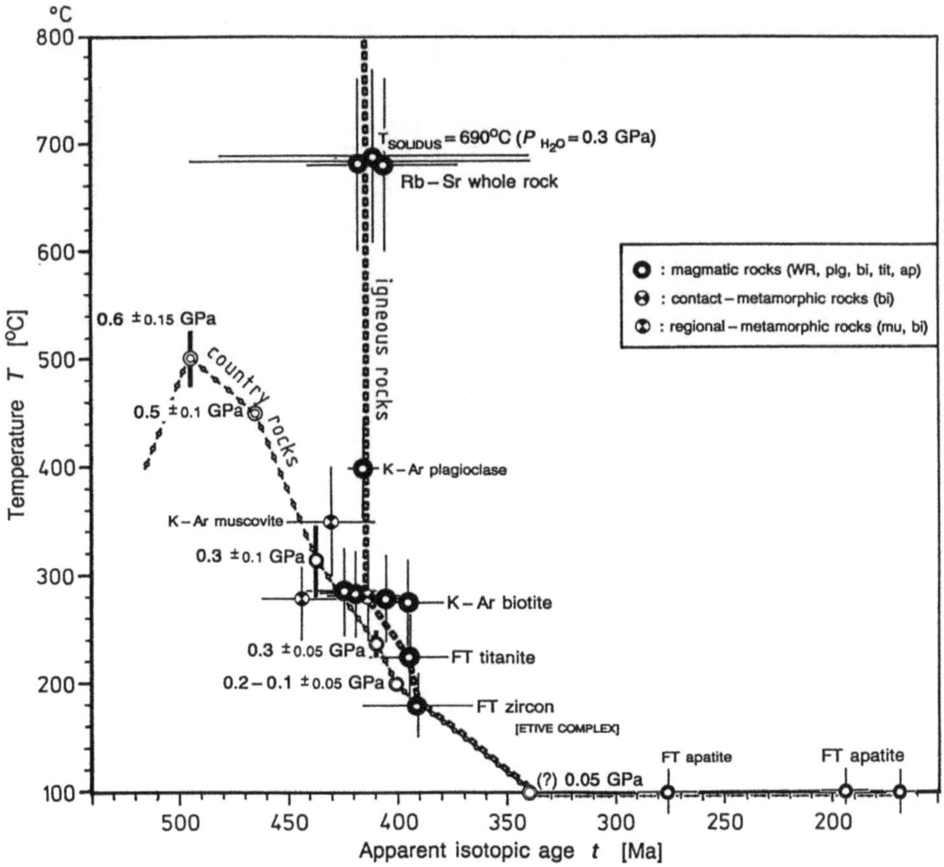

Fig. 3.14. Conceptual P-T-t paths for metasedimentary and igneous rocks in the Ballachulish area. The diagram is based on P-T model data (Dempster 1985; Pattison 1985), and on apparent isotopic ages for whole rocks (Rb-Sr), muscovite-biotite (K-Ar) and sphene-zircon-apatite (fission track FT), compiled from various literature sources cited in the text

suggests a country rock temperature of 250 ± 50 °C prior to plutonic emplacement in Ballachulish, as included within the conceptual P-T path of Fig. 3.14. In this diagram, discontinuous uplift rates, geothermal gradients and the pressure gradient (0.1 GPa/3.6 km) were accepted from Dempster's model for the Perthshire geotraverse (1985, Fig. 10), to simulate the *P-T* path for Ballachulish country rocks using muscovite-biotite isotopic ages (Harper 1967; Dewey and Pankhurst 1970).

An average Rb-Sr whole-rock isotopic age of 409 ± 48 Ma for 2 diorite and 14 granite samples (Harmon and Halliday 1980; Haslam and Kimbell 1981) may be interpreted as the age of emplacement of the relatively small igneous complex, since its thermal history was mainly controlled by the hot and "dry" diorite pluton, which obviously cooled from ca. 1100 to 700 °C within 0.9 ± 0.1 Ma, creating a thermal aureole within ~0.2 Ma (Weiss and Troll, Sect. 4.3; Buntebarth, Chap. 18, this Vol.). Regarding the 690–280 °C isotopic closure range for igneous and contact-metamorphic rocks and minerals from Ballachulish, including all K-Ar biotite data (Miller and Brown 1965; Hurford 1977) and omitting an ill-defined, impure biotite-amphibole separate of suspect 387 ± 6 Ma closure age (P.E. Brown et al. 1968), as discussed in Weiss (1986, p. 199), a best estimate for the magmatic event in Ballachulish may be given as 412 ± 28 Ma.

Fission track data for sphene and apatite from Ballachulish quartz diorites, and for zircon from the neighbouring Etive Complex (Harper 1967; Hurford 1977), indicate that, 10–20 Ma after plutonic emplacement, the T-t paths of metasediments and magmatic rocks converged near 200 ± 30 °C. The convergence falls into a period of strong regional uplift, attaining ~0.5 mm/year at 400–390 Ma (Dempster 1985). The severely discordant mesozoic FT-ages for magmatic apatite (Hurford 1977) imply a temporary cessation of regional uplift at 350–300 Ma, when the complex was still at a depth at least 1.5–2 km (even assuming a high geothermal gradient of 40–50 °C/km), referring to a minimum temperature of 80 °C which is needed for the apparent thermal resetting of apatite (Hurford 1977).

4 Thermal Conditions and Crystallization Sequence, as Deduced from Whole-Rock and Mineral Chemistry

S. Weiss and G. Troll

This chapter presents new data on the chemistry of Ballachulish igneous rocks (Sects. 4.1.1, 4.1.2), considering the effects of assimilation and hybridization (Sect. 4.1.3). A section on solidus temperatures and the granite system (Sect. 4.1.4) introduces several condensed paragraphs on mineral chemistry, phase relationships and geothermometry (Sect. 4.2; see Weiss and Troll 1989 for further details). The implications for fluid activity and the order of crystallization (Sect. 4.3, Fig. 4.11) are given ahead of a comprehensive summary on the petrogenesis of the Ballachulish pluton (Sect. 4.4).

4.1 Whole-Rock Chemistry

Except for a detailed study on the chemistry and mineralization of the small central leucogranite and its proximity by Haslam and Kimbell (1981), previous chemical work on Ballachulish diorite or granite mainly involved reconnaissance studies and was based on only a few samples; Stephens and Halliday (1984) classified the calc-alkaline, strongly Sr-Ba-K enriched magmatic rocks of Ballachulish mainly as 'Caledonian I (igneous)-type', representative of the post-tectonic Argyll Suite (410–415 Ma). Two quartz diorite samples from the pelite-contaminated NW margin of the complex gave $^{87/86}Sr_I$ of 0.7043 (Harmon 1983), suggestive of an ill-constrained age near 410 Ma (Harmon and Halliday 1980). Haslam and Kimbell (1981) presented a Rb-Sr isochron of 11 porphyritic granites with $^{87/86}Sr_I$ of 0.7040, yielding a whole-rock age of 406 ± 34 Ma.

The interpretation of new chemical data in this section, including the variation diagrams and maps in Figs. 4.2–4.7 is based on 99 WDS-XRF analyses of Weiss (1986, Table III) and further 21 analyses from various literature sources (compiled by Weiss 1986, Fig. 76). Nineteen representative analyses are listed in Table 4.1 including C.I.P.W.-normative and modal data; sampling localities are shown in Figure 4.1.

G. Voll, J. Töpel, D.R.M. Pattison, F. Seifert
Equilibrium and Kinetics
in Contact Metamorphism

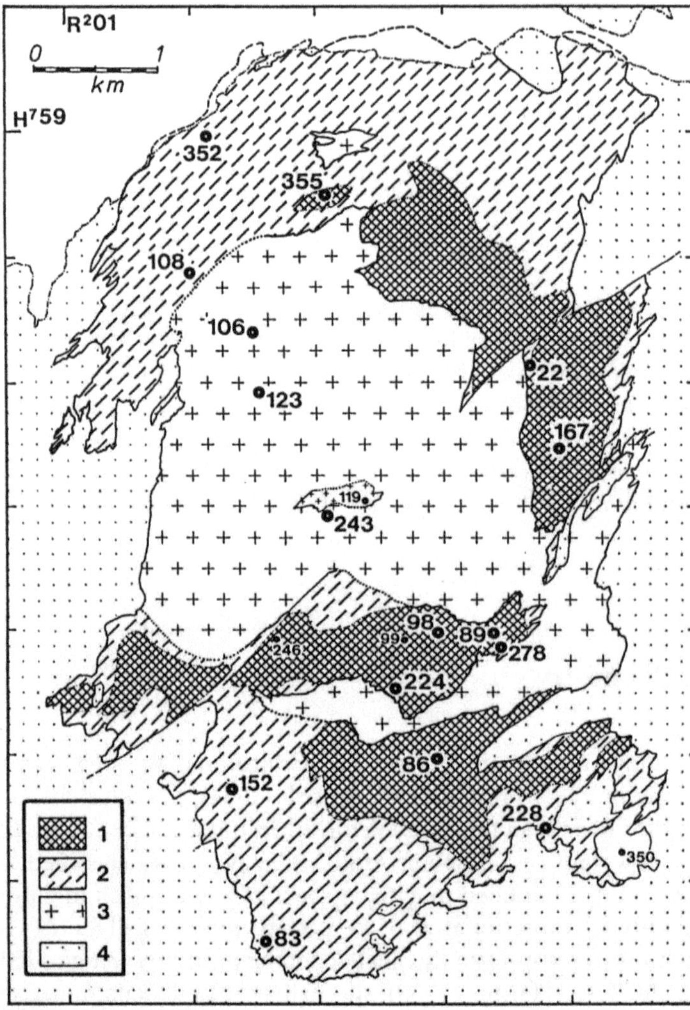

Fig. 4.1. Simplified map of Ballachulish igneous complex (cf. Fig. 3.1) *Numbers* indicate the locations of representative samples in Table 4.1, selected for detailed microprobe study and whole-rock chemistry (No. 350 = mafic appinite). *1* Monzodiorites; *2* quartz diorites; *3* granites; *4* metasediments

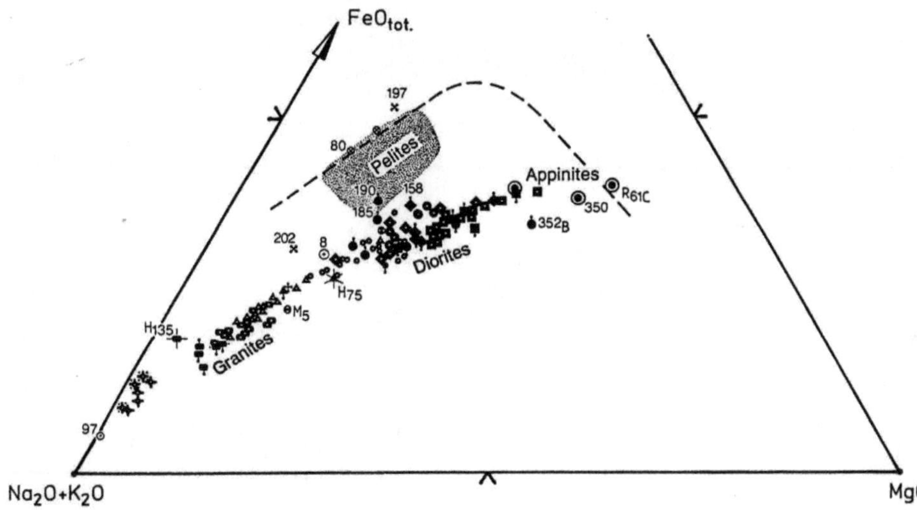

Fig. 4.2. AFM variation diagram of Ballachulish igneous rocks (Weiss 1986, Tables IIIa,b). For symbols, see Table 4.1. The *dashed line* separates calc-alkaline from tholeiitic suites (Irvine and Baragar 1971)

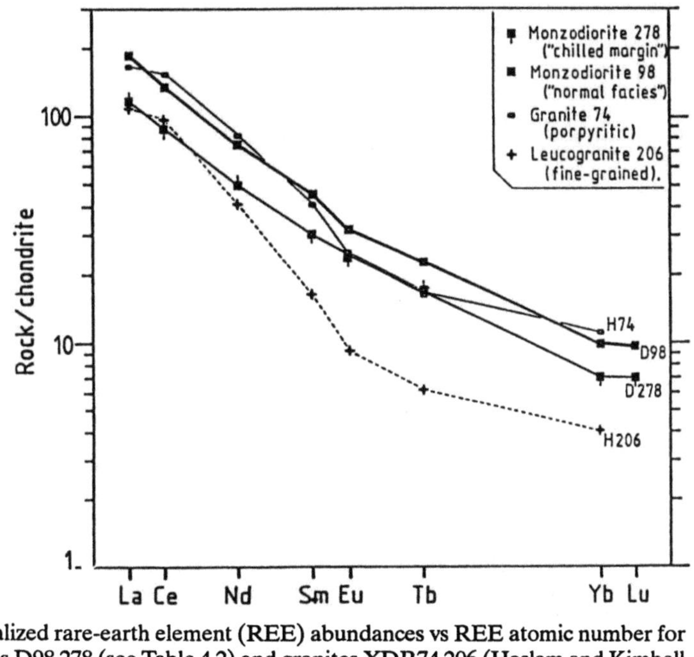

Fig. 4.3. Chondrite-normalized rare-earth element (REE) abundances vs REE atomic number for Ballachulish monzodiorites D98,278 (see Table 4.2) and granites XDR74,206 (Haslam and Kimbell 1981)

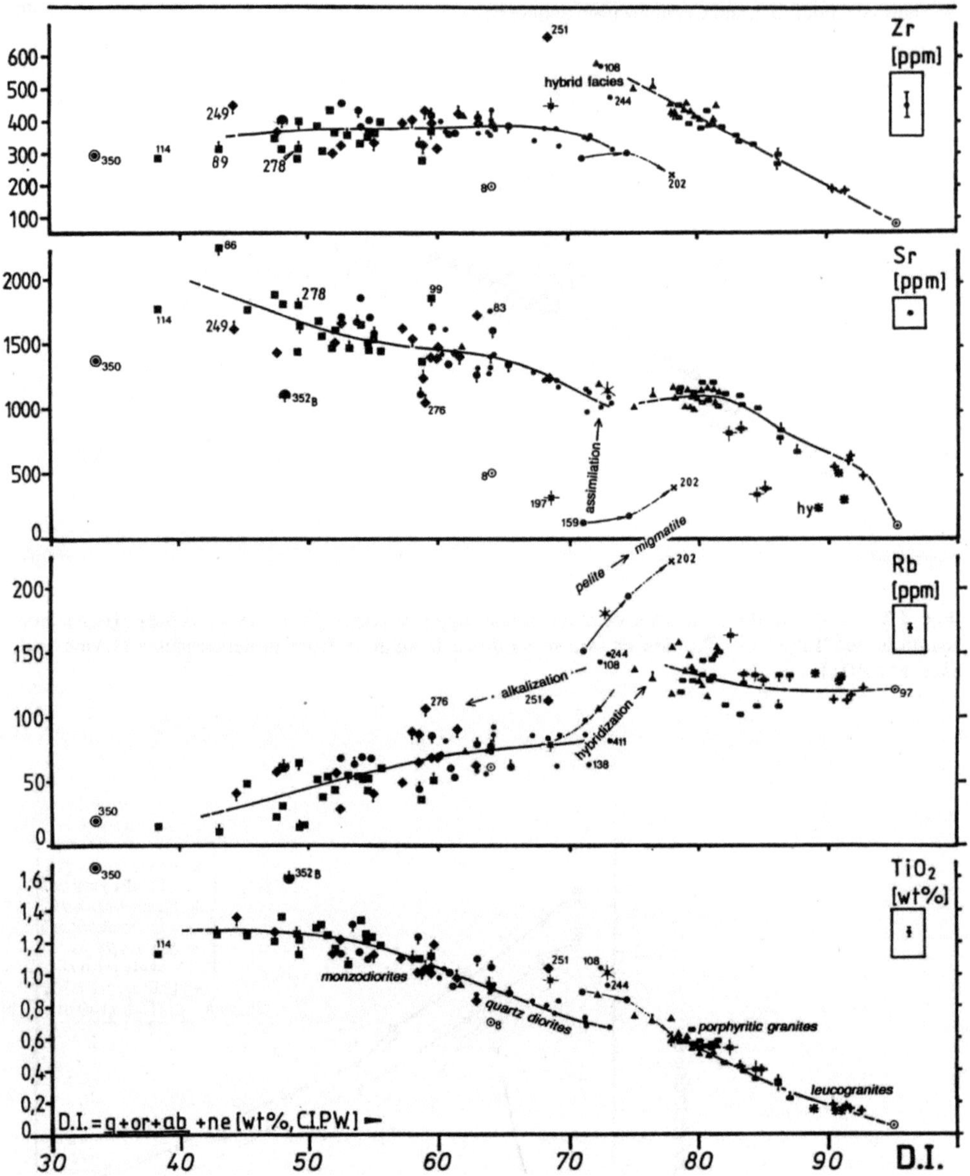

Fig. 4.4. Zr-Sr-Rb-Ti patterns for Ballachulish igneous rocks (Weiss 1986, Tables IIIa,b), plotted versus the differentiation index D.I. (Thornton and Tuttle 1960) to evaluate the effects of fractional crystallization, assimilation and hybridization during the magmatic evolution of the suite monzodiorite-quartz diorite and granite. For rock symbols, see Table 4.1 and Fig. 3.9

Fig. 4.6. **a** Areal isoline pattern of the differentiation index D.I. (Thornton and Tuttle 1960), used to model the zonal fractionation sequence for the diorites (D.I. < 55/ > 65) and the central granites (D.I. < 80/ > 90). **b** Areal distribution of rubidium within the diorite envelope, tracing the contours of mesotype monzodiorites (Rb < 60-70 ppm) and leucocratic quartz diorites (Rb > 70 ppm). The sample distribution is shown in Fig. 4.5; for lithologies, see Fig. 3.1

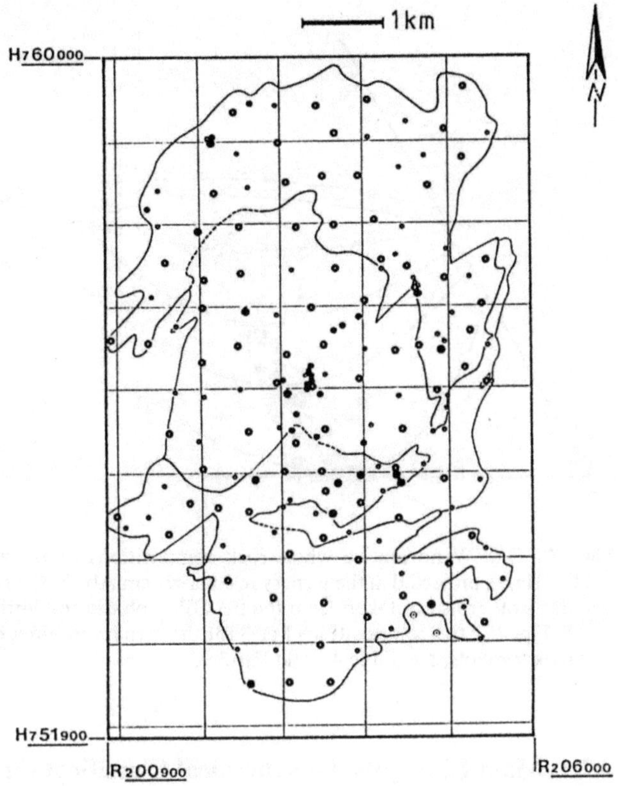

Fig. 4.5. Sample distribution across Ballachulish Complex, used to generate the areal isoline patterns for geochemical data (Fig. 4.6, *large dots*) and for modal data (Fig. 4.11, *large* and *small dots*). For lithologies, see Fig. 3.1

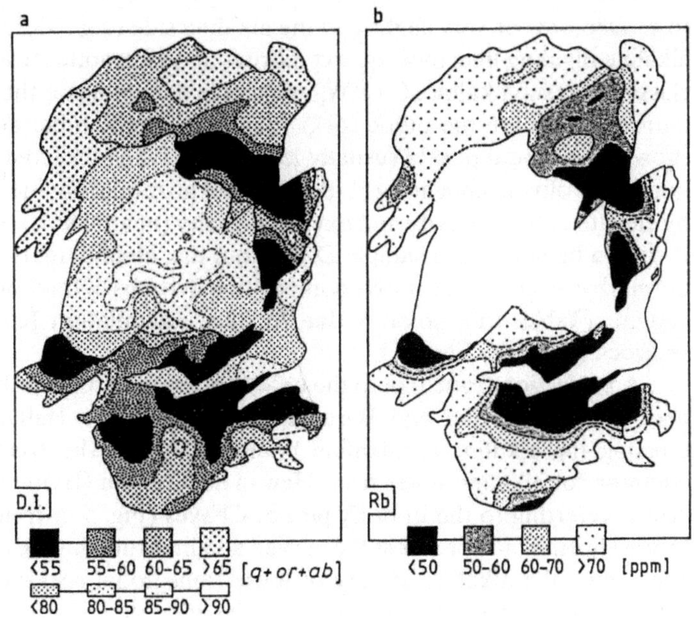

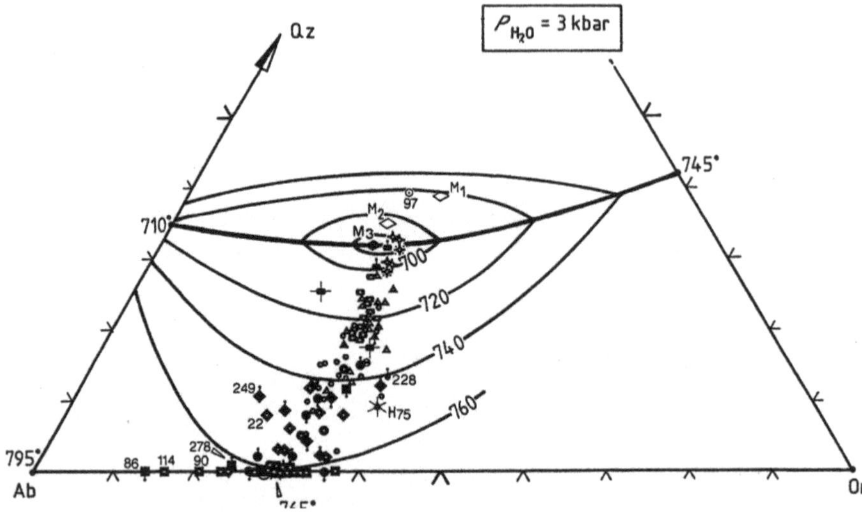

Fig. 4.7. C.I.P.W.-normative whole-rock compositions of Ballachulish magmatites (Weiss 1986, Tables IIIa,b), projected in the ternary residua system Ab-Or-Qz-H_2O (haplogranite). H_2O-saturated liquidus isotherms are taken from the 0.3 GPa isobaric equilibrium diagram of Tuttle and Bowen (1958, Fig. 25); the compositions M1-3 for the ternary minima refer to Johannes (1985, Fig. 2.1). For rock symbols, see Table 4.1 and Fig. 3.9

4.1.1 Main Elements: Geochemical Classification

In a total alkalies/silica diagram (Irvine and Baragar 1971), monzodiorites plot on (e.g., D86,278) or very slightly to the alkaline side (e.g., 98,224) of the alkaline-subalkaline dividing line; most quartz diorites and all granites from Ballachulish may be classified as subalkaline. C.I.P.W.-normative data indicate that hypersthene-bearing monzodiorites are mesotype rocks (> 30 wt% mafic minerals); although they may show some modal quartz (usually less than 0.5–2 vol%), the rocks are q-free, with normative olivine appearing (1–8 wt%). Yet no modal olivine is present, conditioned by elevated alkali activity and oxygen fugacity (near NNO; see Sect. 4.2.3 and 4.2.4). It should be noted that sample D278 from the 'chilled' margin (Fig. 3.4c) is quartz-normative despite its primitive composition in terms of fractionation stage and Rb-Sr content (Table 4.1), possibly due to silica assimilation from adjoining quartzite wallrock (compare Fig. 5.1).

A projection from di onto the ne'-ol'-q' plane of the basalt tetrahedron of Yoder and Tilley (in Irvine and Baragar 1971) shows the Ballachulish monzodiorites crossing the alkaline-subalkaline boundary within the triangle ab-ol'-hy, as it is common for alkaline suites from Hawaii and Tristan (Irvine and Baragar 1971, Fig. 4c,d). Referring to the ol-di-hy plot of Chayes (Fig. 5 in Irvine and Baragar 1971), monzodiorites cluster across the curved alkaline-subalkaline discriminant line, again confirming the alkaline affinity of these transitional rocks; effusive equivalents of most of these potassic rocks were to be classified as trachybasalts of the alkali olivine basaltic series (Irvine and Baragar 1971).

In di-ol-ne-q space (Yoder and Tilley 1962), monzodiorites and quartz diorites follow a silica trend from the pl-di-ol-hy subtetrahedron (olivine basaltic, e.g., D86) across the plane of silica saturation (hypersthene basaltic, e.g., D278) into the pl-di-hy-q subtetrahedron (quartz tholeiitic). Such an andesitic 'silica trend' contrast with the 'nephelinic trend' of critically SiO_2-undersaturated, ne-normative appinites (e.g., D350), crossing both joins En-Jd and Fo-Ab within the system Ne-Fo-SiO_2 (Yoder and Tilley 1962; O'Hara 1968). This suggests that the evolution of CO_2-H_2O rich appinitic magmas was controlled by pressures in the range of c. 1–2 GPa, provided that the members of this suite are predominantly related by equilibrium crystal fractionation and do not represent cumulate rocks; this is indicated by the presence of frequent chilled margins bordering potassic olivine gabbros and pyroxene diorites (Wright and Bowes 1979). By contrast, the silica trend of the monzodiorite magma was obviously controlled by crystal fractionation (and addition of silicic metasediments) mainly on upper crustal level, since the hypersthene gabbro thermal divide Plg-Hy-Cpx (O'Hara 1968) prevents the development of q-normative magmas at pressures above c. 0.5 GPa.

Alkalic monzodiorites, subalkalic quartz diorites and the granites define a nearly straight calc-alkaline trend lacking any significant iron enrichment on a conventional AFM-Plot (Fig. 4.2). This Bowen-type 'line of liquid descent', which clearly differs from the differentiation paths of tholeiitic suites (above the dashed divider in Fig. 4.2; compare Irvine and Baragar 1971), was apparently controlled by a relatively high oxygen fugacity (near NNO). This favoured the fractionation of Ti-Fe oxides and Mg-rich mafic silicates at an early differentiation stage (Haggerty 1976).

Using the mole fraction $X_{ALK} = \{Al_2O_3/(Na_2O + K_2O + 0.5 * CaO)\}$ in the geotectonic classification of Pitcher (1982) to aid a discrimination of 'igneous (I)-type' and 'sedimental (S)-type' magma sources, only the ilmenite-rich hypersthene diorites ($X_{ALK} = 0.98$–1.03) fit the meta-aluminous, 'Caledonian I (igneous)-type', whereas pelite-contaminated quartz diorites ($X_{ALK} = 1.03$–1.10) are transitional to peraluminous, crustal-anatectic magmas. With $X_{ALK} = 1.10$–1.35, the corundum-normative granites of Ballachulish appear to be related to 'S (sedimental)-type' granitoids, which were largely derived by crustal melting.

4.1.2 Rare Earth Elements: Possible Source Regions for the 'Hot and Dry' Diorites

Figure 4.3 shows the chondrite-normalized rare earth element (REE) patterns for two monzodiorites (Table 4.2) and for two granites (Haslam and Kimbell 1981). The lack of a major Eu anomaly both in the dioritic and the granitic units of Ballachulish Complex suggests that the sources of these Sr-Ba-K-enriched rocks were feldspar free or had a high oxygen fugacity (Drake and Weill 1975). The slight Eu depletion of monzodiorite D98 may account for a restricted low-pressure, fractional crystallization of plagioclase, compared to the 'chilled' sample D278. Despite the less evolved differentiation stage of D278, the overall REE-level for this quartz-normative sample is lower. This may be either due to the observed plagioclase enrichment of D278 (Table 4.1), or to a dilution with

Table 4.1. Chemical and modal analyses of representative igneous rocks from the Ballachulish Complex[a]

Rock type	Appinite	Monzodiorite						Quartz Diorites								Granites			
Sample	B350	D86	D278	D167	D98	D224	D89	D152	D355	D22	D83	D228	D352	D141	D108	G106	G123	G243	G119
Symbol[b]	◎	■	■	■	■	■	■	●	◆	◆	•	•	•	•	•	■	▲	■	✦
SiO_2	48.03	52.22	55.92	55.11	55.65	55.47	56.29	56.30	57.68	59.36	58.25	60.10	61.77	63.39	60.60	65.18	67.06	70.31	72.33
TiO_2	1.67	1.25	1.12	1.30	1.33	1.24	1.22	1.09	1.00	0.83	0.90	0.93	0.83	0.72	1.04	0.59	0.56	0.30	0.17
Al_2O_3	13.74	16.48	16.18	16.54	16.28	16.20	16.19	17.57	17.26	17.10	18.32	15.41	16.33	16.24	16.64	16.11	16.46	15.26	14.54
Fe_2O_3*	2.37	2.91	1.75	2.18	2.16	2.14	1.92	1.97	1.81	1.74	2.35	1.80	1.57	1.39	1.60	1.28	1.29	0.84	0.69
FeO*	6.40	5.13	4.77	4.74	4.68	4.31	4.64	4.53	3.90	3.20	3.22	3.16	3.18	2.72	3.22	1.80	1.69	1.06	0.70
MnO	0.14	0.14	0.11	0.11	0.10	0.11	0.11	0.13	0.10	0.11	0.13	0.08	0.09	0.08	0.07	0.06	0.04	0.02	0.03
MgO	9.29	6.61	5.74	4.88	5.11	5.00	4.86	4.00	3.25	3.35	2.97	3.66	2.93	2.53	2.47	1.44	1.21	0.73	0.27
CaO	10.00	8.92	7.08	7.09	6.89	6.57	6.52	5.59	5.27	5.26	4.24	5.03	4.29	3.03	3.87	2.43	2.43	1.31	0.86
Na_2O	2.87	4.43	4.37	4.25	4.38	4.27	4.51	3.95	4.24	4.94	4.85	3.76	4.38	3.99	5.25	4.51	5.16	4.13	4.05
K_2O	1.61	0.99	1.96	2.59	2.92	3.06	2.86	3.03	3.03	2.72	3.30	3.93	3.51	3.89	4.40	3.99	4.11	4.07	4.47
P_2O_5	0.50	0.55	0.36	0.43	0.46	0.46	0.37	0.42	0.39	0.34	0.36	0.40	0.28	0.27	0.36	0.18	0.13	0.09	0.05
H_2O^+*	n.d.	1.04	0.59	0.63	0.67	0.70	n.d.	n.d.	n.d.	0.92	1.36	n.d.	0.81	n.d.	n.d.	0.71	n.d.	n.d.	0.64
F*	0.13	n.d.	0.03	0.06	0.09	0.06	0.07	0.08	n.d.	0.06	0.07	0.07	0.06	n.d.	0.13	n.d.	0.08	0.03	n.d.
L.O.I.#	2.78	0.38	0.24	0.70	0.10	0.32	0.86	1.54	1.44	0.09	0.17	1.07	0.37	1.68	0.95	0.51	0.67	0.81	0.59
Total	99.53	100.79	100.05	100.42	100.62	99.91	100.25	100.20	99.36	100.02	100.51	99.40	100.40	99.95	100.57	98.79	100.90	98.94	99.33
Trace elements [ppm]								Trace elements [ppm]								*Trace elements [ppm]*			
Rb	23	13	17	39	52	45	55	70	69	61	88	76	87	99	143	131	154	110	115
Sr	1374	2254	1805	1555	1666	1447	1543	1707	1416	1729	1766	1412	1288	986	1019	1186	1131	844	558
Zr	292	320	313	308	329	359	367	403	396	400	437	395	339	351	573	411	449	294	192
Ba	729	1320	1410	1449	1792	1535	1362	1714	1450	1642	1789	2209	1472	1523	1288	1446	1361	1374	1850
Ni	112	71	77	66	51	77	64	50	53	63	11	49	38	31	55	34	29	34	10
V	412	240	205	208	281	266	237	174	152	124	146	140	118	105	149	43	30	36	n.d.
Zn	80	99	85	80	88	72	76	80	n.d.	68	92	64	61	n.d.	n.d.	57	n.d.	n.d.	16
C.I.P.W. Norms [wt%]								C.I.P.W. Norms [wt%]								*C.I.P.W. Norms [wt%]*			
Q	—	—	0.7	—	—	—	—	3.3	5.4	5.5	3.6	8.6	9.5	14.4	2.5	16.2	14.3	26.8	29.7
or	9.6	5.8	11.6	15.3	17.2	18.1	16.9	17.9	18.1	15.8	19.4	23.3	20.7	23.0	25.9	23.8	24.1	24.3	26.2
ab	23.0	37.3	37.0	35.9	36.8	36.3	38.1	33.4	36.2	47.1	40.9	32.0	37.0	33.8	44.2	38.6	43.3	35.3	34.5
an	20.0	22.0	18.8	18.4	16.1	16.0	15.5	21.3	19.3	15.8	18.4	13.6	14.5	13.5	8.8	11.3	9.5	6.2	4.3
cor	—	—	—	—	—	—	—	—	—	—	—	—	—	0.5	—	0.3	0.1	1.8	1.5
di- [wo]	11.2	7.8	5.5	5.8	6.3	5.7	6.0	1.6	1.9	3.1	0.1	3.7	2.1	—	3.3	0.8	—	0.8	—
di- [en]	7.8	5.4	3.6	3.8	4.2	3.9	3.9	1.0	1.2	2.1	0.1	2.6	1.3	—	2.1	3.6	3.0	1.8	0.6
di- [fs]	2.5	1.7	1.4	1.6	1.6	1.4	1.7	0.5	0.6	0.8	0.1	0.8	0.6	—	1.0	1.4	1.2	0.8	0.4
hy- [en]	—	2.7	10.8	8.3	5.9	7.8	7.5	9.0	7.0	6.3	7.3	6.6	6.0	6.3	4.0	—	—	—	—
hy- [fs]	—	0.8	4.3	3.4	2.2	2.7	3.2	4.6	3.6	2.5	2.7	2.1	2.7	2.8	2.0	—	—	—	—
ol- [fo]	10.9	5.9	—	0.1	1.9	0.8	0.5	—	—	—	—	—	—	—	—	—	—	—	—
ol- [fa]	3.8	2.1	—	0.1	0.8	0.3	0.2	—	—	—	—	—	—	—	—	—	—	—	—

C.I.P.W Norms [wt%]																			
ne	0.7	—	—	—	—	—	—	—	—	—	—	—	—	—	—	—	—	—	—
mt	3.5	4.2	2.5	3.1	3.1	3.1	2.8	2.8	2.6	2.5	3.4	2.6	2.3	2.0	2.3	1.9	1.8	1.2	1.0
il	3.2	2.3	2.1	2.5	2.5	2.4	2.3	2.1	1.9	1.6	1.7	1.8	1.6	1.4	2.0	1.1	1.0	0.6	0.3
ap	1.2	1.3	0.8	1.0	1.1	1.1	0.9	1.0	0.9	0.8	0.8	0.9	0.7	0.6	0.8	0.4	0.3	0.2	0.1
D.I.	32.6	43.1	49.3	51.2	54.0	54.4	55.0	54.6	59.7	63.0	63.9	63.9	67.2	71.2	72.6	78.6	81.7	86.4	90.4
Modal proportions [vol%]																			
Plg	42.9	64.6	65.2	60.0	60.1	56.8	55.5	52.6	53.8	60.5	60.5	60.8	53.5	49.5	50.8	53.1	45.2	41.6	43.5
Akf	0.5	0.7	7.6	15.4	13.9	11.3	14.9	7.9	8.8	8.1	8.9	8.9	11.8	17.8	21.8	15.5+4.1	16.9+6.5	17.2+13.6	27.6
Qz	0.2	0.6	0.2	0.3	0.2	0.2	0.6	10.2	10.8	9.0	8.3	11.6	12.2	16.1	9.9	16.1	19.9	23.8	25.0
Opx	—	3.9	11.6	5.7	8.1	9.4	9.9	—	—	0.2	—	—	—	—	—	—	—	—	—
Cpx	1.9	7.3	11.8	13.1	13.1	13.6	10.8	0.9	2.9	1.4	tr.	0.9	0.4	0.2	0.5	—	0.3	—	—
Am-I	46.9	10.1	0.9	1.1	0.7	0.5	0.5	9.6	5.5	8.3	9.1	6.4	6.4	5.9	5.1	0.7	2.1	—	—
Am-II	—	6.5	—	—	—	—	—	—	0.3	—	—	—	—	—	—	—	—	—	—
Bi	4.8	3.9	1.4	2.2	2.4	3.5	5.9	17.7	15.1	10.8	10.7	10.3	14.1	7.3	10.0	8.6	7.8	3.2	1.7
Mu	—	—	—	—	—	—	—	—	—	—	—	—	—	0.3	—	0.1	—	—	0.9
OPQ	0.3	2.2	0.8	1.4	1.0	1.5	1.1	0.5	0.9	0.7	1.1	0.7	0.5	1.1	0.8	0.6	0.5	0.4	0.5
Ap	0.7	0.4	0.5	0.6	0.4	0.8	0.7	0.2	0.4	0.4	0.4	0.3	0.3	0.4	0.1	0.3	tr.	tr.	0.2
Zi	—	—	—	—	—	—	—	—	0.1	tr.	tr.	0.1	tr.	0.2	tr.	0.1	tr.	tr.	tr.
Sph	1.1	—	—	—	—	tr.	—	0.1	0.2	tr.	0.3	tr.	0.4	0.5	0.6	0.3	0.7	0.1	tr.
Ep	0.7	—	—	—	—	—	—	0.2	0.9	—	0.3	—	0.2	0.3	0.1	0.4	—	tr.	0.3
No. of sections	1	3	4	2	4	4	3	2	1	2	2	3	1	1	2	1	1	1	1
Total counts	1496	7712	5930	5000	6222	6571	5962	2567	2784	5034	5000	3763	1957	2101	5002	1927	1688	103	1445

[a]Symbols and abbreviations: * = Determined by wet chemical analysis (see text); n.d. = not determined. L.O.I # = Loss on ignition, corrected for oxidation of FeO to Fe_2O_3 [+] and for F + H_2O content [-]. If F + H_2O is determined, L.O.I. # mainly corresponds to the loss of CO_2O. D.I. = Differentiation index (Thornton and Tuttle 1960). Akf = Alkali feldspar ('+' = megacrysts); Am-I = amphibole, replacing Cpx; Am-II = amphibole, replacing Opx; Ap = apatite; Bi = biotite (including secondary chlorite); Cpx = clinopyroxene (augite); Ep = epidote (and zoisite-prehnite); Mu = muscovite; OPQ = opaques (mainly ilmenite and magnetite); Opx = orthopyroxene (hypersthene); Plg = plagioclase; Qz = quartz; Sph = sphene; Zi = zircon; tr. = trace (< 0.1 vol%); — = phase not present.

[b]The rock symbols refer to Fig. 3.9.

Table 4.2. Trace-element analyses by INAA[a]

Element	La	Ce	Nd	Sm	Eu	Tb	Yb	Lu	Ta	Th
D278[b]	38	75	31	6.2	1.9	0.8	1.5	0.2	1.1	1.4
D98	60	117	47	9.3	2.6	1.1	2.1	0.3	1.7	10.8
XDR74[c]	55	132	51	8.4	1.9	0.8	2.5	n.d.	n.d.	{34}
XDR206	36	83	26	3.3	0.7	0.3	0.9	n.d.	n.d.	{26}

[a]Instrumental Neutron Activation Analysis; {.} give XRF data;
[b]D278/D98 = monzodiorites ('chilled'/'normal' facies), this chapter.
[c]XDR74/206 = porphyritic granite/leucogranite (see Haslam and Kimbell 1981).

resorbed quartzitic material. The primitive REE-pattern of the porphyritic granite XDR74 from Ballachulish clearly contrasts with the more evolved, low-Sr members of the large Etive Complex where the inner granites show a large negative Eu anomaly (Harmon et al. 1984, Fig. 4). Although the Ballachulish leucogranite was obviously derived from the porphyritic granites by Plg-Bi-Zi fractionation (see Figs. 4.4, 4.6a, 4.7), europium was apparently not fixed in plagioclase as Eu^{2+}. This prevented the development of a negative Eu anomaly, despite considerable plagioclase fractionation and HREE depletion towards the leucogranite. This effect is referred to fairly oxidizing conditions within the late-stage, fluid-enriched leucogranitic liquid (f_{O_2}-range ~ MNO-HM), rather stabilizing Eu^{3+} instead of Eu^{2+} (Sect. 4.3.1).

Strikingly, the actual REE concentrations in Ballachulish granites are, apart from a slight depletion in the Eu-Tb range (Fig. 4.3), similar to rocks of rather more basic compositions, like hypersthene monzodiorites (Ballachulish) and quartz diorites-tonalites (Etive, Foyers, Strontian; see Fig. 4 in Harmon et al. 1984). This is suggestive both of a higher initial REE level and of less effective fractionation in the parental magma. This observation is consistent with the interpretation of Harmon et al. (1984), that the high Sr-Ba-K characteristics of the low-SiO_2 diorite end members are more likely to have been imposed by the magma source region than by fractional crystallization. This involved the selective metasomatic enrichment of mantle material by aqueous fluids derived from the dehydration of oceanic crust during subduction in a destructive plate margin environment.

The incompatible element normalization scheme of Pearce et al. (1984) places the post-orogenic, zoned Ballachulish suite within the field of 'post-collision granites', which comprise several different source regions. Here, batch melting of a significant lower crustal component is indicated, possibly consisting of intermediate Lewisian granulites (e.g., B41 in Drury 1980), specifically in terms of negative Nb-Hf anomalies, a large positive Ce anomaly and low Y-Yb (Weiss 1986, Fig. 89). Likewise, the extremely low Rb/Ba ratios of 0.01–0.03 and Ni-Zn contents of 65 ± 15 and 75 ± 25 ppm for monzodiorites (see Table 4.1 and Weiss 1986, Table IIIa) are nearly identical with respective average values for more than 540 Lewisian granulites from NW Scotland (Rb/Ba — 0.01–0.02; Ni = 70 ± 30, Zn = 80 ± 5 ppm; Holland and Lambert 1975); principal assemblages of these primitive andesites consist of andesine + augite + hypersthene + green amphibole (+ Mt + Ilm ± Bi ± Qz), strikingly similar to K-depleted fine-grained monzodiorites in Ballachulish (e.g., D86,249).

4.1.3 Trace-Element Variation

4.1.3.1 Effects of Fractionation, Assimilation and Hybridization

Thornton and Tuttle (1960) introduced the concept of a differentiation index (D.I.) comprising the C.I.P.W. normative residua components q + or + ab [+ le + nc + ks] and hence relating rocks with a larger differentiation index to a higher fractionation stage. Figure 4.4 shows a plot of the elements Zr-Sr-Rb-Ti versus D.I. The large differentiation range of the diorite suite (D.I. ~ 45–70) is obviously related to a rather effective fractionation favoured by magma convection and adcumulate consolidation, whereas the restricted compositional range of porphyritic granites (D.I. = 80 ± 2.5) is attributed to a 'crystal mush' consolidation (Troll and Weiss Sects. 3.2.3 and 3.4.2, this Vol.), involving the incomplete separation of cumulus phases and viscous intercumulus liquid within a magmatic suspension (McCarthy and Hasty 1976). Omitting the appinite D350, the mafic enclave D352b, the microdiorite dyke DP8 and several cataclastic, hydrothermally altered (mineralized) granites (see Weiss 1986, Fig. 76), the data sets 'diorite' and 'granite' were fitted to a fifth-degree polynomial function in order to model the magmatic evolution within both suites. A schematic 'pelite → migmatite' trend, based on samples P159–P80–M202 (Weiss 1986, Table IIIc), is shown to indicate the effects of pelite assimilation.

The level of zirconium remains nearly constant throughout the D.I. range of monzodiorites (~ 380 ppm); leucocratic quartz diorites (D.I. > 65) show significantly lower values (~ 350 ppm) which suggests the incorporation of up to 30 wt% metasedimentary material (with Zr ~ 290 ppm). Although this proportion appears quite high in terms of magmatic heat capacity and thermal balance, it should be kept in mind that incorporation does not necessarily imply assimilation by complete melting, but may actually involve the mechanical resorption and disintegration of country rock by melting only Qz + Kf ± Plg, thus exceeding a rheologically critical melt fraction of ca. 30–35 vol%, and leaving a more refractory schlieren residue of Bi ± Crd ± Plg (Troll and Weiss, Sect. 3.2.2, this Vol.). The granites display a strongly linear trend of Zr-depletion, obviously related to the early fractionation of zircon within hybrid rocks and porphyritic granites (near 800–750 °C), severely restricting its modal amount in the leucogranites (Table 4.1).

Apparently related to plagioclase fractionation, strontium shows a slight decrease with differentiation, with a distinct hiatus at D.I. ~ 75 separating the diorite suite from the granites. The strongly contrasting Sr levels of uncontaminated quartz diorites (~ 1350 ppm at D.I. 74) and of pelitic metasediments (~ 150 ppm) suggests an upper limit for the admixture of pelitic material to leucocratic quartz diorites (1050 ± 50 ppm) at 25 ± 5 wt% ('assimilation' in Fig. 4.4), referring to the increasingly non-linear Sr depletion in the D.I. range 70–73. A slight Sr depletion of marginal granites (D.I. 75–80) may be due both to pelite contamination and to hybrid mixing with leucocratic quartz diorites. At D.I. > 85, leucogranites are strongly Sr-depleted, specifically the mineralized samples ('hy') and the rhyolite R97; even the unmineralized aplitic granites (e.g., G119, Table 4.1) are severely depleted in Sr-Zn-F-Li, obviously as a result of the separation of an aqueous fluid from the central, residual leucogranite magma (Haslam and Kimbell 1981).

Although rubidium exhibits only minor variations within the diorite suite and the granites, a Rb level near 100 ppm clearly separates both intrusive phases. Notable exceptions are presented by several anomalously Rb-enriched monzodiorites (D276,251) and quartz diorites (D108,244). These alkalized rocks, which are confined to internal contacts or hybrid transition zones to the granites (Troll and Weiss, Sect. 3.2.1), this Vol.), were obviously metasomatized during emplacement of the granite magma. This favoured the crystallization of poikiloblastic alkali feldspar and fluorine-rich biotite (e.g., D108); both phases reach abundances of 15–22 and 14 vol%, respectively, hence constituting a large Rb reservoir (referring to distribution coefficients $K_{D\ (Sr)}$ of 0.34–6.36 for Akf-Bi in intermediate magmatic rocks; Tindle and Pearce 1981, Table II).

Titanium shows a near-linear decrease with increasing fractionation, related to the crystallization of ilmenite (in the diorites) and of biotite (mainly in the granites).

4.1.3.2 Isoline Maps and Areal D.I.-Rb Patterns

Chemical and modal variation within the igneous complex was modelled on the basis of 120 chemical and 150 modal analyses (Weiss 1986). Figure 4.5 shows the position of the samples used to produce the selected distribution patterns of Figs. 4.6 (below) and 4.10 (Sect. 4.3). The data sets 'diorite' and 'granite' were treated separately as to their two-stage emplacement; isolines were generated by bivariate polynomial interpolation (Akima 1978) using the IQHSCV Fortran routine from the IMSL library at the Leibniz processing centre (Munich). The produced isolines strictly refer to the actually analyzed concentration level above each data point or its grid-condensed derivatives (minimizing their mutual elastic energy by smooth surface fitting); by contrast, conventional 'least square' methods (which are frequently used in geochemical studies) create trend lines, hence severely abstracting the original data base.

With respect to the differentiation index D.I. (Fig. 4.6a), the diorite suite exhibits a somewhat patchy isoline zoning; nevertheless, the more primitive compositions (D.I. < 55–60) largely fit the zone of monzodiorites, grading into more fractionated units towards the S-N-W periphery of the complex where tongue-shaped areas with D.I. > 60–65 contain abundant pelite xenoliths; these zones may continuously grade into the marginal, hybrid granites (D.I. < 80). Despite their apparent petrographical and structural homogeneity (see Figs. 3.1–3.8), the porphyritic granites are well subdivided by several concentric D.I. zones, with progressively increasing fractionation towards the central leucogranite (D.I. > 90).

The areal distribution of rubidium (Fig. 4.6b) largely coincides with the petrofacial subdivision of the diorite envelope (Figs. 3.1, 3.10). High Rb zones (> 70 ppm) comprise the heavily pelite-contaminated areas of the NW-N coastal sectors and of Glen Duror, where fine-grained, biotite-rich quartz diorites are frequent. Negative Rb anomalies near Kentallen and at the southern contact of the igneous complex might be due to the contamination with low-Rb basic rocks (e.g., appinites).

4.1.4 Solidus Temperatures for P_{H_2O} = 0.3 GPa, Deduced from Normative Rock Compositions

4.1.4.1 The Granite System An-Ab-Or-Qz-H_2O at 0.3 GPa

In order to place thermal constraints on the late-magmatic crystallization at excess fluid conditions, C.I.P.W.-normative ab-or-qz components for Ballachulish magmatic rocks were projected via an* (an* + ab + or + q = 100 wt%) into the ternary residua system Ab-Or-Qz for P_{H_2O} = 0.3 GPa (Tuttle and Bowen 1958, Fig. 25). Figure 4.7 shows this haplogranite system, with q-free alkaline monzodiorites (an* = 28–22 wt%) plotting immediately above the binary Ab-Or minimum (765 °C at P_{H_2O} = 0.3 GPa, Tuttle and Bowen 1958, Fig. 17). Quartz diorites and granites follow a thermal valley stretching from the binary towards the ternary minimum, suggestive of a differentiation line produced by plagioclase fractionation within the quartz diorites (an* = 25–15 wt%), joined by alkali feldspar in the porphyritic granites (an* = 12–8 wt%). The considerable scatter within the low-silica portion of this trend is probably related to variable modal amounts of biotite (in any case causing a slight shift towards or, referring to the anhydrous C.I.P.W.-norm) as well as to adcumulus-trapping or filter-pressing of a residual pore liquid. The position of the leucogranites on the cotectic line (an* = 5–4 wt%), very near the 0.3 GPa ternary minimum, infers the simultaneous crystallization of sodic plagioclase and alkali feldspar together with quartz near 680 °C, producing the aplitic texture of these microgranites. With isobaric crystal fractionation, the composition of the residual liquid is expected to move into the minimum M3 at 665 °C which defines the solidus for P_{H_2O} = 0.3 GPa (Johannes 1985, Fig. 2.1). Consequently, the cotectic, near-minimum composition of the Ca-poor leucogranites suggests water-excess crystallization at a lithostatic pressure near 0.30–0.25 GPa, since lower water pressures (e.g., 0.1–0.2 GPa) would shift the cotectic/ minimum intersection towards Ab-depleted compositions M1 and M2 in Fig. 4.7 (cf. Johannes 1985, Fig. 2.1). This is consistent with fluid-pressure estimates of 0.3 ± 0.05 GPa derived within the thermal aureole of the igneous complex (see Pattison, Chap. 16, Masch and Heuss-Aβbichler, Chap. 10, this Vol.).

Experimental data on the granite system An-Ab-Or-Qz-H_2O suggest that the solidus temperature rises with respect to the An-free haplogranite system by less than 5–10 °C at P_{H_2O} = 0.3 GPa, referring to plagioclase compositions of 30–15 mol% An (Johannes 1984, Fig. 4) which corresponds to the An range observed in quartz diorites and granites (Fig. 4.10). Plagioclase of An_{28-25} coexists with late-stage alkali feldspar in q-free monzodiorites (containing less than 1–2.5 vol% modal Qz); the solidus temperature T_s which is inferred from Fig. 4.7 (765 ±10 °C), may be corrected for An considering the T/X_{An}-dependence in Qz-free anorthositic rocks (Johannes 1984, Fig. 8), since the liquids-solidus geometry of the T-X_{An} diagram does not appear to change significantly if Or is added to the Ab-An system (Yoder et al. 1957). Correcting for the X_{An}-dependence of T_S and assuming equilibrium crystal fractionation (as to the adcumulate growth of Plg; see Sect. 4.2.2), an effective bulk solidus temperature of 780 ± 10 °C at P_{H_2O} = 0.3 GPa is inferred for q-free compositions, neglecting small amounts of residual hydrous melt; even if 2–3 vol% modal quartz were present at this temperature, the volume proportion of residual syenogranitic liquid would probably not exceed 5 vol% for fluid-excess conditions. This is well in accordance with the results of melting experiments by Robertson and Wyllie (1971b, Fig. 12) on the alkaline pyroxene diorite 130A from the Late Caledonian Debouillie Stock in northern Maine (Appalachians). This syenodiorite is augite- and biotite-rich (with Qz < 0.5 vol%), has its solidus at 755 ± 15 °C and shows 'Qz-out' near 760–770 °C at P_{H_2O} = 0.3 GPa; the fractionation stage of this rock (D.I. = 52) fits the D.I. 49–55 range for Ballachulish monzodiorites quite well (Fig. 4.4).

4.1.4.2 Dependence of the Solidus Temperature T_S on the Fractionation Stage D.I., with Application to the Ballachulish Suite

Summing q + or + ab for intermediate and acid rocks, the differentiation index D.I. constitutes an effective measure of fractionation for quartz dioritic to leucogranitic compositions, with CaO strongly linear correlated to D.I. (Thornton and Tuttle 1960, Fig. 13), as does the C.I.P.W. normative an*-value (Weiss 1986, Fig. 82b). Since the CaAl-NaSi activity in tonalitic and granitic systems obviously controls both X_{An} in plagioclase *and* the solidus temperature T_s (Johannes 1984, Figs. 4,8), it appears reasonable to relate T_s to the D.I. composition of the host rock. This concept of an isobaric D.I./T diagram was first introduced by Piwinski and Wyllie (1968, Fig. 11) to relate variations in phase stabilities and solidus temperature to normative rock compositions (D.I. > 45; see compilation in Weiss 1986, Figs. 81, 82). The resulting solidus estimates for the D.I. fractionation range of Ballachulish rocks (Fig. 4.4, compare Table 4.1) comprise 765 (±20) °C for q-free monzodiorites (D.I. = 50–55), 685–680 (± 15) °C for quartz diorites (D.I. = 55–73), and 680–670 (±15) °C for the granites (D.I. = 79–91), at P_{H_2O} = 0.3 GPa. Nevertheless, the above derived solidus temperatures for fluid-excess conditions appear to represent *lowest* T_S-limits, as to the consolidation of monzodiorites, for reasons outlined next.

The observed high thermal stability for biotite and alkali feldspar (980–910 °C and 940–860 °C; Sects. 4.2.2 and 4.2.4.) in coarse- to medium-grained monzodiorites, where crystallization was mainly equilibrium-controlled (Weiss, Chap. 5, this Vol.), clearly contrasts with textural evidence that suggests a late-magmatic to near-solidus growth for both phases (Troll and Weiss, Sect. 3.3.2, this Vol.). This points to a 'dry' crystallization in a system which was water-deficient even at late stages. Robertson and Wyllie (1971a) have shown that the solidus temperature for diorites is raised by ~ 150 °C to ca. 900 °C at fluid-absent conditions. Hence an effective bulk solidus temperature (see above) of 900 (± 50) °C at P_{total} = 0.3 GPa seems likely for the least altered 'dry' hypersthene diorites containing less than 1 vol% amphibole and 5 vol% biotite, with H_2O in the rock not exceeding 0.7 wt% (see Table 4.1 and Sect. 4.3.1). A comparison of modal data from Weiss (1986, Tab. IIa) with the thermal stability range for primocrysts (Fig. 4.11) indicates that these alkaline rocks (with Qz < 1 vol%) probably contained *less* than 5–15 vol% residual liquid at temperatures of 900–950 °C and P_{total} = 0.3 GPa.

4.2 Phase Relationships and Geothermometry

Mineral analyses were carried out using an ARL microprobe linked to a λ-dispersive system, with beam diameters ranging from 1–30 μm (usually 6 μm; for the complete analytical data set see Figs. 57–63 and Table IV in Weiss 1986). Representative averaged phase analyses are reported in Weiss and Troll (1989, Table II).

4.2.1 Pyroxenes

The composition of pyroxene pairs from monzodiorites is shown in Fig. 4.8. Both pyroxenes are usually Mg-richer in the marginal, finer-grained facies (D99,246,278

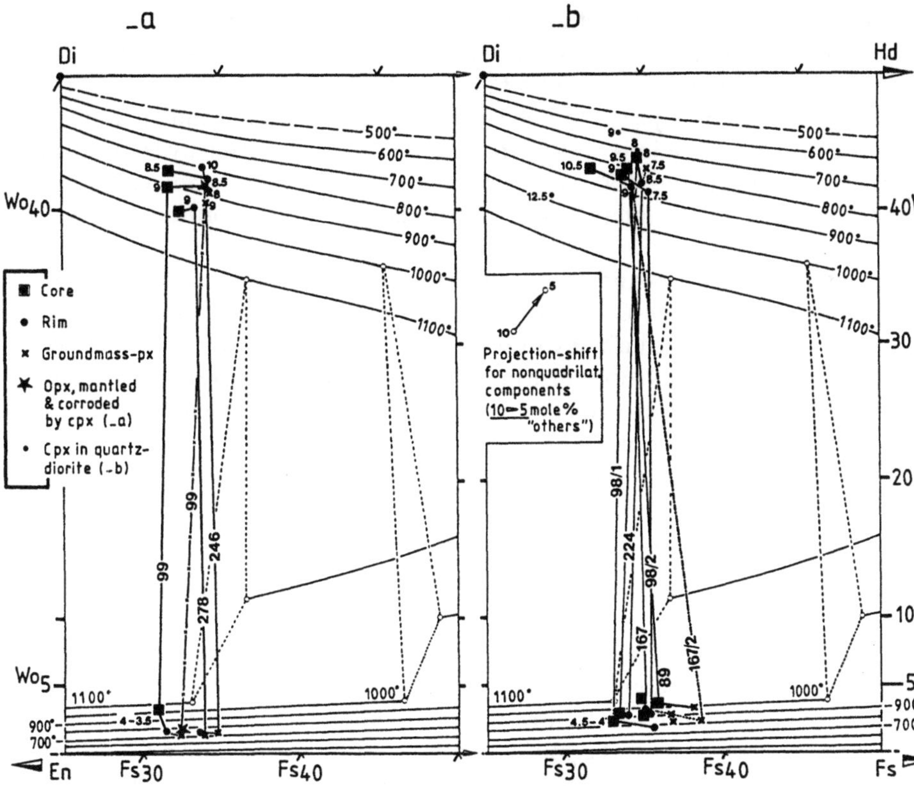

Fig. 4.8. Portion of the pyroxene quadrilateral showing coexisting ortho- and clinopyroxenes from **a** the fine-grained and **b** the medium- to coarse-grained facies of Ballachulish monzodiorites; *tie-lines* are labeled for Opx-Cpx pairs of Weiss and Troll (1989, Table II). The data points are projected from non-QUAD components using the correction procedure of Lindsley and Andersen (1983) onto the graphical thermometer of Lindsley (1983), adjusted to 0.3 GPa; *numbers* give mol% non-QUAD components (others)

in Fig. 4.8a). Clinopyroxene falls in the salite augite field, displaying a fairly narrow compositional range in monzodiorites, where it contains 7.5 to 10.5 mol% non-QUAD components (mainly Ti-Al-Fe Tschermak's molecules and acmite). Compositional zoning is weak to moderate, generally combined with slight Fe-enrichment (less than Fs_{2-4}). Orthopyroxene is a magnesio-hypersthene with compositional zoning weak or absent; having only 3.5–5 mol% non-QUAD components, it displays a moderate chemical variation over the diorite suite (Fs_{29-37}).

4.2.1.1 Pyroxene Geothermometry

Lindsley (1983) presented a graphical two-pyroxene thermometer which critically depends on Ca partitioning between pure Ca-Fe-Mg pyroxenes. Hence, a special projection onto the pyroxene quadrilateral is required (Lindsley and Andersen 1983) to correct the effect of non-QUAD components ('others') thus approximating the bulk activities of Wo, En and Fs. Following the procedure of Lindsley and Andersen, components Wo, En, Fs and 'others' were calculated using fixed Fe^{3+}/Fe^{2+} ratios obtained by wet-chemical analysis and Möβbauer spectroscopy (Weiss 1986; Weiss and Troll 1989). The data were plotted in a portion of the quadrilateral in Fig. 4.8 (with solvus isotherms adjusted to P = 0.3 GPa using the X_{Fe}-dependent T_P-term of Lindsley 1983). Modifications of the pyroxene solvus, as suggested by Nabelek et al. (1987), mainly concern the 1050–1100 °C augite isotherms at intermediate X_{Fe}, hence not critically affecting the interpretation of the stability data for orthopyroxene (see below).

A use of the Lindsley-Andersen geothermometer may be severely restricted by granule exsolution of high-Ca pyroxenes in plutonic environments (Lindsley and Andersen 1983). Al_2O_3-rich clinopyroxenes present further problems, since the complete subtraction of CATS molecules in the largely empirical projection scheme for augite is probably an overcorrection (Lindsley 1983); this underestimate of Wo activity results in unreasonably high apparent temperatures (rising by ~ 100 °C per 1 wt% Al_2O_3; Andersen and Lindsley (1984). The interpretation of equilibration temperatures for Ballachulish pyroxenes is hence based mainly on the data for orthopyroxene, as follows:

1. Texturally and chemically, hypersthene is much more homogenous than coexisting augite.
2. Orthopyroxene only contains 3.5–5 mol% non-QUAD components compared to 7.5–10.5 mol% 'others' in coexisting augite (which results in large potential errors of more than 70–90 °C for Cpx equilibration temperatures).
3. Hypersthene appears to be much less affected by retrogressive Ca-diffusion, critically lowering the apparent equilibration temperature of coexisting augite.

In Figure 4.8, tie-lines were drawn only between coexisting pyroxenes, i.e., omitting zonal Cpx primocrysts with Cr-rich cores (D278,246,167), since these Mg-rich Cpx compositions were probably not saturated with Opx. Although the high Wo content of augite in the slowly cooled, medium- to coarse-grained diorites D98,224

(Fig. 4.8b) indicates a low-T composition which is not at present in equilibrium with coexisting hypersthene (yielding 1100–770 ± 60 °C), the slope of the tie lines is in very reasonable agreement with quadrilateral pyroxene pairs if one considers the possible overcorrection for CATS molecules (Andersen and Lindsley 1984). If an empirical CATS recorrection is applied ('projection shift' in Fig. 4.8b; Weiss and Troll 1989), the revised steeply positive slope of the tie line and the relatively broad X_{Fe}-range in orthopyroxene suggest a former 3-pyroxene assemblage near 1100 °C, with pigeonite and augite subsequently modified by granule exsolution and Ca diffusion.

Combining textural evidence and geothermometrical data (Weiss and Troll 1989), hypersthene crystallization in coarse- to medium-grained diorites obviously started near 1100–1050 (± 60) °C at 0.3 GPa, with augite (and pigeonite?) co-crystallizing, followed by extensive re-equilibration of augite to subsolidus temperature (below 900 °C). Referring to the calculated pyroxene-melt equilibria for H_2O-absent conditions (Weiss and Troll 1989), and to the experimental results on H_2O-CO_2 bearing andesitic-granodioritic systems (Eggler 1972; Naney 1983), the addition of ca. 1 wt% H_2O to a dry melt of intermediate composition would lower the Opx-Cpx liquidus by approximately 50–70 °C; considering intrinsic water contents of 0.6–0.7 wt% (Table 4.1), which supposedly approach the amount of initial magmatic water in the monzodiorites (Sect. 4.3.1), liquidus temperatures of 1140–1130 (± 80) °C (Opx) and 1120–1090 (± 60) °C (Cpx) at P = 0.3 GPa are suggested. These data correspond reasonably well with the maximum equilibration temperatures obtained by geothermometry for orthopyroxene primocrysts from the normal-facies diorites (e.g. D89,98 in Fig. 4.8b). The anomalously low equilibration temperatures for Opx (700–800 °C, in Fig. 4.8a) and the morphology of both pyroxenes in fine-grained to 'chilled' diorites D246,278 (Weiss, Sect. 5.2, this Vol.) suggest the metastable crystallization from a liquid, which appeared strongly 'supercooled' with respect to the Opx-liquidus; here, hypersthene did probably *not* attain a complete exchange equilibrium with augite (which forms skeletal primocrysts; see Figs. 3.6b,5.5).

4.2.2 Feldspars

4.2.2.1 Plagioclase

The compositional range of plagioclase (Fig. 4.9) is fairly restricted in monzodiorites (An_{38-23}, usually < 6–7 mol% An variation per rock sample), large in quartz diorites (An_{46-19}, even in single crystals) and moderate in the granites (An_{29-14}). The liquidus temperature for plagioclase An_{34-29} in monzodiorites was estimated by Weiss and Troll (1989), inferring a liquidus temperature of 1090 ± 60 °C (for ~ 1 wt% H_2O in the melt); this is consistent with the textural evidence that plagioclase joins pyroxene crystallization below 1140–1090 °C.

A genetic interpretation of the zoning profiles of plagioclase in the normal facies of monzodiorites, quartz diorites and granites (Weiss and Troll 1989, Fig. 13) is based on the model for isobaric plagioclase fractionation presented by Loomis (1982),

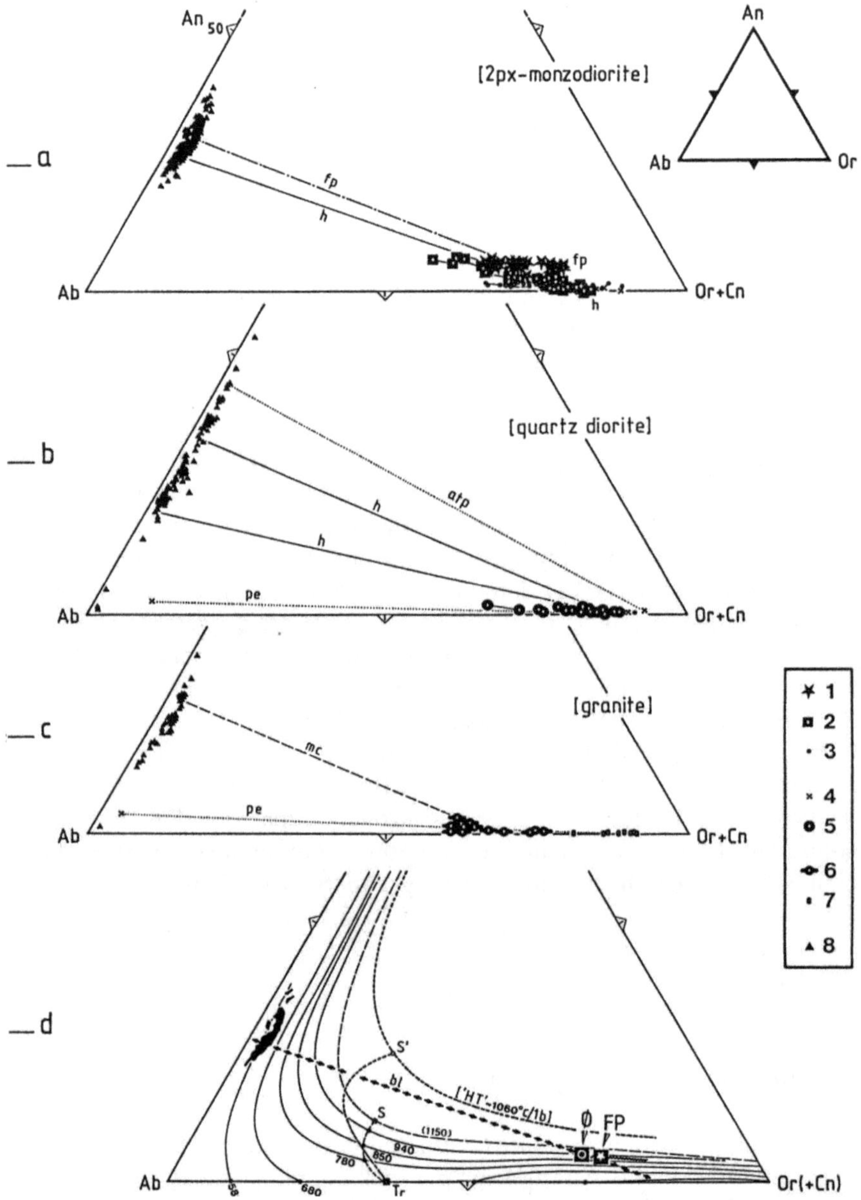

Fig. 4.9. Microprobe analyses of coexisting feldspars in **a** monzodiorites; **b** quartz diorites; and **c** granites from the Ballachulish Complex (Weiss and Troll 1989, Table II). *1,2,3* Film-perthitic Akf (*fp*), homogenous Akf (*h*) and matrix between symplectites, in monzodiorites; *4* antiperthites (*atp*) and perthitic Akf (*pe*) *5* homogenous Akf, in quartz diorites; *6,7* Akf megacrysts (*mc*) and groundmass Akf, in granites; *8* plagioclase. **d** The ternary feldspar system with the solvus isotherms of Seck (1971), adjusted to 0.3 GPa following Ghiorso (1984). T_c-S'/S" are alternative critical solution curves taken from W.L. Brown and Parsons (1985); HT is a possible high-temperature isotherm suggested by Parsons and Brown (1983). O and FP are averaged compositions of symplectite intergrowths (*line bl*) and of film-perthitic areas

showing that, equivalent to an undercooling of the magma, even a minor reduction of a_{H_2O} in a granodiorite melt would considerably reduce X_{An} in fractionating plagioclase, as follows:

1. In monzodiorites, which only contain late- to post-magmatic hydrous phases, the very weak compositional normal zoning of plagioclase (< 3–5 mol% in single primocrysts) indicates a near-equilibrium growth, with low crystallization rates preventing disequilibrium fractionation. The late, near-solidus crystallization of alkali feldspar partly compensated a fractional enrichment of Na-Si in the liquid, favouring An-richer plagioclase rim compositions than inferred from the single phase (Plg) fractionation model of Loomis (1982).
2. In quartzdiorites, corroded plagioclase cores indicate equilibrium fractionation, continuing throughout a period of isothermal crystallization where no hydrous phases are stable. Therefore, water accumulates in the melt; this depresses the Plg liquidus and causes a reverse zoning ($An_{35\rightarrow40}$). Numerous non-periodic oscillations may have formed at this stage, if the host rock shows evidence of extensive pelite contamination (e.g., D228). Finally, the crystallization of biotite and amphibole reduces the water content in the melt thus favouring a strong normal zoning ($An_{45\rightarrow20}$).
3. In the granites, a disequilibrium fractionation trend is initiated following the resorption of Plg xenocrysts from the diorites. The early, H_2O-consuming crystallization of biotite nevertheless prevented a large depression of the plagioclase liquidus. The subsequent growth of alkali feldspar megacrysts (with biotite) and of groundmass alkali feldspar (with quartz) temporarily reduced the Na-Si concentration in the melt favouring the formation of Ca-Al-enriched, inner plateau zones in plagioclase.

4.2.2.2 Alkali Feldspar

In monzodiorites, the compositional range of alkali feldspar (associated with oligoclase An_{30-26}; Fig. 4.9a) comprises homogenous feldspar Ab_{30-15} Or_{65-85} Cn_1 An_{5-0} and film-perthitic feldspar Ab_{30-20} Or_{65-75} Cn_{1-2} An_{6-3}; its composition is nearly identical to a cryptoperthite which crystallized at ~ 950 °C/P ≤ 0.1 GPa in an alkali gabbro from Klokken (Parsons and Brown 1983, Fig. 3a). An-poor homogenous Akf is associated with andesine-oligoclase An_{44-30} in quartz diorites (Fig. 4.9b). Like in the monzodiorites, lobate corrosion-type boundaries between solidus plagioclase rims and solvus interstitial alkali feldspar (Weiss and Troll 1989, Figs. 8,10) might account for a disequilibrium relationship between the two feldspars. In the porphyritic granites, Akf megacrysts Ab_{39-23} Or_{54-76} $An_{2.9-0}$ $Cn_{3.8-1.8}$ coexist with oligoclase An_{26-13} (Fig. 4.10c).

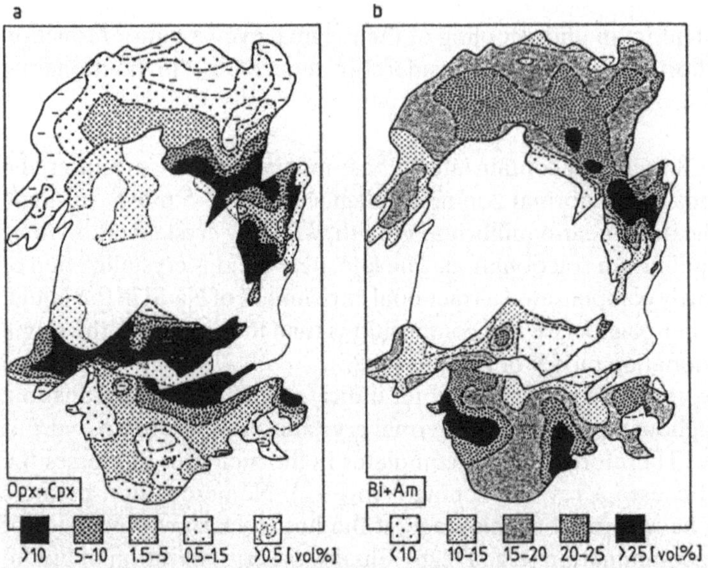

Fig. 4.10.a Areal distribution pattern of pyroxenes (modal augite + hypersthene) in Ballachulish magmatites. Clinopyroxene also occurs as xenocrysts in marginal granites (> 0.5 vol%), due to hybrid mixing leucocratic quartz diorites. **b** Areal distribution of hydrates (modal biotite + amphibole) within the diorite envelope, replacing both pyroxenes. The sample distribution is shown in Fig. 4.5; for lithologies, see Fig. 3.1

4.2.2.3 Thermal Stability of Feldspars

Since andesine-oligoclase always started to crystallize *before* alkali feldspar in Ballachulish rocks (even in the granites), with a single feldspar phase (Plg) continuously forming from the dioritic liquids (except for late-magmatic stages), the weak to moderate fractionation of plagioclase follows solidus paths (Fig. 4.9). Even the most sodic plagioclase An_{27-19}, which obviously coexists with late-stage alkali feldspar in the hypersolvus diorites, contains only 2–4 mol% Or. Hence any application of a two-feldspar solvus thermometer (Ghiorso 1984; W.L. Brown and Parsons 1985; Fuhrman and Lindsley 1988) on these plutonic rocks is restricted, because the activity of the T-critical Or component in plagioclase is severely underestimated. The apparent Or loss from plagioclase might reflect granule exsolution phenomena at magmatic temperatures, or subsolidus alkali exchange with a fluid phase (Fuhrman and Lindsley 1988; Weiss and Troll 1989). Nevertheless, as suggested by Parsons et al. (1986), the compositions of the most ternary, i.e., An-Ab enriched alkali feldspars can be used to estimate *minimum* temperatures of feldspar crystallization.

Referring to the texture and compositional range of two-feldspar symplectite intergrowths (Weiss 1986, Fig. 70d; Fig. 4.9d, this Vol.), which are best developed and coarsest in the more fractionated central portions of the monzodiorites (Weiss and Troll 1989, Figs. 8,9), a simultaneous crystallization of oligoclase blebs and alkali feldspar matrix in the presence of a Ab-An-Or cotectic residual liquid and/or a hydrous fluid is suggested, with H_2O enrichment favouring interface instability and cellular growth of the symplectites (J.V. Smith and Brown 1988). A low-T limit for a late-magmatic co-crystallization is given by the normative composition ab_{70} or_{30} of a hydrous alkaline melt, which is stable above 765 ± 15 °C on the binary minimum for P_{H_2O} = 0.3 GPa (Fig. 4.7); compare Tuttle and Bowen 1958; Fig. 17).

Since the solvus isotherms in the ternary feldspar system (Fig. 3.9d) trend nearly parallel to the Ab-Or binary for K-rich alkali feldspars (Ab_{35-15} Or_{65-85}), a linear T-dependence for An solubility in alkali feldspar is suggested. This implies that Akf equilibration temperatures are raised by 50–70 °C per each mol% An added in excess of a solubility of $An_{1.5-2}$ at 700 °C and 0.3 GPa. Thus the maximum An contents of 7–9 mol% might be attained at temperatures near 1060–1070 °C, referring to the high-*T* solvus isotherms of Parsons and Brown (1983) and Fuhrman and Lindsley (1988).

Compiling the composition-stability data for Ballachulish feldspars (Weiss and Troll 1989), their crystallization range in diorites and granites may be reconstructed as follows:

In monzodiorites, ternary alkali feldspar appeared on the solvus near 940–910 (± 40) °C, joining the crystallization of sodic plagioclase rims. With consolidation and fractionation proceeding, symplectite intergrowths crystallized in the presence of a residual, progressively H_2O-enriched pore liquid (below 900 to 765 °C). Al-Si ordering of this relatively K-Ca-Ba-rich alkali feldspar stranded at intermediate states, while K-poor sodic plagioclase finally attained a largely Al-Si-ordered state. In the quartz diorites, homogenous alkali feldspar only formed at near-solidus temperatures of 700 (±40) °C. Apparent thermal stabilities of 820–740 °C for the cores of Akf megacrysts in the granites suggest minimum crystallization temperatures. This considers the deuteric coarsening of microperthites Ab_{91} Or_5 An_4 (below ~ 550–500 °C; see Parsons 1978, Fig. 6), which produced Ab- and An-poor megacryst compositions (Fig. 4.9c).

4.2.3 Fe-Ti Oxides

Coexisting ilmenite-magnetite pairs were used for oxygen barometry and geothermometry (Weiss and Troll 1989, Fig. 15). Oxide equilibration in monzodiorites was initially controlled by f_{O_2} near NNO, involving the assemblage biotite-sanidine-magnetite-quartz-pyroxene (Weiss 1986, Figs. 22,58); a fairly low maximum equilibration temperature of 880 ± 60 °C indicates that external granule exsolution (Buddington and Lindsley 1964) was efficient over a fairly large T-range (< 880–600 °C). Oxides from three fine-grained diorites plot near the MNO buffer (oxidized samples D99,246,228). This might be due to the assimilation of 'wet' pelitic xenoliths (stabilizing anomalously Mg-rich biotites within these samples). Likewise, the trend for monzodiorites approaches the MNO buffer below 750 °C. This suggests an external buffer control due to the fractional enrichment of water in the residual liquid, as no hydrous phases were stabilized in the main crystallization range (1100–900 °C). Oxides in the more differentiated quartz diorites and the granites follow a reducing trend to NNO-QFM, probably caused by the circulation of hydrous fluids favouring the Fe-Ti exchange between oxides and mafic hydrosilicates. Composite oxidizing-reducing trends for Opx-Qz bearing diorites (transitional between monzodiorites and quartz diorites, e.g., D22) indicate a system which appears closed with respect to 'magmatic' water. Here, the oxidizing portion of the trend (versus MNO) represents the progressive saturation of a residual liquid in H_2O. Excess fluids are subsequently consumed during subsolidus reactions with pyroxenes (uralitization) and plagioclase (sericitization), now buffering the reducing portion of the trend.

4.2.4 Biotite

Biotite is unusually Mg-rich, showing only a moderate increase of iron content with progressive differentiation. Here, X_{Mg} = [Mg/Mg + $Fe^{2+}{}_{total}$] decreases from the monzodiorites [0.69–0.61] to the quartz diorites and the granites [0.57–0.50], following the increasing fractionation stage of the host rocks (D.I. = 51 to 67–73; Table 4.1). Similar 'type II' trends in T-X-f_{O_2} space reported by Wones and Eugster (1965, Fig. 13). indicate 'dry' magma systems, in which the melt is undersaturated with respect to H_2O.

The ternary composition annite-phlogopite-oxyannite of biotites synthesized at various oxygen fugacities by Wones and Eugster (1965, Fig. 1) was used to estimated f_{O_2}-conditions during biotite crystallization, referring to $X_{Fe^{2+}}$-X_{Mg}-$X_{Fe^{3+}}$ ternary biotite compositions in Ballachulish rocks (Weiss and Troll 1989, Table II and Fig. 16). Monzodiorites contain the assemblage Bi-Akf-Mt-Qz-Opx(-Cpx) buffered near NNO. An upper stability limit for biotite is hence given by the univariant, P-insensitive reaction:

3 orthopyroxene + sanidine + H_2O (V) → biotite + 3 quartz (Wones and Eugster 1965).

Referring to the T-X-f_{O_2} plane of Wones and Eugster (1965, Fig. 4), f_{O_2}-estimates and ternary composition of biotites were used to quantify its thermal stability in the Ballachulish diorites and granites (Weiss and Troll 1989):

In monzodiorites D98,167, biotite Ann_{32-34} Oxy_{3-5} became stable at 980–960 (±50) °C at f_{O2} near NNO. In pyroxene-bearing quartz diorites D22,152, biotite Ann_{39-43} Oxy_{4-5} crystallized at 910–870 (± 40) °C and f_{O_2} near NNO. Despite their Fe-richer composition Ann_{44-46} Oxy_{3-2}, a similar thermal stability of 900 (± 50) °C is inferred for reduced, QFM-buffered biotites in the leucocratic quartz diorites D352,108, where augite is extensively replaced by amphibole thus restricting (buffering) the activity of magmatic fluids. The water enrichment by fractional crystallization in the salic granite G123, or the mechanical admixture of 'wet' pelitic material to the marginal acid quartz diorite D228, apparently produced oxidized biotites Ann_{41} Oxy_7 and Ann_{37} Oxy_6, respectively. These compositions were buffered between NNO and MNO, which places an upper thermal stability limit near 850–900 (± 40)°C.

4.2.5 Amphiboles

Amphibole analyses, which are reported in Weiss (1986) and Weiss and Troll (1989, Tab. II) were classified by composition, adopting the IMA nomenclature compiled by Leake (1978). Am-I is a magnesio-hornblende with $X_{Mg} = 0.73–0.60$, showing a substantial substitution to edenite only in leucocratic quartz diorites (D108,352), where Am-I has $(Na+K)_A = 0.50–0.60$. The low-alumina actinolitic amphibole Am-II and the actinolite Am-X have 7.30–7.80 Si and $(Na + K)_A < 0.35$, with X_{Mg} varying between 0.68 and 0.78. Hence, the calcic amphiboles of Ballachulish are Al-poor and relatively Mg-rich, showing a very restricted Mg-Fe compositional variation.

P-T Stability of Amphiboles. In contrast to the 'normal', Al-rich amphiboles which are a common near-liquidus phase in most calc-alkaline rocks (Wones and Gilbert 1982), amphibole textures in Ballachulish suggest a late- to post-magmatic formation *subsequent* to biotite, mainly at the expense of the earlier formed pyroxenes.

Nabelek and Lindsley (1985), Hammarstom and Zen (1986, Fig. 7b) and Hollister et al. (1987) showed that the solubility of total Al-cations Al^T and of tetrahedral alumina Al^{IV} in amphibole may be used to estimate its P-T stability, provided that the amphibole is aluminous (i.e., a magmatic hornblende) and coexists with an igneous six-phase assemblage of Plg-Bi-Mt-Sph-Akf-Qz. Aluminous, edenitic-tschermakitic hornblende suitable for geobarometry only occurs in quartz diorites D108,152,352 (Table 4.1). The Al^T-T equation of Hollister et al. (1987) yields a pressure estimate of 0.19–0.21 (± 0.1) GPa for these edenite compositions. This suggests a minimum value for Ballachulish, considering the P-calibrations of 0.3 (± 0.05) GPa from the thermal aureole and a lower P-limit of 0.25–0.30 GPa based on the cotectic composition of the fractionated leucogranite (Fig. 4.7).

The thermal stability of Ballachulish amphiboles was estimated in a plot of the T-sensitive cations Ti and Al^{IV} (Weiss and Troll 1989, Fig. 17). The deduced magmatic crystallization interval for edenite in quartz diorites ranged from 770–720 (± 70) °C, whereas uralitic magnesio-hornblende appeared at near-solidus temperatures of 710–680 °C. In the hybrid granite G123, edenite crystallized below ca. 700 °C at the expense of augite xenocrysts (derived from the quartz diorites). The actinolitic hornblende Am-II and the actinolite Am-X exclusively formed by subsolidus uralitization of hypersthene and augite below 650–550 °C. The Ti-Al^{IV} distribution pattern in coexisting amphiboles Am-I,II,X from quartz diorites and granites is apparently related to a change in oxygen fugacity from near NNO to HM (stabilizing Ti-poor amphiboles; Helz 1982, Fig. 69), due to the subsolidus circulation of hydrous fluids (< 650–700 °C).

4.3 Fluid Activity and Crystallization Sequence

4.3.1 Fluid Activity

The areal distribution of anhydrous and hydrous phases, respectively modal augite + hypersthene and biotite + amphibole (Fig. 4.10a,b), may be used as a first approach to constrain the effects of magmatic and postmagmatic hydration in Ballachulish Complex. It is obvious that least hydrated areas are almost entirely confined to the inner, hypersthene-rich monzodiorites (compare Figs. 3.1 and 4.10). The areas of maximum hydration either show a prevalence of biotite, within the augite-bearing quartz diorites to the south, or of amphibole, within the alkaline monzodiorites to the northeast (Weiss 1986, Fig. 56c). Whereas the *ratio* Bi/Am appears hence to be controlled by the bulk composition of the host rock (Wones and Gilbert 1982), the *degree* of hydration is not simply a function of the fractionation stage (Fig. 4.6a); this is referred to the assimilation of 'wet', i.e., biotite-rich pelitic material favouring pyroxene-out reactions (Troll and Weiss, Sects. 3.2.2 and 3.3.2, this Vol.), thus producing the low-pyroxene areas (< 0.5–1.5 vol%) in the southern and coastal sectors of the igneous complex (Fig. 4.10).

The largely anhydrous mineralogy (Fig. 4.10) and high-T phase stabilities (Fig. 4.11) suggest a strong water deficiency throughout the magmatic crystallization range of hypersthene-rich monzodiorites. Vapor-absent conditions apparently continued even to the subsolidus range, metastably preserving intermediate structural states for

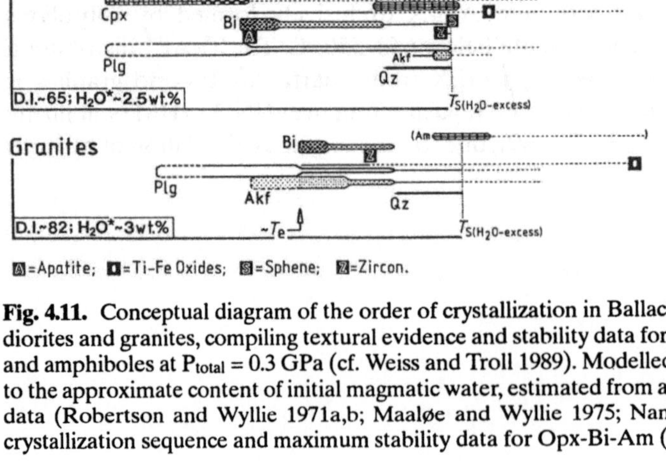

Fig. 4.11. Conceptual diagram of the order of crystallization in Ballachulish monzodiorites, quartz diorites and granites, compiling textural evidence and stability data for pyroxenes, feldspars, biotite and amphiboles at $P_{total} = 0.3$ GPa (cf. Weiss and Troll 1989). Modelled liquidus temperatures refer to the approximate content of initial magmatic water, estimated from a comparison of experimental data (Robertson and Wyllie 1971a,b; Maaløe and Wyllie 1975; Naney 1983) with the observed crystallization sequence and maximum stability data for Opx-Bi-Am (see Sect. 4.2.2)

alkali feldspar (Fig. 3.12b). This suggests that the host rocks had mainly dry histories when they entered the thermal stability range of microcline near 500–450 °C/0.3 GPa (Parsons 1978). Further evidence is given by the granulite-type perthite unmixing of alkali feldspars (Fig. 3.11) and the minimum degree of hydrothermal sericitization of plagioclase in monzodiorites (Fig. 3.10). Except for the pelite-contaminated diorites, the close link between the normal petrofacial sequences shown (Figs. 3.1,3.10) and the degree of igneous fractionation of the host rock (Fig. 4.6a) is obvious. This indicates that, if any late-stage fluids were present, they were released during magmatic evolution. This included the generation and subsequent consumption of magmatic fluids by hydrous buffering. Nevertheless, the anhydrous mafic assemblage was not entirely hydrated in monzodiorites. Magmatic amphibole is not stable in the monzodiorites; even in the quartz diorites it occurs late, following titaniferous biotite. Hence, Bowen's magmatic reaction principle, which postulates the stabilization of hornblende after augite and ahead of biotite in calc-alkalic liquids with increased fractionation and H_2O enrichment (Wones and Gilberts 1982), is not applicable to the H_2O-undersaturated, K_2O-enriched mesotype rocks of Ballachulish Complex. Naney (1983) showed that amphibole stability is not only dependent on the CaO-content of a magma, but is mainly controlled by a minimum H_2O content (~ 3 wt% in a granodiorite melt at 0.3 GPa). In the 'dry' alkaline monzodiorite, phlogopite-rich biotite was hence stabilized at a distinctly higher temperature than magnesio-hornblende.

Besides hypersthene and augite, diorites contain 5–20% modal biotite and subsolidus amphibole. These hydrous phases have ca. 2–3 wt% H_2O, equivalent to 0.15–0.50 wt% structurally bound H_2O. Since whole-rock analyses give 0.6–1 wt% intrinsic H_2O^+ (Table 4.1), the difference presumably represents H_2O trapped within melt-fluid inclusions, fixed along intergranular boundaries or associated with secondary sericite and talc. Nevertheless, most of these pyroxene diorites have *not* attained their maximum hydration at subsolidus conditions, as would be expected in the presence of excess water. We therefore suggest that intrinsic water contents of 0.6–1 wt% closely correspond to an amount of probably 1 ± 0.5 wt% initial water dissolved in an alkaline diorite melt at near-liquidus temperature. This considers the close correspondence between the observed crystallization sequence Opx-Cpx-Plg in normal-facies diorites with the maximum equilibration temperature for hypersthene (1100 ± 60 °C), and the liquidus temperatures, calculated for 1 wt% H_2O (Sect. 4.2.2): 1140–1130 ± 80 °C (Opx), 1120–1090 ± 60 °C (Cpx) and 1090 ± 60 °C (Plg).

Robertson and Wyllie (1971a), Maaløe and Wyllie (1975) and Naney (1983) showed that, at water-undersaturated conditions, the sequence of crystallization is strongly dependent on the initial H_2O content of quartz dioritic — granodioritic and granitic magmas (at fractionation stages D.I. = 72–77 and 83–87, at P = 0.2–0.8 GPa and f_{O_2} buffered near NNO). Extrapolating the experimental data to P_{total} = 0.3 GPa for leucocratic quartz diorites D352,108 (D.I. = 67–73), and regarding the observed crystallization sequence Plg- -Bi-Am (~ 770-720 °C) - -Qz-Akf (~ 700 °C), an initial total H_2O content of 2–3 wt% is suggested for these most fractionated members of the diorite suite. The granites G123,243 have D.I. = 82–86, with alkali feldspar and biotite crystallizing nearly simultaneously; this generates the sequence Plg- -Akf-Bi(850 °C)- - Qz, for an estimated initial H_2O-content of 2.5-3.5 wt% at P_{total} = 0.3

GPa. Progressive fractionation within the porphyritic granite pluton produced a hydrous cotectic melt of rhyolitic composition D.I. = 91–92, presumably saturated with 7.5–8 wt% H_2O at P = 0.3 GPa (Tuttle and Bowen 1958, Fig. 28). Consequently, the crystallization interval for Plg + Akf + Qz is severely reduced, favouring the development of an aplitic texture in this late leucogranite (G119 in Table 4.1).

4.3.2 Order of Crystallization: Compilation of T-X-f_{O_2} Conditions

Figure 4.11 compiles the thermal stability estimates for rock-forming minerals based on the quantitative interpretation of phase compositions in the normal facies of monzodiorites, quartz diorites and granites from Ballachulish (Weiss and Troll 1989; and Sect. 4.2.2). The conceptual diagram also regards the probable extent of late- to post-magmatic recrystallization and Ca-K-Ti re-equilibration, mainly affecting augite, plagioclase and oxides. Modelled thermal data at P_{total} = 0.3 GPa refer to the fractionation stage D.I. of the host rock, the suggested content H_2O^* of initial total water in the system, the inferred equilibrium liquidus data T_L, geothermometric estimates of phase stability T_X within the buffered f_{O_2}-range, and the solidus limit T_S (as presented in the preceeding sections of this chapter).

Initially, a 'dry' alkaline, convecting monzodiorite magma with D.I. 49–53 and H_2O^* ~ 0.5–1.5 wt% was emplaced at a largely liquid state, at T ~ 1100 °C. Hypersthene, augite and sodic andesine lead the crystallization sequence at T_L = 1140–1090 °C ($T_{X\,(Opx)} \leq 1100$ °C), followed at f_{O_2} ~ NNO by Fe-Ti oxides (T_L ~ 1000 °C, $T_X \leq 880$ °C), biotite (T_X = 980–910 °C), apatite (T_L ~ 950 °C) and alkali feldspar (T_X = 940–860 °C). The crystallization interval for Opx-Cpx-Plg primocrysts was obviously restricted to ~ 1100–950 °C. If H_2O^* exceeds ~ 1 wt%, which is the probable limit of maximum subsolidus hydration at vapor-absent conditions, some minor crystallization of biotite, quartz and symplectitic alkali feldspar may continue below T_S ~ 900 °C ('dry') to T_S = 760 °C (H_2O excess). Actinolitic amphiboles are exclusively subsolidus at T_X ~ 720–550 °C and f_{O_2} near HM, while augite, oxides and plagioclase preferentially re-equilibrated to submagmatic temperatures near 650 °C.

A quartz diorite melt with D.I. 59–72 and H_2O ~ 2–3 wt% evolved from the monzodiorite magma, as crystal fractionation, flowage differentiation and assimilation of metasediments proceeded. Augite and plagioclase primocrysts formed at T ~ 1050–900 °C, followed by oxides, apatite, biotite (T_X = 910–870 °C) and edenitic hornblende (T_X = 770–680 °C), at f_{O_2} ~ NNO-QFM. Late-magmatic sphene, quartz, alkali feldspar (T_X = 700 °C) and zircon (T_X ~ 700 °C) crystallized from a hydrous alkalic melt at f_{O_2} ~ NNO-HM. A release of fluids is first to be expected near 700–690 °C, closely above T_S = 680 °C, for H_2O-saturated conditions. It should be noted that these fluids were largely used up in the anatectic melting of pelitic country rocks (Pattison and Harte 1988).

The granite magma already contained primocrysts of sodic andesine and alkali feldspar when it was emplaced at T_e ~ 900–850 °C (D.I. ~ 80, H_2O^* ~ 3 ± 0.5 wt%).

Hybrid mixing with leucocratic quartz diorites and continuing fractionation within the granite magma significantly expanded the D.I.-range (< 75–93) and modified the suggested H_2O^*-contents (~ 2.5–6 wt%). Following the crystallization of oxides, biotite (T_X = 850) and zircon (T_X ~ 800–750 °C) at f_{O_2} ~ NNO-MNO, the appearance of quartz and uralitic actinolite (replacing augite xenocrysts ≤ 710 °C) introduces a static crystallization stage ('pseudoplastic consolidation'; Troll and Weiss, Sect. 3.4, this Vol.). A hydrous fluid was released near 700 °C, before the granitic rocks finally reached their solidus at T_S = 680–670 °C for P_{H_2O} = 0.3 GPa. It must be taken into account, that contact-migmatite zones might also have released subsolidus fluids below 650–680 °C (Pattison and Harte 1988), eventually invading 'dry' marginal portions of the igneous complex. This possibly contributed to the late-stage formation of the hydrated envelope, enclosing pyroxene-rich diorites specifically in the SW-SE-E periphery of Ballachulish Complex (Fig. 3.1).

It is evident from the position of petrofacial zones within the Ballachulish Pluton (Figs. 3.1, 4.1), and from the mineral stability data on the magmatic crystallization sequence (compiled in Fig. 4.11), that the diorite magma may have attained various emplacement temperatures on its present level of exposure, as follows:

1. 1100 °C near the eastern contact (mainly bordered by dry, hypersthene-bearing monzodiorites)
2. 1050 °C near the southern contact (bordered by hydrated monzodiorites and quartzdiorites)
3. 1000 °C near the western margin (with quartzdiorites and granites bordering and underlying, both releasing late-stage fluids)

In a first generalized approach, i.e., disregarding convective flow within the pluton, magma temperatures would imply *minimum* contact temperatures of 800–750 °C (data base of Buntebarth, Chap. 18, this Vol.), referring to a pre-intrusion country-rock temperature of 250 (± 50) °C (Pattison and Voll, Chap. 2, Troll and Weiss, Sect. 3.4.3, this Vol.). A simplified cylindrical model can be derived from the areal distribution of igneous rocks (Weiss 1986, Fig. 94). This may serve as a base reference for the thermal model of Buntebarth (Chap. 18, this Vol.), using an average emplacement temperature of 1050 °C for the diorite magma.

4.4 Summary: Some Aspects of the Petrogenesis of Igneous Rocks in Ballachulish

Watson and Dunning (1979) presented a model for the orogenic evolution of the Grampian Highlands, involving continental collision and early island-arc magmatism at an active plate margin. Subsequent to compaction, thickening and deformation of Moinian-Dalradian units within the Grampian orogen, Late Caledonian granitoids were generated. This group cannot be defined in terms of a single, well-defined magma source (Pearce et al. 1984; Harmon et al. 1984), since it may originate both from the melting of thickened lower crust due to thermal relaxation following a collisional event, and from the melting of upper mantle due to adiabatic decompression (England and Thompson 1984), accompanying discontinuous post-collisional uplift and erosion (Dempster 1985). If the thickening of continental crust involves thrusting at an active plate margin, with substantial mantle heat flow contributing to crustal radiogenic heat supply, the P-T-t paths for crustal units modelled by England and Thompson (1984) are suggestive of very high temperatures near 1050–1150 °C ('Heat Source Distribution III' at P ~ 1.5–1.7 GPa) attained with a thermal relaxation period of ca. 30 Ma for tectonic inactivity at 440–410 Ma (Dempster 1985, Table 3), referring to the structure and position of granulitic lower crust at depths of 50–60 km during the peak of Caledonian regional metamorphism (Shackleton 1979).

The alkaline, hypersthene-bearing monzodiorites are of special relevance to the petrogenesis of Ballachulish Complex, since their trace element patterns of mineralogy show striking affinities both to andesitic granulites and continental lithospheric mantle (Sects. 4.1.2, 4.1.3); compare Harmon et al. 1984). Experimental data of Eggler and Burnham (1973) and Grant (1986) indicate that CO_2-H_2O volatiles might have substantially aided the melting of Lewisian andesitic granulites (having 0.3–0.4/1.2–1.4 wt% CO_2-H_2O; Holland and Lambert 1975) at T ~ 1050–1150 °C and P ~ 1.5–1.7 GPa (as suggested above) within the Caledonian lower crust, which hence likely constitutes a major magmatic contribution to the hot and 'dry' Ballachulish monzodiorites (T ~ 1100 °C, initial H_2O ~ 0.5–1 wt%; Sect. 3.1). Generation and ascent of the intermediate magma was favoured by large-scale faulting of the lithosphere, e.g., along the Great Glen structure (Brown 1979), with the first 'hot trails' set up by CO_2-H_2O-rich mafic precursors of the Appinite Suite originating from an alkalized mantle source (Wright and Bowes 1979). The alkaline diorite magma of Ballachulish was modified on its upper crustal emplacement level (~ 10 km) to produce a monzodiorite-quartz diorite suite by in situ convective fractionation and assimilation of 'wet' Dalradian metasediments, yet on a smaller scale than in the neighbouring Ben Nevis Complex (Haslam 1968) where the occurrence of alkaline, hypersthene-rich diorites is severely restricted (at least on the exposed near-roof level).

In the Ballachulish Complex, coarse- to medium-grained monzodiorites of the central and normal facies (D96,98,223,224; Fig. 5.1) form typical wallrock adcumulates (Tait et al. 1984), consisting of igneously laminated euhedral phases with very weak compositional zoning. These rocks solidified at a relatively low cooling rate from a convecting magma, involving the continuous production and loss of latent heat from the system. Crystallizing more or less simultaneously under isothermal, near-equilibrium conditions, hypersthene, augite and andesine were continuously concentrated and fixed in the 'crystal mush zone' (Morse 1986), which developed along a slowly advancing, subvertical cooling front (Troll and Weiss, Sect. 3.4.1, this Vol.). In medium- to fine-grained rocks of the marginal facies (D99,167,246), however, augite is the prevailing liquidus phase, showing compositional zoning and subophitic intergrowths with plagioclase. Strongly reduced size, modified shape and low-T composition of unzoned hypersthene (associated with skeletal augite in the 'chilled' rocks D88,278) suggests a metastable suppression of Opx-nucleation, favouring disequilibrium growth of hypersthene from a melt whichappearedstronglysupercooledwithrespecttotheOpx-liquidus(Fig.4.8a and Weiss, Sect. 5.2, this Vol.). Tracing the petrofacial units from the centre to the margin of the diorite envelope, the more significant zoning of primocrysts and the decreasing proportion of interstitial minerals (~ 15–20 to < 25 vol%, compare Table 4.1) indicates a transition from adcumulate to orthocumulate growth, due to progressively faster cooling and solidification (Tait et al. 1984). This involves a more effective crystal fractionation at large accumulation rate, and a faster crystallization of residual liquid in the pore space, thus preventing a marked diffusional exchange between primocrysts and trapped liquid. The convective transport of solids and liquids obviously acted throughout the main thermal crystallization range for Opx-Cpx-Plg primocrysts (1100–950 °C), constituting a maximum volume proportion of ca. 50–60% in monzodiorites of the normal facies. With progressive capture of anhydrous cumulus phases in the advancing 'mush zone' and with clearing away the H_2O-enriched rejected solute during convective fractionation (Sparks et al. 1984), the activity of water was probably reduced with respect to the monzodioritic adcumulate (hence constituting an 'open system'), as H_2O and residual components became concentrated in a more differentiated, still-liquid magma fraction which at that time occupied the centre of Ballachulish Complex. This quartzdioritic magma later on expanded towards its western metasedimentary margin and roof, where it became chemically modified due to mechanical resorption and partial assimilation of 'wet' pelitic material, initially involving some high-T dehydration melting of biotite, and due to subsequent hybridization with the younger granite (Troll and Weiss, Sects. 3.2.1.2, 3.4.2, this Vol.). During these later stages of upwelling, bulk contamination of the quartzdiorites with Dalradian metasediments may have attained maximum values of ~ 20–30 wt% (Sect. 4.1.3; compare Frost and O'Nions 1985). When the mesotype portions of the diorite envelope were largely consolidated at T ~ 950–850 °C, a viscous granite magma was emplaced as a 'crystal mush' diapir at T ~ 900–850 °C. Subsequent to the ductile deformation of monzodiorites during in-situ expansion ('ballooning') of this pseudoplastic granite magma (Troll and Weiss, Sect. 3.4.2, this Vol.), consolidation and differentiation proceeded towards its core where a hydrous leucogranite melt formed. Possibly related to tectonic fracturing and the cone-sheet emplacement of rhyolite dykes under conditions of severely reduced fluid-pressure, the residual granite magma underwent a rapid loss of volatiles at T ~ 680 °C, as it was observed for the central cordierite microgranite of the Loch Doon Complex (Tindle and Pearce 1981, Fig. 10).

The petrogenic model proposed for Ballachulish monzodiorites is consistent with the parental role assigned by Nockolds (1949) to similar pyroxene mica diorites in several other plutons belonging to the Scottish suite of Caledonian Younger Granitoids (Halliday and Stephens 1984), including Glen Loy, Ben Nevis, Glen Tilt, Glen Scaddle, Glen Doll, Garabal Hill and Arrochar (see compilation of Pankhurst et al. 1982, Fig. 12.1).

5 Nucleation and Growth of Pyroxene in the Hypersthene Diorites

S. Weiss

5.1 Selection of Reference Area and Delimitation of Thermal Conditions

For a detailed textural study on intermediate igneous rocks of Ballachulish complex, an area in the southeastern part of the pluton was selected, where hypersthene-bearing monzodiorites are nearest to the contact. These relatively 'dry' diorites are least affected by late- to post-magmatic hydration southeast of Sgorr Dhonuill, where a very fine-grained chilled marginal facies bordered quartzitic country rocks, before the emplacement of granitic rocks partially separated the diorites from the metasediments. The term 'chilled margin' is used here in a general sense, defined as a zone of "fine-grained rock, enclosing and grading into the relatively coarser inner parts of an intrusion" (Cox et al. 1979). Figure 5.1 attempts a reconstruction of the monzodiorite main zone before the emplacement of younger granites, omitting the large ring-dyke granite apophysis and correcting for later sinistral fault displacement (compare Fig. 3.13c).

Cooling rates within the Opx-bearing monzodiorite main zone were estimated via the *halbraum* computer program which was kindly supplied by G. Buntebarth in 1987. This program calculates the propagation of a cooling front, relative to a boundary plane separating two media with different thermal properties. Referring to the experimental work of Buntebarth on Ballachulish diorites and quartzites (Chap. 18, this Vol.), input parameters for heat conductivity K = 1.5; 2.3 [W/m×grd C] and temperature conductivity A = 0.0868; 0.1042 [m^2/day] were used to model the cooling rate within the monzodioritic rocks (at T_l = 1100 °C), for varying distances z from the igneous contact which borders quartzitic country rocks (at T_l = 250 °C). The simplified *halbraum* model does not consider convective heat transport which evidently acted during primocryst formation in the diorites (see Figs. 3.4, 3.8). As long as the crystal fraction in an intermediate magma is below 40–50 vol% (Troll and Weiss, Sect. 4.4.1, this Vol.), convection would effectively tend to reduce the temperature variation in an igneous body (Kirkpatrick 1981). Consequently, the cooling rates $\Delta T/\Delta t$ [°C/yr] evaluated below for the marginal and the central facies of hypersthene diorites represent maximum and minimum values, respectively. Besides this, absolute distances of igneous facies types from the tectonically re-arranged igneous contact in Fig. 5.1 can only be given to ±50–100 m. Moreover, the contact to the metasediments is not always vertical, regarding local subsidences of quartzite screens and pelite rafts from a blocky, irregularly structurated roof zone in this southeastern part

G. Voll, J. Töpel, D.R.M. Pattison, F. Seifert
Equilibrium and Kinetics
in Contact Metamorphism

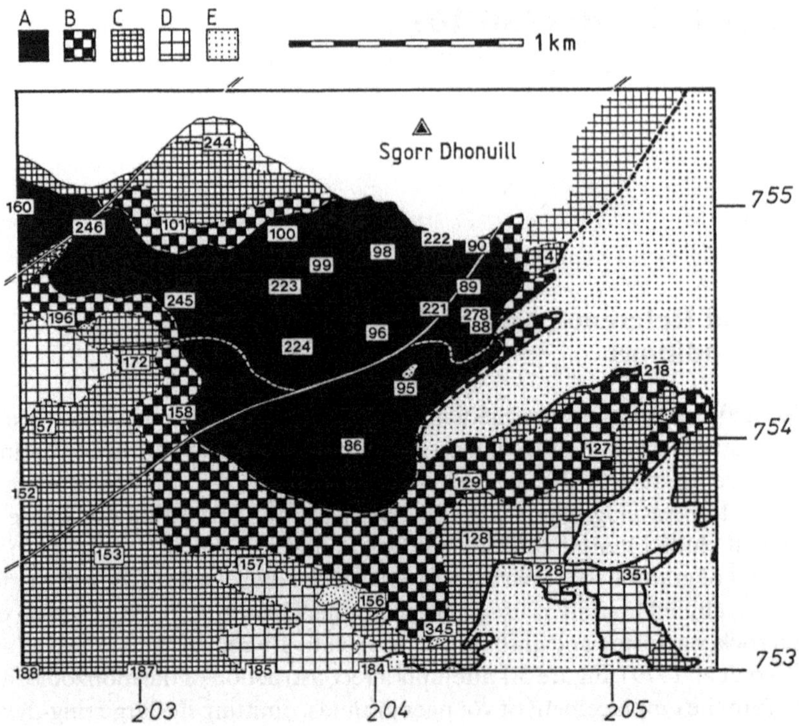

Fig. 5.1. Tectonic reconstruction of the monzodiorite main zone (before the emplacement of younger granites; cf. Figs. 3.1 and 3.13a,b). The actual positions of selected samples (*numbered*) are shown in Fig. 4.1. *A 'Dry' Opx-Cpx monzodiorite; B* 'hydrated' Opx-Cpx-Qz monzodiorite; *C* Cpx-Am-Qz diorite; *D* Bi-Am-Qz diorite; *E* quartzitic country rocks

of the igneous complex (see Figs. 3.1, 3.2). Within these uncertainties, the *halbraum* model approximates cooling rates $\Delta T/\Delta t$ of $\leq$ 4–0.8 and $\geq$ 0.03–0.02 [°C/yr] for marginal and central portions of hypersthene diorites, referring to contact distances of ~50–100 m and ~500–700 m, and to a main crystallization interval of ~1100–900 °C (see Fig. 4.11).

5.2 Nucleation Density, Facies Subdivision and Variation of Crystal Morphology

The nucleation density of both pyroxenes and plagioclase in Ballachulish monzodiorites is estimated via the crystal index (C.I.) of Maaløe (1978): C.I. = $[n/A^{3/2}]/(M \times 0.01)$, where n/A is the number of crystals per area of thin section (A = 1.5 × 2 cm), related to the mode M [vol%] of a mineral (in 3–5 sections per sample).

The C.I. thus only depends on the average size of a given mineral, defining the number of crystals in 1 cm^3 of that mineral. A fine-grained rock hence contains minerals with a larger C.I. than a coarse-grained rock. In Ballachulish samples, uralitic amphiboles Am-I (replacing Cpx) and Am-II (replacing Opx) were counted as Cpx and Opx, respectively.

Textural evidence for hypersthene diorites suggests that phases with the smallest C.I. generally lead the crystallization sequence, followed by phases with larger C.I. The relative sequence of crystallization may hence be modelled as a function of the nucleation density. Secondary effects, like the granule exsolution of augite, may superpose this relation in coarse-grained, slowly cooled diorites; likewise, plagioclase may exhibit slightly increased C.I. values, due to the following:

1. Subgrain formation during ductile deformation of primocrysts (Fig. 3.7a)
2. Late-stage crystallization of symplectite intergrowths, with plagioclase appearing as numerous groundmass platelets in alkali feldspar (Weiss and Troll 1989, Fig. 8.9)
3. Mechanical incorporation of plagioclase xenocrysts and heterogeneous nucleation on spinel phases, originating from partially resorbed pelitic material (Fig. 3.6a)

On the basis of average grain-size relationships (Fig. 3.4 and Weiss 1986, Table 1a) and of nucleation density for the main rock-forming cumulus phases C.I.* = $C.I._{Opx + Cpx + Plg}$, the diorites can be subdivided into three facies types (Fig. 5.2), each type displaying a different crystallization sequence (Fig. 5.3): (1) coarse-grained 'central facies', with C.I.* = 5–15 × 10^3 and Opx-Cpx- -Plg; (2) medium-grained 'normal facies', with C.I.* = 15–35 × 10^3 and Cpx- -Plg -Opx; (3) fine-grained 'marginal facies', with C.I.* = 35–150 × 10^3 and Cpx-Plg- -Opx.

This facies classification is largely independent of the fractionation stage of the host rock (at D.I. < 55–60), comparing D.I. zones and Rb isolines (Fig. 4.6) with the areal distribution of facies types in Fig. 5.2, and hence seems to be mainly related to variable cooling rates. Figures 5.2 and 5.3 show that a coarse-grained central facies of hypersthene diorites is best developed in the bottom parts of the slope south of Sgorr Dhonuill (D96,223,224). The fine-grained sample D99 belongs to a narrow schlieren facies occurring near the top of the monzodiorite main zone (exposed vertically over more than 600 m). The tongue-shaped marginal facies of D99 apparently represents a near-roof portion, extending subhorizontally eastwards into the 'chilled' contact zone (D221,278,88); this facies with a maximum nucleation density C.I.* of 150–330 × 10^3 is restricted to the contact area of 'dry' quartzites to monzodiorites, which are poor in hydrous phases (Bi + Am < 3 vol%). At this eastern contact zone, the high heat-conductivity of quartzites apparently favoured a very rapid removal of latent heat from the supercooled margin of the diorites. To the southeast, a hydrated marginal facies with a somewhat lower C.I.* of 50–150 × 10^3 occurs near the contact to the less heat-conductive pelites. Here, furthermore, the resorption of 'wet' pelitic material increases the oxygen fugacity (Weiss and Troll, Sect. 4.2.3, this Vol.), consequently lowering the liquidus temperature of the diorite and thereby decreasing the degree of supercooling ΔT (Lofgren 1974). This reduces

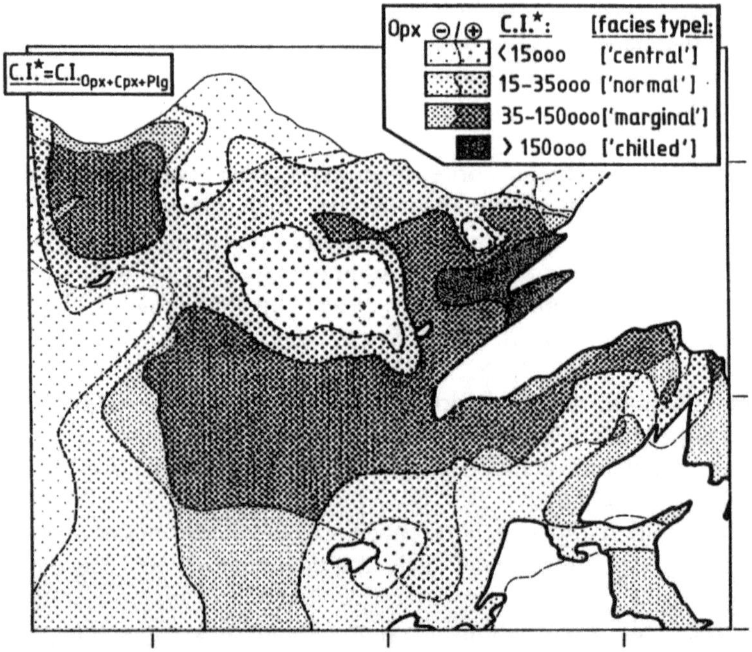

Fig. 5.2. Igneous facies subdivision for the map sector of Fig. 5.1; the estimated nucleation densities are based on the measured crystal index (Maaløe 1978), with C.I.* referring to rock-forming cumulus phases in Opx-free and Opx-bearing units. Calculation of the C.I. isolines followed the method described by Weiss and Troll (Sect. 4.1.3, this Vol.)

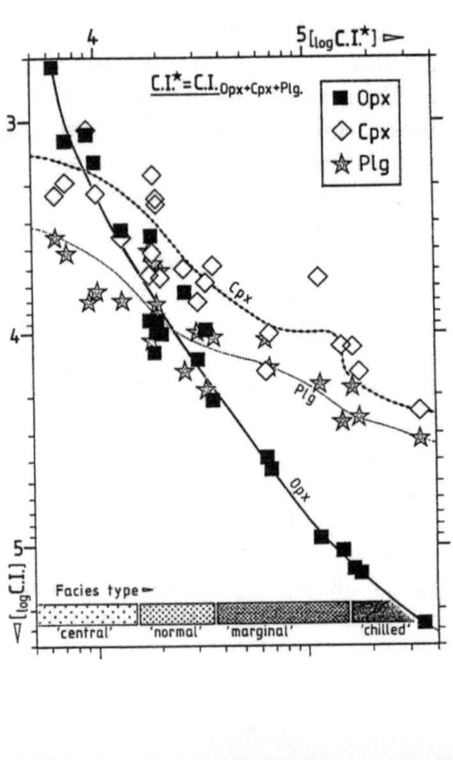

Fig. 5.3 Logarithmic nucleation density diagram for Opx-bearing monzodiorites, with C.I.* summed for Opx-Cpx-Plg cumulus phases and plotted versus the single-phase C.I., which critically varies according to the relative position of Opx (±Cpx-Plg) within the sequence of crystallization. Based on these changes in the order or crystallization, a conceptual subdivision of facies types is depicted in the base frame; it also refers to the areal facies subdivision in Fig. 5.2

the nucleation rate, producing an oxidized *fine-grained*, however not 'chilled', marginal facies (e.g., D86,158 in Fig. 5.1).

Reconnaissance studies on the crystal morphology of hypersthene, augite and plagioclase in monzodiorites showed (Weiss, unpubl. data), that the relatively simple shape of orthopyroxene allows an easy detection even of minor variations in crystal habit; augite and andesine are preferably affected by granule exsolution and recrystallization, which modifies the form of subhedral primocrysts and produces ill-defined subgrains, thus obscuring primary textures. Frequent (100)-twinning of augite and Karlsbad twinning of plagioclase further complicates the relationship of crystal growth and crystal habit.

By contrast, Opx generally develops well-defined, euhedral to subhedral (100)-flattened crystal plates which are untwinned. Their habit is strongly dependent on the average crystal size (see insets of Fig. 5.4). This is best seen in rectangular cross sections perpendicular to the b-axis. These elongate to stubby index sections are non-pleochroic, have parallel extinction and are characterized by lowest first-order interference colours, while narrow Cpx lamellae exsolved along (100) are visible sharp and parallel. The ratio of length to thickness (the aspect ratio LTR) for platy Opx crystals is then derived from the (010) index sections by measuring the ratio of the longest to the shortest crystallographic axis (c/a), with elongate crystals having a larger aspect ratio LTR than stout forms.

Figure 5.4 shows the variation of crystal morphology versus absolute crystal size, plotted as the aspect ratio LTR vs the length of the c-axis [mm] of Opx in the three facies types of fine- to coarse-grained hypersthene diorites. The plots figure all grains of orthopyroxene cut along (010), 20–210 of which were detectable in 3–5 thin sections per rock sample. The considerable scatter of morphological data may be the result of minor deviations from ideal (010)-orientation of Opx, or of growth inhibition by mutual crystal impingement, which is not always obvious in a nearly two-dimensional thin section.

It is evident from the diagrams and the schematic (010)-sections (to the right of Fig. 5.4) that the morphology of Opx is most variable in the 'chilled' marginal facies D88,278. This may be related to a continuously decreasing supercooling ΔT of a melt, where the most non-equilibrium growth form (here: *elongate* subhedral Opx) represents the stable morphology for the initial ΔT-value, and succeeding more equilibrium forms (here: *stout* anhedral Opx) then represent stable morphologies for the succeedingly smaller ΔT-values (Lofgren 1974). The initially high aspect ratio of subhedral Opx (LTR > 3) and the skeletal morphology of Cpx-primocrysts in the 'chilled' monzodiorites D88,278 (Fig. 5.5) apparently represent very large departures from equilibrium crystal growth for both pyroxenes. In contrast, the average aspect ratio of euhedral Opx in the coarse-grained central facies D96,224 constantly remains between 2 and 3 (Fig. 5.4c), suggesting near-equilibrium growth at small cooling rate. Comparing variations of Opx size and morphology in Fig. 5.4, the normal facies obviously takes a status which is transitional between the central and marginal facies types. Epitaxially-related growth pairs consisting of Mg-rich hypersthene crystals mantled with augite (Weiss and Troll 1989, Fig. 11) seem to prefer inhomogenous schlieren portions of normal facies monzodiorites, containing partially-resorbed pelitic

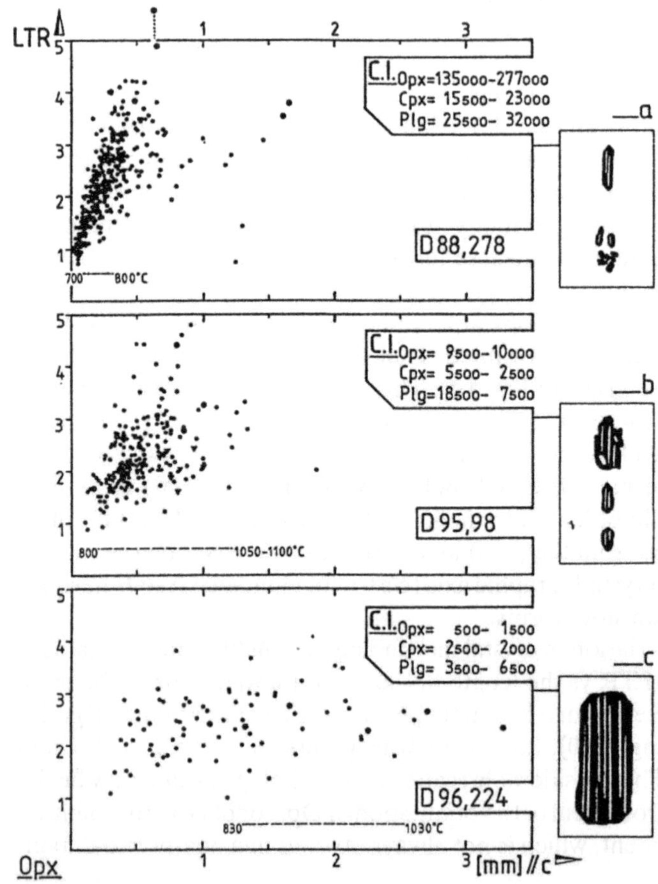

Fig. 5.4 a-c. Orthopyroxene crystal morphology (length/thickness ratio *LTR*), plotted versus absolute crystal size (length ||c) for orthopyroxene in the three facies types of fine- to coarse-grained monzodiorites from Ballachulish. The *insets* to the right give schematic (010)-sections of Opx, to indicate a possible change of crystal habit during growth (which is most significant in the 'chilled' samples D88,278). The single-phase crystal index C.I. for coexisting Opx-Cpx-Plg is tabled to allow a comparison of nucleation density with the aspect ratio LTR (c/a axis). The geothermometer scale refers to apparent thermal stability range for hypersthene in samples D278,98,224 (Fig. 4.8)

xenoliths (Fig. 3.6a). These Cpx-mantled Opx crystals (shown as triangles in Fig. 5.4) may consequently represent the product of metastable crystallization involving stepwise (super)cooling, i.e., a multistep crystallization (Naney and Swanson 1980), due to the incorporation and sinking of xenoliths.

It should be noted that those Opx primocrysts, which are found in fluidal alignment together with Plg-Cpx (Fig. 3.4) are usually well developed with an euhedral shape, forming elongate crystals with a high to maximum aspect ratio LTR (shown as large circles in Fig. 5.4). These faceted crystals evidently formed at very

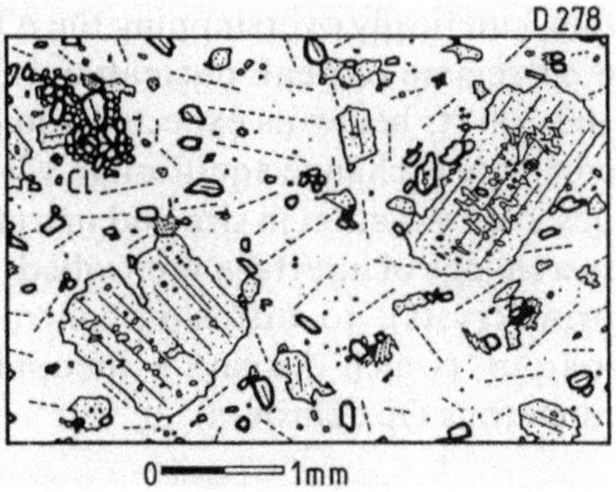

Fig. 5.5. Texture of the 'chilled' marginal facies (monzodiorite D278). Skeletal augite primocrysts (*stippled*) contain trails of *drop-shaped* plagioclase growth inclusions; pyroxene clusters (*Cl*) consist of numerous hypersthene crystals showing synneusis-type intergrowths. *Dashed lines* mark plagioclase laths (in random orientation). Schematic thin section drawing

small degrees of supercooling by interface reaction controlled growth (Maaløe 1985), with relatively high growth rates favoured by convective transport and movement of the Opx crystals in a low-viscous diorite magma (Troll and Weiss, Sect. 3.4.1, this Vol.). By contrast, anomalously low equilibration temperatures for Opx in the 'chilled' facies D278 (Fig. 4.8a) suggest that hypersthene formed in a considerably supercooled melt, under conditions of diffusion reaction controlled growth. This assumption is backed by the skeletal habit of zonal augite primocrysts in the 'chilled' rocks (Figs. 5.5, 3.6b; compare Maaløe 1985, Fig. XI-54). Non-laminar magmatic flow and very high nucleation rates favoured the formation of Opx clusters in the marginal facies (Figs. 5.5, 3.4c). Attributed to the synneusis process of Vance (1969), the drifting together and mutual impingement of numerous crystals suspended in a supercooled viscous melt probably resulted in severely reduced growth rates for those Opx grains, which became closely attached to the clustered groups during consolidation.

5.3 Thermal Controls on Pyroxene Crystallization: Summary

From the centre to the margin of the hypersthene diorite main zone, the nucleation density of both pyroxenes and plagioclase increases dramatically, combined with significant changes in the crystallization sequence. An increase in cooling rates from ≥0.03–0.02 °C/year (centre) to ≤4–0.8 °C/year (margin) may be responsible for this variation. Despite the similar fractionation stage of monzodiorites (which are Na,K-rich) and basaltic andesites (which have higher Ca,Mg; see e.g., Eggler and Burnham 1973), the crystallization of hypersthene in the *alkaline* diorites of Ballachulish seems to be particularly sensitive to nonequilibrium effects, favoured by cooling rates exceeding ~0.1–0.5 °C/year. This results in a metastable suppression of Opx nucleation, kinetically overstepping the ΔT-range of maximum growth rate and entering a ΔT-regime of peak nucleation rate, with hypersthene apparently crystallizing 100–300 °C below its expected equilibration temperature, probably due to lack of complete exchange equilibrium with augite (compare T-scales in Fig. 5.4a,b). The observed variation in size and morphology of both pyroxenes is apparently related to a change of interface-controlled, near-equilibrium growth (producing euhedral primocrysts), to diffusion-controlled disequilibrium growth at the 'chilled margin' (which displays a subophitic Cpx-Plg texture with skeletal augite and numerous Opx clusters).

6 Microstructures and Thermal Behaviour of Igneous Pyroxenes

H. Feuer, L. Schröpfer, and H. Fuess

6.1 Introduction

The cooling history of rocks is manifested in the microstructures of their minerals, which are most conveniently observed by TEM. To help in their interpretation, X-ray diffraction of heated single crystals is applied.

The specimens of this study are from two monzodiorite samples of the 'normal', i.e., coarse- to medium-grained facies, and from the 'chilled' marginal, i.e., very fine-grained facies (D98 and D278, respectively). Mineral chemistry, petrography and crystallization sequence of these rocks are described by Troll and Weiss and Weiss and Troll (Chaps. 3 and 4, this Vol.). The pyroxene crystals from the two facies types are of nearly identical composition but have experienced different cooling histories (Weiss, Chap. 5, this Vol.). Two pyroxenes occur in both facies types, orthopyroxene and clinopyroxene, the average compositions being $Wo_3En_{63}Fs_{34}$ and $Wo_{41}En_{45}Fs_{14}$ respectively. Both pyroxenes produced submicroscopic exsolution lamellae. In some cases, faint lamellae can also be observed by optical microscopy.

The temperatures of exsolution, which define the lower thermal limit of pyroxene crystallization temperature, can be estimated by comparing observed orientations of clinopyroxene lamellae with calculated orientations. In order to calculate these orientations of lamellar interfaces, the lattice constants of pyroxene exsolution pairs as a function of temperature must be known.

The high-temperature behaviour of these igneous pyroxenes is determined by X-ray diffraction of single crystals on a precession camera: (1) in situ up to 1150 °C; (2) after the samples had been annealed in a furnace at 1070 °C and 1170 °C for 2 weeks. All samples were enclosed in evacuated quartz glass capillaries. The aim was in (1) to measure the lattice parameters as a function of temperature, and the temperature of transformation from low to high pigeonite ($P2_1/c \rightarrow C2/c$); in (2) to pursue exsolution reactions. The lattice constants were obtained by the method of triangulation, measuring the reciprocal distances on the films by means of an optical device, the camera being calibrated with single crystals of NaCl.

Samples for transmission electron microscopy (TEM) were prepared from petrographic thin sections by ion thinning.

G. Voll, J. Töpel, D.R.M. Pattison, F. Seifert
Equilibrium and Kinetics
in Contact Metamorphism

6.2 Observation of Microstructures by Transmission Electron Microscopy

6.2.1 Exsolution in Orthopyroxene

The orthopyroxene (hypersthene) crystals from the two monzodiorite facies types contain heterogeneously nucleated augite lamellae parallel to (100); (Fig. 6.1). Augite occurs in both (100) twin orientations possible, sharing its b- and c-axis with the host. The a-axes form an angle of 106° either with +c and -c of the host. The c-lattice constant of augite is adapted to that of the host. The lamellae thickened by the propagation of ledges along the (100) interface (cf. Champness and Lorimer 1973; Kohlstedt and Vander Sande 1976). Augite lamellae up to a width of ≈20 nm have planar (100) interfaces. Lamellae exceeding this width show ledges at their (100) interfaces (Fig. 6.2 and Table 6.1). The abundance of precipitates is rather small in all samples. In general, the (100) lamellae are continuous and of constant thickness, but a few were observed to terminate within the orthopyroxene. Neither pigeonite nor Guinier-Preston zones have been observed in the orthopyroxene hosts.

6.2.2 Exsolution in Clinopyroxene

Most augite grains from the two monzodiorite facies types contain '001' and '100' pigeonite and (100) orthopyroxene lamellae (Fig. 6.3, Table 6.2). Similar complex assemblages have been observed by Jaffe and Jaffe (1973) and Jaffe et al. (1975) in metamorphic augites, and by Rietmeijer (1979) and Rietmeijer and Champness (1982) in iron-rich igneous pyroxenes. The symbol 'h0l' indicates that the lamellae are approximately parallel to (h0l). Some augite crystals from the 'normal' monzodiorite facies are devoid of orthopyroxene. All precipitates nucleated heterogeneously and thickened by the propagation of ledges.

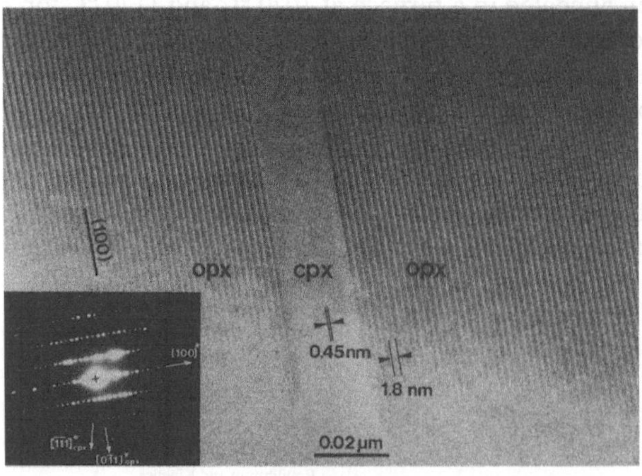

Fig. 6.1. Multibeam lattice image of an augite lamella parallel to (100) of the orthopyroxene host (*opx*), projected along [011]; [200 kV]. The lamella is ≈20 nm thick. The specimen is from the monzodiorite 'normal' facies

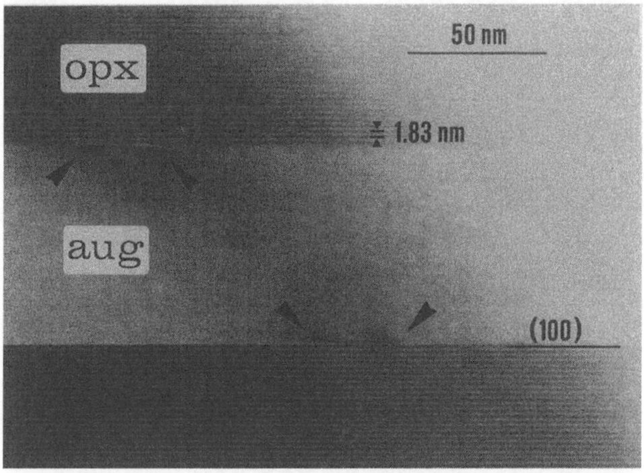

Fig. 6.2. Lattice image of an augite lamella (*aug*) in orthopyroxene host (*opx*), projected along [010]; Brightfield, [200 kV]. The (100) fringes of orthopyroxene are imaged. Growth ledges (marked by *arrows*) which are 1.83 nm high occur at the matrix/lamella interface. This corresponds to one layer (d_{100}) of orthopyroxene. The strain fields indicate that the lattices are elastically strained at these ledges. The specimen is from the 'chilled' monzodiorite

Table 6.1. Thickness and spacing of augite lamellae in orthopyroxene host from monzodiorites of the 'normal' facies (D98) and from the 'chilled margin' (D278)[a]

		D98	D278
Lamellae	Thickness [nm]	8–100	1.83–100
	Spacing [nm]	50–1000	500–2000
Ledges	Distance [nm]	≈20–100	≈100–200
	Height [nm]	1.83–7.32	1.83–3.66

[a]The values for distance and height of growth ledges are also given.

The orthopyroxene lamellae are exactly parallel to $(100)_{aug}$, continuous and of almost constant width. Growth ledges are visible at the interfaces of lamellae which exceed a width of 50 nm. The orthopyroxene exsolution lamellae themselves contain thin (100) lamellae that are an odd number of 0.9 nm units wide. The most common width is ≈2.7 nm. These clinopyroxene lamellae occur in both (100) twin orientations possible. The abundance of these (100) lamellae and of (100) stacking faults within the orthopyroxene lamellae is much higher in specimens from the 'chilled margin' rock than from the 'normal' monzodiorite. This is shown by the occurrence of streaks parallel to a* in the diffraction patterns of orthopyroxene within augite hosts from the 'chilled margin' and their absence in samples from the 'normal' facies.

In both facies types, late-stage pigeonite nucleated at the interfaces of (100) orthopyroxene lamellae. This pigeonite grew along the (100) lamellae, locally forming thin rims around the orthopyroxene lamellae. These rims are common in augite

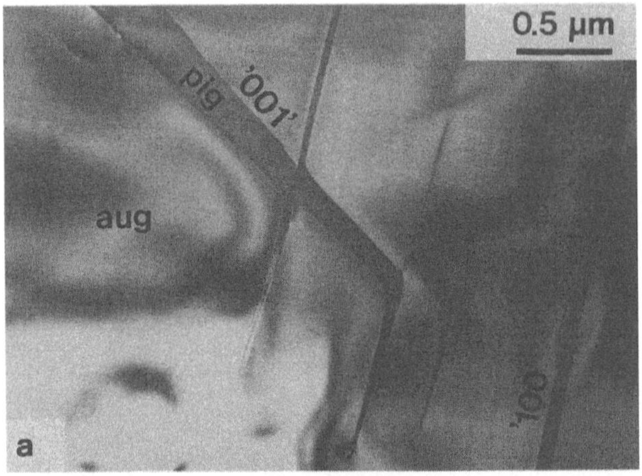

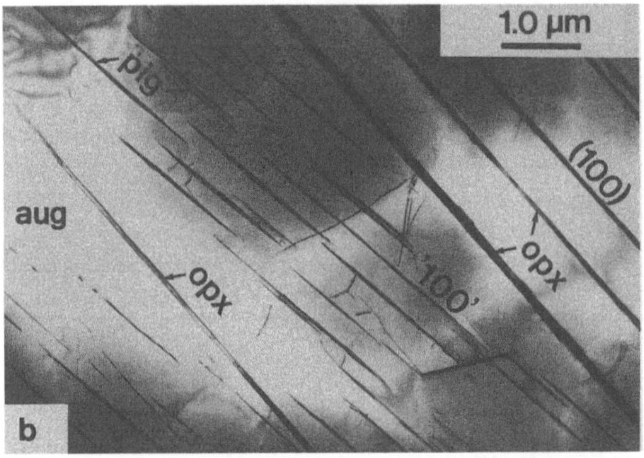

Fig. 6.3 a. Pigeonite lamellae (*pig*) in an augite host (*aug*) from the monzodiorite 'normal' facies. The '001' lamellae change their habit plane from '001' to '100' at their tips. A few growth ledges occur at the '001' interface. **b** Exsolution lamellae in augite host from the 'chilled' monzodiorite. The orthopyroxene lamellae (*opx*) on which late stage pigeonite (*pig*) has nucleated are exactly parallel to (100) of the host. The '100' pigeonite lamellae occur in areas between the orthopyroxene lamellae, forming an angle of 10° with the c-axis of augite (*aug*)

crystals from the 'chilled margin', but are rare in specimens from the 'normal' monzodiorite facies.

The pigeonite exsolution lamellae deviate several degrees from the (001) and (100) planes. In D98, the '100' lamellae commonly form an angle of 7–8° with $(100)_{aug}$ ('100' Λ $c_{aug} \approx$ 7–8°), some lamellae deviating from the (100) plane only by 2–4°. In specimens from the 'chilled margin', the '100' lamellae form an angle of 10° with the c-axis of augite. The '001' lamellae have phase boundaries which deviate from the (001) plane of augite by 10–12° ('001' Λ $c_{aug} \approx$ 116–118°). The angles refer to the traces of the lamellae. The reported values were determined by TEM.

Only few data for comparison of such angles exist. Jaffe et al. (1975) reported values obtained by optical microscopy from metamorphic samples of similar Fe/(Fe + Mg) ratio (≈0.24), but higher Al content, which are considerable 5–10° larger. When

Table 6.2. Size and distribution of the different exsolution lamellae in augite hosts from monzodiorites of the 'normal' facies (D98) and from the 'chilled margin' (D278)[a]

		D98	D278
(100)opx	Thickness [μm]	0.04–0.3; (0.1)	0.1–0.2
	Spacing [μm]	1–20	0.03–10; (0.05–0.5)
'001'pig	Thickness [μm]	0.05–0.5; (0.2)	0.05–0.7
	Spacing [μm]	0.3–10.0	0.5–10.0
	$\wedge a_{Aug}$ [°]	10–12	10–12
'100'pig	Thickness [μm]	0.02–0.1; (0.05–0.08)	0.02–0.1
	Spacing [μm]	0.1–4.0	1–3
	$\wedge c_{Aug}$ [°]	7–8 and 2–4	10

[a]The most frequently observed values are given in parentheses. The orientation of the phase boundaries of the pigeonite exsolution lamellae is also given. The '001' lamellae deviate from the a-axis of augite by 10–12° ($\hat{=}$ '001' $\wedge c_{aug}$ = 116–118°).

comparing the morphological orientation of exsolution lamellae in clinopyroxenes of different origins, one should be aware that this orientation is dependent on the lattice parameters of host and lamellae, and therefore depends on both the chemical omposition and the temperature at which exsolution takes place (see Sect. 6.4).

The '001' boundaries are composed of alternating (001) and probably (100) faces, the '100' boundaries of (100) and most likely (001) faces (Fig. 6.4). The '001' pigeonite lamellae with a width above 50 nm usually contain some (100) stacking faults.

The '001' pigeonite lamellae change their habit plane from '001' to '100' at their tips (Fig. 6.3). Similar observations were reported by Kitamura et al. (1981), and by Reitmeijer and Champness (1982). Some lamellae change their orientation up to eleven times (Fig. 6.5). This is observed in specimens from both host rock types. With a few exceptions, the '100' segments form obtuse angles with the '001' lamellae. In general, the change in the habit plane of the pigeonite lamellae continuously takes place from the '001' to the '100' segment, corresponding to a relative lattice rotation

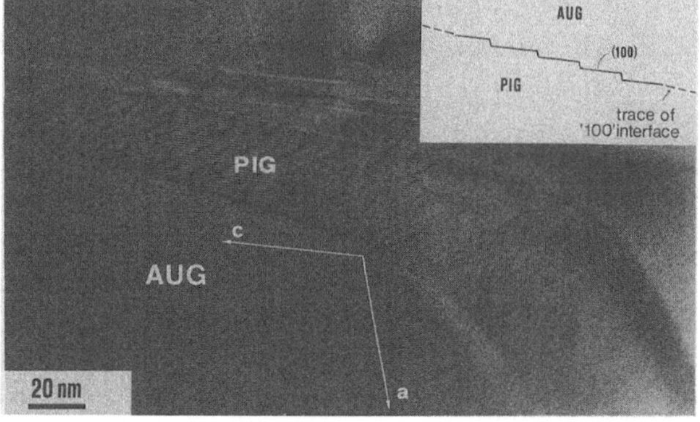

Fig. 6.4. Lattice image of a pigeonite lamella (*PIG*) in an augite host (*AUG*). The phase boundary (trace) of the '100' lamella is parallel to *b*, but forms an ≈8° angle with *c* of augite. The *inset* shows a schematic drawing of the stepped interface

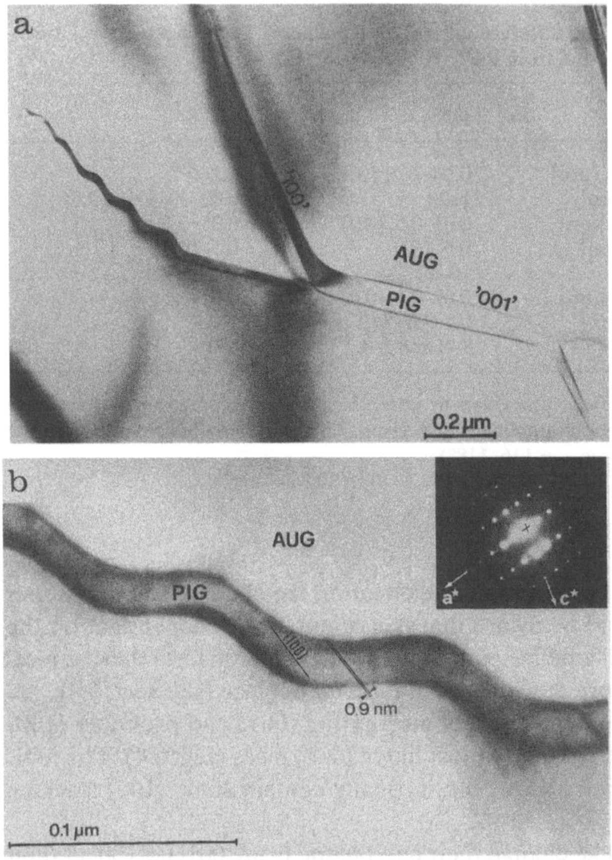

Fig. 6.5 a. The step-like pigeonite lamella (*PIG*) in an augite host (*AUG*) results from the repeated change of its orientation. The lamella probably nucleated at the tip of the coarser '001' lamella. **b** At higher magnification, the (100) lattice fringes of pigeonite can be seen in the lamella. The fringes slightly change their orientation when passing from '001' segments into '100' segments, and vice versa

of 2.8° around the b-axis; exceptionally, some lamellae show an irregular boundary between the two pigeonite segments.

A common feature of all the pigeonite lamellae observed in this study is the absence of antiphase boundaries. This indicates that pigeonite formed as low pigeonite ($P2_1/c$). This is further discussed in Sections 6.3.1 and 6.5.

6.2.3 Alteration of Pyroxene to Amphibole, Biotite and Talc

The primary pyroxenes from the two monzodiorite facies types are partly replaced by amphibole, biotite and talc. These alteration reactions are predominantly topotactic.

In augite, the amphibole occurs as incoherent blebs (3–10 µm), as thick semi-coherent (010) lamellae (0.1–0.2 µm) and as thin coherent (010) lamellae (≤10 nm; average = 5 nm); it is homoaxial with augite ($c_{amph} || c_{aug}$, $b_{amph} || b_{aug}$, $a_{amph} \approx || a_{aug}$). The abundance of amphibole lamellae is highly variable, both within and between samples. The amphibole lamellae appear to have nucleated at the interfaces between

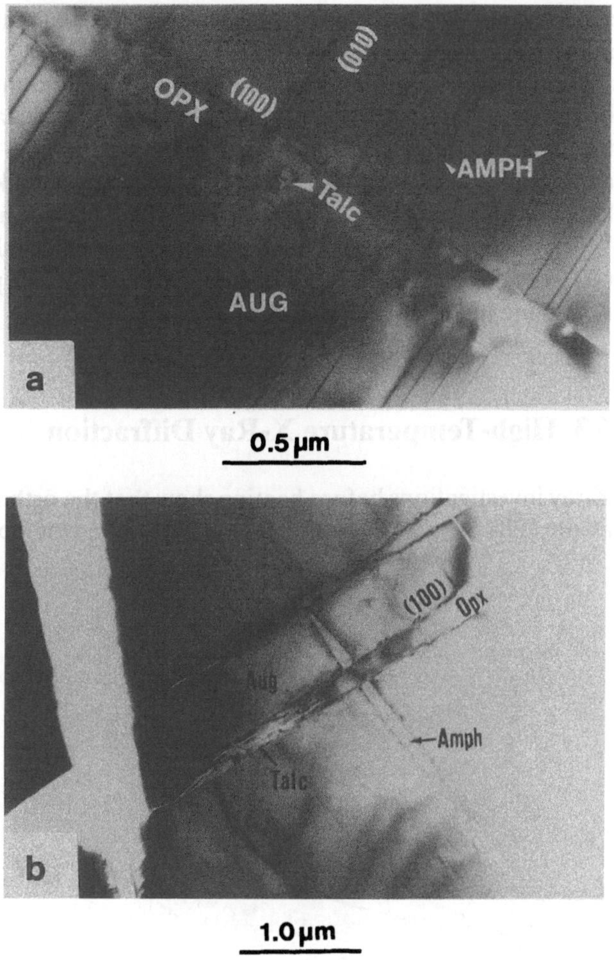

Fig. 6.6 a. The (010) amphibole lamellae (*AMPH*) nucleated at the interfaces between orthopyroxene exsolution lamellae (*OPX*) and the host augite (*AUG*). The 5-10-nm-wide lamellae are coherent with the augite host. Brightfield; [001]-projection; [100 kV]. **b** The orthopyroxene exsolution lamella is partly replaced by talc. Brightfield: [100 kV]

augite and orthopyroxene exsolution lamellae, and to have grown predominantly along the c- (chain-) and a- (layer-) directions in the host augite (Fig. 6.6a). In general, the orthopyroxene and pigeonite exsolution lamellae are not replaced by amphibole. Absence of alteration of (100) orthopyroxene lamellae is common (P.P.K. Smith 1977; Nakajima and Ribbe 1981; Otten 1985).

Biotite also partially replaces the augite at grain boundaries. The (001) sheets of biotite are almost parallel to $(100)_{aug}$. Biotite is locally deformed and shows complex bending and kinking. The layers are bent and split into chevron folds. Although most of the biotite is a 1M polytype, there may be random intercalations of a two- and a three-layer polytype ($2M_1$ and 3T, respectively). Complex interlayering of different mica polytypes is common in biotite (Iijima and Buseck 1978; Czank and Schumacher 1986). Biotite exhibits stacking disorder originating from random layer orientations of n(120°) in layer stacking sequences (cf. Iijima and Zhu 1982) yielding diffuse streaking along c* in all reciprocal lattice rows with $k \neq 3n$. Different superlattices which are due to regularly spaced stacking faults have been observed in biotite (cf. Bell and Wilson 1977).

All pyroxene grains studied contain small amounts of fibrous talc along grain boundaries and fractures. Talc also nucleated at dislocations and other faults within the crystals. In general, the talc sheets are parallel to (100) of host pyroxene, additionally in augite parallel to (001). They are often kinked and buckled. Partially bent and split layers occur as 10–100 nm thick packets at pyroxene grain boundaries. When the augite crystals contain exsolution lamellae, talc predominantly replaces the (100) exsolution lamellae, i.e., orthopyroxene and '100' pigeonite, the adjacent augite remaining unaffected (Fig. 6.6b).

6.3 High-Temperature X-Ray Diffraction

X-ray investigations before heating show that the orthopyroxene host only exsolved augite (Fig. 6.7), always parallel to (100) and in twin positions of equal volume. This

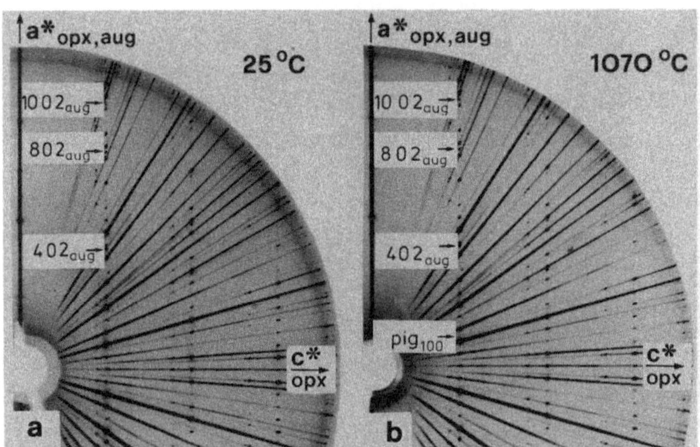

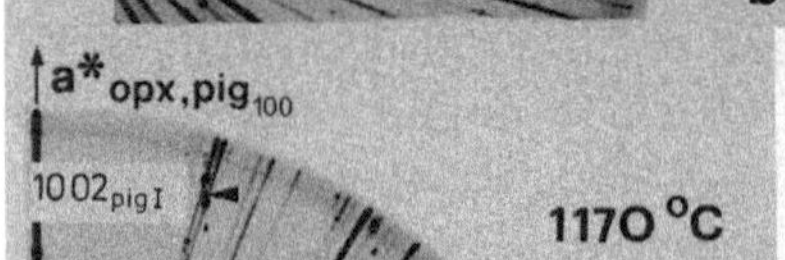

Fig. 6.7. Precession photographs (Mo unfiltered, $\mu = 30°$) of host orthopyroxene showing exsolved augite. **a** Before heating: there are two augites in (100) twin position of nearly equal volumes. $h0l_I$ and $(-h-2)0l_{II}$ reflections are superposed; only aug_I is indexed. The reflection 602_I is extremely weak; the one observed in its place is -802_{II}, which is of medium intensity. **b** After annealing at 1070 ± 10 °C for 2 weeks: orthopyroxene and augite are beginning to react, forming pigeonite. **c** After annealing at 1170 ± 10 °C for 2 weeks: augite is consumed, with orthopyroxene and the new pigeonite phase coexisting. Pigeonite is twinned after (100); the twins are of equal volume. **a** 160 mA/50 kV, 20 h; **b** 160 mA/45kV, 42 h; **c** 160 mA/45 kV, 33.5 h)

equal twinning is strong evidence that orthopyroxene was the primary phase, not a pigeonite, which might have inverted to orthopyroxene after exsolution: only the orthorhombic pyroxene provides two equivalent orientations for the exsolving augite.

In the diffraction diagrams all phases exhibit sharp reflections; hardly any streaks are observed, the involved phases scattering rather incoherently.

Within the limits of error, augites have the same lattice constants in both facies types (Table 6.3). Many augite grains from the 'normal' monzodiorite and all of the 'chilled margin' contain orthopyroxene as well as two pigeonites. The intergrown phases have their lattices in strict common crystallographic orientation, apart from a slight relative lattice rotation (≈0.1°) around the b-axes of augite and pigeonite. Again, the reflections are sharp, no streaks being observed in samples from the 'normal' facies (Fig. 6.9). However, the exsolved phases are not free of strain (see below).

The lattice parameters of exsolved '001' and '100' pigeonite are not adapted to the augite host (Fig. 6.8), their a differing from a_{aug} by $\Delta a = 0.09$ and 0.10 Å respectively; the corresponding differences in c are 0.06 and 0.03 Å. These Δa are larger than the values for pyroxenes from plutonic rocks, as given by Buseck et al. (1980).

6.3.1 Transformation of Pigeonite, Thermal Expansion of Host Augite and Exsolved Pigeonite

On heating to 1150°C, the exsolved pigeonites do not transform from low to high temperature structure, since b-reflections ($h + k = 2n + 1$) can still be observed in precession photographs; this is absolutely unexpected. If the chemical composition did not change on cooling, this would mean that the pigeonite already formed as low

Table 6.3. Lattice constants of augite hosts and exsolved pigeonites from monzodiorites of the 'normal' facies (D98), the 'chilled margin' (D278), and from the Whin Sill (Settlingstone), N England (each data set is an average of five grains)

		D98	D278	Whin Sill
Augite host	a [Å]	9.742(5)	9.747(6)	9.742(4)
	b [Å]	8.919(3)	8.917(3)	8.921(5)
	c [Å]	5.257(4)	5.251(3)	5.258(3)
	β [°]	106.00(4)	106.01(2)	106.00(6)
	V [Å^3]	439.1	438.7	439.3
'100' Pigeonite	a [Å]	9.639(5)	9.645(12)	—
	b [Å]	8.919(3)	8.917(3)	8.921(5)
	c [Å]	5.227(2)	5.218(7)	—
	β [°]	108.70(7)	108.83(8)	108.52(15)
	V [Å^3]	425.6	424.8	
'001' Pigeonite	a [Å]	9.652(4)	9.653(14)	9.718(11)
	b [Å]	8.919(3)	8.917(3)	8.921(5)
	c [Å]	5.196(3)	5.192(3)	5.207(4)
	β [°]	108.57(4)	108.50(9)	108.90(7)
	V [Å^3]	424.0	423.8	427.1

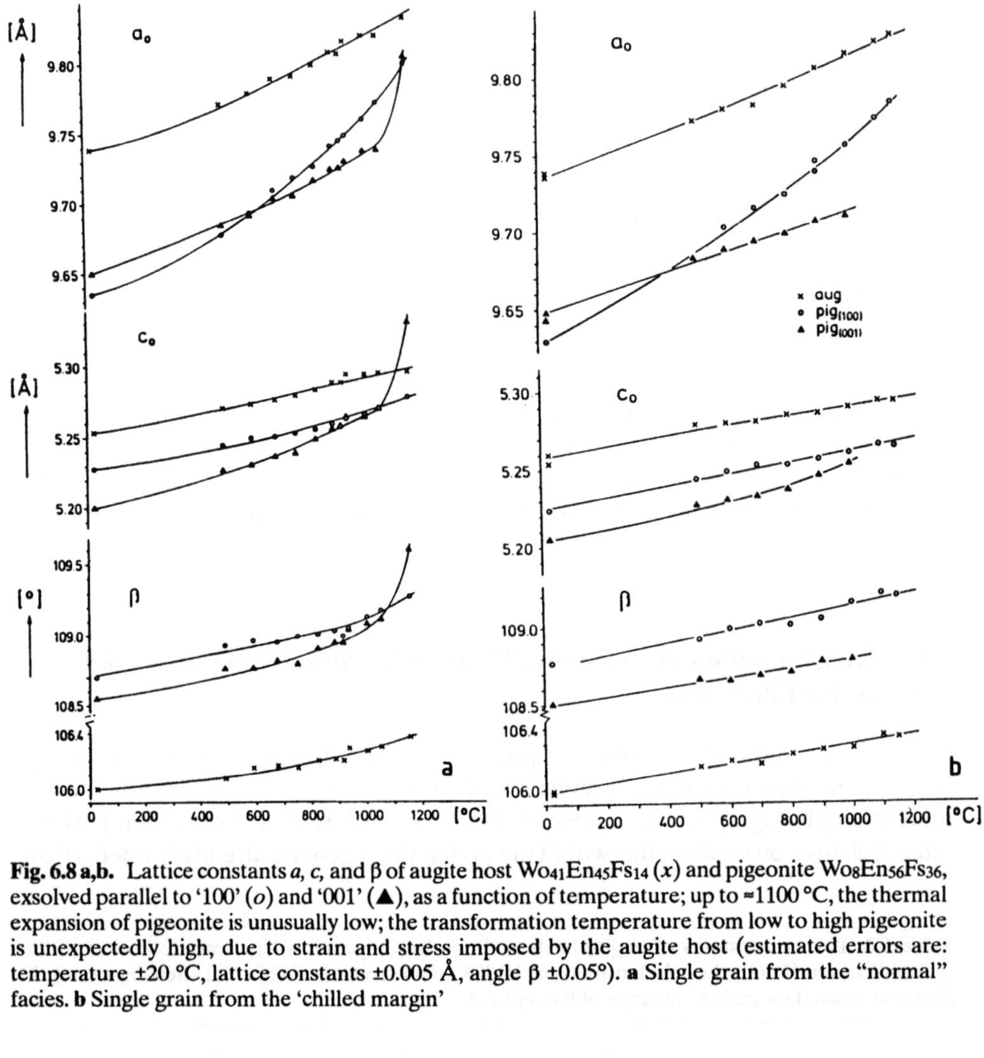

Fig. 6.8 a,b. Lattice constants *a, c,* and β of augite host $Wo_{41}En_{45}Fs_{14}$ (*x*) and pigeonite $Wo_8En_{56}Fs_{36}$, exsolved parallel to '100' (*o*) and '001' (▲), as a function of temperature; up to ≈1100 °C, the thermal expansion of pigeonite is unusually low; the transformation temperature from low to high pigeonite is unexpectedly high, due to strain and stress imposed by the augite host (estimated errors are: temperature ±20 °C, lattice constants ±0.005 Å, angle β ±0.05°). **a** Single grain from the "normal" facies. **b** Single grain from the 'chilled margin'

pigeonite, because the stable liquidus temperature of the host augite is estimated to the 1120–1090 ± 60 °C (Weiss and Troll, Sect. 4.2.1, this Vol.).

However, when annealed for two weeks at 1070 °C and then quenched, the lattice parameters a of '001' pigeonite and c of '100' pigeonite are adapted to the host augite (Table 6.4), most probably caused by a change in composition due to homogenization. These pigeonites, which are richer in Ca, presumably represent the original compositions during the onset of exsolution. Although higher Ca contents lower the temperature of the high-low transformation, this does not a priori exclude that these pigeonites initially formed as low pigeonite. Here, it is inferred from the absence of antiphase boundaries, that pigeonite formed as low pigeonite. Stress imposed by the host augite may be an important parameter, forcing pigeonite to form in the low temperature

Table 6.4. Lattice constants of augite host and exsolved pigeonite before and after annealing at 1070 °C for 2 weeks (data of a single grain)

	Augite host		'100' pigeonite		'001' pigeonite	
	Before	After	Before	After	Before	After
a [Å]	9.735	9.741(5)	9.636	9.694(5)	9.648	9.752(5)
b [Å]	8.920	8.920(5)	8.920	8.920(5)	8.920	8.920(5)
c [Å]	5.258	5.259(5)	5.227	5.262(5)	5.196	5.244(5)
β [°]	106.00	106.08(10)	108.80	108.49(10)	108.60	108.65(10)
V [Å^3]	438.9	439.1	425.3	431.5	423.8	432.2

structure at high temperatures, because the low temperature phase has the smaller unit cell volume.

The thermal expansion in a and c of exsolved pigeonite, calculated from Fig. 6.8, is about one-half the expansion of free pigeonite, and up to 1050 °C, rather corresponds to the nearly linear expansion of the host augite, which shows normal values (from 25 to 1000 °C: $\Delta a/a = 8.6(4)$, $\Delta c/c = 8.0(4)$, $\Delta\beta/\beta = 2.6(3)$, all $\times 10^{-6} \times °C^{-1}$). For comparison: The corresponding values of a synthetic pigeonite host of quite similar composition ($Wo_8En_{53}Fs_{39}$) are $\Delta a/a = 17.7(4)$, $\Delta c/c = 18.8(4)$, $\Delta\beta/\beta = 11.7(3)$; hence the lattice constants of the exsolved pigeonite, though not adapted to the host, cannot be 'free'. Note that the thermal expansion in a, c, and β of free low pigeonite is by far not linear (e.g., Smyth 1974).

6.3.2 Strain and Stress Imposed on Pigeonite Exsolution Lamellae

In order to evaluate the strain and stress exerted upon the exsolved pigeonite lamellae, their strain-free equilibrium lattice constants must be known. Following Morimoto and Tokonami (1969) they can be calculated from phases exsolved approximately parallel to (100) and (001) which have a or c respectively 'free'; b is equal in all three phases. Thus we must apply the lattice parameters of the annealed pyroxenes (two weeks at 1070 °C), thereby taking into account not only the strain and stress caused by different thermal expansion, but also by change in chemical composition.

The 'free' parameters are given in Table 6.5; the values of the strain ellipsoids referring to these 'free' parameters are shown in Table 6.6. Applying the elastic moduli of augite, given by Aleksandrov et al. (1964), to the pigeonite, stress ellipsoids are obtained (Table 6.6) and strain energies E^i are estimated. Values of two different grains are presented in Table 6.6.

It should be noted that in the case of the Ballachulish pyroxene lamellae, the strain energy for '100' pigeonite is smaller than for '001' pigeonite, as can readily be expected by comparing the volume differences of the unit cells $\Delta V = V(pig\ \text{'free'}) - V(pig\ \text{'before'})$. From Tables 6.4 and 6.5 we get $\Delta V(pig_{100}) = 5.7$ Å^3 and $\Delta V(pig_{001}) = 7.2$ Å^3. This contrasts with the findings of Morimoto and Tokonami (1969) for augite

Table 6.5. Lattice constants of 'free' pigeonite calculated from those of the two annealed pigeonites in Table 6.4

a [Å]	b [Å]	c [Å]	β [°]	V[Å^3]
9.706(10)	8.920(5)	5.251(10)	108.57(10)	431(2)

Table 6.6. Strain (ε_i) and stress (σ_i) ellipsoids of exsolved pigeonite, ε_2 and σ_2 are parallel b. θ is the angle between ε_1 or σ_1 and c, measured from +c to +a in (010). Data of two different samples are shown. The estimated errors are ≤10 ° for θ, ≤20% for the other values

	'100' pigeonite			'001' pigeonite	
	[kbar]	[GPa]		[kbar]	[GPa]
$\varepsilon_1 = -0.0041$	$\sigma_1 = -13.7 =$	−1.37	$\varepsilon_1 = -0.0060$	$\sigma_1 = -17.1 =$	−1.71
$\varepsilon_2 = -0.0000$	$\sigma_2 = -8.4 =$	−0.84	$\varepsilon_2 = -0.0000$	$\sigma_2 = -8.4 =$	−0.84
$\varepsilon_3 = -0.0091$	$\sigma_3 = -22.9 =$	−2.29	$\varepsilon_3 = -0.0107$	$\sigma_3 = -30.0 =$	−3.00
θ = −19°	θ = −37°		θ = −77°	θ = −66°	
	$1/3\Sigma\sigma_i = -15 =$	−1.50		$1/3\Sigma\sigma_i = -19 =$	−1.90
$E^a = 12$	J/cm^3		$E^c = 20.5$	J/cm^3	
=0.38	kJ/mol		= 0.63	kJ/mol	
$\varepsilon_1 = -0.0043$	$\sigma_1 = -13.4 =$	−1.34	$\varepsilon_1 = -0.0056$	$\sigma_1 = -15.9 =$	−1.59
$\varepsilon_2 = 0.0000$	$\sigma_2 = -7.7 =$	−0.77	$e_2 = 0.0000$	$\sigma_2 = -7.6 =$	−0.76
$\varepsilon_3 = -0.0082$	$\sigma_3 = -20.5 =$	−2.05	$\varepsilon_3 = -0.0098$	$\sigma_3 = -27.2 =$	−2.72
θ = −9°	θ = −35°		θ = −83°	q = −67°	
	$1/3\Sigma\sigma_i = -14$	= −1.4		$1/3\Sigma\sigma_i = -17$	= −1.70
$E^a = 11$	J/cm^3		$E^c = 17$	J/cm^3	
=0.34	kJ/mol		= 0.54	kJ/mol	

lamellae in pigeonite hosts from the Moore County meteorite. We think that the smaller strain energy of '100' pigeonite in Ballachulish augite becomes the dominant term on cooling, causing the observed change in the habit plane of pigeonite lamellae from '001' to '100' (Fig. 6.5).

6.3.3 Three-Pyroxene Field

Though a reaction takes place when augite hosts ("3-px grains") were annealed at 1070 °C for 2 weeks, three pyroxene phases still coexist (Fig. 6.9), indicating that the 3-px field for Ballachulish pyroxenes lies at or near this temperature. Heating 2 weeks at 1170 °C results in a completely homogenized augite (Fig. 6.9).

When, on the other hand, an orthopyroxene host containing exsolved augite undergoes the same heating procedure, both phases react to form pigeonite at 1070 °C,

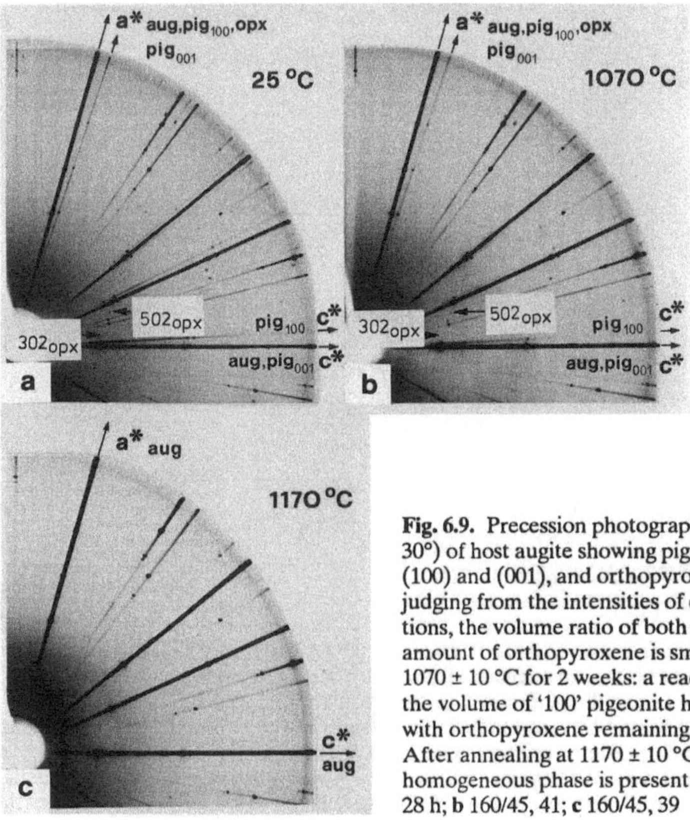

Fig. 6.9. Precession photographs (Mo unfiltered, μ = 30°) of host augite showing pigeonite exsolved parallel (100) and (001), and orthopyroxene. **a** Before heating: judging from the intensities of corresponding reflections, the volume ratio of both pigeonites is ≈1:1; the amount of orthopyroxene is small. **b** After annealing at 1070 ± 10 °C for 2 weeks: a reaction has taken place; the volume of '100' pigeonite has clearly diminished, with orthopyroxene remaining rather unchanged. **c** After annealing at 1170 ± 10 °C for 2 weeks: only one homogeneous phase is present: augite. **a** 160 mA/40 kV, 28 h; **b** 160/45, 41; **c** 160/45, 39

but augite is not markedly consumed; all three phases seem to coexist at this temperature (Fig. 6.7). When heated for 2 weeks at 1170 °C, augite has completely reacted and the orthopyroxene host coexists with pigeonite only (Fig. 6.7).

It is inferred from both experiments that the 3-px field for Ballachulish pyroxenes lies at or near 1070 °C, and is determined by the triangle augite $Wo_{41}En_{45}Fs_{14}$, orthopyroxene $Wo_3En_{63}Fs_{34}$, and pigeonite $Wo_8En_{56}Fs_{36}$. This temperature is in very good agreement with the one derived from Ishii's (1975) lower stability reaction line or "pigeonite eutectoid reactions line" at low pressure, which gives 1060 °C for pigeonite of the above composition. The chemical composition of this pigeonite was estimated by means of the nomogram of Turnock et al. (1973) using the measured lattice parameters b and β.

6.4 Calculation of Exsolution Temperatures in Augite

We have used the orientation of the interfaces of pigeonite exsolution lamellae, as proposed by Robinson et al. (1977), to estimate the temperatures of exsolution. The

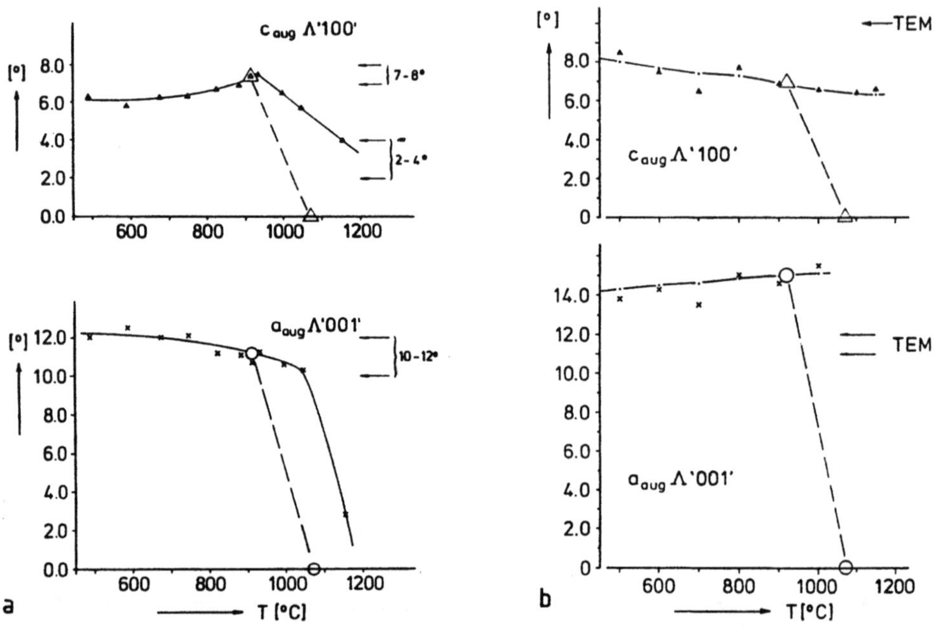

Fig 6.10 a,b. Calculated orientations of the '001' (x) and '100' (▲) pigeonite lamellae in augite as a function of temperature; *solid lines* are calculated from data of unequilibrated phases (not corrected for changes in composition on cooling), *dashed lines* from equilibrated phases. The ranges of angles obtained by TEM are indicated by *arrows*. **a** Single grain from the "normal" facies. Estimated exsolution temperatures of pigeonite are: first generation '100' pigeonite at 1000 ± 30 °C, second '100' pigeonite at 920 ± 50 °C, and '001' pigeonite at 935 ± 50 °C. **b** Single grain from the 'chilled margin'. Estimated exsolution temperature of '001' pigeonite is 950 ± 50 °C. Values of '100' pigeonite cannot be determined (see text)

orientation for the '001' and '100' lamellar interfaces ("exact phase boundaries") were calculated as a function of temperature (Fig. 6.10, solid lines) using the measured lattice parameters of augite host and exsolved pigeonite plotted in Fig. 6.8. The observed values are compared with the calculated model angles. The calculated curves are rather unspecific below 1070 °C and can only give an upper limit of exsolution temperatures.

In specimens from the 'normal' facies, the observed angles of a first generation '100' pigeonite yield unreasonable high temperatures of exsolution; the upper limits for the other pigeonites appear very high, too. The explanation for this is the change of composition and the associated change of lattice constants on cooling. For the calculation of 'exact phase boundaries', the lattice constants of the exsolved pigeonite and augite host must be known as a function of temperature and composition. A reasonable approach is to use the lattice constants of the pyroxene annealed at 1070 °C which is thought to be nearly equilibrated, and of a pyroxene held at 925 °C for 4 weeks, which did not undergo any measurable changes. The lattice constants of pigeonite annealed at 1070 °C yield orientations for the '001' and '100' lamellar interfaces which are close (±1°) to the (100) and (001) planes. Using these parameters, the dashed lines of Fig. 6.10 are obtained, which are the realistic continuation of the solid lines above 925 °C. Finally, estimates of the temperatures of exsolution obtained from the orientations of the exsolved pigeonite lamellae were 1000 ± 30 °C for first generation '100', 920 ± 50 °C for the second generation '100', and 935 ± 50 °C for '001'. These data are consistent with the 1100–950 °C main crystallization range of augite suggested by Weiss and Troll (Chap. 4, this Vol.).

In augites from the 'chilled margin' (D278), the orientation of the '001' lamellar interfaces suggests that '001' pigeonite nucleated at 950 ± 50 °C, which represents a reasonable value. However, the range of angles of the '100' lamellae obtained by TEM lies beyond the calculated values which are based on the cell dimensions of a single grain only (Fig. 6.8b). It must be taken into account that the values of model angles (i.e., calculated orientations) scatter; the average angle of five grains shows a standard deviation of 2.4°. Thus the observed 10° angle (±1°) of the '100' lamellae is within the limits of error to be expected in Fig. 6.8. If the heated single grain or the TEM samples were not representative, the angles cannot compare favourably.

6.5 Discussion and Conclusions

The pyroxenes from both monzodiorite facies types show heterogeneously nucleated exsolution lamellae. Primary orthopyroxene contains the equilibrium precipitate, augite parallel to (100), and may therefore be called "orthopyroxene of the Bushveld type" (Hess and Phillips 1938). Despite the different cooling histories, orthopyroxene displays nearly identical microstructures in both rock facies types. Likewise, the augite hosts have quite similar lattice constants (Table 6.3), hence most probably indicating the same chemical composition. Nevertheless, they differ in the quantity of their precipitates: augite from the 'chilled' marginal monzodiorite always contains exsolved orthopyroxene and pigeonite, being less frequent, and present in smaller amounts, within augite from the 'normal' monzodiorite facies. There, again, is only a slight variation concerning the lattice constants of the exsolved '100' pigeonite. The difference $\beta_{(\text{'100' pig})}$-$\beta_{(\text{'001' pig})}$ is larger in the 'chilled margin' samples (≈0.33°) than in the 'normal' (≈0.14°), because the latter samples are nearer to equilibrium. When compared to clinopyroxenes from the Whin Sill, augite hosts show the same lattice parameters, but the lattice parameters of exsolved '001' pigeonite evidently differ (Table 6.3). The Δa of the Whin Sill '001' pigeonite and augite host is only 0.02 Å. Values of 0–0,02 Å are characteristic of quickly cooled pyroxenes.

The absence of antiphase boundaries within pigeonite lamellae suggests that pigeonite already nucleated as low pigeonite ($P2_1/c$). This is well supported by the X-ray heating experiments, which yield unexpectedly high transformation temperatures ≥1150 °C, due to the stress imposed by the augite host. The maximum values of stress obtained in pig(100) and pig(001) are $\sigma_3 = -23$ and -30 kbar, respectively (Table 6.6); if we take overall values $1/3\Sigma\sigma_i$ we get 15 and 19 kbar. Our values $\Sigma\sigma_i$ are 5–4 times the internal stress $\Sigma\sigma_i = 9$ and 14 kbar, respectively, for augite lamellae in a pigeonite host from the Moore County meteorite given by Morimoto and Tokonami (1969). That case is inasmuch different as those augite lamellae have adapted their lattice parameter a or c to the host pigeonite. Since pigeonite exsolved already as low pigeonite, the spinodal decomposition mechanism was not effective, because of the different space groups of augite (C2/c) and pigeonite ($P2_1/c$). This is consistent with the absence of homogeneously distributed lamellae, which usually result from spinodal decomposition.

By comparing the orientation of '001' and '100' lamellar interfaces with calculated orientations of exact phase boundaries as a function of temperature, pigeonite exsolution in "normal" facies samples is estimated to have occurred at ≤1000 °C, constraining the temperature of crystallization of augite to ≥1000 °C. The crystallization temperature of augite from the 'chilled margin' is found to be ≈50 °C lower, about 950 °C. From this data, no strong supercooling of the 'chilled' marginal facies (monzodiorite D278) can be assumed. However, compare Weiss (Chap. 5, this Vol.).

None of the lattice parameters of the exsolved pigeonite is adapted to the augite host ($\Delta a = 0.1$ Å, $\Delta c = 0.03$–0.06 Å), as to be expected for plutonic rocks, but Δa is unusually large.

The upper thermal stability limit for amphiboles from monzodiorites covers a range of ≈710–680 °C, as reported by Weiss and Troll (Sect. 4.2–4.5, this Vol.). This temperature postdates the '100' pigeonite and the (100) orthopyroxene exsolution lamellae, because (010) amphibole lamellae terminate against them.

Though the estimated cooling rate of the rocks was rather small (≤4–0.8 °C/year for the margin, ≥0.02–0.03 °C/year for the centre of the monzodiorite ring; Weiss, Chap. 5, this Vol.), the pyroxenes did not equilibrate since most of the crystals are '3-px grains'. This would be expected from specimens which experienced faster cooling, e.g., in the Whin Sill, N England (>9°/year, J. Töpel, pers. comm.), but no 3-px grains are observed. On the other hand, the Bushveld augites which have undergone extremely slow cooling also appear as 3-px grains (Robinson et al. 1977).

The Ballachulish augites are outstanding because of the pronounced change in habit plane of pigeonite exsolution lamellae from '001' to '100'. This change is explained by the smaller strain energy of '100' pigeonite (≈0.34 kJ/mol compared with ≈0.59 kJ/mol of '001' pigeonite) which becomes the dominant term on cooling.

The orthopyroxene exsolution lamellae in augite hosts from both monzodiorite facies exhibit thin (100) clinopyroxene slabs. These are abundant in the 'chilled' samples. Most likely, they are growth faults that formed by the propagation of 0.9-nm ledges during thickening of the orthopyroxene.

The low density of (100) stacking faults in the '001' pigeonite lamellae indicates that these lamellae did not experience great relief of strain. The (100) stacking faults commonly form within coherent '001' lamellae to relieve the strain caused by the differential contraction of the pigeonite, compared to that of the augite host (Robinson et al. 1977; Nobugai and Morimoto 1979). However, due to the prevailing width of 0.2 μm, the straining of the confined pigeonite may have been too small to cause formation of stacking faults on (100) of pigeonite as a relief for strain imposed by the host. Possibly the faults formed within semi-coherent pigeonite lamellae as a precursor to the inversion of pigeonite to orthopyroxene (e.g., Nobugai et al. 1978) although there is no other evidence for this reaction in the Ballachulish specimens.

7 The Shape of the Intrusion Based on Geophysical Data

W. Rabbel and R. Meissner

7.1 Introduction

Geological information about the geometric shape of a rock body is generally restricted to shallow depths. Concerning the Ballachulish pluton, it is the purpose of the present chapter to obtain information on the geological structure at larger depths by using geophysical methods. The following items will be investigated:

1. Location of metamorphic contacts and of boundaries between different igneous rocks
2. Estimation of their dip angles
3. Mean depth of the Ballachulish pluton
4. The lateral extension of the pluton under the surrounding metamorphic rocks.

The above-mentioned problems will be studied by using the results of our own magnetic surface measurements, as well as the aeromagnetic and gravity data of the British Geological Survey (Institute of Geological Sciences 1979, 1981).

7.2 Surface Magnetics

In the northeastern and sourthern parts of the Ballachulish Complex, nearly 40 profile-km of the magnetic total intensity were mapped in order to study the interior boundaries between different types of igneous rocks and the outer contact zones between igneous and metamorphic rocks. The distance between the measurement points along the profiles was 5 m. Three main questions guided the surface magnetic investigations:

1. Are the geologically mapped metamorphic contacts and interior boundaries reflected in the surface magnetic field? If so:
2. Can the magnetic anomalies be used for mapping in areas not accessable to geologists?
3. What are the dip angles of these contacts and boundaries?

The geological map in Fig. 7.1 contains the location of three magnetic profiles (see Figs. 7.2–7.4). The magnetic anomaly map of Fig. 7.5 is the result of 7000 measure-

G. Voll, J. Töpel, D.R.M. Pattison, F. Seifert
Equilibrium and Kinetics
in Contact Metamorphism

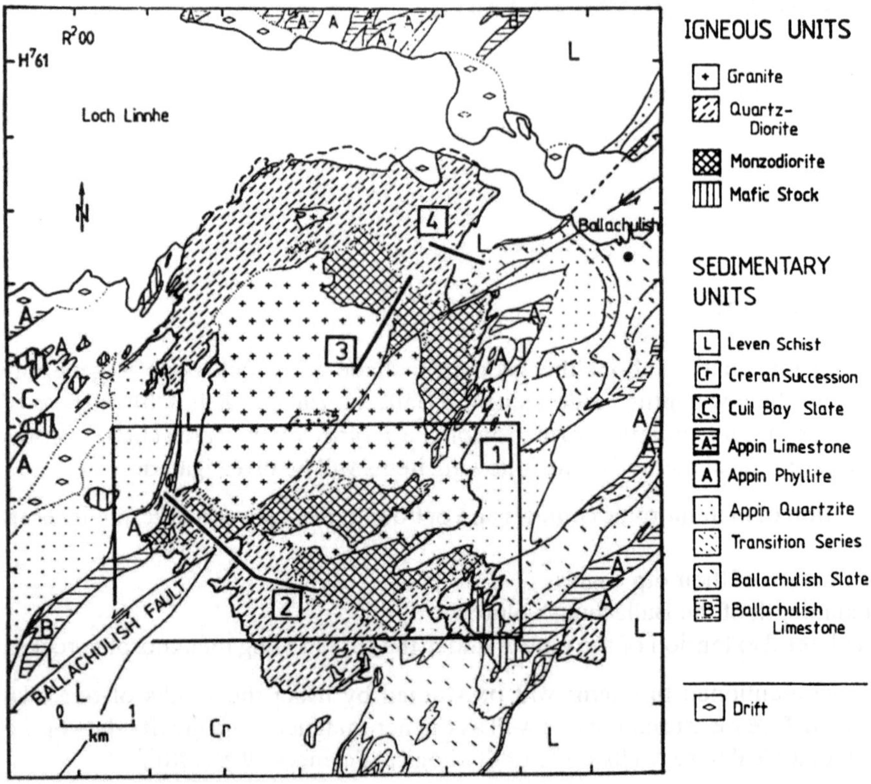

Fig. 7.1. Geology of the Ballachulish area (after Pattison and Harte 1985; Weiss 1986) and location of magnetic profiles and magnetic anomaly map (see also Figs. 7.2-7.5)

ment points in a 5 km^2 area. The magnetic field data is interpreted in the following two sections.

7.2.1 Magnetic Profiles

Profiles 2–4 (Figs. 7.2–7.4) show the typical appearance of surface magnetic data from the Ballachulish area, i.e., a large number of small-scale anomalies superimposed on anomalies of some hundred meters wavelength and considerable amplitude. In particular they cross the following:

1. Three outer contacts between diorite and phylites and the Ballachulish Fault in the SW (profile 2, Fig. 7.2)
2. The inner contact between granite and diorite in the NE (profile 3, Fig. 7.3)
3. The outer contact between diorite and metamorphic country rock in the NE (profile 4, Fig. 7.4)

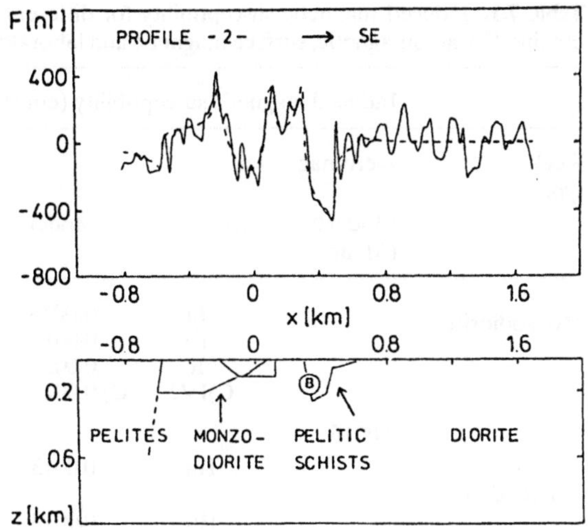

Fig. 7.2. Surface magnetic profile 2. Field anomaly (*solid line*) and modelled anomaly (*dashed line*) with geological interpretation (location 2 in Fig. 7.1; susceptibility values, see Table 7.1. *F* Magnetic intensity; *X*, *Z* spatial coordinates; *B* Ballachulish Fault

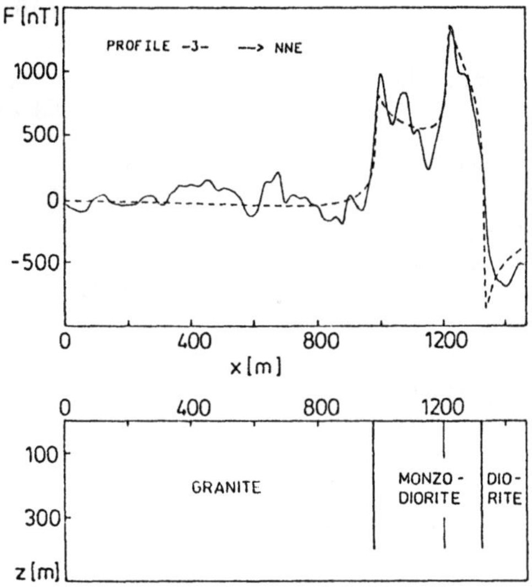

Fig. 7.3. Surface magnetic profile 3. Field anomaly (*solid line*) and modelled anomaly (*dashed line*) with geological interpretation (location 3 in Fig. 7.1; susceptibility values, see Table 7.1). *F* Magnetic intensity; *X*, *Z* spatial coordinates

The selected profiles cross the strike of the geological formations at nearly right angles, allowing a two-dimensional interpretation. The procedure was performed with an interactive computer program (Rabbel 1987) based on the formulae of Talwani and Heirtzler (1962) for horizontally infinite prism-shaped bodies. Remanent magnetization was ignored. The cross sections of the resulting structures are given at the bottom of Figs. 7.2–7.4 together with the geological classification given in Fig. 7.1. The corresponding calculated susceptibility values (see Table 7.1) are in the range of laboratory values obtained by Entwisle (1984) for rock samples from the Ballachulish area. Maximum susceptibilities are found for monzodiorite.

Table 7.1. Induced magnetic susceptibility for different rock types from the Ballachulish pluton obtained by aeromagnetic, surface magnetic and laboratory investigations

	Induced magnetic susceptibility (emu)				
Rock type	Aeromagnetics			Surface magnetics	Laboratory Data
	Model 2 (Mean)	Sector	Model 3	[..] = Profile No.	after Entwisle (1984)
Monzodiorite		Ia	0.0019	0.002–0.005 [3]	0.0002–0.0033
		Ib	0.001		
		Ic	0.0022	0.001-0.003 [2]	
		C)IV?	C) 0.0021		
	0.0016				
Quartz-diorite		IIa	0.0013	0.002 [4]	0–0.0016
		IIb	0.0013		
		IIc	0.0014	0.002 [2]	
		IId	0.0010		
			A) 0.0013		
Granodiorite		III	0.0020		
		C)IV?	C) 0.0021		
Granite		IV	0.0012	0	0–0.0001
			B) 0.0		
Country Rock	0	V	0.0001	0–0.001 [4]	0 (Quartzites)
		ELSE	0.0	0 [2]	0–0.002 (Pelites)
Depth range (km)	0.0–4.2		0.0–4.2	~0.0–0.3	0
			A) 0.25–4.2		
			B) 0.0 –0.5		
			C) 0.5 –4.2		

Generally, the magnetic models (Figs. 7.2–7.4) coincide well with the geologically mapped units. As suggested by the geological profiles of Troll and Weiss (Chap. 3, this Vol.) the dip angles of the metamorphic contacts and interior boundaries determined by the magnetic method are mostly steep, with typical values of 70°-90° estimated with an error of ±15°. The depth of resolution is about 300 m. Referring to Figs. 7.2 and 7.4, we note that the marginal contacts dip slightly away from the pluton. The tongue-shaped pelite screen near the centre of profile 2 (Figs. 7.1, 7.2), which has been displaced by the Ballachulish Fault, seems to be a body of shallow depth. The fault zone itself ('B' in Fig. 7.2) has no pronounced magnetic anomaly, although it represents a 50-m-wide, hydrothermally altered zone of significant veining (Weiss 1986) and of crack-induced, mechnical weakness, verified by shallow seismic data (Rabbel 1987).

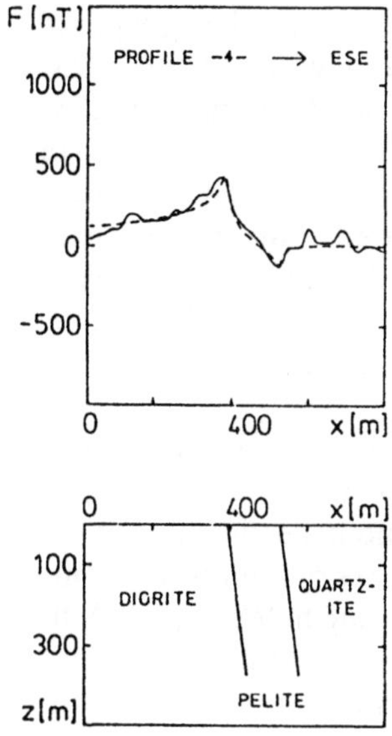

Fig. 7.4. Surface magnetic profile 4. Field anomaly (*solid line*) and modelled anomaly (*dashed line*) with geological interpretation (location 4 in Fig. 7.1; susceptibility values, see Table 7.1). *F* Magnetic intensity; *X,Z* spatial coordinates

7.2.2 Magnetic Anomaly Map

Figure 7.5 shows a map of the magnetic total intensity for the southern part of the Ballachulish Complex (location see Fig. 7.1). The map is based on 20 mainly parallel magnetic profiles. Before the two-dimensional interpolation was carried out, the individual profiles were low-pass filtered with a cut-off-wavelength of 250 m.

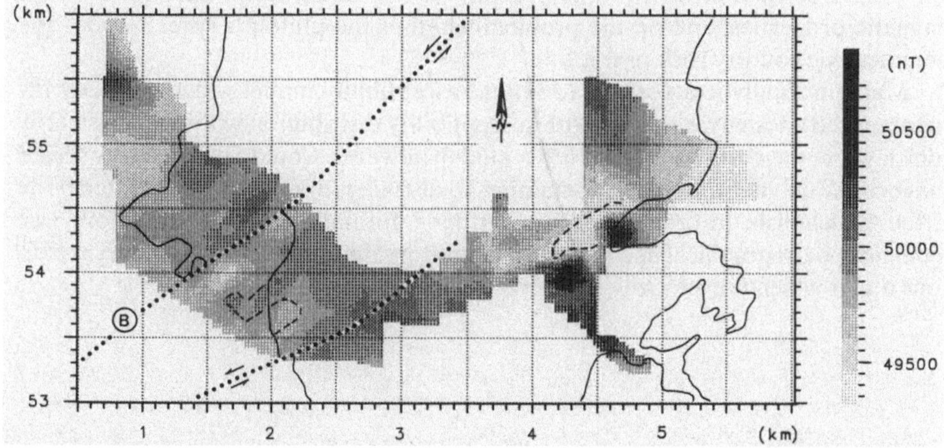

Fig. 7.5. Surface magnetic map of the southern Ballachulish Complex (location 1 in Fig. 7.1; *B* Ballachulish Fault). Outcrop contour from geological mapping indicated by *solid line* (after Weiss 1986), suggested modifications indicated by *dashed line*

The map can be divided into areas of low and high magnetization, being separated by the surface contact contour line of the Ballachulish pluton which has been taken from Weiss (1986; solid line in Fig. 7.5). The geological boundaries usually fit well to the magnetic high and low total intensities, with one exception in the east. Here the country rock (here: quartzite) extends into the pluton and only a few outcrops are exposed. The observed low magnetic intensity values, as well as the distinct susceptibility contrast between quartzite and diorite, based on laboratory data (Table 7.1, last column), suggests that the contour of the pluton must be modified. The modification is indicated by the dashed line and now included in the actual geological map presented in this volume (map 2).

Apart from the low levels of the magnetic total intensities which follow the fault-associated pelite 'tongue' into the interior of the complex, further striking features of the map are the strong positive magnetic anomalies in the NW- and SE-parts. They originate from spherical or prismatic bodies of high susceptibility, with a geometrical centre at a depth of 100–200 m. According to Weiss (1986), they can be identified as basites of the Appinite Suite (amph-diorite) which have been emplaced near the margins of the complex without reaching today's earth's surface.

The question whether the northwestern appinitic body is isolated or whether the pluton as a whole continues under the northwestern metamorphic rocks will be discussed in the following section.

7.3 Aeromagnetics

According to the aeromagnetic map of the British Geological Survey (Institute of Geological Sciences 1981) the Ballachulish Complex reveals a remarkable anomaly with a maximum of 500 nT (Fig. 7.6). The interpretation of the aeromagnetic data will yield information on the mean depth of the Ballachulish pluton, its mean magnetic properties, and on the problem whether the pluton is spread under the metamorphic country rock or not.

Mathematically it is possible to find a susceptibility model which perfectly fits the observed anomaly, e.g., a smooth susceptibility distribution within an arbitrarily thick layer at the earth's surface. Such a model, however, would only be a new scaled image of the magnetic anomaly, containing no distinct properties of the isolated body of the Ballchulish. In order to obtain further information we will, therefore, use geological, petrographical and magnetic laboratory data as definite boundary conditions under which the anomaly is analyzed.

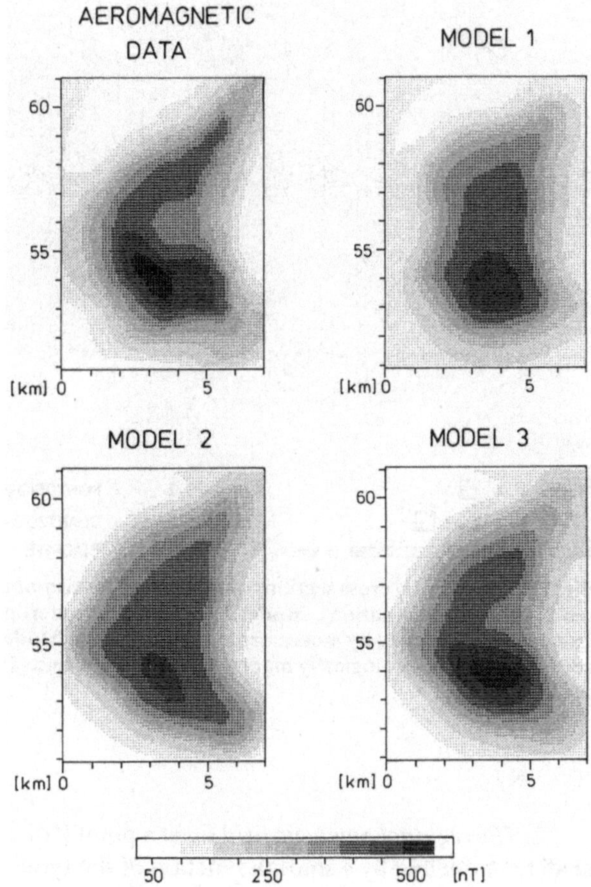

Fig. 7.6. Aeromagnetic data (after Institute of Geological Sciences 1981) and synthetic anomalies of *Models 1-3* (see Fig. 7.7)

7.3.1 The Linear Inversion Method

The calculations were carried out with an iterative interpretation procedure of modelling and linear inversion. The a-priori assumptions and their consequences are listed below:

1. *The pluton* and its petrographic units are approximated by vertical prisms of quadratic cross-section and constant depth (Fig. 7.7). Since the dip angles of the contacts are steep, between 70–90° (see Sect. 7.2), the error arising from this assumption is small. The corresponding model anomalies are computed due to the formulae of Bhattacharyya (1964). The anomaly contribution F_B of N given bodies at a point of observation P is then

$$F_B(P) = f_1(P)S_1 + f_2(P)S_2 + ... + f_n(P)S_n, \quad (1)$$

where S_k (K = 1..N) are susceptibility values and f_K are partial derivatives of the magnetic field with respect to susceptibility (to be determined numerically).

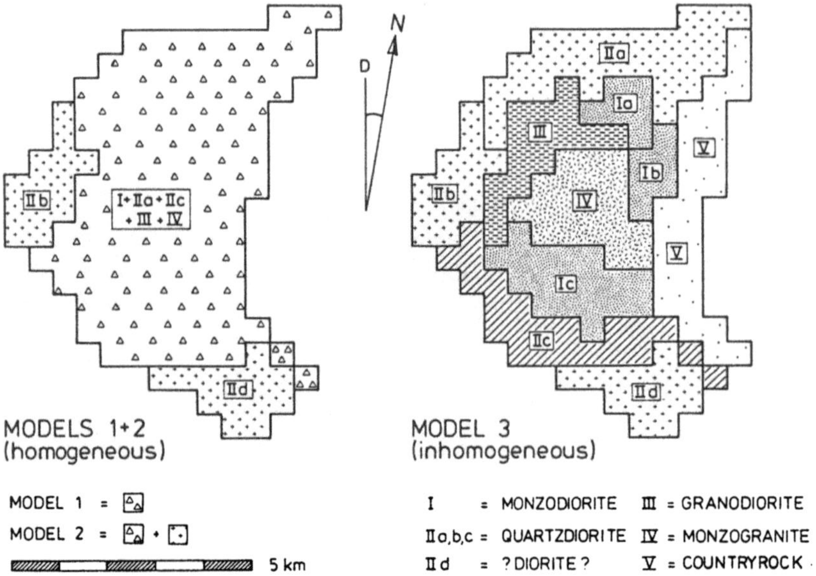

Fig. 7.7. Horizontal cross sections of *Models 1-3 (susceptibility values, see Table 7.1; anomalies, see Fig. 7.6; D* declination). **Model 1** = geological outcrop contour; **Model 2** = Model 1 plus igneous units covered by metamorphic country rock; **Model 3** = Model 2 subdivided in sectors corresponding to geologically mapped petrographic units (after Weiss 1986) plus country rock sector V.

2. *The regional magnetic field* F_R at a point P of a horizontal plane of observation shall be modelled by a smooth function of the type

$$F_R(P) = r_0 + r_1X + r_2Y + r_3XY, \tag{2}$$

with horizontal spatial coordinates X,Y and unknown constants r_K (K = 0..3) within the observation area.

3. We assume that the *height of observation* (terrain clearance) is constant over ground and neglect topographic corrections. For a given model body and a point of observation P(X,Y,H) the change of the anomaly F_H with height H is linearized:

$$F_H\,(X,Y,H+h) = f_H(P)h \tag{3}$$

with height increment h and the corresponding partial derivative f_H (to be determined numerically).

4. Effects of *remanent magnetization* are neglected. The *residuum of the magnetic field anomaly* F, i.e., the difference between the observed anomaly and the anomaly of an initial model body, can now be described as a system of M linear equations

$$F(P_K) = F_B(P_K) + F_R(P_K) + F_H(P_K), \tag{4}$$

for K = 1..M points of observation P_K. They are solved simultanously for the unknown susceptibility values S_J (J = 1..N), the regional field coefficients r_L (L = 0..3) and the height correction h in a least squares sense (inversion theory outlined, e.g., by Parker 1977, or Zschau 1979). Equation (4) applies if the height of observation is known only approximately. If height corrections can be neglected Eq. (4) reduces to

$$F(P_K) = F_B(P_K) + F_R(P_K). \tag{5}$$

As a data base for the inversion, 315 (15 × 12) intensity values with 0.5 km spacing were taken from the aeromagnetic map.

7.3.2 Interpretation Steps

The sequence of interpretation comprises four steps. They are illustrated in Figs. 7.6–7.8 and will be described below.

1. *Homogeneous initial model.* At first a homogeneous pluton model is designed, of which the contours approximate the geologically mapped outcrop line of the Ballachulish Complex (Fig. 7.7: Model 1). With fixed values for the height of observation (0.5 km) and for the pluton's depth range (4 km) the formal model parameters are optimized according to Eq. (5). The resulting magnetic anomaly pattern, however, deviates significantly from the aeromagnetic data in the W- and SE-parts of the observation area (Fig. 7.6: Model 1).

2. *Improved contours for the homogeneous model.* The initial model can be improved significantly by a modification of the pluton's contour line (Fig. 7.7: Model 2). Inverting Eq. (4) yields an improved observation height (0.58 km) and an anomaly pattern which approximates the gross characteristics of the field data (Fig. 7.6: Model 2). At this stage, it is already evident that the pluton extends under the western and southeastern country rocks.

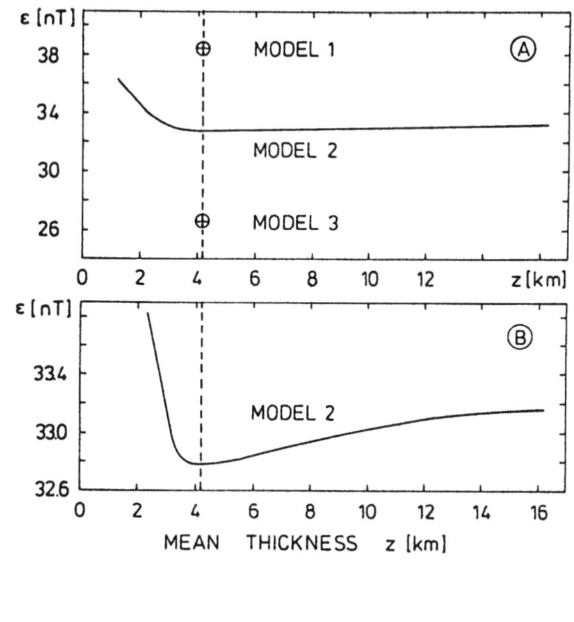

Fig. 7.8 A. Average absolute residuum ε for *Model 1-3* of Fig. 7.7 as a function of depth extent *Z*. **B** Detail of **A** Notation: $\varepsilon = (|F_1| + |F_2| + ... + |F_K| + ... + |F_N|)/N$. N = 315 = number of equally spaced observation points from the area of Fig. 7.6. F_K = residuum = observed minus modelled anomaly. |..| = absolute value

3. *The depth problem.* In the linear equation system (4) there is no variable with depth dependence of the Ballachulish pluton. Introducing such a term [formally corresponding to Eq. (3)] as a trial into Eq. (4) indicates via an eigenvalue analysis that the aeromagnetic data contains little information on the pluton's depth range. Figure 7.8 displays average absolute residuals of the aeromagnetic anomaly for different depth extents of the pluton. Varying the depth range of Model 2 (Step 2) while optimizing its formal parameters [Eq. (5)] yields a significant decrease of the residuum only between $Z = 1$ km and $Z = 4$ km thickness. A slight residual minimum is obtained for $Z = 4.2$ km. We therefore conclude that the minimum-depth extent of the Ballachulish pluton is about 4 km while a maximum-estimate cannot be derived from the data.

4. *Inhomogeneous model.* In the previous steps, observation height and depth extent were optimized for Model 2 and fixed. Now the pluton's interior and country rocks can be divided into segments approximating the contours of its mapped petrographical units (Fig. 7.7: Model 3). The linear inversion procedure [Eq. (5)] yields mean susceptibility values for the ten different sectors (see Table 7.1). The improvement of the interpretation is indicated by the remarkable decrease of the average absolute residuum from 38 nT (Model 1; Step 1) to 26 nT (Model 3; see Fig. 7.8). The corresponding anomaly pattern indeed displays all basic features of the observed aeromagnetic data (Fig. 7.6: Model 3). In general, the calculated susceptibility values fit well to the values obtained by laboratory measurements on rock samples (see Table 7.1). The details of this interpretation step are outlined below.

5. *Discussion on Step 4.* The inhomogeneous Model 3 (Fig. 7.7) includes three sectors outside the geologically mapped outcrop contour, i.e., IIb, IId and V. The corresponding susceptibility values obtained by the least squares inversion method provide information on their petrographic composition and help to decide if these sectors belong to the pluton or not (see Table 7.1).

Sector IIb (western part) has the same susceptibility as the diorite sector IIa and will therefore consist of diorite enclosing screens of metamorphic country rock.

The susceptibility of sector V (eastern part) is by a factor 0.1 smaller than the smallest value of the igneous units I-IV. We conclude that sector V is a metamorphic country rock (mainly quartzitic).

The susceptibility of sector IId (southeastern part) is smaller than that of the interior sectors but clearly higher than that of the country rocks. Thus it is reasonable to assume that this sector IId belongs to the diorite sectors IIa-c. Without significant loss of modelling precision the one-layer sector IId of 0.001 emu can be replaced by a two-layer sector representing a 250-m-thick country-rock coverage of 0.0 emu, under which the diorite of typical 0.0013 emu (cf. sectors IIa-c) spreads.

A similar argument can be applied to the central sector IV (granite). In contrast to the remaining areas, the magnetic susceptibility here deviates significantly from the laboratory data for rock samples from the Ballachulish (Table 7.1). The one-layer sector IV of 0.0012 emu can be split into a pure granite package of typical 0.0 emu at the surface, extending to a depth of 0.5 km and covering a second package of monzodiorite- or diorite-contaminated rocks of 0.002 emu (cf. sectors Ic, III and Sect. 7.4).

Table 7.2. Density values of rock samples from the Ballachulish area (after Entwisle 1984)

Rock Type	Density [Mg/m^3]		
	Minimum	Maximum	Middle of interval
Monzo-diorite	2.81	2.84	2.825
Quartz-diorite	2.73	2.75	2.740
Granite	2.61	2.64	2.625
Pelitic schists	2.77	2.82	2.795
Quartzite	2.62	2.65	2.635

7.4 Gravity

Relative to the surrounding area, the Ballachulish complex has only a weak gravity anomaly of +5 mGal (Institute of Geological Sciences 1979). This observation suggests a small density contrast between the pluton and its surroundings. It is compatible with the laboratory-derived density values of Entwisle (1984) for rock samples from this area (see Table 7.2) if (1) the country rock mainly consists of metapelites and (2) the pluton is mainly dioritic.

Since a pronounced gravity anomaly is absent, the pluton's mean total depth cannot be estimated and compared to the aeromagnetic results. The gravity signature, however, contains information on the depth extent of its pure granitic centre: assuming the density values given in Table 7.2, the granitic sector (IV in Fig. 7.7) should yield a central anomaly of –7 to –9 mGal, if it would reach the aeromagnetically estimated minimum depth of 4 km. This differs remarkably from the observed positive (!) anomaly and indicates that the central granite of low density must be contaminated with or graded into diorite at shallow depth (as suggested in the previous section).

7.5 Conclusions

Based on surface magnetic, aeromagnetic, gravity and laboratory data, the following conclusions concerning the structure of the Ballachulish Complex are derived:

1. The pluton has a miminum depth range of 4 km.
2. The western and southeastern diorites of the complex spread laterally 0.5–1.5 km below the metamorphic country rocks. In the eastern part, its lateral boundary coincides with the geologically mapped surface contours.
3. Within the upper 0–400 m, the contacts dip steeply (close to the vertical).
4. The pluton as a whole consists mainly of diorite or of diorite-contaminated granitic rocks. The central porphyritic granite layer is graded into or contaminated by diorite at shallow depth.

Part III
The Contact Aureole and Its Rocks

8 Petrography and Mineral Chemistry of Pelites

D.R.M. Pattison and B. Harte

8.1 Introduction

This chapter describes the petrography and mineral chemistry of an exceptionally well-developed sequence of prograde mineral zones in pelitic and semipelitic rocks in the Ballachulish aureole. Two schematic petrogenetic grids are derived: the first is for mineral assemblages below the onset of partial melting, which define the mapped isograds in Maps 1 and 2 and Figure. 8.1; and the second is for high-grade mineral assemblages which occur sporadically within the zone of partial melting (Harte et al., Chap. 9, this Vol.) and within pelitic screens within the igneous complex. The two grids, when linked, provide a continuous petrogenetic grid from the lowest to highest grade in the aureole. In Pattison (Chap. 16, this Vol.) the continuous grid is calibrated in P-T space.

Sixty-eight specimens were selected from the prograde mineral zones for mineral chemical analysis. For the specimens below the zone of partial melting, emphasis is place on: the variation with grade of Mg/(Mg + Fe) in coexisting chlorite, biotite and cordierite; (Fe,Mg)Si = 2Al, the Tschermak exchange, in chlorite, muscovite and biotite; K/K + Na in muscovite and biotite; and F and Ti in biotite. For the high grade specimens, emphasis is placed on the compositional parameters that are required for the geothermometric and geobarometric calculations in Pattison (Chap. 16, this Vol.).

The chapter closes with a short description and interpretation of textural evidence for localized retrogression in the aureole.

8.2 Previous Work

The earliest descriptions of contact metamorphism around the Ballachulish Igneous Complex were by MacCulloch (1817, p. 126), who noted the induration of the pelitic schists near the contacts. Around the turn of the century, routine mapping and initial petrographic analysis of the contact metamorphism by geologists of the Scottish Geological Survey, including J.S.G. Wilson, J.J.H. Teall, J.S. Flett, and E.G. Bailey, established the extent and principal petrographic features of contact metamorphism (Bailey and Maufe 1960). Apart from studies by Neumann (1950) on the transition of pyrite to pyrrhotite in the aureole, and by Muir (1953a,b) on the petrography of

G. Voll, J. Töpel, D.R.M. Pattison, F. Seifert
Equilibrium and Kinetics
in Contact Metamorphism

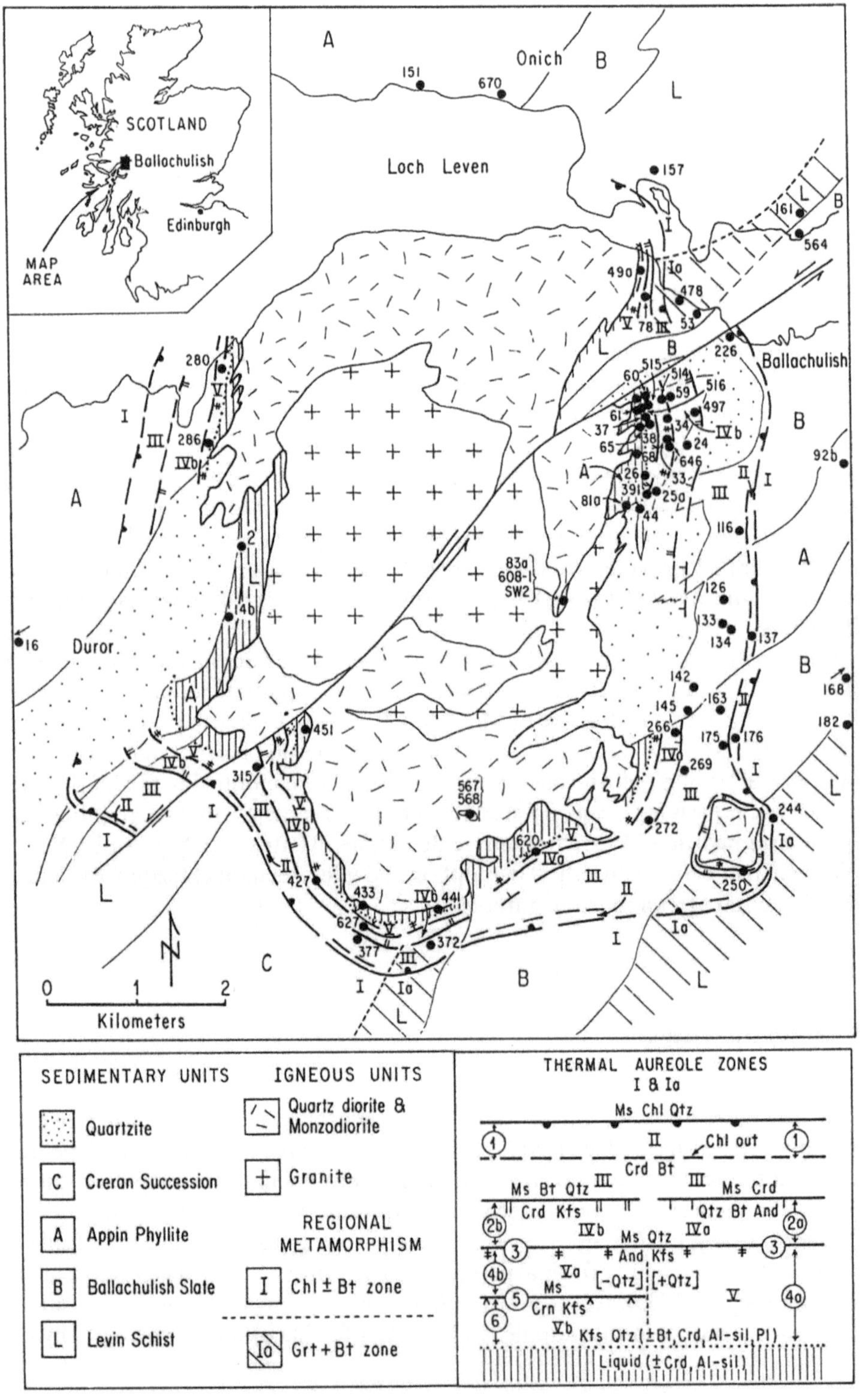
SCOTLAND
Ballachulish
Edinburgh
MAP AREA
Onich
Loch Leven
Ballachulish
Duror
Kilometers
0 1 2
SEDIMENTARY UNITS
Quartzite
C Creran Succession
A Appin Phyllite
B Ballachulish Slate
L Levin Schist
IGNEOUS UNITS
Quartz diorite & Monzodiorite
Granite
REGIONAL METAMORPHISM
I Chl ± Bt zone
Ia Grt + Bt zone
THERMAL AUREOLE ZONES
I & Ia
Ms Chl Qtz
II
Chl out
Crd Bt
Ms Bt Qtz III
III Ms Crd
Crd Kfs
Qtz Bt And
IVb Ms Qtz IVa
And Kfs
Va [-Qtz] [+Qtz]
Ms
V
Crn Kfs
Vb Kfs Qtz (±Bt, Crd, Al-sil, Pl)
Liquid (± Crd, Al-sil)

xenoliths in the igneous complex, no further work was done on the aureole until the commencement of this integrated study in 1981. Much of the material for this chapter comes from Pattison (1985, 1987, 1989, 1991) and Pattison and Harte (1985, 1988).

8.3 Petrography of Metamorphic Zones Up to the Onset of Partial Melting

Figure 8.1 shows the location of the prograde metamorphic zones around the igneous complex. Figure 8.2 is a photograph of Fraochaidh, in the southern part of the aureole, showing the distribution on the ground of the mineral zones. The metamorphic zones were delineated by mineral identification in the field and by the examination of about 400 thin sections.

Fig. 8.2. Distribution of pelitic mineral zones on the mountain Fraochaidh, southern margin of the igneous complex (see Map 2). Ornaments are the same as in Fig. 8.1. The broken squiggly line represents the onset of partial melting. Note the more resistant, massive weathering outcrops in Zone IV compared to Zone III. This is due to the production of large amounts of modal K-feldspar by reaction P2b

---

Fig. 8.1 Simplified geology of the Ballachulish Igneous Complex and contact aureole, showing the mineral zones and location of specimens discussed in the text. For clarity, only quartzite and the four main pelitic units have been shown. *B* represents both the Ballachulish Slate and Transition Series (see Map 2). Specimen localities are given as *black dots* with adjacent specimen number

Table 8.1. Mineral abbreviations (Kretz 1983) and formulae

Phase	Abbreviation	Model formulae
Albite	Ab	$NaAlSi_3O_8$
Andalusite	And	Al_2SiO_5
Biotite	Bt	$K(Fe,Mg)_3AlSi_3O_{10}(OH)_2$
Calcite	Cal	$CaCO_3$
Chlorite	Chl	$(Fe,Mg)_5Al_2Si_3O_{10}(OH)_8$
Cordierite	Crd	$(Fe,Mg)_2Al_4Si_5O_{18}{\cdot}0.5H_2O$
Corundum	Crn	Al_2O_3
Diopside	Di	$CaMgSi_2O_6$
Dolomite	Dol	$CaMg(CO_3)_2$
Forsterite	Fo	Mg_2SiO_4
Garnet	Grt	$(Fe,Mg)_3Al_2Si_3O_{12}$
Graphite	Gr	C
Hydrous vapour	H_2O or V	H_2O
Hypersthene	Hy	$(Fe,Mg)_2Si_2O_6$
Ilmenite	Ilm	$FeTiO_3$
K-feldspar	Kfs	$KAlSi_3O_8$
Magnesite	Mgs	$MgCO_3$
Granitic melt (liquid)	L	—
Muscovite	Ms	$KAl_3Si_3O_{10}(OH)_2$
Periclase	Per	MgO
Plagioclase	Pl	$NaAlSi_3O_8$-$CaAl_2Si_2O_8$
Quartz	Qtz	SiO_2
Sillimanite	Sil	Al_2SiO_5
Spinel	Spl	$(Fe,Mg)Al_2O_4$
Talc	Tlc	$Mg_3Si_4O_{10}(OH)_2$
Tremolite	Tr	$Ca_2Mg_5Si_8O_{22}(OH)_2$
Wollastonite	Wo	$CaSiO_3$

In the following sections, the outcrop characteristics and thin-section petrography of the different metamorphic zones in Fig. 8.1 are described. Reactions are inferred between different metamorphic zones, based on mineralogical changes and on textural relationships in the rocks. Mineral abbreviations are listed in Table 8.1, following Kretz (1983). Mineral reactions are balanced according to the ideal mineral formulae in Table 8.1. Reaction numbers correspond to those in Pattison and Harte (1985).

The model pelitic system K_2O-MgO-FeO-Al_2O_3-SiO_2-H_2O (KFMASH) has been adopted for the initial petrographic analysis below. All minerals which show significant changes in distribution with grade in the aureole can be satisfactorily modelled in this system. It is additionally assumed that up to the onset of partial melting, there was a hydrous fluid phase present during metamorphism (see discussion in Pattison and Harte 1985).

In the following sections, Roman numerals are used to describe both mineral assemblages and metamorphic zones and subzones. Zones and subzones represent differences in metamorphic grade. Mineral assemblages with Roman numerals refer to assemblages developed at the same grade, but in rocks of different bulk composition. For example, Zone IV contains two assemblages, IVa and IVb, developed in different units at the same grade. In contrast, subzones Va (lower grade) and Vb (higher grade) refer to two subzones developed in quartz-absent rocks in Zone V. Table 8.2 summarizes these relationships.

8.3.1 Zones I and Ia (Marginal Zones)

Zones and I and Ia include regional grade schists and phyllites that have experienced no apparent contact metamorphic effect. Zone I refers to assemblages containing Ms + Qtz ± Chl ± Bt, whilst subzone Ia refers to assemblages containing Ms + Qtz + Grt + Bt ± Chl (see Pattison and Voll, Chap. 2, this Vol.).

In some specimens from Zones I and Ia, it is possible that biotite is of contact metamorphic origin. For example, in DP151, coarse, green, randomly orientated biotite porphyroblasts overgrow small tabular, brown, schistosity-parallel biotites; the former may represent contact metamorphic growth. In a number of specimens, regional grade chlorite pseudomorphs after garnet are converted to biotite as one approaches the igneous complex (see Sect. 8.3.4).

Apart from these rare examples, however, the presence of biotite in Zones I and Ia appears to be unrelated to proximity to the igneous complex. The biotite is usually orientated parallel to regional metamorphic planar fabrics (e.g., D_1 schistosity or later crenulations), which suggests that it is predominantly of regional metamorphic origin. Consequently, a contact metamorphic biotite zone is not considered to be an important feature.

8.3.2 Zone II (Lower Cordierite Zones; Chlorite Zone)

The first pelitic mineral change that constitutes a mappable thermal metamorphic zone in all stratigraphic units is the first appearance of cordierite. In outcrop, it occurs in 0.5–2-mm ellipsoidal 'spots' that overgrow pre-existing fabrics, producing classic 'spotted slates' (e.g., Harker 1954; Fig. 8.3). The cordierite ellipsoids are elongated parallel to the dominant pre-existing regional fabric (normally the D_1-schistosity). On cleavage surfaces, the ellipsoids are randomly orientated. Accompanying the development of cordierite is a general decrease in fissility of the slates and phyllites.

In thin section, the incipient cordierite occurs in 0.5–1.5-mm elongate shapeless poikiloblasts that overgrow the cleavage (Fig.8.4). Some crystals show patchy extinction. The cordierite contains abundant inclusions of muscovite, quartz, accessory minerals and sometimes pre-existing regional biotite, but no primary chlorite. Most of the cordierite spots are completely altered to grey-brown cryptocrystalline aggregates of sericitic material, or to orange-brown 'films' that are isotropic in crossed nicols (pinite); (see Section 8.8.1). Sometimes the cordierite is partially altered at its margins, giving a zoned appearance (Fig. 8.4). Maresch et al. (Chap. 14, this Vol.) discuss the ordering state of cordierite in these assemblages in more detail.

Table 8.2. Summary of mineral zones, diagnostic assemblages, and inferred model reactions in the Ballachulish aureole

Zone			Subdivision	Occurrence	Diagnostic assemblage	Model reaction	Reaction No.
Marginal zones (regional assemblages)		I[a]	I: Chl ± Bt subzone	All units	Ms + Chl + Qtz ± Bt	—	—
			Ia: Grt + Bt subzone	Leven Schist Creran Succession	Ms + Chl + Qtz + Bt + Grt	Ms + 3Chl + 3Qtz = 4 Grt + Bt + 12 H_2O	R5
Ms + Qtz stable zones of contact meta-morphism	Lower cordierite zones	II:	Crd + Chl zone	All units	Ms + Chl + Qtz + Crd + Bt	Ms + Chl + 2 Qtz = Crd + Bt + 3.5 H_2O	P1
		III:	[Chl] zone	All units	Ms + Qtz + Crd + Bt	Chl consumed by reaction P1	—
	Upper cordierite zone	IV[b]	IVa: And assemblage	Ballachulish Slate	And + Bt + Qtz + Crd + Ms	2 Ms + 3 Crd = 7 Qtz + 8 And + 2Bt + 1.5 H_2O	P2a
			IVb: Kfs assemblage	Creran Succession Leven Schist Appin Phyllite	Kfs + Bt + Qtz + Crd ± Ms	6 Ms + 2 Bt + 15 Qtz = 3 Crd + 8 Kfs + 6.5 H_2O	P2b
Al_2SiO_5 + Kfs stable zones of contact metamorphism		V[c]		All units	Ms + Bt + Qtz + Crd + And (Sil) + Kfs	Ms + Qtz = And (Sil) + Kfs + H_2O	P3
			V: Qtz assemblage	Qtz-bearing pelites	Bt + Qtz + Crd + Kfs + And (Sil)	9 Qtz + 2 Bt + 6 And (Sil) = 3 Crd + 2 Kfs + 0.5 H_2O	P4a
			Va: Ms subzone	Qtz-absent pelites	Ms + Bt + Crd + Kfs + And (Sil)	9 Ms + 3 Crd = 2 Bt + 15 And (Sil) + 7 Kfs + 8.5 H_2O	P4b
			Vb: Crn subzone		[Ms + Bt + Crd + Kfs + And (Sil) + Crn] Bt + Crd + Kfs + And (Sil) + Crn	Ms = Crn + Kfs + H_2O 2 Bt + 15 And (Sil) = 9 Crn + 3 Crd + 2 Kfs + 0.5 H_2O	P5 P6
			Subzone of partial melting	All units	Same as Zones V but containing textures and structures suggestive of anatexis. At highest grade, Grt ± Hy may be present in Qtz-bearing assemblages, and Spl may be present in Qtz-absent assemblages.	Kfs + Qtz + H_2O ± Bt ± Pl = L ± Crd ± Al_2SiO_5	—

[a]Zone I represents regional grade assemblages, developed prior to contact metamorphism.
[b]Zone IV contains two assemblages, IVa and IVb, which are developed at approximately the same grade in rocks of different bulk composition (see text).
[c]Zone V assemblages may be Qtz-bearing or Qtz-absent. The same Qtz-bearing assemblage occurs throughout zone V. In Qtz-absent rocks, a lower grade (Va) muscovite subzone, and a higher grade (Vb) corundum subzone may be defined.

Fig. 8.3. Zone II. Ballachulish Slate (collected east of DP116 on Fig. 8.1) showing incipient development of cordierite spots on cleavage surfaces

Fig. 8.4. Zone II. Ballachulish Slate (DP116) showing ellipsoidal cordierite poikiloblasts elongate parallel to the preexisting slaty cleavage. The apparent zonation in the cordierite is due to selective pinitization. Long dimension of photo is 8 mm

Accompanying cordierite in all specimens are small grains of biotite. These normally occur adjacent to or intergrown with the cordierite, but sometimes occur discretely away from the cordierite. In rocks that contain no pre-existing regional grade biotite, the newly-formed biotite often envelopes ilmenite grains or occurs on the edges of chlorite grains. The biotite is typically orientated parallel to the pre-existing schistosity.

Primary chlorite decreases in abundance abruptly as one enters Zone II. Primary chlorite is distinguished texturally from secondary chlorite by its occurrence in well-defined, tabular grains of uniform grain size (100–200μ) that lie parallel to the schistosity, in contrast to secondary chlorite, which is of more variable and coarser grain size, is less tabular and is often a conspicuous alteration product of biotite or cordierite.

The appearance of Crd + Bt, coupled with the modal decrease in Chl, suggests the operation of the model divariant KFMASH reaction (balanced according to ideal mineral compositions in Table 8.1, and treating Fe and Mg as one component):

$$\mathrm{Ms + Chl + 2\ Qtz = Crd + Bt + 3.5\ H_2O.} \qquad \text{(P1)}$$

The full Ms + Chl + Qtz + Crd + Bt assemblage has been found in 17 specimens, mostly in the graphitic Ballachulish Slate. This assemblage characterizes Zone II and in some areas (Fig. 8.1) can be mapped over a narrow (<100 m) interval, which constitutes the outermost contact metamorphic zone in the aureole.

The first appearance of cordierite varies in distance from the contact from <400 m in the north to 1700 m on the east and southwest contacts; the widest zones are adjacent to or incorporate large expanses of quartzite. On the east flank, cordierite first occurs slightly farther from the contact in the Ballachulish Slate than in the adjacent Appin Phyllite (Fig. 8.1).

8.3.3 Zone III (Lower Cordierite Zones; Chlorite-Absent Zone)

Zone III is distinguished from Zone II by the absence of primary chlorite, which gives the assemblage Ms + Qtz + Crd + Bt in all pelitic and semi-pelitic units. In no specimens has either of Ms or Qtz been consumed. Zone III is also characterized by the increase in size (1–3 mm) and modal abundance of cordierite (up to 30%), which accompanies a hardening of the slates and phyllites (Fig. 8.5).

In thin section, cordierite in Zone III occurs in inclusion packed poikiloblastic crystals, often with irregular, patchy, lobe-like margins (Fig. 8.6). Biotite occurs in discrete tabular grains, whilst muscovite retains its schistosity-parallel, tabular habit from lower grades. Plagioclase in both Zones II and III is albitic.

Zone III varies in width from <200 m in the north to 700 m on the east and southwest flanks, and is the widest individual zone in the aureole.

Fig. 8.5. Zone III. Appin Phyllite (DP126), showing an overall reduction in fissility and more abundant and larger cordierite poikiloblasts than in Zone II

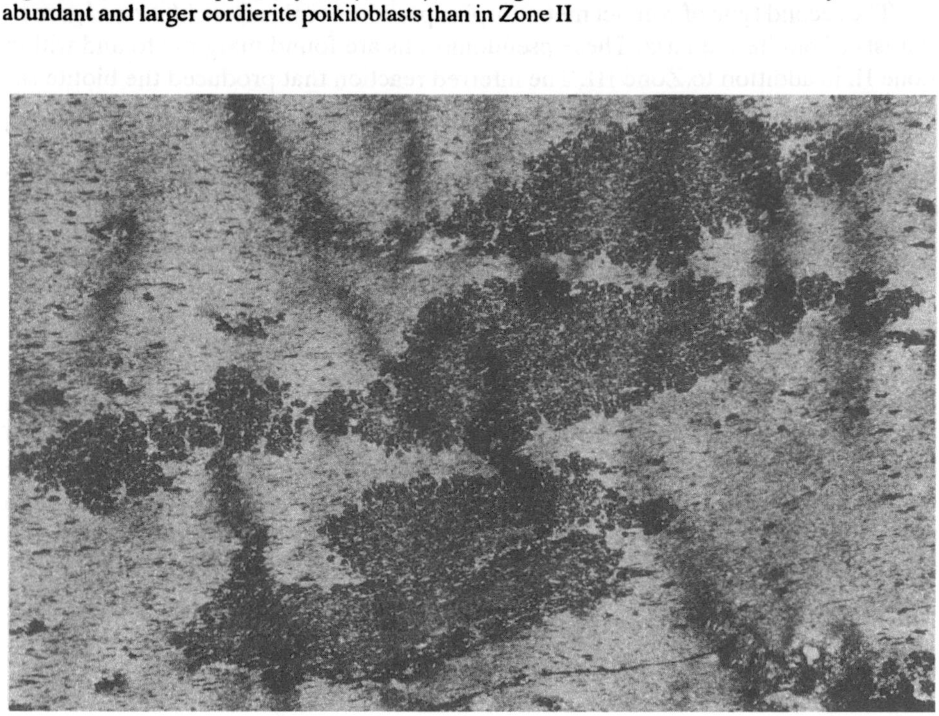

Fig. 8.6. Zone III. Ballachulish Slate (DP163), showing the development of cordierite poikiloblasts with irregular, patchy margins overgrowing the preexisting crenulated schistosity. Long dimension of photo is 9 mm

8.3.4 Overprinting of Regional Garnet-Bearing Assemblages in Zones II and III

Two main types of pseudomorphs after regional grade garnet are found in the contact aureole: one arising from the replacement of garnet itself, the other from the replacement of regional retrograde chlorite that had partially or wholly pseudomorphed garnet prior to contact metamorphism. The pseudomorphs are most clearly developed in Zones II and III.

The pseudomorphs of garnet itself are not present in Zone II, where garnet texturally appears inert in assemblages in which contact metamorphic cordierite has grown. However, in Zone III and above, regional garnet is replaced by an intergrowth of cordierite and biotite. This texture suggests the progress of the model KFMASH reaction

$$4\ \mathrm{Grt} + 3\ \mathrm{Ms} + 3\ \mathrm{Qtz} + 1.5\ \mathrm{H_2O} = 3\ \mathrm{Crd} + 3\ \mathrm{Bt}. \qquad (\mathrm{P1'})$$

A notable feature of this reaction is that it is a prograde vapour-consuming (hydration) reaction rather than the more usual vapour-producing (dehydration) reaction. Water is a reactant because cordierite at this grade is a hydrous mineral (e.g., Mirwald and Schreyer 1977). In the Ballachulish aureole, the release of water due to the widespread progress of dehydration reactions such as Ms + Chl + 2 Qtz = Crd + Bt + 3.5 H_2O (reaction P1) may have provided the water required for reaction (P1′) to proceed.

The second type of contact metamorphic pseudomorph, after *chloritized* garnet, consists of biotite ± quartz. These pseudomorphs are found marginal to and within Zone II, in addition to Zone III. The inferred reaction that produced the biotite is:

$$3\ \mathrm{Ms} + \mathrm{Chl} = 3\ \mathrm{Bt} + 4(\mathrm{Fe,Mg})_{-1}\mathrm{Si}_{-1}\mathrm{Al}_2 + 7\ \mathrm{Qtz} + 4\ \mathrm{H_2O}. \qquad (\mathrm{R4})$$

Evidence from mineral chemistry for the progress of this reaction is discussed in more detail in Section 8.6.4 below.

8.3.5 Zone IV (Upper Cordierite Zone)

Upgrade of Zone III, pelites acquire either of the assemblages *And* + Qtz + Bt + Crd + Ms (assemblage IVa) or *Kfs* + Qtz + Bt + Crd + Ms (assemblage IVb). Either of these assemblages is diagnostic of Zone IV. Mineral assemblages IVa and IVb are developed in rocks of different bulk composition at the same grade. Assemblage IVa has only been found in the graphitic Ballachulish Slate, whilst assemblage IVb has only been found in the Appin Phyllite, Leven Schist and pelitic Creran Succession.

8.3.5.1 Assemblage IVa (Andalusite Assemblage)

In the Ballachulish Slate, andalusite appears in the Zone IV assemblage And + Ms + Crd + Qtz + Bt (assemblage IVa). At the outer margin of Zone IV, andalusite crystals are typically small and ill-formed, sometimes producing a slightly knotted ap-

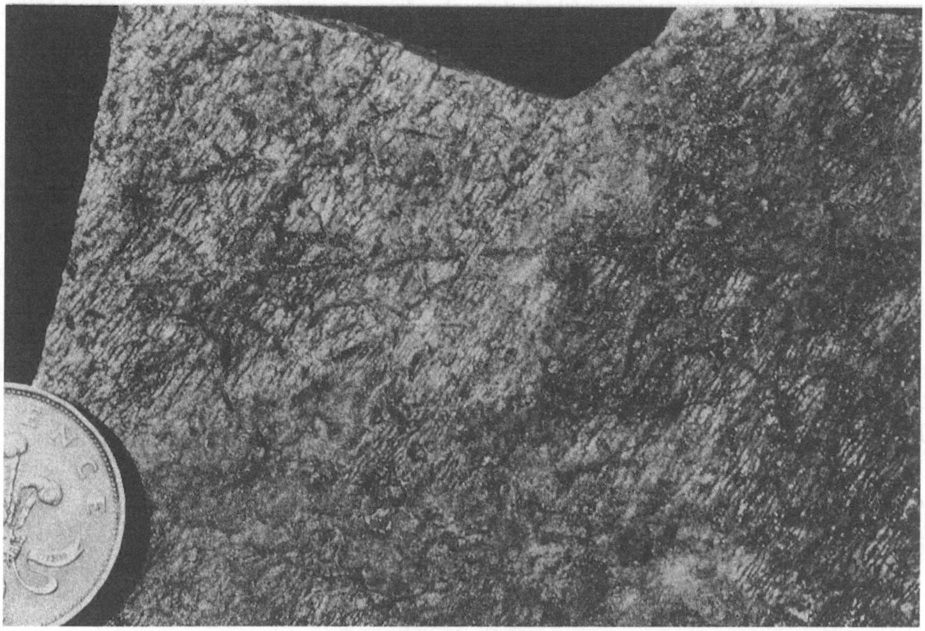

Fig. 8.7. Zone IV, assemblage IVa. Ballachulish Slate (SE contact) showing the random development of andalusite prisms within crenulated cleavage planes. Note that the slate has retained its cleavage from lower grades

pearance on cleavage planes in the slates. In specimens further within Zone IV, andalusite occurs more conspicuously in prisms up to 8 mm in length, randomly orientated within the cleavage planes (Fig. 8.7).

In this section, the andalusite prisms typically have one or two sharp crystal edges, but the other edges, and the core of the crystals, are ragged and intergrown with quartz. Chiastolite texture is rare. The andalusite tends to occur adjacent to cordierite poikiloblasts, with biotite and quartz typically partially surrounding the andalusite or intergrown in its ragged edges. In some specimens, andalusite occurs in aggregates of small ($<50\ \mu m$) prisms, intergrown with quartz and biotite. In all specimens, andalusite is unaltered. Sillimanite has not been observed in any of the specimens.

The development of And + Bt + Qtz intergrowths upgrade of the Zone III assemblage Ms + Crd + Bt + Qtz suggests the progress of the model divariant KFMASH reaction:

$$2\ \text{Ms} + 3\ \text{Crd} = 8\ \text{And} + 2\ \text{Bt} + 7\ \text{Qtz} + 1.5\ H_2O. \qquad \text{(P2a)}$$

All specimens with andalusite from Zone IV contain the full IVa assemblage Ms + Crd + Bt + And + Qtz.

Zone IV in the Ballachulish Slate and Transition Series, characterized by the IVa assemblage, is well developed in the southeast part of the aureole (Fig. 8.1), where it varies from 200–300 m in width. It has also been tentatively identified from field observations in Transition Series outcrops on the east flank (Fig. 8.1; Map 2).

8.3.5.2 Assemblage IVb (K-Feldspar Assemblage)

Above Zone III in the Leven Schist, Creran Succession and Appin Phyllite, K-feldspar appears in the second of the Zone IV assemblages, Kfs + Ms + Bt + Qtz + Crd (assemblage IVb). The development of K-feldspar produces a dramatic textural change in the pelites: relatively fissile Zone III phyllites pass upgrade into massive hornfels in Zone IV.

Going upgrade through Zone IV, muscovite decreases markedly in abundance. Abundant ellipsoidal cordierite poikiloblasts typically weather down relative to the surrounding K-feldspar rich matrix, producing a mesh- or net-like texture (Fig. 8.8). The volumetric proportion of cordierite in the hornfels ranges up to 50%.

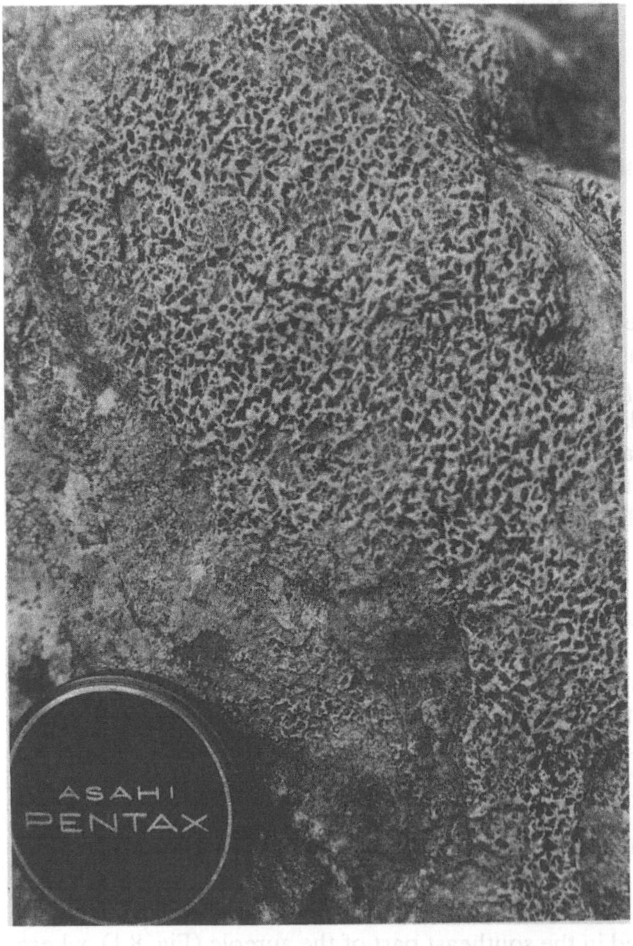

Fig. 8.8. Zone IV, assemblage IVb. Appin Phyllite, 100 m east of DP34, showing the massive recrystallized hornfels characteristic of this zone. Note the mesh or network texture, defined by randomly orientated, weathered cordierite poikiloblasts in a resistant K-feldspar-rich matrix

In thin section, cordierite occurs in two main habits: (1) 2–5-mm ellipsoidal, inclusion-packed poikiloblastic crystals; and (2) 1–3-mm, more prismatic, less inclusion-filled crystals (see also Maresch et al., Chap. 14, this Vol.). Both textures are found together in some specimens. K-feldspar occurs both in the matrix surrounding the cordierite and within the cordierite poikiloblasts, along with inclusions of muscovite, biotite, quartz and accessory minerals. In the Bt + Qtz ± Pl ± Ms matrix, the K-feldspar occurs in small (50–100 μm) subhedral, crypto-perthitic grains. Biotite occurs in small (<150 μm), reddish-brown, tabular, subhedral crystals. Plagioclase in Zone IVb has decreased modally from Zone III, and is oligoclase rather than the albite found at lower grades.

In about 2/3 of specimens, cordierite is partially or wholly altered. The cordierite typically alters to either fine-grained sericitic pinite, or to intergrowths of relatively coarse (100–250 μm) muscovite, biotite and quartz (Sect. 8.8.2).

The development of intimately associated Crd + Kfs upgrade of Ms + Bt + Qtz-rich phyllites suggests progress of the divariant KFMASH reaction

$$6\,\mathrm{Ms} + 2\,\mathrm{Bt} + 15\,\mathrm{Qtz} = 3\,\mathrm{Crd} + 8\,\mathrm{Kfs} + 6.5\,\mathrm{H_2O} \qquad \text{(P2b)}$$

In semipelitic specimens in Zone IV, the most common assemblage is Bt + Crd + Qtz + Kfs, suggesting that Ms was consumed by reaction (P2b). On the east flank, a muscovite-out boundary in semipelites may be mapped (see Map 2). One very muscovite-rich pelitic specimen contains the assemblage Ms + Bt + Crd + Kfs; in this case, quartz was consumed in reaction (P2b).

Zone IV, defined by Assemblage IVb, varies in width from <100 m in the northeast to 300 m in the east and southwest. The width of Zone IV is typically ~1/4–1/3 the combined width of Zones II and III.

8.3.5.3 Distinction Between Primary and Secondary Muscovite

The distinction between primary and secondary muscovite is critical to the interpretation of assemblages in Zone IVb and at higher grades, because at least some secondary muscovite is found in *all* samples from these zones. In the passage from Zone III muscovite-rich phyllites into muscovite-poorer Zone IVb hornfels, primary muscovite is subhedral, tabular and fine grained (<100 μm), and continues to define pre-existing regional schistosity and small scale crenulations. Sometimes it is concentrated in continuous layers, which are thought to reflect original primary compositional layering. In specimens of appropriate bulk composition, these features are observed to continue into Zone Va (see below).

Secondary muscovite occurs in coarser (200–1000 μm), isolated grains that typically vary in grain size in a single thin section. They are more euhedral, cross cut or partially replace other minerals, especially cordierite, and are randomly orientated. Secondary muscovite is commonly intergrown with biotite and late chlorite.

The above textural distinction is the only reliable way to distinguish between primary and secondary muscovite in Zone IVb and above, because chemically they are very similar (Sect. 8.64 below).

8.3.6 Muscovite + Quartz Breakdown

The boundary between Zones IV and V occurs where primary muscovite + quartz no longer coexist, and give way upgrade to the assemblage K-feldspar + Al_2SiO_5:

$$Ms + Qtz = Al_2SiO_5 + Kfs + H_2O \qquad (P3)$$

This is a degenerate univariant reaction in the model KFMASH system. In the field, the model univariance of this reaction is manifested in the abrupt change in texture and mineralogy of the pelites across the boundary, and the scarcity of the full univariant assemblage Ms + Qtz + Bt + Crd + Kfs + Al_2SiO_5 (four samples out of about 400).

In the Ballachulish Slate, Al_2SiO_5 + Kfs + Bt + Qtz + Crd hornfelses occur in resistant, massive-weathering outcrops a few tens of metres upgrade of low-weathering outcrops of knotted slates containing the IVa assemblage Ms + Qtz + Bt + And + Crd. Andalusite occurs in euhedral prisms 5–10 mm in length, sometimes up to 6 cm (Fig. 8.9). In the Appin Phyllite, Leven Schist and Creran Succession, massive pelitic hornfels with the IVb assemblage develop conspicuous prisms of andalusite for the first time in Zone V, and most specimens have lost any pre-existing micaceous schistosity.

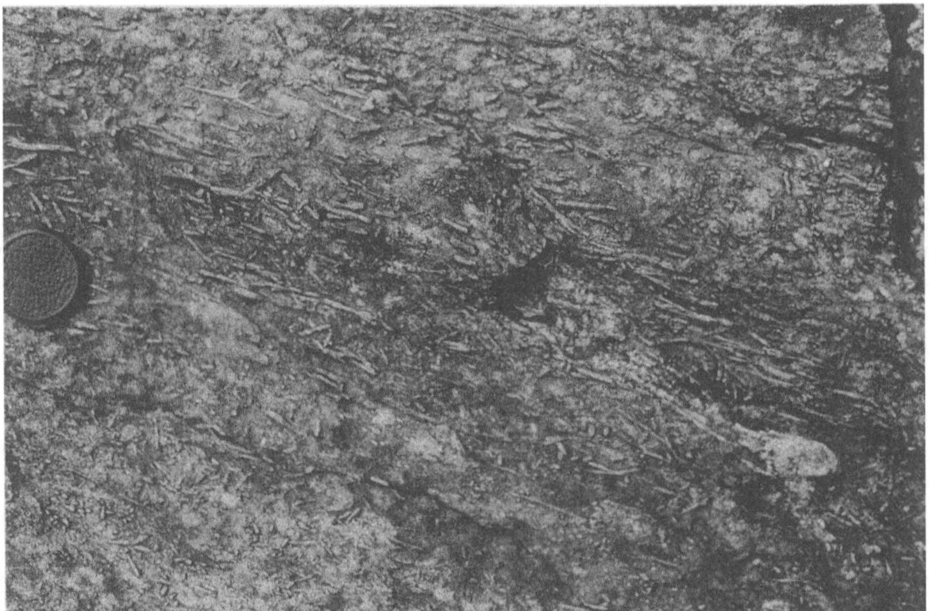

Fig. 8.9. Andalusite + K-feldspar-rich hornfels near DP620. Large (5-6 cm) andalusite prisms are developed on the surface of a massive weathering outcrop, only 200 m upgrade of the Zone IV slate in Fig. 8.7. The proliferation of And + Kfs is due to reaction (P3), Ms + Qtz = And + Kfs + H_2O

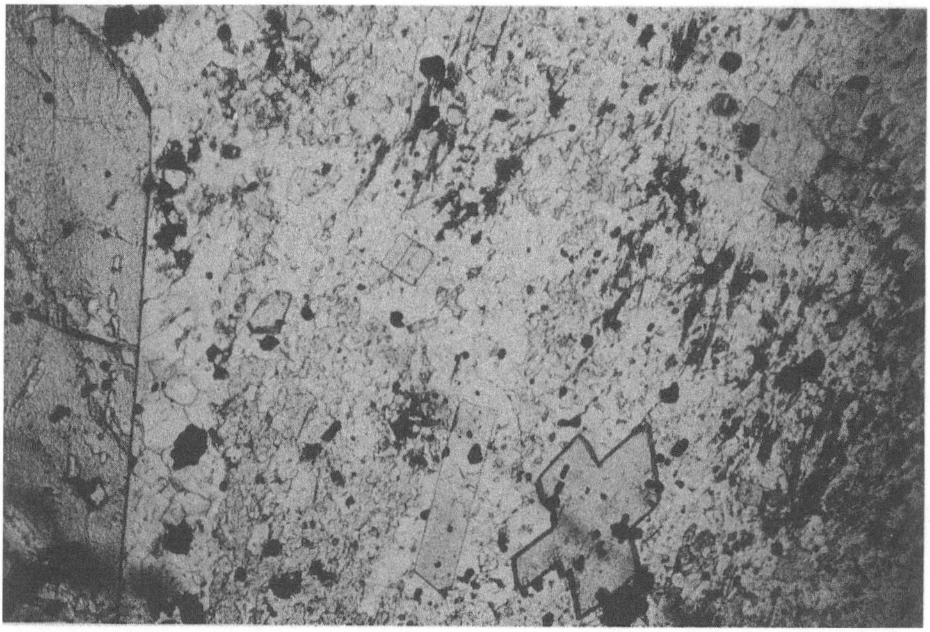

Fig. 8.10. Zone V. Appin Pelite (near DP515) showing the development of euhedral andalusite prisms in a matrix of K-feldspar, quartz, cordierite (diffuse, medium-relief poikiloblast to the *right* of the large andalusite prism), sillimanite (fibrous sprays nucleating on opaque grains), and ilmenite. Long dimension of photo is 9 mm

8.3.7 Andalusite and Sillimanite

The first development of sillimanite closely coincides with reaction (P3) (Pattison 1991). Below reaction P3, andalusite is the only Al_2SiO_5 polymorph present. Sillimanite first occurs in 0.5–2.0 mm long prisms in cordierite and fibrous sprays that nucleated on andalusite and accessory minerals (Figs. 8.10 and 8.11). The thickness of sillimanite prisms ranges from < 0.5 μm (fibrolite) to 18 μm (sillimanite) (Pattison 1991).

Upgrade of reaction (P3), most Al_2SiO_5-bearing specimens contain both polymorphs. Sillimanite becomes more abundant and coarser (up to about 50 μm) as the contact is approached. Even at these higher grades, however, most of the Al_2SiO_5 is andalusite.

There is no evidence for the *reaction* of andalusite to sillimanite, even when there is evidence for sillimanite having nucleated *on* andalusite crystals (Fig. 8.11). This suggests that sillimanite formed from the reaction of matrix minerals (Pattison 1991). The common occurrence of andalusite in rocks upgrade of the first occurrence of sillimanite is probably due to the sluggish kinetics of the andalusite-sillimanite polymorphic inversion reaction.

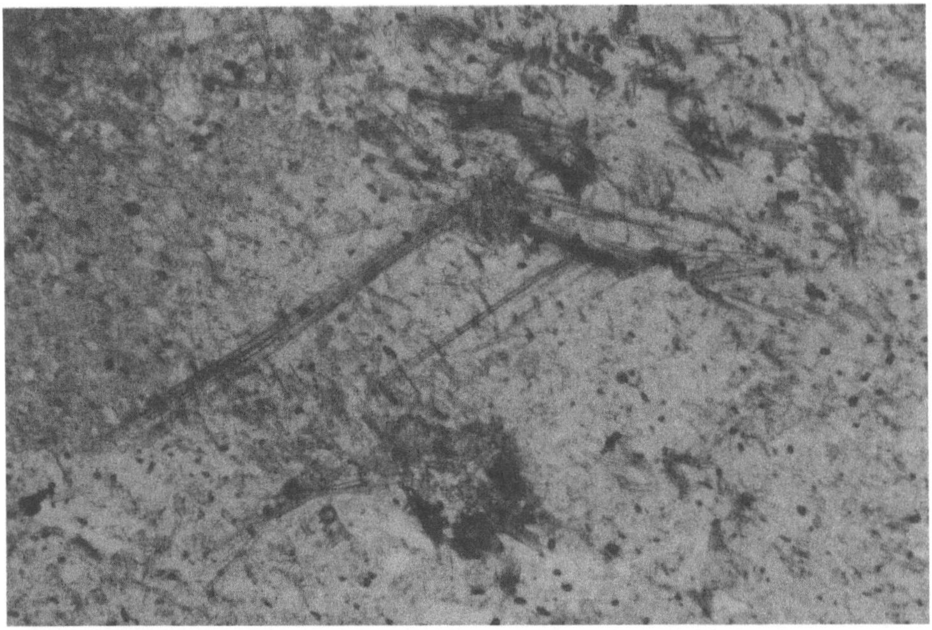

Fig. 8.11. Subzone Va. Appin Pelite (DP25a) showing the nucleation of silliminate on an andalusite prism (*top centre*) and on zircon (*above* andalusite prism). There is no evidence for the *reaction* of andalusite to silliminate. The matrix comprises cordierite, muscovite and biotite. The long dimension of photo is 5.5 mm

8.3.8 Zone V (Al_2SiO_5 + K-Feldspar Zone)

Pelites upgrade of the reaction (P3) isograd contain either muscovite or quartz, resulting in quartz-bearing or quartz-absent assemblages. Separate prograde sequences of assemblages are developed in each of these assemblages. Up to the onset of partial melting, quartz-bearing pelites maintain a common situations. In quartz-absent pelites, a lower grade subzone Va and higher grade subzone Vb may be distinguished.

8.3.8.1 Quartz-Bearing Assemblages (Zone V)

The majority of pelites in the Ballachulish aureole upgrade of reaction (3) have lost their muscovite, giving the quartz-bearing assemblage Bt + Crd + Qtz + Al_2SiO_5 + Kfs. This assemblage is characteristic of Zone V in quartz-bearing rocks. The rocks are massive and preserve no trace of schistosity. Andalusite typically occurs in prominent euhedral-subhedral prisms intergrown with quartz and sometimes biotite, surrounded by K-feldspar ± biotite (Fig. 8.10). Cordierite generally retains its textures from assemblage IVb, occurring both in prismatic crystals and ellipsoidal poikiloblastic crystals containing inclusions of quartz, biotite and K-feldspar, and sometimes silliminate (see above).

The Zone V assemblage is involved in the model divariant KFMASH reaction:

$$6\ Al_2SiO_5 + 2\ Bt + Qtz = 3\ Crd + 2\ Kfs + 0.5\ H_2O. \quad \text{(P4a)}$$

Zone V has as its lower boundary the Ms + Qtz breakdown. The upper boundary of Zone V is the igneous contact, although as described below and in Pattison and Harte (1988) and Harte et al. (Chap. 9, this Vol.), migmatitic structures and textures suggest the onset of partial melting in the upper portion of Zone V. The width of Zone V below the onset of partial melting is 100–200 m.

8.3.8.2 Quartz-Absent Assemblages; Subzone Va (Muscovite Subzone)

Subzone Va is defined by the assemblage Ms + Bt + Crd + Al_2SiO_5 + Kfs. In the field, subzone Va is characterized by light-coloured, cordierite-spotted hornfelses that contain schistose, fine-grained muscovite and small, stubby, anhedral andalusite crystals.

In thin section, andalusite occurs in two main textural associations. In about half the specimens, it occurs in 2–4 mm rectilinear, skeletal crystals and crystal aggregates substantially intergrown with K-feldspar (And:Kfs ≈ 1:1 by vol.; Fig. 9.7b); this is in contrast to andalusite in quartz-bearing Zone V assemblages, which occurs in euhedral prisms, often intergrown with quartz, and surrounded by K-feldspar (Fig. 8.10). In the second association of the Va assemblage, andalusite occurs in subhedral, relatively inclusion-free crystals, surrounded by a fringe of K-feldspar and small tabular biotite grains (Fig. 8.12).

The subzone Va assemblage is involved in the model divariant reaction:

$$9\ Ms + 3\ Crd = 2\ Bt + 15\ Al_2SiO_5 + 7\ Kfs + 8.5\ H_2O. \quad \text{(P4b)}$$

The 15:7:2 stoichiometric ratio of And:Kfs:Bt corresponds roughly to the mode of the second of the andalusite textures described above. In several specimens the presence in one thin section of both textures suggests the passage through both reaction (P3) (skeletal And + Kfs texture) and P4b (And + Kfs + Bt texture).

Subzone Va is only present as a mappable zone in very muscovite-rich pelites belonging to the Appin Phyllite lithology in the small syncline on the northeast flank, where it is about 100 m in width.

8.3.8.3 Quartz-Absent Assemblages; Subzone Vb (Corundum Subzone)

In quartz-absent assemblages, the next major mineralogical change above subzone Va is the development of corundum, which is diagnostic of subzone Vb. In the field, corundum occurs in 0.5–1-mm rounded grains that weather above the matrix, in contrast to the larger, more prismatic andalusite crystals (Fig. 8.13). The corundum grains are usually surrounded by K-feldspar.

In thin section, corundum normally occurs in 0.3–2-mm, high relief, ill-shaped, ragged crystals, intergrown with and rimmed by K-feldspar. Rarely, stubby, barrel-shaped hexagonal prisms with [001] parting are developed (Fig. 8.14). Corundum sometimes envelopes ilmenite or sulphide grains.

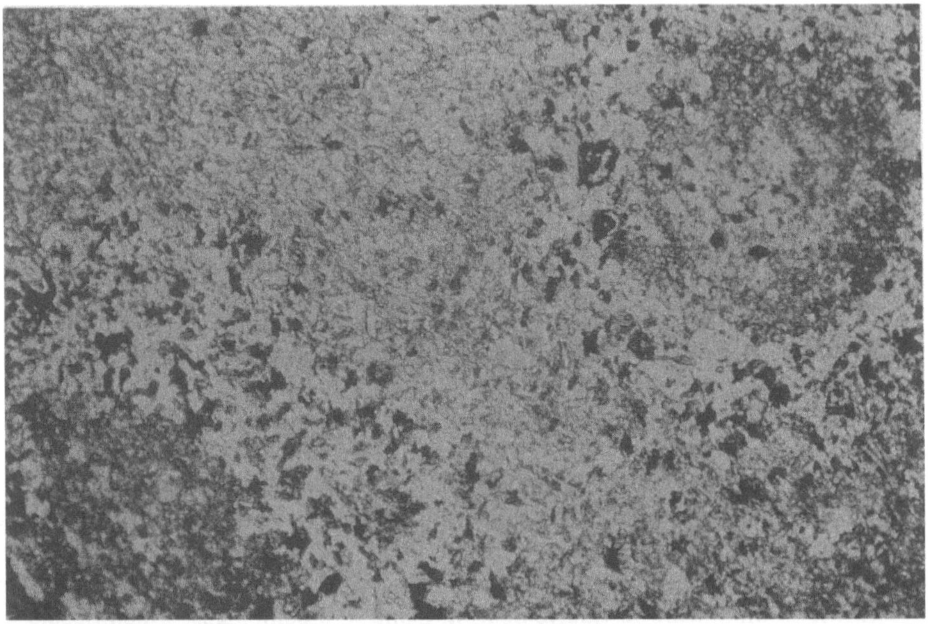

Fig. 8.12. Subzone Va. Appin Pelite (DP514) showing the development of andalusite crystals (high relief) surrounded by clear K-feldspar and biotite. The matrix consists of cordierite (*bottom right*) and abundant fine-grained, schistosity-parallel muscovite with no quartz (*centre*). Long dimension of photo is 5.5 mm

Fig. 8.13. Subzone Vb. Appin Pelite (near DP60) showing abundant small, rounded, corundum grains and an andalusite prism (*bottom left*) in a masive K-feldspar- and cordierite-rich hornfels

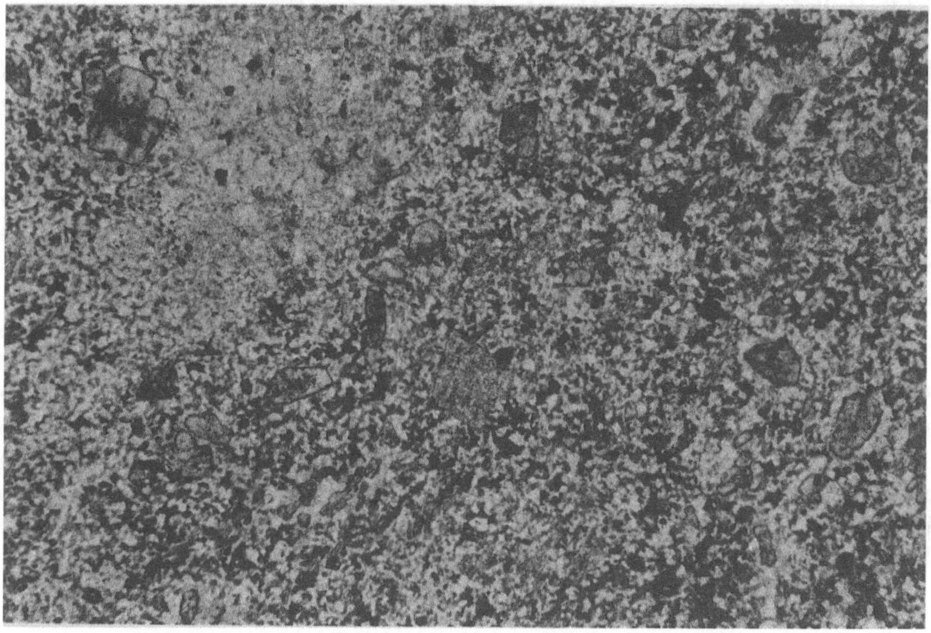

Fig. 8.14. Subzone Vb. Appin Pelite (near DP37) showing the full Subzone Vb assemblage Bt + Crd + And(+Sil) + Kfs + Crn: stubby, high-relief corundum grains, medium-relief andalusite prism (*centre*), low-relief cordierite poikiloblast (*upper left*), biotite and abundant clear K-feldspar. Long dimension of photo is 5.5 mm

In corundum-bearing assemblages, muscovite is usually absent, but in two specimens, fine-grained, schistosity parallel muscovite that is thought to be primary occurs in the assemblage Ms + Crn + Al_2SiO_5 + Kfs + Bt + Crd. This assemblage suggests the occurrence of the degenerate KFMASH univariant reaction:

$$\mathrm{Ms = Crn + Kfs + H_2O.} \qquad \text{(P5)}$$

Reaction (P5) therefore marks the boundary between subzones Va and Vb.

Above reaction (P5), subzone Vb is defined by the assemblage Bt + Crn + Crd + Al_2SiO_5 + Kfs. This assemblage is involved in the model KFMASH divariant reaction:

$$\mathrm{2\ Bt + 15\ Al_2SiO_5 = 9\ Crn + 3\ Crd + 2\ Kfs + 0.5\ H_2O.} \qquad \text{(P6)}$$

Overgrowths of And ± Bt by Crd, and the nucleation of corundum in andalusite prisms, suggest the progress of reaction (P6).

Subzone Vb, like subzone Va, is only mappable in very aluminous pelitic layers in the Appin Phyllite on the northeast flank, where it occurs at the same grade as the upper parts of quartz-bearing Zone V. Evidence for partial melting in quartz-bearing assemblages occurs at about the same distance from or slightly nearer the igneous contact than reaction (P5) (see Fig. 8.1). In all other areas in Map 2 and Fig. 8.1, the reaction (P5) isograd has been drawn to highlight the presence of corundum, whether formed by reaction (P5) or (P6).

8.3.9 Bulk Composition of Quartz-Absent Assemblages

The occurrence of quartz-absent assemblages in Subzones Va and Vb is the result of normal prograde metamorphism of aluminous pelitic rocks, rather than indicating metasomatic removal of SiO_2 at high grades to eliminate quartz. This is shown by the SiO_2:Al_2O_3 ratios of specimens DP137, DP424, DP59, DP25a and DP433 from Zones II, III, IVb, Va and Vb, respectively: 2.24, 2.38, 2.14, 2.36 and 2.36 (raw data from Pattison 1985). These data show that there is no significant decrease in SiO_2:Al_2O_3 in the high grade quartz-absent assemblages (DP59, 25a and 433) compared to the lower grade quartz-bearing assemblages (DP137 and 424).

The development of quartz-absent assemblages in the aureole is due to a prograde sequence of reactions that consumed large modal amounts of quartz and produced K-feldspar and/or cordierite, both relatively Si-rich minerals (see Table 8.2). The most important of these is reaction (P2b), which consumes quartz and muscovite in the ratio 3:1. In some muscovite-rich pelites, muscovite outlasted quartz during passage through reactions (P2b) and (P3), allowing the development of subzone Va and Vb assemblages. This is further confirmed by the mineral chemistry of the different zones (see below).

8.4 High-Grade Assemblages in the Zone of Partial Melting

In this section, we describe the high-grade mineral assemblages in the subzone of partial melting, which form the highest grade rocks of Zone V. The occurrence of partial melting is inferred from migmatitic textures and structures in the rocks (Pattison and Harte 1988; Harte et al., Chap. 9, this Vol.), rather than from any sudden change in mineral assemblages. The subzone of partial melting varies in width from <50 m (northeast) to 350–400 m (west flank).

8.4.1 Reaction at the Beginning of Anatexis

Pattison and Harte (1988) and Harte et al. (Chap. 9, this Vol.) argue that migmatization and associated phenomena were due primarily to anatexis. Initial melting reactions at the migmatite boundary were of the form:

$$\text{Kfs} + \text{Qtz} + H_2O \pm \text{Bt} \pm \text{Pl} = \text{L} \pm \text{Crd} \pm Al_2SiO_5$$

This generalized reaction emphasizes the essential participation of Kfs, Qtz and H_2O to produce liquid, and the selective participation of Bt and Pl, depending on their presence or absence in the protolith (Pattison and Harte 1988). Cordierite and andalusite may be products of melting, depending on the exact reaction. The approximate position of this melting boundary, as indicated by textures and structures, is shown on Maps 1 and 2 and Fig. 8.1.

A major consequence of vapour-consuming melting is that all free hydrous vapour will have been consumed at the anatectic boundary. In place of the free vapour phase present downgrade of the partial melting boundary, a silicate melt phase will have been present in quartz-bearing assemblages. In contrast, however, quartz-absent assemblages may not have melted until substantially higher temperatures were reached, (e.g., Johannes 1983 Fig. 7), so that although quartz-absent reactions occurred within the zone of partial melting, they may have released free vapour rather than producing a melt phase. To acknowledge this possibility, 'L or H_2O' is written in the quartz-absent reactions below. Free vapour released by reactions in quartz-absent layers may have provided a local flux for vapour-consuming partial melting in interbedded quartz-bearing layers (see more detailed discussion in Pattison and Harte 1988).

8.4.2 Quartz-Bearing Assemblages

Throughout most of the zone of anatexis, quartz-bearing assemblages are the same as in subsolidus Zone V: Bt + Crd + Kfs + Qtz + Ilm ± Pl in semipelitic rocks, and Bt + Crd + Kfs + Qtz + Al_2SiO_5 + Ilm ± Pl in pelitic rocks. Garnet-bearing assemblages occur within two coherent pelitic screens surrounded by the igneous complex (SW2 and DP568) and along the southern contact near DP433: Bt + Crd + Qtz + Grt + Kfs + Ilm ± Pl (Fig. 8.1). The garnet occurs in fresh, red, 1–3-mm crystals that show a rectilinear parting. In thin section, the garnet is euhedral and unaltered, and contains inclusions of quartz, ilmenite, cordierite and biotite (Fig. 8.15).

The development of garnet upgrade of the Zone V assemblage Bt + Crd + Qtz + Kfs + Al_2SiO_5 suggests the operation of the model KFMASH divariant reaction:

Bt + Crd + Qtz = Grt + Kfs + L. (P11)

The stoichiometry of this reaction is uncertain because of the uncertain composition of the melt.

The highest grade quartz-bearing specimen in the aureole (DP568) contains hypersthene in the assemblage Bt + Crd + Qtz + Grt + Hy + Ilm + Pl. The hypersthene occurs in 0.5–2-mm, subhedral, strongly pleochroic, well-cleaved crystals containing inclusions of quartz and ilmenite (Fig. 8.16).

The occurrence of hypersthene in the above assemblage suggests the progress of the model KFMASH univariant reaction

Bt + Grt + Qtz = Crd + Hy + Kfs + L. (P12)

In DP568, biotite occurs in large, poikiloblastic, cross-cutting flakes that do not appear to be in textural equilibrium with rest of the assemblage (Fig. 8.16), and may therefore be of late-stage origin (Sect. 8.7.1 below). The sporadic occurrence of the garnet- and hypersthene-bearing assemblages do not allow any separate mineral zones to be mapped in the migmatite zone.

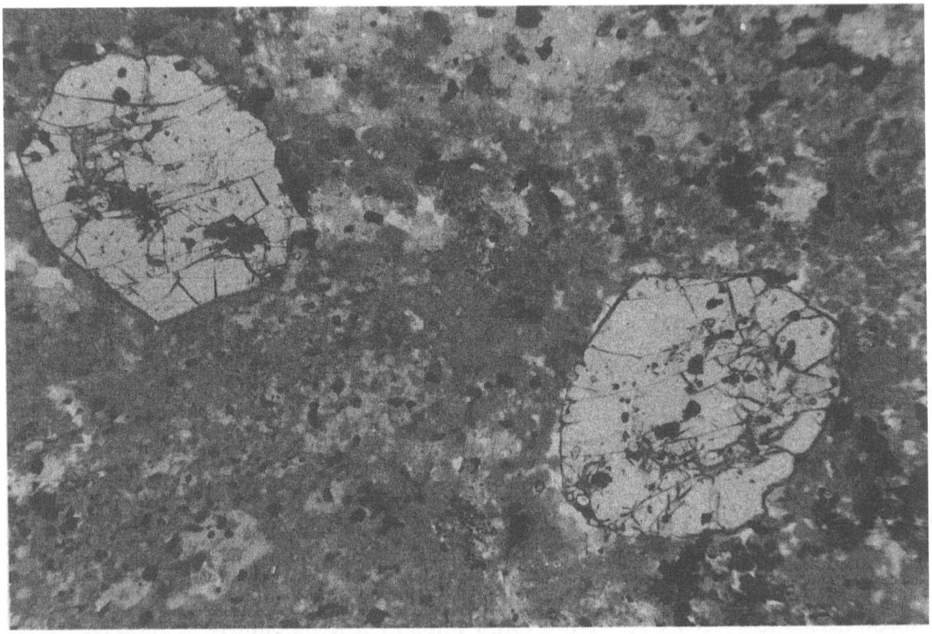

Fig. 8.15. Zone of partial melting, pelitic screen, DP567. Fresh euhedral garnets, containing inclusions of biotite, cordierite, ilmenite and quartz in a matrix of altered cordierite (dirty grey), plagioclase, quartz and ilmenite. Long dimension of photo is 6 mm

8.4.3 Quartz-Absent Assemblages

At the low-grade margin of the melting zone, quartz-absent assemblages are the same as in Zone Vb (Bt + Al_2SiO_5 + Crd + Crn + Kfs). In a number of specimens closer to the igneous contact, the assemblage Crn + Crd + Kfs + Ilm ± Pl with *either* Bt or Al_2SiO_5 is common. These assemblages suggest that the divariant reaction

$$\text{Bt} + Al_2SiO_5 = \text{Crd} + \text{Crn} + \text{Kfs} + \text{L or } H_2O, \qquad \text{(P6)}$$

proceeded until one of Bt or Al_2SiO_5 was consumed.

Along the northeast and southern contacts, and within the more northerly of the two pelitic screens (SW2), pleonaste spinel is found sporadically in quartz-absent assemblages. Along the southern contact, it is present in the assemblage Spl + Bt + Crn + Al_2SiO_5 + Crd + Kfs + Ilm ± Pl. It typically occurs in 50–300 μm, equidimensional, deep green, isotropic grains, commonly occuring adjacent to corundum (Figs. 8.17, 8.18), or within the centres of cordierite crystals and crystal aggregates. The development of spinel in the above assemblage suggests the KFMASH univariant reaction:

$$\text{Bt} + Al_2SiO_5 = \text{Spl} + \text{Crn} + \text{Crd} + \text{Kfs} + \text{L or } H_2O \qquad \text{(P8)}$$

In a number of specimens, typically in the more northerly pelitic screen, spinel is developed in the assemblages Bt + Crn + Kfs + Crd + Spl + Ilm ± Pl and Al_2SiO_5 + Crn + Crd + Spl + Ilm ± Pl. In these assemblages either of Bt or Al_2SiO_5 may have

Fig. 8.16. Zone of partial melting, pelitic screen, DP568. High-relief, blocky, subhedral crystals of hypersthene with well-developed cleavage, in a matrix of quartz, plagioclase, cordierite and ilmenite. Note the large poikiloblastic biotite crystals that overgrow ilmenite and quartz, which therefore appear texturally to be of late-stage origin. Long dimension of photo is 6 mm

been consumed during passage through reaction (P8). The Bt-bearing assemblage is involved in the model divariant reaction

$$\text{Bt} + \text{Crn} + \text{Crd} = \text{Spl} + \text{Kfs} + \text{L or } H_2O \qquad (P7)$$

Overall, spinel is more abundant in quartz-absent assemblages than garnet or hypersthene is in quartz-absent assemblages. Spinel has been found in quartz-absent assemblages in the country rocks on the south west, south and east margins of the igneous complex. With the exception of one garnet-bearing locality on the south flank, garnet and hypersthene have only been found in pelitic screens within the igneous complex. These observations suggest that the Spl-producing reactions (P7) and (P8) occur at lower temperatures than Grt- and Grt + Hy-producing reactions (P11) and (P12).

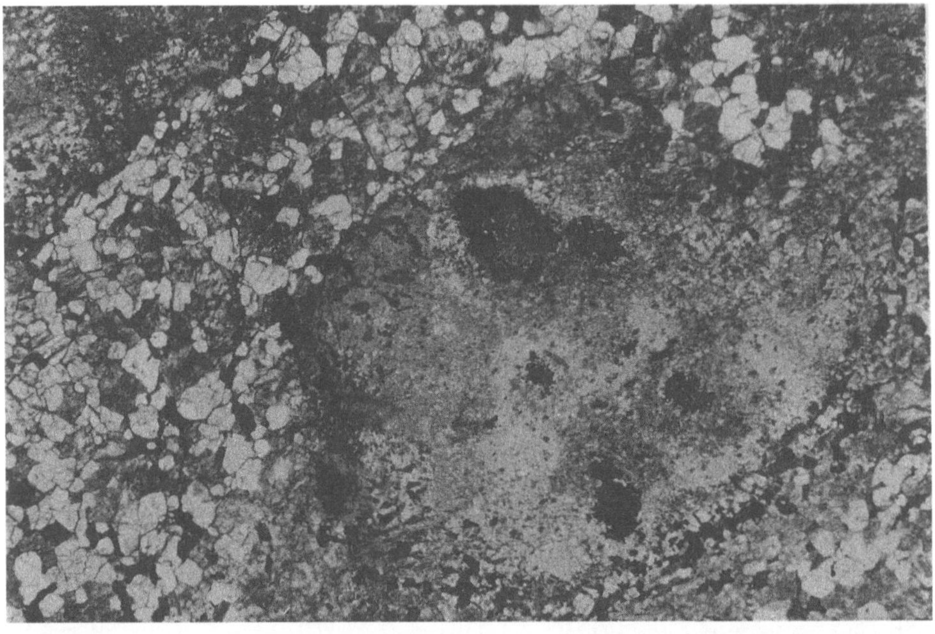

Fig. 8.17. Zone of partial melting, pelitic screen, DP608-1; quartz-absent fragment surrounded by Type D leucosome (terminology as in Pattison and Harte 1988; and Harte et al., Chap. 9, this Vol.). Within the fragment is spinel (black aggregates) that nucleated on corundum (high-relief mottled crystals) in a matrix of cordierite, K-feldspar, biotite and ilmenite. Long dimension of photo is 15 mm

8.5 Petrogenetic Grids for the Mineral Assemblages and Reactions

8.5.1 Schematic Petrogenetic Grid for Metamorphic Zones Below the Onset of Partial Melting

The subsolidus mineral zones and model KFMASH reactions are summarized in Table 8.2. All of the model reactions involve combinations of the minerals Qtz, Chl, Ms, Bt, Crd, Al_2SiO_5, Crn, Kfs and inferred hydrous vapour. Schreinemakers' analysis of these phases in the model six-component KFMASH system is summarized in Table 8.3. The exclusions of Qtz + Crn and Chl + Crn are inferred from a wide range of assemblages, and imply that co-existence of eight phases could only occur at a corundum-absent invariant point, which is considered to be metastable (Pattison and Harte 1985). This means that there are no stable KFMASH invariant points. Thus, *for the minerals considered*, all KFMASH univariant reactions must terminate at KFASH and KMASH invariant points. The numbered reactions in Table 8.3 refer to the reactions deduced from the assemblages and textures in the Ballachulish aureole. It can be seen that there is excellent agreement between the theoretically predicted range of reactions and those inferred above from the mineral assemblages and textures.

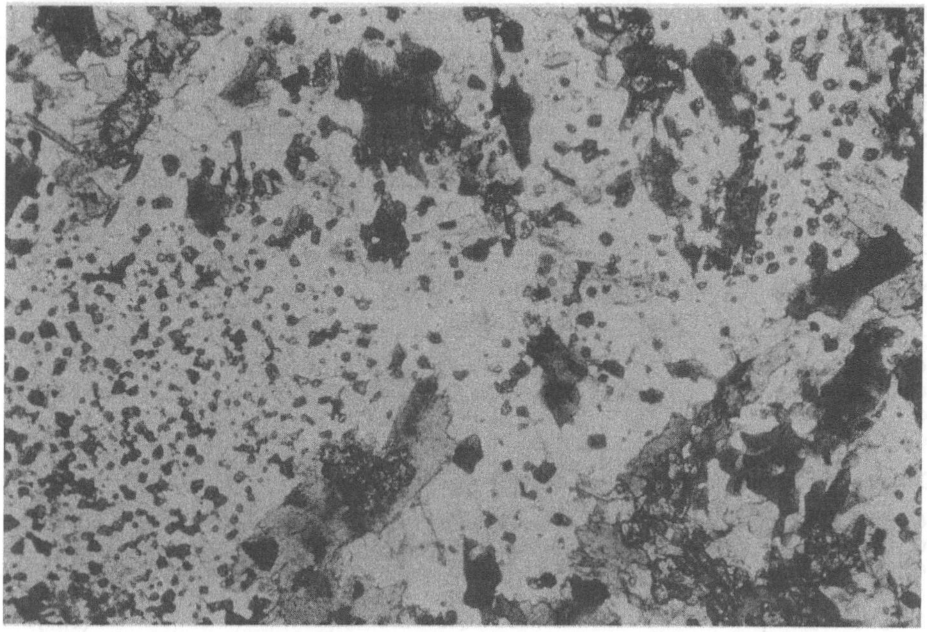

Fig. 8.18. Quartz-absent xenolith, near DP451. Small, isolated, dark grey grains of spinel (*bottom left, top right*), and high relief corundum grains in a matrix of K-feldspar, plagioclase and biotite. Long dimension of photo is 2.3 mm

Figure 8.19 illustrates a schematic Schreinemakers' grid for the reactions in the end member KMASH or KFASH systems. Figure 8.20 shows the reactions in the full KFMASH system; invariant points in KMASH become univariant curves in KFMASH, univariant curves become divariant fields, etc. These figures include high-grade reactions in the melting zone, as discussed below. Figures 8.19 and 8.20 have been orientated schematically in P-T space to be consistent with the experimental data listed in Pattison (1989) and Pattison and Harte (1985).

The reactions tabulated in Table 8.3 occur downgrade of the melting boundary. The ornaments on the reaction boundaries in Fig. 8.20 correspond to those in Fig. 8.1. The isobaric trajectory in Fig. 8.20 is consistent with the sequence of metamorphic zones in the aureole.

From Fig. 8.20, it is apparent that reactions (P2a) and (P2b), and (P4a) and (P4b), overlap in P-T space. Assuming constant pressure, a schematic T-X(Fe-Mg) diagram has been constructed in Fig. 8.21. It can be seen that above Zone III, either reaction (P2a) or (P2b) can be entered, depending on the Mg/(Mg + Fe) ratios of coexisting cordierite and biotite (assuming fixed activity of all other components). Above reaction (P3), the presence or absence of quartz determines whether reaction (P4a) or (P4b) is entered. Tests of the applicability of this diagram to the assemblages are provided below and in Pattison (Chap. 16, this Vol.).

Table 8.3. Summary of KFMASH reactions below the onset of partial melting

Phases:	Ms Chl Qtz Bt Crd Crn Al_2SiO_5 Kfs V = 9
Components:	K_2O-FeO-MgO-Al_2O_3-SiO_2-H_2O (KFMASH) = 6
Exclusions:	Qtz-Crn, Chl-Crn
Other Restrictions:	V assumed present in all assemblages
	Qtz and Ms present in all assemblages below reaction P3
	Kfs present in all assemblages above reaction P3
Invariant Points:	O[a]
Univariant curves:	(Invariant pts in KFASH or KMASH): 4[b]

Name	Reaction		Reaction No.
[Crn,Al_2SiO_5]	Ms + Chl + Qtz + Kfs = Bt + Crd + V		
[Crn,Kfs]	Ms + Chl + Qtz = Crd + Bt + Al_2SiO_5 + V		(P1b)
[Crn,Chl]	Ms + Qtz = Al_2SiO_5 + Kfs + V	Degenerate	(P3)
[Qtz,Chl]	Ms = Crn + Kfs + V	Degenerate	(P5)
Divariant curves (univariant curves in KFASH or KMASH): 13[c]			
[Crn,Kfs,Al_2SiO_5]	Ms + Chl + Qtz = Crd + Bt + V		(P1)
[Crn,Kfs,Chl]	Ms + Crd = Bt + Al_2SiO_5 + Qtz + V		(P2a)
[Crn,Kfs,Ms,Bt]	Chl + Al_2SiO_5 + Qtz = Crd + V		
[Crn,Kfs,Crd]	Ms + Chl = Bt + Al_2SiO_5 + Qtz + V		(P1a)
[Crn,Al_2SiO_5,Chl]	Ms + Bt + Qtz = Crd + Kfs + V		(P2b)
[Crn,Al_2SiO_5,Ms]	Chl + Kfs = Bt + Crd + Qtz + V		
[Crn,Al_2SiO_5,Bt]	Ms + Chl + Qtz = Crd + Kfs + V		
[Crn,Al_2SiO_5,Crd]	Chl + Kfs = Ms + Bt + Qtz + V		
[Crn,Chl,Bt,Crd]	Ms + Qtz = Al_2SiO_5 + Kfs + V		(P3)
[Crn,Chl,Ms]	Bt + Qtz + Al_2SiO_5 = Crd + Kfs + V		(P4a)
[Crn,Chl,Qtz]	Ms + Crd = Bt + Al_2SiO_5 + Kfs + V		(P4b)
[Qtz,Chl,Al_2SiO_5,Crd,Bt]	Ms = Crn + Kfs + V		(P5)
[Qtz,Chl,Ms]	Bt + Al_2SiO_5 = Crn + Crd + Kfs + V		(P6)

KFMASH univariant curves (invariant pts in KFASH or KMASH) are designated by the phases absent. Reaction curves are numbered as in the figures. V refers to hydrous vapour.

[a]Given only the above exclusions, it is possible that one invariant point is stable, [Crn], but this is presumed to be metastable (see Pattison and Harte 1985).
[b]This assumes that muscovite is pure $KAl_3Si_3O_{10}(OH)_2$, which makes the [Crn,Crd], [Crn,Bt] and [Crn,Chl] unvariant reaction degenerate and identical. The [Crn,Ms] and [Crn,Qtz] reactions are absent for the same reason as the [Crn] invariant point.
[c]With the exception of (P1a), the numbered reactions are seen in the Ballachulish aureole. Only cordierite bearing reactions are considered (see text). Chl-Al_2SiO_5 and Chl-Kfs have not been observed, and cordierite always coexists with biotite.

8.5.2 A Petrogenetic Grid for High-Grade Assemblages

The limited number of high grade assemblages in the Ballachulish aureole makes it difficult to construct a Schreinemakers' net in a similar manner to the lower grade zones. Therefore, high-grade assemblages containing the same collection of minerals as at Ballachulish were examined from several other aureoles, including the Comrie Aureole (Tilley 1924), the Lochnagar aureole (Chinner 1962) and the Belhelvie aureole (Stewart 1946; see Pattison and Harte 1985). All assemblages involve combinations of Qtz, Crd, Bt, Al_2SiO_5, Crn, Hy, Grt, Spl, Kfs and inferred melt.

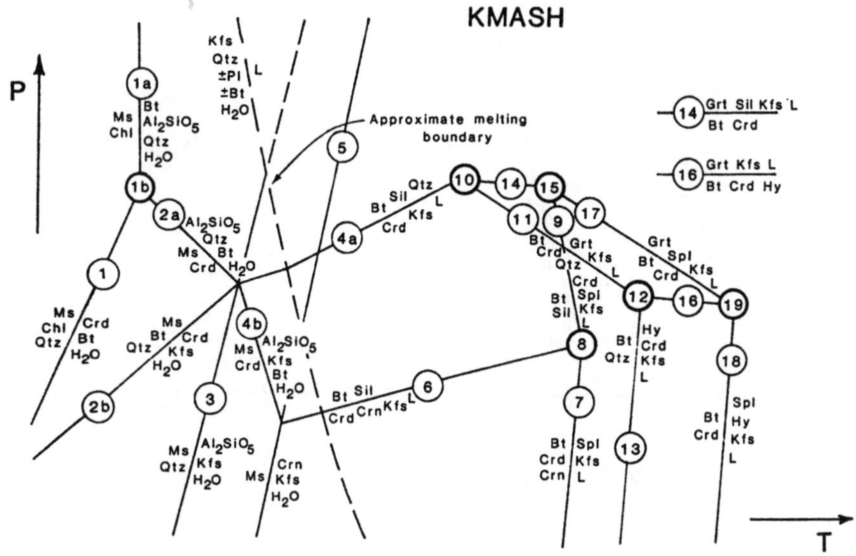

Fig. 8.19. Schematic KMASH (or KFASH) petrogenetic grid (not to scale) for reactions in the Ballachulish aureole. Subsolidus reactions from Table 8.3 are to the *left* of the approximate melting boundary (represented by Kfs + Qtz ± Pl ± Bt + H_2O = L). Reactions from Table 8.4 are to the *right* of the melting boundary. With the beginning of melting, reactions may generate melt (L) rather than H_2O, and most assemblages cease to be saturated in H_2O (see text for further discussion)

Table 8.4 lists the possible invariant points, univariant reactions and divariant reactions involving these minerals in the model system KFMASH, assuming that Bt + Crd + Kfs + melt were present in all assemblages and that the following mineral pairs were unstable: Qtz + Crn, Hy + Al_2SiO_5, Hy + Crn, Grt + Crn and Spl + Qtz (Pattison and Harte 1985). These exclusions appear to hold for the above contact aureoles, although it is recognized that associations such as Hy + Al_2SiO_5, Spl + Qtz and Grt + Crn are well known in higher pressure settings (e.g., Droop 1989; Hensen and Harley 1990).

The resultant Schreinemakers' nets in KMASH (or KFASH) and KFMASH are illustrated in the higher temperature parts of Figs. 8.19 and 8.20, respectively (i.e. to the right (upgrade) of the dashed vapour-saturated solidus curve, 'approximate melting boundary'). The topology of Figs. 8.19 and 8.20, and the balancing of reactions P8, P12 and P15 in these figures and in Table 8.4, represent corrections from Figs. 9 and 10 of Pattison and Harte (1985). Hensen and Harley (1990) independently arrived at the same topology for the reactions, strengthening the interpretation here.

The high-grade part of the Schreinemakers' nets have been orientated in P-T space to be consistent with observed assemblages, limited experimental data and with the following general volume and entropy considerations: biotite is likely to be on the low temperature side of the reaction, melt on the high temperature side, cordierite on the low pressure side, and garnet on the high pressure side (Pattison and Harte 1985). Reaction (P8), Bt + Al_2SiO_5 = Crd + Crn + Spl + Kfs + L, has been located on low temperature side of reaction, P12, Bt + Grt + Qtz = Crd + Hy + Kfs + L, because Spl-bearing quartz-absent assemblages are more widely distributed in the country

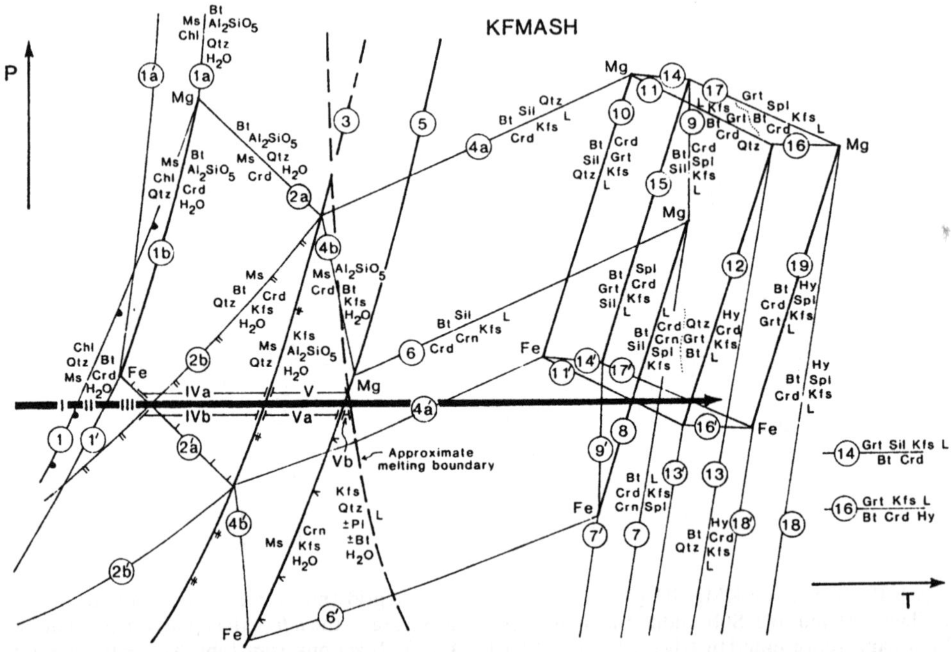

Fig. 8.20. Schematic KFMASH petrogenetic grid (not to scale) showing the correlation between mineral zones and model reactions in the Ballachulish aureole. This diagram corresponds to the grid in Fig. 8.19 for the end member KMASH or KFASH subsystems. Zones and ornaments correspond to those in Fig. 8.1. Fe-Mg divariant reactions are labelled on the bounding Mg-end member curves. 18 and 18' refer to the Mg-rich and Fe-rich bounding curves for reaction 18, respectively. Subsolidus reactions from Table 8.3 are to the *left* of the dashed vapour-saturated solidus curve Kfs + Qtz + H_2O ± Pl ± Bt = L; reactions in the zone of partial melting from Table 8.4 are to the *right* of the solidus. Univariant *curves* have been drawn parallel to each other for simplicity, although some may intersect at higher pressures (e.g., Grant 1985). The *arrow* represents the isobaric trajectory at Ballachulish

rocks than Grt- or Grt + Hy-bearing quartz-present assemblages (Sect. 8.4.3). This therefore constrains reaction P15, Bt + Al_2SiO_5 + Crd = Spl + Grt + Kfs + L, to be on the low temperature side of reaction P12 (Fig. 8.20).

The bold arrow in Fig. 8.20 is an isobaric trajectory consistent with the progression of assemblages of Ballachulish. Other aureoles contain assemblages characteristic of higher and lower pressure trajectories within the same grid. For example, abundant Grt + Sil-bearing assemblages in the Belhelvie aureole suggest the occurrence of reactions (P10) and (P15) and a higher pressure trajectory than at Ballachulish. On the other hand, in the Comrie aureole, the complete absence of garnetiferous assemblages and extensive development of Crd + Crn + Spl and Crd + Hy + Spl assemblages suggest the predominant involvement of reactions (P13) and (P18), and therefore lower pressures (Pattison and Harte 1985).

It should be noted that in the high-grade portion of the grids in Figs. 8.19 and 8.20, that a(H_2O) will have been internally buffered to low values by melting reactions [see Pattison, Chap. 16, this Vol., for quantitative estimates of a(H_2O)]. This is in contrast to the uniform and high a(H_2O) assumed in the subsolidus portion of Figs. 8.19 and 8.20.

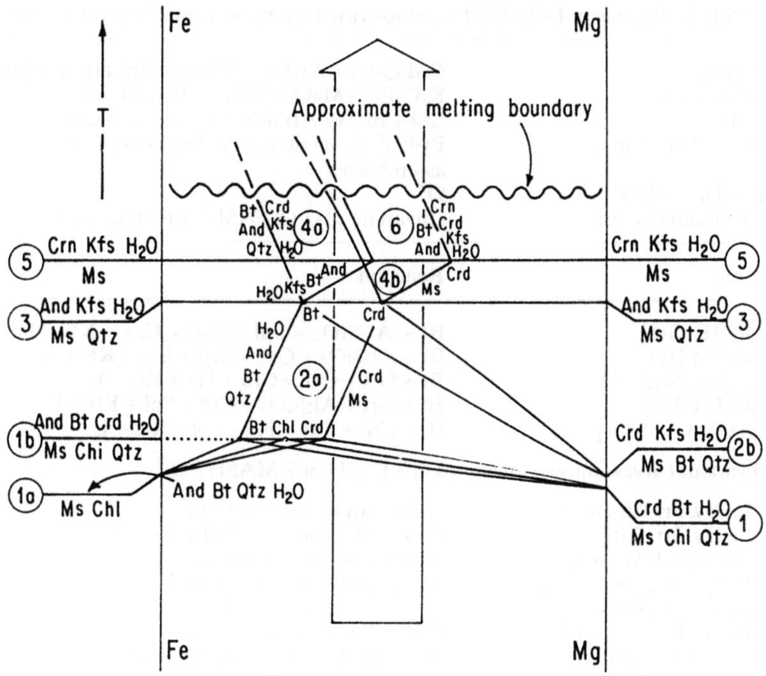

Fig. 8.21. Schematic T-X(Fe-Mg) diagram, corresponding to the isobaric trajectory in Fig. 8.20. The broad *arrow* encompasses the bulk of the Mg/(Mg + Fe) compositional variation in the aureole

8.5.3 Linking of the Low-Grade and High-Grade Grids

Separate Schreinemakers' analysis was made for assemblages below the onset of anatexis (Table 8.3) and for assemblages involving melt (Table 8.4). The lowest grade at which migmatitic and textural features suggest the occurrence of partial melting is where pelitic rocks contain the Zone V assemblages Bt + Al_2SiO_5 + Qtz + Crd + Kfs (quartz-present) or Bt + Al_2SiO_5 + Crd + Crn + Kfs (quartz-absent). These assemblages undergo the KFMASH divariant reactions:

$$\text{Bt} + \text{Al}_2\text{SiO}_5 + \text{Qtz} = \text{Crd} + \text{Kfs} + \text{H}_2\text{O or L} \qquad \text{(P4a)}$$

$$\text{Bt} + \text{Al}_2\text{SiO}_5 = \text{Crd} + \text{Crn} + \text{Kfs} + \text{H}_2\text{O or L} \qquad \text{(P6)}$$

At the beginning of melting it is inferred that melt (L) replaces H_2O on the product side of these reactions, although this is less certain for quartz-absent reactions (see above). Strictly speaking, this change from H_2O-bearing to L-bearing assemblages involves an invariant point on the KFASH and KMASH univariant curves for reactions (P4a) and (P6) in Figs. 8.19 and 8.20 (e.g., Abbott and Clarke 1979). However, given that limited H_2O availability (Pattison and Harte 1988) precludes substantial co-existence of H_2O 'vapour' with melt, only the vapour-absent melting curves are relevant above the anatectic boundary. Thus we have not shown reaction

Table 8.4. Summary of KFMASH reactions for high-grade assemblages in the zone of partial melting

Phases:	Crd Qtz Bt Al_2SiO_5 Kfs Crn Spl Hy Grt (melt) = 10	
Components:	K_2O-FeO-MgO-Al_2O_3-SiO_2-H_2O = 6	
Exclusions:	Qtz-Crn, Hy-Al_2SiO_5, Grt-Crn, Spl-Qtz[a]	
Other restrictions:	Bt-Crd-Kfs-melt assumed present in all assemblages[b]	
Invariant points:	O[a]	
Univariant curves:	(Invariant pts in KFASH or KMASH): 5[a]	
Name	**Reaction**	**Reaction No.**
[Qtz,Hy,Grt]	Bt + Al_2SiO_5 = Spl + Crd + Crn + Kfs + L	(P8)
[Spl,Crn,Hy]	Bt + Al_2SiO_5 + Qtz = Crd + Grt + Kfs + L	(P10)
[Spl,Crn,Al_2SiO_5]	Bt + Grt + Qtz = Crd + Hy + Kfs + L	(P12)
[Qtz,Crn,Hy]	Bt + Grt + Al_2SiO_5 = Crd + Spl + Kfs + L	(P15)
[Qtz,Crn,Al_2SiO_5]	Bt + Crd + Grt = Hy + Spl + Kfs + L	(P19)
Divariant curves (univariant curves in KFASH or KMASH): 20[b]		
[Qtz,Hy,Grt,Kfs,Bt,L]	Crd + Crn = Al_2SiO_5 + Spl	
[Qtz,Hy,Grt,Crd]	Bt + Al_2SiO_5 + Crn = Spl + Kfs + L	
[Qtz,Hy,Grt,Al_2SiO_5]	Bt + Crd + Crn = Spl + Kfs + L	(P7)
[Qtz,Hy,Grt,Crn]	Bt + Al_2SiO_5 = Crd + Spl + Kfs + L	(P9)
[Qtz,Hy,Grt,Spl]	Bt + Al_2SiO_5 = Crd + Crn + Kfs + L	(P6)
[Spl,Crn,Hy,Kfs,Bt,L]	Crd = Grt + Al_2SiO_5 + Qtz	
[Spl,Crn,Hy,Crd]	Bt + Al_2SiO_5 + Qtz = Grt + Kfs + L	
[Spl,Crn,Hy,Al_2SiO_5]	Bt + Crd + Qtz = Grt + Kfs + L	(P11)
[Spl,Crn,Hy,Qtz]	Bt + Crd = Grt + Al_2SiO_5 + Kfs + L	(P14)
[Spl,Crn,Hy,Grt]	Bt + Al_2SiO_5 + Qtz = Crd + Kfs + L	(P4a)
[Spl,Crn,Al_2SiO_5,Kfs,Bt,L]	Grt + Qtz = Hy + Crd	
[Spl,Crn,Al_2SiO_5,Crd]	Bt + Qtz = Grt + Hy + Kfs + L	
[Spl,Crn,Al_2SiO_5,Qtz]	Bt + Crd + Hy = Grt + Kfs + L	(P16)
[Spl,Crn,Al_2SiO_5,Grt]	Bt + Qtz = Crd + Hy + Kfs + L	(P13)
[Qtz,Crn,Hy,Kfs,Bt,L]	Crd + Spl = Grt + Al_2SiO_5	
[Qtz,Crn,Hy,Crd]	Bt + Al_2SiO_5 = Grt + Spl + Kfs + L	
[Qtz,Crn,Hy,Al_2SiO_5]	Bt + Crd = Grt + Spl + Kfs + L	(P17)
[Qtz,Crn,Al_2SiO_5,Kfs,Bt,L]	Grt = Hy + Spl + Crd	
[Qtz,Crn,Al_2SiO_5,Crd]	Bt + Grt = Spl + Hy + Kfs + L	
[Qtz,Crn,Al_2SiO_5,Grt]	Bt + Crd = Spl + Hy + Kfs + L	(P18)

[a]Spinel and quartz do not coexist in assemblages from the aureoles listed in the text. At higher pressures, the only one of these four exclusions that continues to apply is Qtz-Crn (Hensen and Harley 1990)
[b]If all assemblages contain Bt-Kfs-L, then there are 15 divariant curves. If all assemblages contain Bt-Kfs-Crd-L, then there are 10 divariant curves (numbered).

curves involving vapour in the region of partial melting and have omitted the invariant points for simplicity in Figs. 8.19 and 8.20. From the viewpoint of the natural assemblages, this treatment is justified by the continuity of the assemblages across the beginning of melting, and the fact that the parageneses in the melting region can be well modelled with the above assumptions. Thus the (P4a) and (P6) reaction curves in Figs. 8.19 and 8.20 may be seen as showing a transition from vapour-bearing assemblages to melt-bearing assemblages, the latter stable above the approximate melting curve shown.

The combined grids in Figs. 8.19 and 8.20 provide a single continuous grid for *all* observed pelitic assemblages from the lowest to highest grade in the aureole.

8.6 Mineral Chemistry of Metamorphic Zones Below the Onset of Anatexis

In this section, some of the major chemical variations of pelitic minerals in the subsolidus metamorphic zones are discussed. Full mineral assemblages and numerical chemical data are listed in Tables 2 and 3 of Pattison (1987), along with details of sample selection and microprobe operating conditions.

8.6.1 Mg/(Mg + Fe) Variations

In Fig. 8.20, the divariant reaction bands, bounded by univariant KMASH and KFASH reactions, carry implications for variations in mineral Mg/(Mg + Fe) with grade. These are more explicitly portrayed in the T-X(Fe-Mg) diagram in Fig. 8.21. Using Fig. 8.21, it should be possible to predict Mg/(Mg + Fe) variations of coexisting minerals in the different zones. For example, in Zone IV the Mg/(Mg + Fe) ratios of minerals with assemblage IVa [reaction (P2a)] should all be more Fe-rich than those with assemblage IVb [reaction (P2b)], and they should converge upgrade at reaction (P3) where a common assemblage is formed. The same applies to Zone V with the assemblages of reactions (P4a) and (P4b), except that their convergence of mineral compositions at reaction (P3) is on the low-grade side of Zone V. Mg/(Mg + Fe) ratios of assemblages IVb and Va [reactions (P2b) and (P4b)] should all be more Mg-rich than assemblages on reaction (P3); similarly, the assemblages IVa and V [reactions (P2a) and P4a)] should be more Fe-rich than assemblages on reaction (P3).

Figure 8.22 illustrates in a series of AFM-type diagrams the compositions of coexisting cordierite, biotite ± chlorite seen in the different contact metamorphic zones, together with regional chlorite, biotite and garnet in Zones I and Ia (see Pattison 1987, for a description of the various projections used). In Zone assemblages II and Zone III assemblages carrying retrograde chlorite, chlorite plots to the Fe-rich side of the Crd-Bt tie line (Fig. 8.22b). As reaction P1 proceeds, the Crd + Chl + Bt triangle moves to more Fe-rich compositions, which is supported by a weak but consistent Mg/(Mg + Fe) zonation in cordierite in DP269 and DP377 to slightly more Fe-rich rim compositions.

In the IVb assemblage of Zone IV, all specimens with the exception of DP78 have Mg/(Mg + Fe) ratios lower than or equal to those of the reaction (P3) univariant assemblage Ms + Qtz + Bt + Crd + Al_2SiO_5 + Kfs, which is consistent with Fig. 8.21. In the IVa assemblage of Zone IV, however, the Mg/(Mg + Fe) ratios of Crd and Bt are *more* Mg-rich than the univariant assemblage, which contradicts the predicted

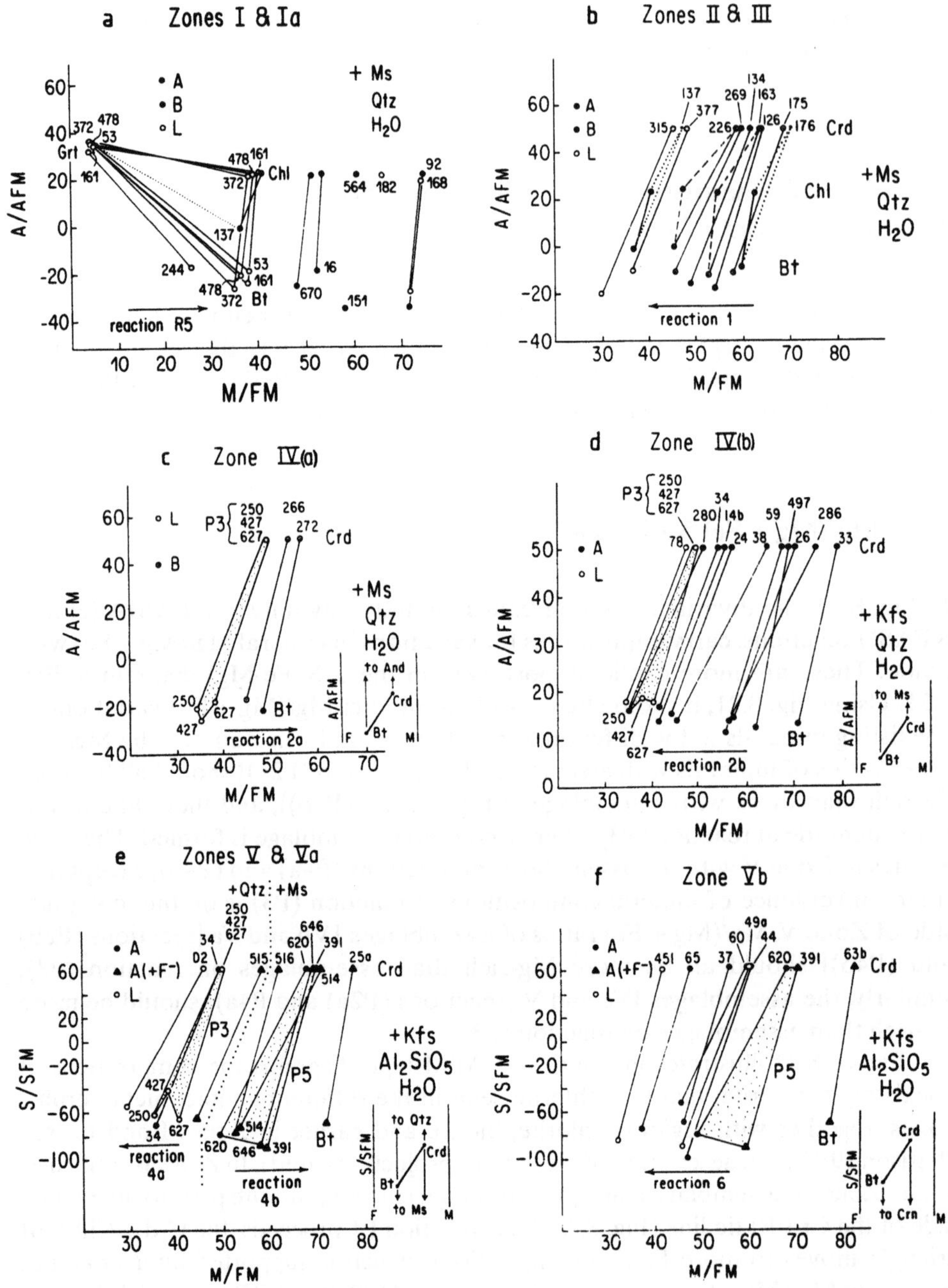
a Zones I & Ia
A/AFM
M/FM
• A
• B
○ L
+ Ms
Qtz
H2O
Grt
Chl
Bt
reaction R5
b Zones II & III
Crd
Chl
Bt
+Ms
Qtz
H2O
reaction 1
c Zone IV(a)
P3
Crd
Bt
+Ms
Qtz
H2O
to And
reaction 2a
d Zone IV(b)
+Kfs
Qtz
H2O
to Ms
reaction 2b
e Zones V & Va
+Qtz : +Ms
• A
▲ A(+F⁻)
○ L
S/SFM
P3
P5
+Kfs
Al2SiO5
H2O
to Qtz
to Ms
reaction 4a
reaction 4b
f Zone Vb
to Crn
reaction 6

pattern from Fig. 8.21. It is shown in Pattison (Chap. 16, this Vol.) that this effect is due to the presence of graphite in the slates, which introduces C-bearing fluid species to the hydrous vapour, thereby lowering a(H_2O) and displacing reaction P2a to more Mg-rich compositions. Notably, however, there is a weak but consistent zonation to more Mg-rich rim compositions in cordierite in DP266 and DP272, consistent with the positive slope of reaction (P2a) in Fig. 8.21.

Of the quartz-bearing assemblages of Zone V, two specimens are more Fe-rich than the reaction-(P3) univariant assemblages, consistent with the predicted trend, but DP515 is more Mg-rich, which contradicts the pattern. In the Zone Va assemblage (with muscovite and no quartz) the minerals are more Mg-rich than the reaction-P3 univariant assemblage, consistent with Fig. 8.21, but three of the four are *more Mg-rich* than the reaction-(P5) univariant assemblage Ms + Bt + Crd + Kfs + Al_2SiO_5 + Crn, which contradicts Fig. 8.21. Notably, all of the V and Va assemblages from Zone V that contradict the predicted pattern contain biotite with substantial (0.6–1.4 wt%) F (Pattison 1987; Table 3). In assemblage Vb of Zone V, all of the specimens, with one exception, are more Fe-rich than the reaction-(P5) univariant assemblage, consistent with Fig. 8.22. DP63b, the anomalously Mg-rich specimen, contains 1.5 wt% F.

In summary, if one omits the graphitic assemblages IVa and the specimens containing F-rich biotite in Zone V, the ranges in Mg/(Mg + Fe) of coexisting minerals in the different zones fit well with the predicted ranges in Fig. 8.21.

8.6.2 F in Biotite

Specimens containing biotite with >0.6 wt% F generally have Mg-rich compositions relative to those that contain little or no F. This is consistent with the phenomenon of Fe-F avoidance in biotite (Munoz 1984), in which Mg-biotite is stabilized by F. The F-contents of biotites represent up to 0.35 out of 2.0 hydroxyl sites (Pattison 1987).

Fig. 8.22 Compositions of Chl, Bt and Crd through the prograde mineral zones at Ballachulish, plotted in a variety of AFM-type projections (from Pattison 1987). *A* Appin Phyllite; *B* Ballachulish Slate; *L* Leven Schist and Creran Succession. The *arrows* indicate the predicted movement of Mg/(Mg + Fe) tie lines going through the number reaction. **a** Zone I and Ia (regional assemblages). Zone I refers to garnet-absent assemblages, Zone Ia to garnet-bearing assemblages. **b** Zones II and III. For specimens 137 and 176 in Zone II, *dotted lines* join analyzed chlorite + biotite compositions to *inferred* cordierite compositions (because the cordierite in these specimens is pinitized); the inferred cordierite compositions were determined by drawing a line from Bt parallel to the Crd + Bt tie lines of the other specimens. In specimens 163 and 226 from Zone III, *dashed lines* connect Crd + Bt compositions to Chl that appears texturally to be secondary. **c** Zone IV, assemblage IVa. P3, *Stippled pattern*, is for the range in compositions of Crd and Bt from the three univariant assemblages corresponding to reaction P3. **d** Zone IV, assemblage IVb. P3 *Stippled pattern* — see **c**. **e** Zones V and Va. P3, P5, *stippled patterns* are for the range of compositions of Crd and Bt in univariant assemblages corresponding to reactions (P3) and (P5). A(+ F) — Appin Phyllite specimens with biotite that contains >0.6 wt% F. **f** Zone Vb. See **e** for symbols

It appears that the relatively high F-contents in some biotites are the cause of their magnesian compositions. If, in contrast, high magnesium were the cause of high F-contents, then one would expect to see in the most Mg-rich biotites the highest F-content. This pattern is not seen: for example, biotite in DP620 (Zone Vb, Mg/(Mg + Fe) = 0.68, wt% F = 0.3) is more magnesian than biotite in DP515, (Zone V, Mg/(Mg + Fe) = 0.59, wt% F = 0.8). The Mg-rich composition of biotite in DP620 is explained by the normal movement of Mg/(Mg + Fe) tie lines through reaction (P4b) up to reaction P5.

Figure 8.23 shows the variation with grade of F-content of biotite through the aureole. There is no evidence for significant fractionation to more F-rich compositions at higher grades; the within-zone variation is greater than the variation with grade. This is probably because biotite remains modally abundant even in the highest grade specimens, so that any fractionation effects would have been masked. The lack of correlation of biotite F-content with grade indicates that original bulk compositional variation is the main control on biotite F-content in the aureole.

8.6.3 K/(K + Na) in Muscovite, Biotite and K-Feldspar

The variation of K/(K + Na) in biotite and muscovite is illustrated in Figs. 8.23 and 8.24. K/(K + Na) in muscovite drops from an average of 0.93 in Zone I to 0.89 in Zones II and III. The higher muscovite K/(K + Na) ratio in assemblage IVb (0.93)

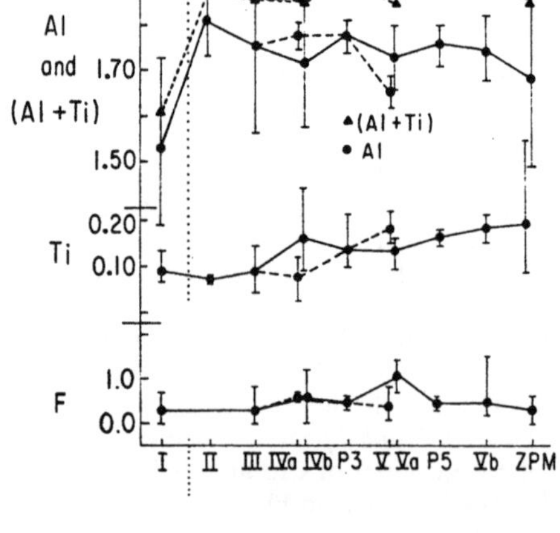

Fig. 8.23. Variation with grade of K/(K + Na), Si, Al, Ti, Al + Ti (cations) and F (wt%) in biotite from all units. For each zone, the mean is indicated by a *point*, and the absolute range is indicated by the *brackets*. *Roman numerals* denote metamorphic zones of assemblages from Fig. 8.1 and elsewhere in the text. P3, P5 = Reactions (P3) (Ms + Qtz = Al_2SiO_5 + Kfs + H_2O) and (P5) (Ms = Crn + Kfs + H_2). *R* Regional metamorphism; *T* thermal metamorphism; *ZPM* zone of partial melting. The *lines* joining assemblages IVa and V, respectively, to the overall trend have been *dashed* to distinguish them from assemblages IVb and Va, which occur at the same grade

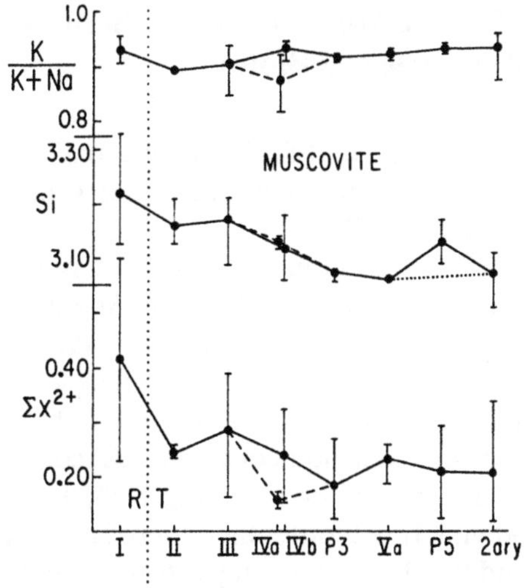

Fig. 8.24. Variation with grade of K/(K + Na), Si, and ΣX^{2+}(= Fe + Mg + Mn + Ca) in muscovite from all units. *2* = secondary muscovite from high-grade zones. All other symbols the same as in Fig. 8.23. The line to assemblage IVa has been *dashed* to distinguish it from assemblage IVb, which occurs at the same grade. The *dotted line* in the Si trend going from Zone Va to 2 is drawn because of doubts about the anomalously high Si values in (P5)

compared to assemblage IVa (0.88) is probably due to the development in assemblage IVb of K-feldspar, which fractionates Na into its structure more strongly than muscovite [typical K/(K + Na) in Kfs is 0.8; see below]. Above Zone IV, in which K-feldspar is always present, K/(K + Na) in muscovite ranges between 0.91–0.93. The absolute abundance of K cations in muscovite is subparallel to K/(K + Na), varying from an average of average of 0.86 in Zone I, 0.78 in Zone II, 0.81 in Zone III, and 0.85–0.92 above Zone III (Pattison 1985).

The pattern of K/(K + Na) in biotite is similar qualitatively to that of muscovite, but at higher absolute K/(K + Na), (Fig. 8.23). The absolute abundance of K cations in biotite also follows that of muscovite, ranging from an average of 0.84 in Zone I, 0.67 in Zone II, 0.80 in Zone III and 0.85–0.93 in and above Zone IV (Pattison 1985). In both muscovite and biotite there is a conspicuous drop in K/(K + Na) and absolute K going from Zone I (regional grade) to Zone II (outermost contact metamorphic zone).

In Zones I-III and in assemblage IVa, plagioclase is albitic (Na/(Na + Ca) = 0.99). The contact metamorphic oligoclase 'isograd' coincides with the lower boundary of assemblage IVb in Zone IV, in which Na/(Na + Ca) in plagioclase is 0.71–0.84. The lower Na content of plagioclase in assemblage IVb is probably due to Na-exchange between plagioclase and the newly formed, Na-bearing K-feldspar, which coexists with oligoclase rather than albite at these grades (e.g., Goldsmith and Newton 1974). Based on spot analyses, plagioclase generally decreases in Na/(Na + Ca) from 0.80 to 0.70 going from assemblage IVb upgrade into the zone of partial melting (Pattison 1985). Spot analyses of cryptoperthitic K-feldspar in and above Zone IVb give a range of K/(K + Na) of 0.75–0.88, clustering around 0.80 (Pattison 1985).

8.6.4 (Fe,Mg)Si = 2 Al (Tschermak Substitution) in Chlorite, Muscovite and Biotite

Figures 8.23 and 8.24 illustrate the variation with grade of Si-content (cations) of biotite and muscovite. In Zones I and II, chlorite shows a similar pattern (Fig. 6, Pattison 1987).

The variation in Si content is the most reliable indication of the extent of Tschermak exchange [(Fe,Mg)Si = 2 Al] in these minerals. This is in contrast to the variation of divalent cations or Al content, which are subject to uncertainties in Fe^{2+}/Fe^{3+} and other substitutions (see Miyashiro and Shido 1984; Pattison 1987, for a fuller discussion). For example, in biotite, if the Tschermak exchange were the main exchange reaction controlling Al content, then Al should show an inverse relationship with Si; referring to Fig. 8.23, this pattern is not seen. However, if one combines Al and Ti in biotite, then there is a very good negative correlation between Al + Ti and Si. This pattern suggests the operation in biotite of the exchange reaction

$$(Fe,Mg)^{2+} + 2Si^{4+} = Ti^{4+} + 2Al^{3+}.$$

Stoichiometrically this reaction bears similarities to the simpler Tschermak exchange

$$(Fe,Mg)^{2+} + Si^{4+} = 2Al^{3+}.$$

8.6.4.1 Variation in Si Content

On average, the Ballachulish Slate contains the most aluminous micas, and the Appin Phyllite the most siliceous, resulting in a primary bulk compositional bias to Figs. 8.23 and 8.24 (Pattison 1987). To remove this bias, Fig. 8.25 plots the variation in Si content of chlorite, muscovite and biotite in samples from the Appin Phyllite, the unit best represented at all grades in the aureole.

A major feature of Fig. 8.25 is that the three minerals show subparallel trends of Si content with grade. Taking muscovite as an example, average Si content shows the following pattern: 3.22 in Zone I, 3.16 in Zone II, 3.18 in Zone III, 3.13 in Zone IV and 3.06–3.08 in Zone V [apart from quartz-absent reaction-(P5) assemblages which have an unexpectedly high average Si content of 3.13]. Secondary muscovite (as

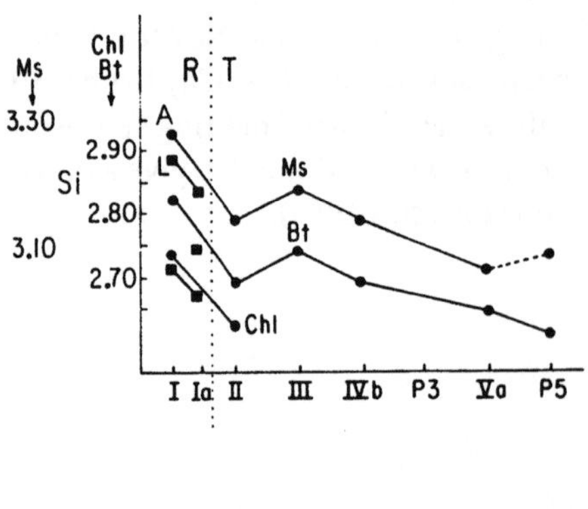

Fig. 8.25. Comparison of the variation with grade of Si in muscovite (*Ms*), biotite (*Bt*), and chlorite (*Chl*) in the Appin Phyllite (contact metamorphism = *heavy dots*) and the Leven Schist (regional metamorphism = *squares*) (from Pattison 1987). The *symbols* are the same as in Figs. 8.23 and 8.24

distinguished texturally using the criteria described earlier) from 26 specimens at different grades has an average Si content of 3.07. This is indistinguishable from primary muscovite upgrade of assemblage IVb, and dictates the need for textural distinction between the two generations of muscovite, as discussed in Section 8.3.5.

Overall, the trend with grade is towards decreasing Si content, with the largest drop going from Zone I (regional grade) to Zone II in the aureole. This is consistent with Velde's (1965) experiments that show that muscovite Si contents decrease with lowering pressure and increasing temperature. Regional metamorphic conditions (Zones I and Ia) were in the range 450–550 °C, 0.6 ± 0.1 GPa (Pattison and Voll, Chap. 2, this Vol.) in contrast to conditions in Zone II of the aureole of 550–560 °C, 0.3 GPa (Pattison, Chap. 16, this Vol.), which represents near isothermal decrease in pressure. Above Zone II in the aureole, which represents isobaric increase in temperature, Si contents also decrease.

A notable departure from this pattern is the apparent increase in Si content of muscovite and biotite going from Zone II to Zone III (Fig. 8.25). Within Zone III, uniquely defined by the KFMASH assemblage Ms + Bt + Crd + Qtz, there is also an increase in Si content of the two minerals with increasing grade (see Fig. 9 of Pattison 1987).

8.6.4.2 The T-X(Al-Si) Diagram

Si variations in Fig. 8.25 suggest that the exchange (Fe,Mg)Si = 2Al operates sympathetically in all of the sheet silicates. The Tschermak exchange has usually been discussed with reference to one mineral, such as muscovite in the following reaction:

phengitic Ms + Chl = less phengitic Ms + Bt + Qtz + H_2O,

(e.g., Mather 1970). This reaction, rewritten in KMASH using $Mg_{-1}Si_{-1}Al_2$(Ts), becomes:

$$\underset{\text{Muscovite}}{3\,KAl_3Si_3O_{10}(OH)_2} + \underset{\text{Chlorite}}{Mg_5Al_2Si_3O_{12}(OH)_8} =$$

$$\underset{\text{Biotite}}{3\,KMg_3AlSi_3O_{10}(OH)_2} + \underset{\text{Tschermak exchange (Ts)}}{4\,Mg_{-1}Si_{-1}Al_2} + \underset{\text{quartz}}{7\,SiO_2} + \underset{\text{water}}{4\,H_2O}.$$

The second reaction shows that the Tschermak exchange is a general exchange component applicable to all amenable phases, rather than being restricted to only one.

Using the same reasoning as for the construction of a T-X(Fe-Mg) diagram, a T-X(Si-Al) diagram is constructed in Fig. 8.26 that illustrates the influence of Al-Si variation on the mineralogy and mineral chemistry of a variety of low-grade assemblages in the aureole (for a fuller discussion, see Pattison 1987). The relative ranking in Si/(Si + Al) is Bt > Ms > Chl. Because all variation in Si/(Si + Al) is assumed to be due to the Tschermak exchange, the relative Si/(Si + Al) ratios between the three minerals remains constant.

Figure 8.26 explains several notable features of assemblages in regional and contact metamorphic rocks. In regional grade specimens outside the aureole, the

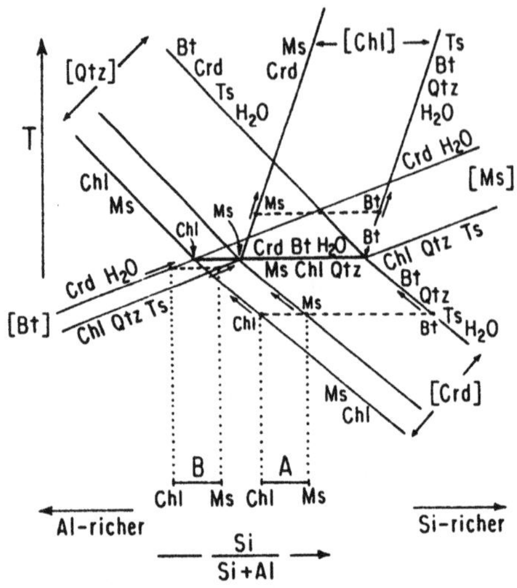

Fig. 8.26. Schematic isobaric T-$X_{Al\text{-}Si}$ diagram for assemblages at Ballachulish (see Pattison 1987, for a discussion of the construction of this diagram). The above topology and movement of the lines has been drawn to be consistent with the observed Si/(Si + Al) variation of coexisting Ms + Bt ± Chl in the Ballachulish assemblages (see Figs. 8.23-8.25). Two pairs of coexisting Chl and Ms with differing Si/(Si + Al) ratios (*A* and *B*) are drawn to illustrate the contrast in evolution of Si contents with increasing grade (see text). The *short dashed lines* represent tie lines between Ms ± Chl ± Bt in different assemblages. The small *arrows* indicate the direction of migration of the Si/(Si + Al) content of the coexisting micas as the different reactions proceed

assemblage Ms + Chl + Qtz is developed in the most Al-rich specimens (typically Ballachulish Slate), whilst the assemblage Ms + Chl + Bt + Qtz is developed in more Si-rich compositions (typically Appin Phyllite or Leven Schist). In Fig. 8.26, these two compositions are labelled B and A, respectively. During prograde regional metamorphism, the more Si-rich Ms and Chl (A) entered the reaction

Ms + Chl = Bt + Ts + Qtz + H_2O,

producing biotite, whilst at the same grade, Ms and Chl (B) did not develop Bt because they were too aluminous.

In cordierite-bearing Zone II assemblages Ms + Chl + Bt + Qtz + Crd, Si contents of Ms, Chl and Bt are intermediate between those of Al-richer Ms + Chl + Qtz assemblages and Si-richer Ms + Chl + Bt + Qtz assemblages (e.g., Ms: 3.18, 3.16, 3.27, respectively; see Table 5, Pattison 1987). This pattern is predicted by Fig. 8.26, in which the more Si-rich Ms, Chl and Bt become more *Al-rich* during passage through the reaction Ms + Chl = Bt + Ts + Qtz + H_2O, whilst the more Al-rich Ms and Chl become more *Si-rich* during passage through the reaction Chl + Ts + Qtz = Crd + H_2O. These two reactions terminate upgrade against reaction (P1), where the Si-contents of Ms, Chl and Bt are predicted to be the same regardless of which of the two divariant reactions they passed through on the way to reaction (P1). Examining regional garnet-bearing specimens, the same relationship holds true, with the Si contents of Ms, Chl and Bt in Ms + Chl + Qtz + Grt + Bt assemblages lying between those of Al-rich Ms + Chl + Qtz assemblages and Si-rich Ms + Chl + Bt + Qtz assemblages found at lower grade (Table 5, Pattison 1987).

Within the Zone III assemblage Ms + Bt + Crd + Qtz, Si contents of Ms and Bt increase with grade. Referring to Fig. 8.26, this may be explained by the progress of the reaction Bt + Ts + Qtz + H_2O = Crd + Ms, in which Bt and Ms become more Si-rich with increasing grade. This reaction, like the garnet pseudomorphing reaction (P1'), Ms + Grt + Qtz + H_2O = Crd + Bt, has water as a reactant, due to the hydrous nature of cordierite.

Significant variation in Si content in muscovite and biotite ceases above Zone IV. There appears to be a lower limit of Si content in muscovite (~3.07) and biotite (~2.68) that is reached by reaction (P3).

8.7 Mineral Chemistry of High-Grade Assemblages in the Zone of Partial Melting

Selected mineral chemical data of four Crd + Grt ± Hy quartz-bearing assemblages and four Crd + Spl + Crn ± Al_2SiO_5 quartz-absent assemblages are listed in Table 16.1 and 16.2 (Pattison, Chap. 16, Part IV, this Vol.).

8.7.1 Quartz-Bearing Assemblages

The garnet chemical data in Table 16.1 are from the cores of generally flat compositional plateaus away from any inclusions. Typical garnet compositions are Fe:Mg: Ca:Mn = 76–78:14–17:2–4:3–5. Fe^{3+} calculated from stoichiometry is negligible. Going from core to rim in SW2a and SW2b, X_{Mn} increases from 0.05 to 0.09, X_{Mg} decreases from 0.14 to 0.09, and X_{Fe} and X_{Ca} remain constant. This zoning pattern is consistent with a process involving garnet resorption (Tracy 1982). In DP567 and DP568, Ca and Mn remain constant going from core to rim, but X_{Fe} increases from 0.76–0.78 to 0.79–0.83, and X_{Mg} decreases from 0.16–0.17 to 0.10–0.14. The amount of departure from the core values depends upon the adjacent minerals (negligible variation beside quartz, maximal variation beside biotite), which suggests a mechanism of diffusive reequilibration rather than resorption.

Cordierite and matrix biotite show no significant variation in Mg/(Mg + Fe) going from core to rim. Biotite inclusions in garnet are more magnesian and more Ti-rich than biotite grains in the matrix [e.g., DP567: Mg/(Mg + Fe) biotite = 0.46 (inclusion) vs 0.38 (matrix); Ti cations = 0.37 (inclusion) vs 0.23 (matrix)].

In DP568, core analyses of hypersthene have an average Fe:Mg:Ca:Mn ratio of 64:35:1:1. Al_2O_3 content is 2.6 wt%, corresponding to 0.12 Al cations for 6 oxygens. Fe^{2+} calculated from stoichiometry is negligible.

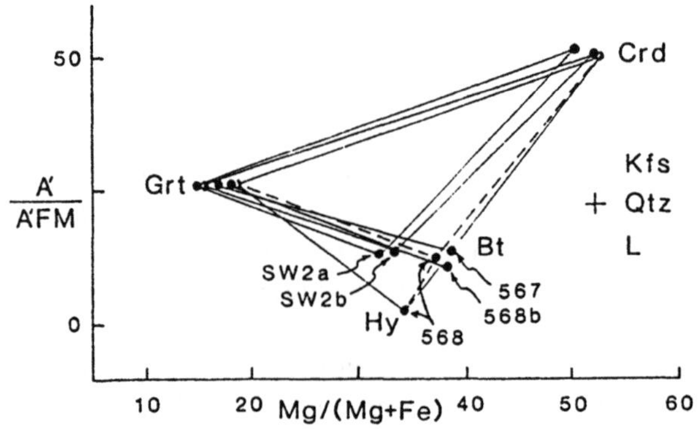

Fig. 8.27. A'FM diagram projected from Kfs, Qtz and L showing the compositions of Crd, Bt, Grt and Hy from high-grade quartz-bearing assemblages

Grt + Crd + Bt ± Hy compositions from the above samples are plotted in an A'FM diagram in Fig. 8.27. In DP568, biotite in the assemblage Bt + Crd + Grt + Hy plots within the Grt + Hy + Crd triangle, in contrast to the more typical topology of this assemblage in which Bt plots to the Mg-side of the Crd + Hy tie line (e.g., Droop and Charnley 1985). This may support the textural evidence that biotite in DP568 is of late stage origin.

8.7.2 Quartz-Absent Assemblages

Pleonaste spinel has been analyzed in four specimens (see Table 16.2, this Vol.). In three specimens, the ratio of Fe:Mg:Mn:Zn is 88–89:10:0–1:0, whilst in DP433, high Zn-content gives a Fe:Mg:Mn:Zn ratio of 77:10:0:13. The presence of Zn in spinel tends to expand the stability field of spinel, thereby shifting reactions such as (P7) and (P8) to lower temperatures (e.g., Montel et al. 1986). Figure 8.28 illustrates the relative compositions of Spl, Crd, Bt and Al_2SiO_5 in an S'FM diagram projected from Kfs, Crn and H_2O.

8.8 Retrograde Reaction in the Aureole

In addition to the prograde features described above, there is a range of textures in pelitic assemblages indicative of retrogression at different stages in the cooling of the aureole (Pattison 1985). Most of these textures indicate hydrous retrogression. The textures are described below according to the prograde mineral zone in which they occur.

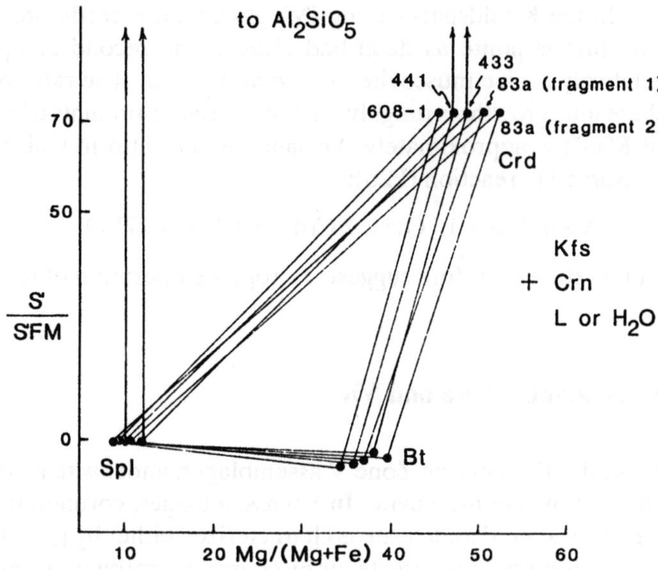

Fig. 8.28. S'FM diagram projected from Kfs, Cor and L or H_2O showing the compositions of Bt, Crd and Spl from high-grade quartz-absent assemblages

8.8.1 Zones II and III

Out of 72 specimens from these zones, 53 contain cordierite that is completely altered. As described in Section 8.3.2, the alteration consists of either an amorphous, isotropic-looking, orange-brown 'film', or a generally coarser intergrowth of muscovite and chlorite. Both of these types of alteration comprise the general term 'pinite'.

Cordierite is more susceptible to alteration than biotite. In rocks in which biotite is wholly or partly chloritized, cordierite is always completely altered, whilst in many rocks in which cordierite is altered, biotite is fresh. In rocks that contain chloritized biotite, cordierite is typically replaced by the coarser Ms ± Chl type of alteration. This suggests the retrograde operation of the reaction

$$\text{Ms} + \text{Chl} + \text{Qtz} = \text{Crd} + \text{Bt} + H_2O \qquad \text{(P1)}$$

The 'orange-brown film' type of alteration may represent a later near-surface weathering.

8.8.2 Zone IV

In the andalusite-bearing assemblage IVa, andalusite is generally unaltered. Rarely it has thin, incomplete rims of muscovite or fine grained sericite at its edges. Cordierite alters similarly to cordierite in Zones II and III.

In the K-feldspar-bearing IVb assemblage, cordierite alters in two main ways. The first is pinite as described above. The second comprises an intergrowth of relatively coarse muscovite, biotite and quartz. The ratio of muscovite to biotite in these intergrowths is roughly 3:1, with variable amounts of quartz. The 3:1 proportion of Ms:Bt is approximately the same as the ratio in which these two minerals are consumed in reaction (P2b):

$$6\ \mathrm{Ms} + 2\ \mathrm{Bt} + 15\ \mathrm{Qtz} = 2\ \mathrm{Crd} + 8\ \mathrm{Kfs} + 6.5\ \mathrm{H_2O}. \qquad \text{(P2b)}$$

This texture therefore suggests retrograde operation of reaction (P2b).

8.8.3 Zones V, Va and Vb

In And + Kfs-bearing Zone V assemblages, andalusite is commonly, but not always, rimmed by late muscovite. In Vb assemblages, corundum shows the same pattern. Severely altered andalusite is characterized either by pseudomorphous replacement by cryptocrystalline sericite, or by coarse intergrowths of muscovite and quartz. The latter texture is suggestive of the retrograde operation of the reaction

$$\mathrm{Ms} + \mathrm{Qtz} = \mathrm{And} + \mathrm{Kfs} + \mathrm{H_2O}. \qquad \text{(P3)}$$

Similar to the lower grade zones, cordierite is the mineral most susceptible to alteration in Zone V assemblages. Many rocks that contain relatively fresh andalusite contain cordierite that has been replaced by Ms + Bt ± Qtz intergrowths or by pinite.

8.8.4 Zone of Partial Melting

Alterations of individual minerals similar to those described above may be found in this subzone, but in addition a turbid alteration of K-feldspar is more widespread. In some cases, an extensive but patchy alteration of the rocks with formation of sericite + chlorite is seen. In these rocks, andalusite is wholly or partially altered to sericite, cordierite is wholly altered to Ms + Bt ± Qtz or pinite, and biotite is largely or wholly altered to chlorite; no K-feldspar is left. These sericite + chlorite patches represent zones of intense, low-grade hydration of the pre-existing high grade assemblage.

Some distinctive textures with somewhat different retrograde characteristics are also seen, most typically in Zone V assemblages in the Chaotic Zone (west flank) of Pattison and Harte (1988) and Harte et al. (Chap. 9, this Vol.). Large (>1 mm), poikiloblastic, cross-cutting biotite flakes surround or occur nearby andalusite prisms; the biotite may also contain large inclusions of quartz and cordierite. In many specimens, this is the only type of biotite present, in contrast to the smaller, tabular biotite crystals seen elsewhere in pelites of equivalent grade. The andalusite prisms may show ragged edges intergrown with quartz. In one specimen, mats of fibrolitic sillimanite growing in quartz are associated with the coarse andalusite-rimming biotite.

The coarse, cross-cutting texture of the biotite, and its conspicuous spatial association with andalusite and rarely fibrolite, are suggestive of the retrograde operation of the reaction

$$Bt + Al_2SiO_5 + Qtz = Crd + Kfs + H_2O \text{ or } L \quad \text{(P4a)}$$

Since these rocks have undergone partial melting, it may be that melt (L) is the appropriate phase on the right-hand side of the reaction, in which case the textures may be a product of crystallization of the melt during cooling. This type of retrogression is, however, quite distinct to that which has given rise to typical chlorite and sericite alteration products in many rocks.

Similar coarse, poikiloblastic biotite crystals occur in high grade Grt ± Hy assemblages in DP567 and DP568 (Fig. 8.16 and discussion in Sect. 8.4.2 above), which are also interpreted to be of late-stage origin.

8.8.5 Interpretation of Retrograde Textures

From the above textures, it is clear that hydrous retrogression proceeded at all grades in the thermal aureole. It is significant that retrograde equivalents of many of the prograde reactions appear to have operated, which indicates that late hydrous fluids were variably present during the cooling history (although not everywhere at the same time and probably not in very large volumes; Harte et al., Chap. 19, this Vol.). In some specimens, more than one type of alteration product indicates that late hydrous fluids percolated through these specimens at different times during cooling (Pattison 1985).

8.8.6 Distribution of Alteration in the Aureole

The part of the aureole that has suffered the most widespread low-grade (chlorite + sericite) alteration is the Chaotic Zone (west flank). Other areas showing extensive alteration include the Ballachulish Bridge area (northeast flank), the southwest contact in Glen Duror, and the aureole around the small stock in the southeast of the area. These distinctions in degree of alteration are based primarily on alteration of rocks in and above Zone IV. Alteration of lower grade rocks farther out in the aureole (mainly pinitization of cordierite) appears to be more uniformly distributed.

It is significant that the Chaotic Zone shows the most intense low-grade alteration in the aureole. This area also shows evidence of the most extensive partial melting. Troll and Weiss (Chap. 3, this Vol.) and Rabbel and Meissner (Chap. 7, this Vol.) argue that the quartz diorite underlies the pelites in this area. During crystallization of the quartz diorite, water will have been released into the overlying pelites, providing a ready source of fluids for melting and later hydrous retrogression (Pattison and Harte 1988). Extensive melting within the rocks will also have stored water which would be released and potentially cause retrogression on cooling. This model is also consistent with the oxygen isotopic data of Hoernes et al. (Chap. 18, this Vol.); specimens from the Chaotic Zone show significant shifts from typical regional pelitic values to more igneous values.

8.9 Summary

A concentric sequence of pelitic and semipelitic mineral assemblage zones has been mapped around the Ballachulish Igneous complex. The width of the aureole, as defined by the outer development of Zone II, varies from 400 m (NE contact) to 1700 m (E and W contacts). The width of individual subzones in the aureole are generally proportional to these variations. The widest zones are adjacent to or enclose large expanses of quartzite.

The prograde reactions in the aureole produce profound textural contrasts in the metapelites. In Zone I-III, the pelites are relatively fissile, whilst in Zone IV and at higher grades, assemblages with substantial modal development of K-feldspar and cordierite are classic massive hornfelses.

The petrography of the mineral assemblages, including contact metamorphic overprints of regional grade assemblages, have been described in detail. Schreinemakers' analysis of the main phases in the model pelitic system KFMASH has been done on the assumption of H_2O-bearing assemblages below the onset of anatexis, and H_2O-absent, melt-bearing assemblages in the anatectic zone. The modelled reactions may be linked to provide a continuous schematic petrogenetic grid that accounts for all of the assemblages in the Ballachulish aureole, and is also applicable to other thermal aureoles.

Above the Ms + Qtz breakdown, separate prograde sequences of quartz-bearing and quartz-absent assemblages are developed. The development of these relatively uncommon quartz-absent assemblages is due to passage through prograde reactions which consume large modal amounts of quartz.

The chemical variations of pelitic minerals from 60 contact and regional specimens below the onset of anatexis, and 8 specimens from the anatectic zone, have been examined. With the exception of specimens from the graphitic Ballachulish Slate and those with F-rich biotite, the changes in Mg/(Mg + Fe) ratios for chlorite, cordierite and biotite at different grades are consistent with those predicted by continuous Fe-Mg reactions in the petrogenetic grid. Specimens that contain F-rich biotite are anomalously magnesian, consistent with the theory of Fe-F avoidance in biotite. Graphitic specimens in assemblage IVa are anomalously magnesian because of the relative shift to more magnesian compositions of reaction P2a due to lowered $a(H_2O)$ (see Pattison, Chap. 16, this Vol.).

The Tschermak exchange, (Fe,Mg)Si = 2 Al, operates sympathetically in coexisting chlorite, muscovite and biotite in the various metamorphic zones. By treating (Fe,Mg)Si = 2 Al as a general exchange vector that operates in all amenable phases, an isobaric T-X(Al-Si) diagram may be constructed that accounts for the distribution of low grades assemblages in the Ballachulish area, and for the variation in Si content of chlorite, muscovite and biotite in a variety of regional and contact metamorphic assemblages.

A range of textural features in the pelites indicate localized hydrous retrogression in the aureole. Several of the textures suggest retrograde operation of a number of the prograde reactions [e.g., (P1), (P2b) and (P4a)], and indicate that hydrous fluids were variably present during cooling of the aureole, although not everywhere at the same time and not necessarily in very large volumes. Pelites on the west flank (the Chaotic Zone) are the most extensively retrogressed as well as showing the greatest melting. Release of water from the underlying quartz diorite may be responsible directly for their extensive melting, and both directly and indirectly responsible for their extensive retrogression.

9 Field Relations and Petrography of Partially Melted Pelitic and Semi-Pelitic Rocks

B. Harte, D.R.M. Pattison and C.M. Linklater

9.1 Introduction

Evidence of partial melting is seen in pelitic and semi-pelitic metasediments within the Ballachulish aureole in many locations immediately adjacent to the intrusive complex (within Zone V of Pattison and Harte 1985, and Chap. 8, this Vol.). In this chapter we summarize that evidence as seen both in the field and under the optical microscope. We endeavour in our presentation to largely keep description and interpretation in separate sub-sections, in an attempt to lay out the evidence for melting as objectively as possible. The chapter is concerned with textures and structures, their interpretation, and their implications for the rheological and crystallization behaviour of the melts. The reader is referred to Pattison and Harte (1985, 1988; Sect. 8.5, this Vol.) for notes concerning the melting reactions.

The evidence of partial melting which we shall present, occurs within 0.5 km of the mapped contact of the intrusive complex; an approximate limit to the occurrence of partial melting is shown on Fig. 9.1 and Maps 1 and 2. This boundary is not well defined because the most widespread migmatitic rocks seen in the field are only well developed in pelites and semi-pelites with strong composition layering, and even within these rocks are only usually abundant within 100-200 m of the intrusive complex. In locating the melting boundary, we have also taken into account microscopic features, especially xenomorphic quartz textures, albite rims around K-feldspar, and cloudy perthitic K-feldspar (Pattison and Harte 1988).

We commence our description by considering migmatitic rocks in which leucosome development is accompanied by fracturing of cordierite-rich hornfels layers. The resultant migmatites, which show pull-apart (boudin) and breccia-like (agmatitic) structures, show their greatest variety and development on the eastern side of the intrusive complex. In general, the scale of leucosome development and disruption in these rocks is < 0.5 m. We then describe leucocratic veins and sheets, which are also widely distributed, but at the same time restricted in thickness and abundance. One part of the aureole stands out in the field as showing evidence of melt development and associated country-rock disruption on much larger scales. This 'Chaotic Zone' is on the western flank of the intrusive complex, and is the subject of the last part of our account.

In the text, the terms leucocratic and melanocratic are used in a purely descriptive and relative sense to indicate modal characteristics. The term *leucosome* is used to indicate a leucocratic domain (extending over 1 cm), which appears to have formed

G. Voll, J. Töpel, D.R.M. Pattison, F. Seifert
Equilibrium and Kinetics
in Contact Metamorphism

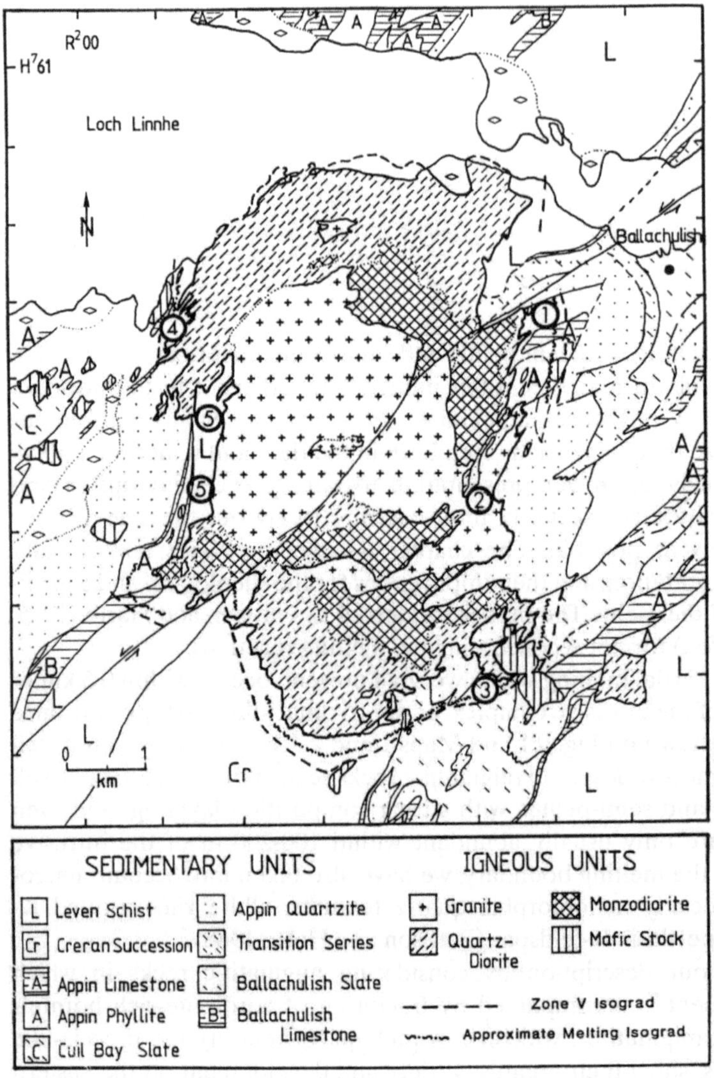

Fig. 9.1. Geological map of the Ballachulish intrusive complex and its aureole. The Zone V isograd and the approximate isograd or outer limit of partial melting (see text) within the aureole are shown close to the contacts with the intrusive complex. Particular localities referred to in the text are marked by numbers: *1* area between intrusive complex and Allt Giubhsachain; *2* metasedimentary enclaves west of Sgorr Dhearg; *3* contact zone (NN053531) by Coire Chaorain; *4* area south and west of Kentallen; *5* 'Chaotic Zone'

as a result of the contact metamorphism, and is therefore part of the neosome (Ashworth 1985), but without implying the process of formation. Although the most widespread migmatites at Ballachulish show boudin and breccia-like structures, unlike those of the typical stromatic migmatites of regional metamorphic terranes, we consider the term *migmatite* to be appropriate for them because they contain leucocratic domains, which are clearly leucosomes, in association with mesosomes and some biotite-rich selvedges which might be termed melanosomes (Ashworth 1985). The xenomorphic quartz texture of Pattison and Harte (1988) is here referred to as *interstitial-xenomorphic* in order to better indicate its character.

Previous publications on these rocks have particularly dealt with the development of boudin and agmatitic structures, and the contrasts between the 'Chaotic Zone' and other parts of the aureole (Pattison and Harte 1988). Partial melting phenomena have also been described around smaller intrusions nearby and of similar age to the Ballachulish complex (Platten 1982, 1983). Where these intrusions occur close to the Ballachulish main complex, the aureoles of the major and minor complexes appear to be continuous. Further work on textures and compositional data in the Ballachulish aureole rocks is presently being undertaken in the course of Ph.D studies by one of us (C. Linklater).

9.2 Migmatites Characterized by Pull-Apart and Breccia-Like Structures

9.2.1 Field Description

Leucosomes associated with boudinage and breccia-like structures in the hornfelses are widespread within about 100-200 m of the contacts of the intrusive complex. They are associated with rocks showing obvious compositional layering (largely relict bedding) on scales of up to 0.5 m, which is characteristic of much of the pelitic/semi-pelitic units in the aureole. Although the migmatitic rocks described here are widespread and involve small-scale disruption, it is a characteristic of rock exposures in the contact zone everywhere other than the Chaotic Zone, that relict bedding and regional metamorphic structures are for the most part clearly preserved.

Particularly good exposures of the types of migmatitic rocks described in this section are found by the contact on the ridge (locality 1, Fig. 9.1) to the west of the Allt Guibhsachain on the eastern flank of the intrusive complex, and to the SW of here (locality 2, Fig. 9.1) where screens of metasediment occur within the diorite (see Weiss and Troll, Chap. 3, Part II, this Vol., for cross sections of the contacts). Other good localities include the immediate contact area near Coire Chaorain and the contact zone to the E and S of Kentallen; localities 3 and 4 respectively, in Fig. 9.1.

The boudinage structures associated with leucosome formation typically involve the pull-apart of cordierite-rich (with or without andalusite) layers, and the infilling of the volume between the dismembered edges of these layers by material which varies from being distinctly leucocratic (K-feldspar and/or quartz rich) to material with more biotite and a more general semi-pelitic appearance. Figure 9.2a shows a typical example in which relict bedding is crosscut by relatively thin quartzo-felspathic leucosomes. The leucosomes are frequently at ca. 90° to bedding, but may be orientated at smaller angles. Sometimes obvious leucosomes also occur parallel to the layering and connect with the crosscutting (boudin-infilling) leucosomes (see below), and in other cases the crosscutting leucosomes connect with the leucocratic

Fig. 9.2. Examples of leucosome development associated with boudinage and breccia structures. **a** Metasedimentary layers (bedding) are seen dipping steeply to the *right*, with cordierite-rich layers (*centre*) showing pull-apart structure infilled by narrow crosscutting leucosomes. Coin is 3 cm across. NN008572, by Kentallen. **b** Cordierite-rich metasedimentary layer to the *right* and area with agmatitic structure in *centre* and *left*. A boudinage-type leucosome crosscuts the cordierite-rich layer and is in continuity with the leucocratic material of the agmatitic structure. NN057575 west of Allt Guibhsachain

matrix of small-scale breccia-like or agmatitic structures. However, the dominant feature of many of these boudinage-related leucosomes is their restricted continuity across the layering, such that they often only form a roughly planar transect of one or two metasedimentary layers (Fig. 9.2a; and Figs. 3, 4 and 7 of Pattison and Harte 1988). This behaviour is in marked contrast to the pervasive vein structures described in Section 9.4.

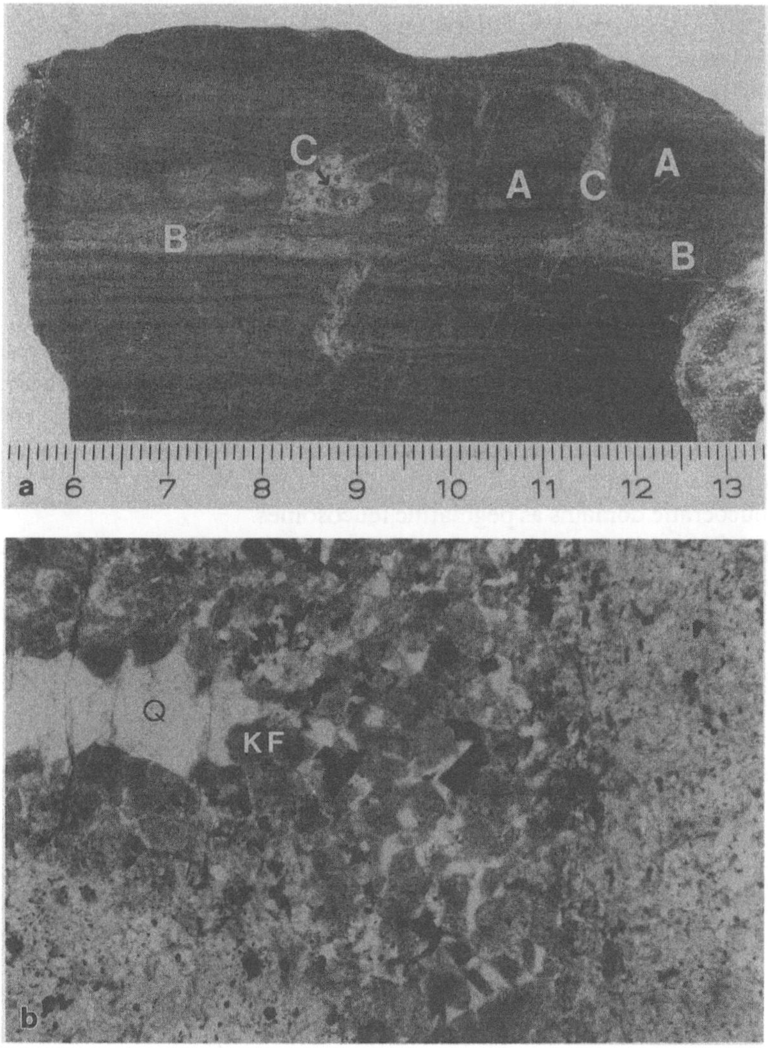

Fig. 9.3. Migmatitic rocks showing boudin-type structures. **a** Interbedded pelitic and semi-pelitic material, showing: rigid, pulled apart, cordierite-rich layers (*A*); ductile, layers (*B*); and well developed leucosome of infilling leucocratic material (*C*); scale in mm; see text for detailed description. **b** Detail of part of a type-C leucosome, similar to those in **a**, showing quartz-core (*Q*) and adjacent zone with cloudy-altered euhedral-subhedral K-feldspar (*KF*) and interstitial-xenomorphic quartz. NN058575, west of Allt Giubhsachain

More detail of the relations seen in migmatitic rocks showing structures of boudin or partial brecciation type are illustrated in Fig. 9.3a. Three main compositional and textural domains may be distinguished in hand specimen:

1. Type (A): Primary compositional layers (beds) that underwent brittle fracture to form millimetre-centimetre-scale rectilinear fragments. The most conspicuous of these layers are typically cordierite-rich.

2. Type (B): Layers that show evidence of ductile (rather than brittle) behaviour and dominantly consist of quartz and K-feldspar, with variable amounts of biotite and plagioclase. These layers may show some internal modal variation with development of leucocratic domains (Fig. 9.3a); their appearance varies from that of straightforward metasedimentary beds of pelitic or more quartzo-felspathic composition, to more granular and leucocratic irregular layers suggestive of leucosomes. They may show evidence of flow or bending towards the gaps between the rectilinear (A) fragments (Pattison and Harte 1988, Figs. 3 and 7).

3. Type (C): Medium- to coarse-grained leucosomes that cross-cut and fill the spaces in between the rectilinear (A) fragments. They consist dominantly of K-feldspar and quartz, with minor plagioclase and biotite, and often pass gradationally both in texture and mode into the ductile (B) layers (Fig. 9.3). The minerals may be evenly distributed in these leucosomes, but quite commonly the central zone of the boudin is occupied by relatively coarse (2 mm) quartz grains with a surrounding volume dominated by K-feldspar (Fig. 9.3b). We refer to such relatively coarse leucocratic domains as pegmatitic leucosomes.

The scale and extent of development of these boudinage-type structures are quite variable. Scale appears to be usually dictated by the nature and thickness of compositional layering associated with the original foliation. In the aureole as a whole, the boudinage structure with sharp crosscutting leucosomes is particularly associated with cordierite-rich layers up to about 10 cm thick, with the leucosomes being up to ca. 1 cm thick.

In some cases, and particularly in the area of metasedimentary screens on the east side of the intrusive complex, the boudinage becomes more pervasive in rocks showing regular thin compositional layering. The result is a very regular rectilinear chocolate tablet or brickwork appearance, in which leucocratic material is developed both along and across the layering (Fig. 9.4a; Fig. 4 in Pattison and Harte 1988). The detailed modal and textural constitution of this leucocratic material (D in Fig. 9.4a) varies considerably, and may show all the characteristics of B and C domains as described above with respect to Fig. 9.3a.

Occasionally, more extensive break-up and disorganization of the metasedimentary layers leads to the formation of breccia or agmatitic structures in which flakes or tablets of hornfels are seen 'floating' with irregular orientations in leucosome. In some cases these agmatitic structures form irregular volumes or patches (Figs. 9.2b and 9.4b), whilst in other cases they form sheet-like bodies crosscutting the metasedimentary layering. Between these extreme cases are agmatites which form irregular bodies, partly conformable with and partly crosscutting the relict bedding structure. In scale the irregular agmatitic patches are usually < 0.25 m across and vary with the scale of the relict bedding; whilst the agmatitic sheets are usually < 0.15 m thick.

The leucocratic matrix in these breccia-like migmatites shows similarities to the ductile/leucocratic material forming (B), (C) and (D) domains in the boudinage structures, and may show physical continuity with such material (Fig. 9-2b). However, the agmatitic leucosomes are often coarser grained and contain more plagioclase. These characteristics also lead to similaries with the medium- to coarse-grained leucosomes (Sect. 9.4.2).

Fig. 9.4 a. Disrupted interbedded pelite/semi-pelite in which cordierite-rich layers show a tablet structure, and leucocratic material *D* both crosscuts and runs parallel to the metasedimentary layering. *D* material grades into semi-pelitic material. **b** Agmatitic structure. Coins approximately 2 cm across. Metasedimentary screen enclosed in intrusive complex west of Sgorr Dhearg

9.2.2 Thin Section Characteristics

The type-C leucosomes of the boudin-type structures show textures that are typically found in granites and pegmatites. These include: (1) euhedral to subhedral perthitic K-feldspar crystals: (2) quartz showing interstitial textures infilling between feldspar crystals; (3) occasional granophyric intergrowths of K feldspar and quartz, and (4) coarse-grained quartz-cored structures. Several of these features may be seen in Fig. 9.3b.

The large areas of quartz occupying the cores of pegmatitic segregations are particularly striking, since the whole area may consist of only one or two optically continuous quartz crystals. Euhedral-subhedral K-feldspar crystals may project into such quartz. These features are clearly similar to the final two of the four stages of pegmatite crystallization described by Mehnert (1968, p. 167).

In parts of these leucosomes where the minerals are more evenly distributed, small single quartz crystals also assume irregular, interstitial and xenomorphic shapes, which are moulded between other phases in the rock and often contain euhedral-subhedral projections of feldspar. Such textures also occur where quartz is not very abundant in the variable domains of types (B) and (D) as described above.

Where quartz occurs extensively in the type (B) and (D) layers described above, its usual aspect is of grains with a dominant overall spherical-rounded shape. At the same time, some locally straight grain boundaries mould the outline of these round shapes and indicate the presence of some euhedral faces (Pattison and Harte 1988, Fig. 8).

Two other distinctive textures are associated with the leucosomes in the boudin-type and agmatitic migmatites: (1) the variable occurrence of intergranular albite rims around K-feldspar crystals; (2) biotite occasionally forms large poikilitic crystals which enclose feldspars and quartz of similar size to those in the matrix. In addition, we may note that the K-feldspar in leucocratic domains in migmatititc rocks is characteristically cloudy and more perthitic than in the high-grade rocks lacking leucosomes (Pattison and Harte 1988). Muscovite in relatively large grains of apparently late-stage origin occurs sporadically in the leucosomes.

Despite the distinctive textural features noted above, a considerable proportion of the leucocratic material of (C) and (D) type, and particularly that forming the closely associated B layers, has 'normal' metamorphic textures. By this we mean that the constituent mineral grains (mostly feldspars, quartz, biotite) show: (1) moderately smooth grain boundaries; (2) regular (usually large) dihedral angles; and (3) little evidence of rational faces (except in biotite). Thus they have the texture of reasonably well-equilibrated solid state aggregates.

The typically hornfelsic type A layers of boudin structured rocks are usually very rich in cordierite, together with a moderate proportion of biotite and K-feldspar, and often some andalusite and/or sillimanite. They may or may not contain quartz, corundum or spinel. Often these layers show a very small-scale leucocratic network structure, and further description of this and the associated textures is given in the next section. At the contact of the fractured cordierite-rich (A) layers with the assorted leucocratic and (B) material in the boudin structures, it is common to see narrow (0.5 mm) selvedges marked by a greater proportion of biotite and/or a lack of K-feldspar.

9.2.3 Interpretation

Both the structure and the detailed textural characteristics of the migmatites described in this section suggest that the migmatites are a product of partial melting. At the textural level, the interstitial-xenomorphic quartz textures, euhedral-subhedral K-feldspar, albitic rims on K-feldspar and occasional granophyric intergrowths are all features associated with granites and pegmatites (Mehnert 1968). We further discuss the interstitial-xenomorphic quartz texture and associated euhedral-subhedral K-feldspar in Sect. 9.3, and suggest that it results from the quartz taking on the shape of pre-existing melt pools in the late stages of melt crystallization. On the other hand, the rounded quartz textures in the quartz-rich layers are believed to result from the actual melting process, with the overall rounded shape reflecting the shape of residual grains generated as the melt develops. The partial development of rational faces on such grains again dates from the crystallization stage, when a coating of quartz precipitated on the residual grains.

As noted, some parts of the (B), (C) and (D) domains show metamorphic textures indicating subsolidus equilibrium. It is uncertain to what extent these textures are a product of achievement of textural equilibrium in the subsolidus after melt crystallization, rather than indicative of an absence of melt at any time. We have no solution to this conundrum, but since many grain boundaries showing good subsolidus equilibrium occur in domains that also show sparse interstitial-xenomorphic quartz, we believe that subsolidus re-equilibration may have been substantial.

On the larger scale, the characteristics of the boudin-type and agmatitic structures clearly suggest the flow of material which can inject itself between fractured hornfels and carry xenoliths. The highly localized occurrence of the leucosomes in these structures is particularly distinctive. Those leucosomes infilling between pulled-apart hornfels fragments frequently have little continuity beyond the edges of the bed they transect, and such connections to adjacent layers as do exist are an apparent merging of the leucosome material (C) with material of more normal semi-pelitic aspect (B). The structures clearly suggest that the pull-apart of the rectilinear layers (A) results in the flow of the ductile layers (B), with the leucosomes (C) being largely derived from the (B) layers. With the agmatitic structures as well, there is typically no sign of a connection to pervasive leucosomes crosscutting a series of metasedimentary units, and melt represented by the leucosome appears to be of local origin. Therefore, and as further argued by Pattison and Harte (1988), we have no doubt that these migmatites are the products of in situ partial melting. Further evidence of the intimate occurrence of melt at the grain scale in many of the highest grade rocks is adduced in the next section.

The boudinage and tablet structures show an extensional fracturing process particularly affecting the cordierite-rich hornfels layers. This may be accounted for if one considers the likely volume change caused by melting, and the development of melt to differing amounts in semi-pelitic and pelitic lithologies. Given that the melting process resulted in the replacement of solid plus a small amount of volatile fluid by solid plus melt (Pattison and Harte 1985, and Chap. 8, this Vol.), the overall change in volume would depend on the relative amounts of solid, volatile fluids and melt, as well as their unit volumes. Assessing the amount of melting, even within one layer, is extremely difficult (see Sects. 9.2.3 and 9.3.3), but the amount of melt in semi-pelitic layers probably exceeded 10% close to the intrusive complex. The volatile fluid phase (with fluid pressure equal to load pressure: $P_F = P_L$) present prior to melting (Pattison and Harte 1985; see also Pattison, Chaps. 16, 19, this Vol.), probably had a volume of < 1%, on the basis that the fluid porosity of the metamorphic rocks concerned was probably less than 1% (Fyfe et al. 1978). Thus expansion must be expected to have accompanied melting, and this would have raised the fluid (meaning volatile fluid or melt) pressure.

Such a build up of fluid pressure with melting would reduce the effective stresses (Hubbert and Rubey 1959). Given pressures during metamorphism of 0.3 GPa (Pattison, Chap. 16, this Vol.), the amount of melting necessary to generate melt (fluid) pressures in excess of this is estimated as below 10% (calculation following Yoder 1976, p. 171). Thus it is readily envisaged that the least principal effective stress would become tensional, and melting would lead to hydraulic fracturing of the rock layers (Nicolas and Jackson 1982) in a manner consistent with that described above. Clearly, this fracturing has been controlled by the strong pre-existing layering anisotropy in the rocks, coupled with the accentuation of these differences by varying amounts of melting in different layers.

9.3 Small-Scale Leucocratic Structures Within and Along Rock Layers

9.3.1 Field Description

Small-scale textures and structures involving separation of leucocratic and melanocratic mineral constituents on scales less than 1 cm, and including the widespread spotted structure of cordierite-bearing hornfelses, are extensively present throughout the aureole, but show some changes in features at different grades of the aureole. At low grades, the spotted structure of the cordierite-rich hornfelses has the typical appearance of dark cordierite spots (porphyroblasts), in a matrix which closely resembles that of the regional schists unaffected by contact metamorphism. In the moderate- to high-grade hornfelses of Zones IV and V, the spotted structure, on weathered surfaces, is defined not so much by the spots themselves as by the presence of an irregular honeycomb or network structure of leucocratic ribs (Fig. 9.5a; Pattison and Harte Chap. 8, Fig. 8.8). In cordierite-rich rocks, the leucocratic ribs are usually < 1 mm thick, whilst the cells between these ribs usually have diameters of 2-4 mm. The ribs defining the network do not usually show a preferred orientation, but sometimes the ribs are longer and perhaps thicker in some directions than others. In some cases, particularly in relatively uniform layers of semi-pelitic composition, the network is defined by coarser leucocratic ribs, 1-2 mm thick, while the cells between the ribs may approach 1 cm across (Fig. 9.5b).

This network type of spotted structure first develops outside the zone of partial melting, as noted, but within the zone of highest grade rocks showing boudin- and breccia-structured migmatites, this network structure may show features which are more suggestive of a vein network. Thus, the leucocratic network structure may become more continuous and less irregular (e.g., Pattison and Harte 1988, Fig. 11); or the network becomes accompanied by and connects with more continuous veinlets of leucocratic material, similar to that forming the ribs of the network (Fig. 9.5d). In

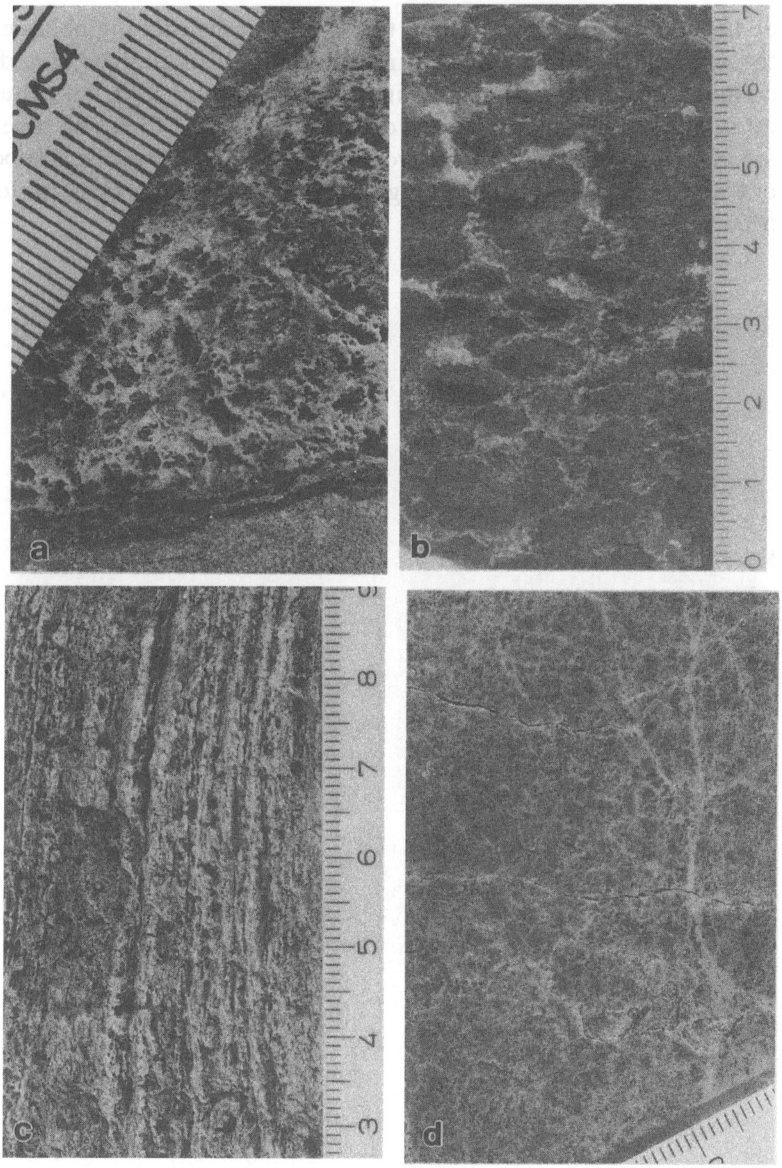

Fig. 9.5. Small-scale structures shown by leucocratic material (small divisions on scales are in mm). **a** Leucocratic network structure in cordierite-rich hornfels, showing typical irregular honeycomb structure found in both Zones IV and V. **b** Coarser network of leucocratic ribs (very rich in potash felspar) enclosing cells rich in cordierite and quartz. **c** Laminated structure running parallel to the original foliation. **d** Rock showing leucocratic network structure, as in **a** and **b**, but with development of veinlets and linking of network structure with more regular and continuous veins

addition, thicker veins (2-3 mm thick) may occur running oblique or parallel to the original foliation. The thin network structure and associated veinlets may connect with these thicker veins (Figs. 9.5d and 9.6), and thus indicate an interconnecting system of veins of varying sizes. At the same time, the relations are complex; sometimes the network and veins of different sizes appear to transect one another (Fig. 9.6b), as though overlapping episodes of vein formation have occurred.

Fig. 9.6. Small-scale leucocratic network structure passing into veinlets and connecting with more continuous veins. In **a** the majority of the leucocratic network shows a predominant orientation from *upper right* to *lower left* and a continuous passage to vein-like structures of similar orientation. In **b** gradual transitions from an irregular network to veinlet structures are seen in many parts of the photograph, but some of the thicker veins also appear to transect this structure. NN053531 by Coire Chaorain (locality 3, Fig. 9.1)

The leucocratic network structures described above are best developed in pelitic and semi-pelitic layers which have relatively uniform petrographic features on the scale of several centimetres in thickness. In laminated rocks (with thin layering of original sedimentary or regional metamorphic origin, or originally strong cleavage structure) the above network structure is often not well developed, but at the highest grades, the laminated structure becomes accentuated. Thus thin seams (typically < 2 mm thick) of more leucocratic, and sometimes vein-like material, alternate with more melanocratic material (Fig. 9.5c) in a very regular, almost platy structure. Biotite selvedges may form along the contacts of these thin seams and add to the sharpness of their appearance. The more leucocratic and vein-like seams in such laminated arrays occasionally thicken very locally in a bead-like structure, or they may become connected by short leucosomes at right angles to the lamination, thus showing aspects of the boudin-like structure (Sect. 9.2) on a very small scale. Within a sequence of rock layers adjacent to the contact with the intrusive complex, it is common to find layers with both the network structure (Fig. 9.5a,b) and layers with the strong laminated appearance (Fig. 9.5c) as well as transitions between the two.

All these fine leucocratic and vein-like structures are most commonly <2 mm in thickness. They are typically of whitish colour and feldspar-rich, rather than the more pellucid quartz-rich veins or seams which often run parallel to foliation in the regional metamorphic rocks. Indeed, as one follows the metasediments towards the contact with the intrusive complex, for example to the west of the Allt Giubhsachain (area 1, Fig. 9.1), one may see that thin quartz veins of regional metamorphic origin become feldspar-rich, whilst the number of veins simultaneously increases. Where cordierite-rich layers with network structure are adjacent to semi-pelitic and laminated material, there are typically thin leucocratic seams along the contacts which may connect with both the plane-parallel veins of the semi-pelite and the thicker veins and network of the pelite. Thin biotite selvedges may occur between the leucocratic seams or veins and the cordierite-rich layers.

Small-scale leucocratic structures with evidence of more continuous vein development, as noted above and shown in Figs. 9.5d and 9.6, are seen well at localities 1 and 3 (Fig. 9.1), and in these areas there is also evidence of these very small-scale leucocratic structures connecting to larger-scale vein structures. Thus a network structure which is showing the initial development of veinlets may pass into one with more strongly developed vein structures, and these in turn pass into conspicuous and regularly orientated veins (Fig. 9.6). On the hilltop west of Coire Chaorain (locality 3), the vein networks are connected in this way to pervasive vein sets of the type described in Section 9.4.

9.3.2 Thin-Section Features

Where it first appears in the aureole in Zone IV, the leucocratic network structure described above commonly consists of K-feldspar, quartz and muscovite, with the latter often showing a preferred orientation appearing to be a relict of the regional schistosity. Within the zone of melting, the leucocratic networks are completely reconstituted texturally and are dominated by K-feldspar, with very variable amounts of aluminosilicate, plagioclase, biotite and quartz. When present, biotite is usually randomly orientated.

K-feldspar is commonly by far the most abundant phase, and in some cases forms more than 75% of the network. Areas in thin section consisting of K-feldspar alone, commonly show a regular equant polygonal grain structure, with approximately 120° angles at triple junctions. When quartz is present, it often has a distinctive interstitial and xenomorphic habit (Fig. 9.7a), with a single crystal of quartz showing a series of low angle penetrations along the grain boundaries of adjacent K-feldspar grains. K-feldspar often shows euhedral extensions into such quartz (Fig. 9.7a), and there are clear similarities to the features described in the larger leucosomes (Sect. 9.2).

In the more aluminous hornfelses, andalusite is often a prominent constituent of the leucocratic networks alongside K-feldspar, and sometimes corundum is also present. In such cases the andalusite crystals often show a skeletal habit, in which variably shaped but sometimes subhedral, tiny segments of andalusite are seen to show optical continuity. K-feldspar occupies the space between the andalusite segments, and there is often a halo of K-feldspar alone around the andalusite (Fig. 9.7b), with biotite becoming a prominent phase beyond this halo (see also Pattison and Harte, Chap. 8, this Vol.).

The leucocratic seams in the laminated rocks, and the extensions into more obvious veinlets and veins, also tend to be dominated by K-feldspar, with variable amounts of quartz, plagioclase and biotite. A variety of textures are seen as with the type (B) domains in the boudin structured rocks (Sect. 9.2). Where limited amounts of quartz are present, it commonly shows the interstitial xenomorphic habit with low dihedral angles already described. Where quartz is more abundant, it commonly shows the both rounded and subhedral habit noted in Section 9.2. Biotite may occur as randomly orientated small laths or larger poikiloblastic grains (similar to those illustrated in Fig. 9.7d, though the specimen illustrated comes from the Chaotic Zone). At the same time, many parts of these rocks may show equigranular metamorphic textures as noted for the (B) domains is Section 9.2.2.

The rock volumes in between the small-scale leucocratic network structure, and forming the more melanocratic seams in the laminated rocks, are typically dominated by cordierite. In addition to cordierite, it is usual to find some biotite and K-feldspar and sometimes quartz, which may all occur both as separate grains and as inclusions in cordierite. In thin section, continuous areas of cordierite in plane polarized light are often seen to be formed by aggregates of crystals in crossed nicols, and in these aggregates there appears to be a restricted variation in optical orientations. In some cases, the aggregates may be formed by interlocking cordierite porphyroblasts. Sometimes the cordierite shows polysynthetic twinning, and there may be segments showing polsynthetic twins orientated at approximately 60° or 120° to one another. Where quartz is present in the cordierite domains, it frequently displays the interstitial-xenomorphic appearance with low-angle penetrations between cordierite grains (Fig. 9.7c). At the margin of the cordierite-rich domains, there is frequently a distinct selvedge in which biotite is relatively abundant and/or K-feldspar is lacking.

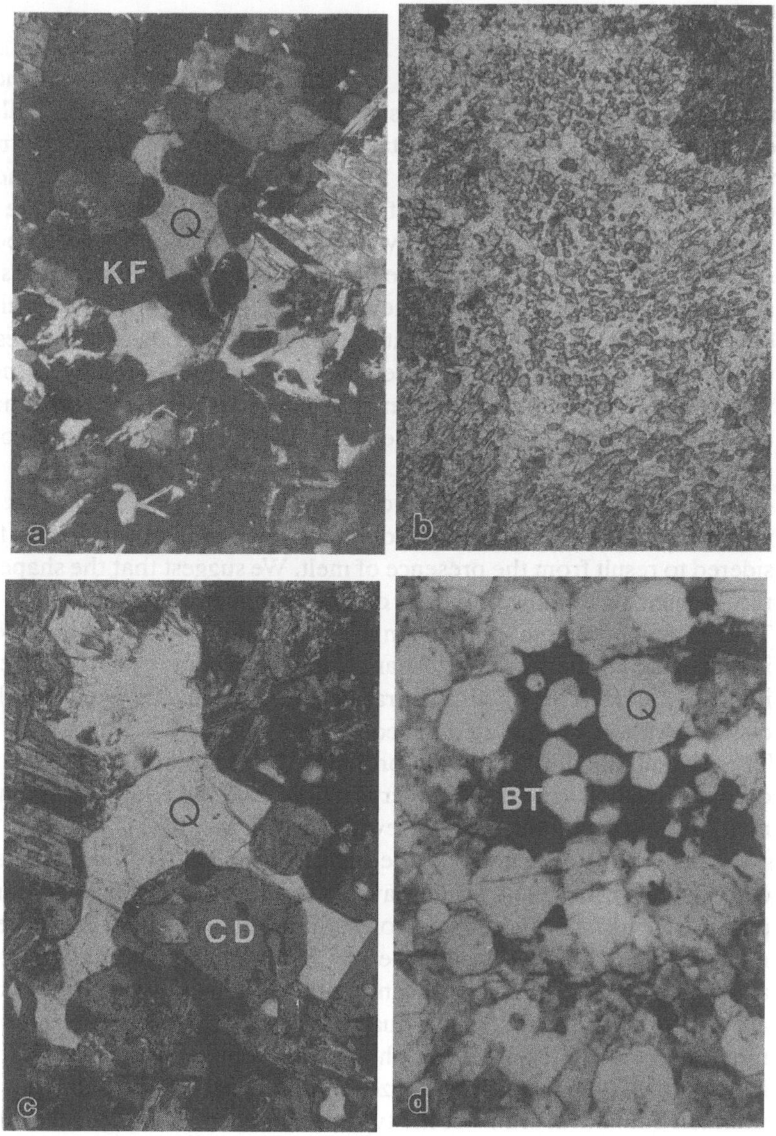

Fig. 9.7a. Interstitial-xenomorphic quartz (*Q*) showing a series of low angle penetrations along the grain boundaries with K-feldspar (*KF*). Some K-feldspar shows euhedral extensions into the quartz (figure is 0.5 mm across). **b** A group of several skeletal andalusite grains embedded in K-feldspar, and tending to have a halo of K-feldspar alone. Biotite is prominent towards the edges of the photograph (figure is 3.0 mm across). **c** Interstitial-xenomorphic quartz (*Q*) within the cordierite-rich (*CD*) domain of Fig. 9.5b (figure is 0.5 mm across). **d** Interstitial-poikilitic biotite (*BT*) enclosing rounded-subhedral quartz (*Q*) grains (figure is 1.0 mm across.) All specimens from Locality 1 (Fig. 9.1) except **d** which is from locality 5 (Chaotic Zone)

9.3.3 Interpretation

The way in which the small-scale irregular leucocratic network structure develops small-scale venous characteristics and locally becomes linked with larger, more continuous veins, close to the contacts with the intrusive complex, strongly suggests the presence of melt. Similarly, in rocks with laminated structure, the development of laminae with more uniformly feldspathic compositions, with the occurrence of biotite selvedges and occasional veinlets crosscutting more melanocratic laminae, suggests melt presence. Clearly the network and laminated structures are not purely a product of melting, because they are developments of structural features seen outside the zone of melting. Thus, it is more a matter of these structures having a local control over the extent of melt development because they represent modal variations within the pre-existing rocks. This makes it extremely difficult to estimate the amount of melt which these small-scale leucocratic structures may have contained in the melting subzone.

The thin section characteristics provide further strong evidence of melting. In particular, the wide development of interstitial-xenomorphic quartz textures is considered to result from the presence of melt. We suggest that the shape of such quartz results from it assuming the shape of small volumes of melt during its crystallization. That a small pool of melt has been partly pseudomorphed by the interstitial quartz, is supported by the way K-feldspar crystals with some euhedral faces often project into the quartz. The small dihedral angles, typically $< 60°$, shown by such quartz contrast with its normal ca 120° equilibrium angles in contact with many silicates. These small dihedral angles are probably the result of both the chance shapes left occupied by melt as feldspar has crystallized, and the equilibrium dihedral angles of melt in contact with feldspar (Jurewicz and Watson 1985).

This is not to suggest that the melt had the composition of quartz. The other components of the melt would have crystallized on pre-existing mineral grains, in particular feldspars. Biotite and cordierite would crystallize from the melt to a very much lesser degree because of the low Fe-Mg melt contents expected (Puzriewicz and Johannes 1988). It is significant that the interstitial-xenomorphic quartz shapes are typical of situations where quartz is a relatively minor constituent and would therefore be a late-crystallizing phase, which could pseudomorph the shape of the last volumes of the melt to crystallize. Where quartz is more abundant it usually forms rounded-subhedral grains of the type discussed in Section 9.2.3, which are thought to largely represent residual grains formed during melting.

It follows from the above that the volume represented by interstitial-xenomorphic material only provides an absolutely minimal estimate of the amount of melt present at the time of maximum melting. The problem of estimating the maximum amount of partial melting is exacerbated by the difficulty of knowing to what extent melt textures have been removed by sudsolidus re-equilibration (see Sect. 9.2.3).

Although it is very difficult to quantify the amounts of melt present in different rock layers, the detailed textural evidence provides important data on the distribution of melt. The occurrence of interstitial-xenomorphic quartz in the cordierite-rich domains in some rocks clearly indicates that they also contained melt. Furthermore, Jurewicz and Watson (1985) provide evidence that silicic melts commonly show dihedral angles against coexisting silicates of < 60°, this is supported in the present case by the low dihedral angles of the interstial-xenomorphic quartz. This means that with a reasonable approach to textural equilibrium, the melt in both melanocratic and leucocratic domains would have formed a connected network, penetrating along the grain edges of the solids even when present in very small amounts (e.g., Beeré 1975; Hunter 1987).

These considerations also have an important bearing on the redistribution of the melt components during melt crystallization. The leucocratic domains of the rocks often have compositions dominated by K-feldspar and not suggestive of expected melt compositions. This is partly a product of their initial subsolidus development, but we believe it has been accentuated during melt crystallization, because the continuity (or connectivity) of the melt network allows diffusion of the K-feldspar component of the melt into domains rich in solid K-feldspar mineral during crystallization (Harte and Linklater 1989). Similar segregation of the quartz component of the melt may also occur, such that in some cases it becomes most abundant as solid quartz in the cordierite domain. The development of the selvedge structures, often relatively rich in biotite, between K-feldspar-rich and cordierite-rich domains is also thought to be a product of the same migration of components through the melt (Harte and Linklater 1989).

The above features of redistribution of melt chemical components during the crystallization of the melt clearly have wider importance with respect to the extent that leucosomes in some migmatites may show bulk melt compositions, melanosomes show restite compositions, and the sum of the two represents the parent rock (Mehnert 1968). With melt connectivity through layers of widely differing melt and solid proportions, and the redistribution of melt chemical components during melt crystallization, no simple identification of originally molten and solid parts of the rock will be possible in the resultant solid migmatite. This may well provide an explanation of why some leucosome and melanosome compositions do not fit a simple melt and restite model, and why mineral compositional distinctions between leucosome and melanosome are not always clear (Johannes 1985).

9.4 More Extensive and Crosscutting Leucosomes

We place into this category those leucosomes which form structures which cut continuously across several compositional layers of the original foliation of the metasediments. Thus they have much more the aspect of externally-derived intrusive material than the leucosomes described in Sections 9.2 and 93. However, we shall suggest in the interpretative section that they also appear to be a product of local partial melting. From the descriptive viewpoint three principal types of extensively crosscutting leucosomes may be distinguished as described below.

9.4.1 Pervasive Crosscutting Vein Sets

The veins forming these sets very largely consist of quartz, K-feldspar, biotite, plagioclase and late-stage muscovite. They have greater continuity and regularity of development than the intimate networks of veinlets and veins described in Section 9.3, and are typically 2-10 mm thick (Fig. 9.8). The larger veins may be traced over several metres or tens of metres and show some consistent orientations. Usually there

Fig. 9.8 a. Pervasive veins of subvertical orientation are cut across by crack-like alteration veins running sub-horizontally. Both sets of veins traverse dominantly granular material of the Chaotic Zone. **b** Pervasive veins (dipping steeply to the *left*) cut across the general foliation (parallel to pen), but branch to form veins running parallel to the foliation. The host rock varies between the laminated and granular types of the Chaotic Zone

is a dominant orientation trend, which crosscuts the metasedimentary layering, but subordinate trends may also occur; and the veins in these different orientations interconnect with on another. Usually one trend is parallel to the metasedimentary layering, so that not all veins are crosscutting, even though the vein sets as a whole transect the general rock foliation.

In detail the thinner veins in these vein sets, may show much branching despite maintaining an overall trend. Sometimes a thicker vein is a composite of subordinate veins which locally divide and then come together again. Usually the veins appear to cut across the network and veinlet structures of Section 9.3, but locally there are clear connections between the two (see Sect. 9.3 and Fig. 9.6).

Conspicuous sets of these veins have a limited occurrence. They are best seen in the Chaotic Zone (see also Sect. 9.5), and in a more limited area of the SE contact zone by Coire Chaorain (locations 5 and 3 respectively, in Fig. 9.1).

9.4.2 Medium- to Coarse-Grained Leucosomes

These leucosomes have a relatively coarse grain size and granular texture. Quartz, potash feldspar and plagioclase are their principal constituent minerals, but biotite and muscovite may also be prominent. The proportions of the minerals vary widely, and in the field local concentrations of otherwise sparse muscovite plates are particularly obvious.

The bodies formed by these leucosomes may be sheet-like (usually 1-10 cm thick) or more irregular. The irregular bodies are usually up to 0.5 m in total width, and consist dominantly of a series of intrusive arms penetrating the hornfelses principally along the direction of the original foliation, but with apophyses extending in other directions. These leucosomes always occur sparsely in any area, and sheet-like bodies are rarely traceable over more than a few metres; thus they do not appear to form regularly orientated sets of sheets or veins. Some of the thicker bodies carry randomly orientated fragments of hornfels.

9.4.3 Crack-Like Alteration Veins

In the field these often appear as regular crack or joint surfaces, in which the immediately adjacent rock surfaces (to a depth of 1 mm or less) are formed of uniform pale grey material, probably formed by alteration of the original rock. Although the amount of leucocratic material may be very small, these crack-like veins may show considerable continuity over several metres (Fig. 9.8a). They are probably the latest of the leucosome and vein structures to be generated in the Ballachulish aureole.

9.4.4 Interpretation

The leucosomes described in this section show greater lateral continuity than those of Sects. 9.2 and 9.3 and in the case of the interconnected vein sets may represent, in aggregate, relatively largely volumes of material. In combination with these features, their extensive crosscutting of the metasedimentary layering and foliation, suggests intrusive material which does not have the same very localized (in situ) derivation as the leucosomes described in Sections 9.2 and 9.3.

However, we have noted that some pervasive veins show connections to the small-scale leucocratic networks and veinlets. Also some co-sanguinity between the medium-to coarse-grained leucosomes and the agmatitic migmatites of Section 9.2 is suggested by: (1) similar variations in mineralogical constitution; and (2) the occasional occurrence of hornfels xenoliths in the medium-coarse leucosomes.

Thus the pervasive veins and the medium-coarse leucosomes may have been derived from melts formed within the metasediments. Their distinction to the leucosomes of Sections 9.2 and 9.3, is therefore associated with their greater mobility and a tendency to be of relatively late age of formation. Perhaps the greater mobility of the pervasive and crosscutting leucosomes results from their acquiring high water contents during the late stages of melt crystallization. Such a situation would also explain why these leucosomes may both cut across and yet connect with the more localized and small-scale leucocratic features.

The crack-like alteration veins appear in many respects to be a later and distinct phenomenon. They seem to represent narrow channelways along which fluid has passed and caused metasomatism to the immediately adjacent rock.

9.5 The Chaotic Zone

9.5.1 Field Description

The diverse phenomena described in Sections 9.2-9.4 are all found widely distributed throughout the inner aureole of the Ballachulish intrusive complex, including the 'Chaotic Zone' on the western side of the complex (Fig. 9.1, area 5). This zone is characterized by disruption of the original foliation on a much more extensive scale than elsewhere. Thus while the overall preservation of the orientation of metasedimentary layering is evident in exposures a few metres across in most parts of the inner aureole, some exposures of this scale in the Chaotic Zone consist of diversely orientated xenoliths dispersed through a granular matrix. However, we must not overemphasize this larger scale disruption; we shall see that considerable continuity of metasedimentary layering and foliation is still present in the Chaotic Zone.

In broad terms the rocks of the Leven Schists of the Chaotic Zone are fine to medium grained, and, whilst transitional types are common, two dominant distinctive types may be recognized in the field:

(a) Obvious metasedimentary rocks showing evidence of original compositional layering (bedding) on scales of centimetres to metres (Fig. 9.9). Many of these rocks possess a fine-scale laminated structure similar to that described in Section 9.3. There are also layers richer in cordierite, which show the fine leucocratic network structure described in Section 9.3; rarely, there are very andalusite-rich layers.
(b) Rocks with a generally granular appearance (Fig. 9.10b), which are much more homogeneous in both composition and structure than those of type (a). Quartz, feldspars and biotite are again prominent, giving the appearance of relatively quartzofelspathic semi-pelitic rocks; however, particularly cordierite- or andalusite-rich layers are lacking except as xenoliths. Some foliated structure may be present but it is usually relatively weak and typically diffuse; sharply bounded units are generally absent. Sometimes the diffuse foliation bends in a manner suggestive of flow (Pattison and Harte 1988).

The junctions between these two rock types may be both sharp and gradational. Where a sharp change occurs, there is usually a leucocratic vein connecting with the pervasive, dominantly crosscutting set (Sect. 9.4.1) present along the contact (Fig. 9.9b). Gradational changes between the rock types may occur both parallel to and at irregular angles to the original metasedimentary foliation.

Vein sets of the type described in Sect. 9.4.1 are common in both of these lithologies (Figs. 9.8 and 9.9); they tend to be thicker and more diffuse when crossing the more granular rock type. Despite commonly branching, the veins usually show considerable general continuity along particular directions and thus slice pervasively through both the laminated and granular lithologies. The veins vary in thickness from around 1 mm to 10 mm, but the thicker veins often appear to be composite and formed by the coming together and running parallel of several thinner veins, which may separate again elsewhere. The average dominant trend of these veins is oblique to the original foliation and dipping very steeply to the N or NNE, but a second common orientation is parallel to the original foliation and dipping steeply to the ENE. However, a wide variety of orientations occur, including sub-horizontal veins.

Breccia and xenolith structures are common throughout the Chaotic Zone. Both structures are marked by the occurrence of disorientated fragments of metasediment, and they vary from situations in which there are a large number of randomly orientated fragments occurring close together (agmatitic migmatites) to situations where relatively isolated fragments (xenoliths) occur (schollen migmatites). The most striking xenolithic occurrences consist of fragments of the layered and laminated lithology 'floating' within a matrix of the granular lithology (Fig. 9.10). Typical occurrences involve xenoliths of laminated rocks, with their foliations in diverse orientations, or xenoliths of cordierite-andalusite hornfels lying within reasonably homogeneous granular lithology.

These structures are not confined to regions with a granular matrix, however. A brecciated appearance may be found within regions of the layered and laminated lithology, whereby closely packed rock fragments lie with their foliations disorientated with respect to one another. Usually the rock fragments in such situations are centimetres or a few tens of centimetres across, but fragments ca. 1 metre across also

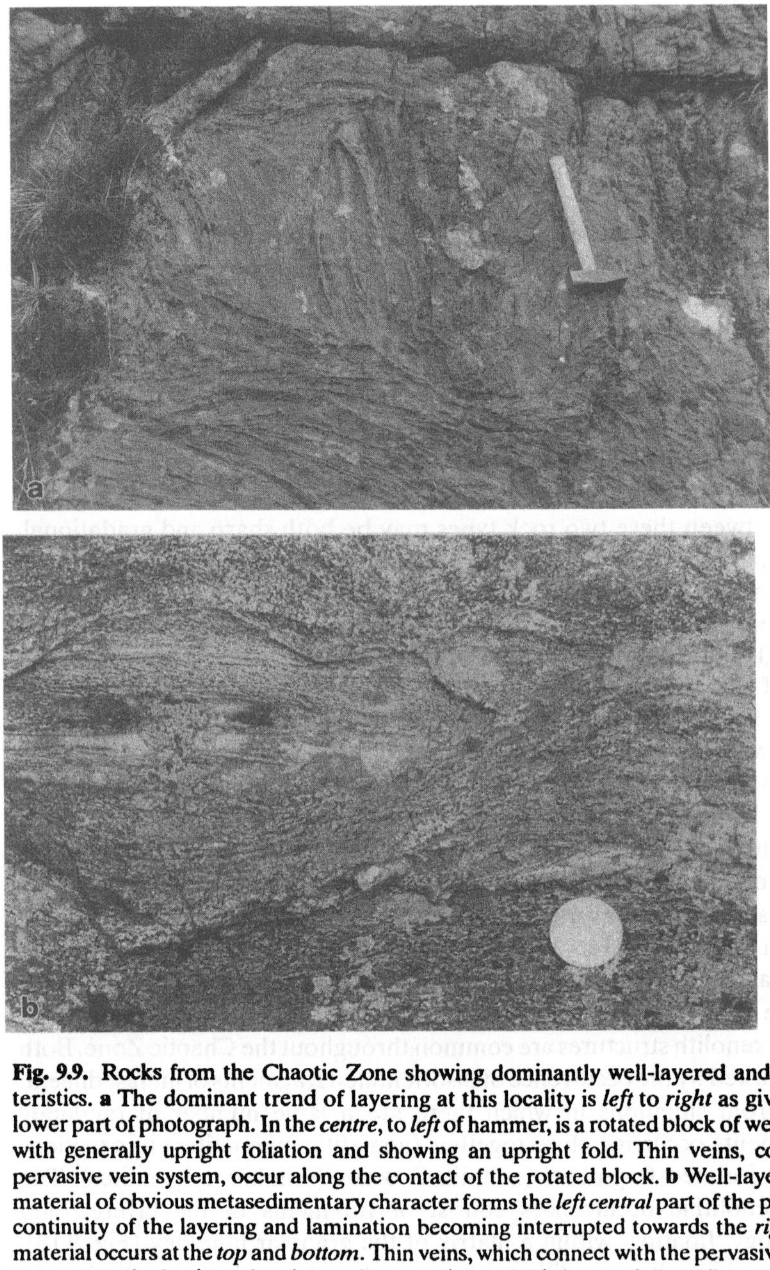

Fig. 9.9. Rocks from the Chaotic Zone showing dominantly well-layered and laminated characteristics. **a** The dominant trend of layering at this locality is *left* to *right* as given by the rocks in lower part of photograph. In the *centre*, to *left* of hammer, is a rotated block of well-layered material, with generally upright foliation and showing an upright fold. Thin veins, connecting with the pervasive vein system, occur along the contact of the rotated block. **b** Well-layered and laminated material of obvious metasedimentary character forms the *left central* part of the photograph with the continuity of the layering and lamination becoming interrupted towards the *right*. More granular material occurs at the *top* and *bottom*. Thin veins, which connect with the pervasive vein system, tend to separate the laminated and granular material, and this is especially well seen at *upper boundary* of the laminated rocks

Fig. 9.10. **Xenoliths in the Chaotic Zone rocks. a A relatively large laminated xenolith dominates** the photograph and is surrounded by dominantly granular material containing many small xenoliths. **b** Granular rock type containing small xenoliths which are all of similar laminated rock type

occur. Occasionally there are striking occurrences where an exposure shows considerable continuity and coherence of foliation in the layered and laminated lithology, but a block of layered rock with different orientation lies embedded in the sequence. Figure 9.9a illustrates such a case where the disrupted block is unusually large (about 1 m across).

Typically, where disrupted fragments of laminated rocks lie in close proximity, they are separated by leucocratic vein material, which usually appears to be continuous with material of the pervasive vein sets. Such vein material anastomoses from the more regularly orientated veins and wraps around the disrupted fragments, but it does not usually cut across the xenolith and breccia fragments themselves.

The occurrence of disruption within the Chaotic Zone is impressive in the field, but the extent of disruption and the distances of transport of rock fragments with respect to one another is not so clear. Where trains of xenoliths occur over several metres within the granular lithology, they frequently appear to represent portions of only one or a few metasedimentary layers. This is particularly obvious with respect to trains of andalusite- and cordierite-rich hornfelses, which are not common rock types in this part of the aurole, and are most frequently seen as xenoliths, presumably because of their relatively rigid and competent nature. Where breccias show a close association of diverse semi-pelitic and pelitic rock types, the rock fragments are usually only a few centimetres across, and may well be derived from close to one another by disaggregating thinly layered (centimetres scale) units containing a variety of rock types; such units may be found within coherent parts of layered and laminated lithology.

9.5.2 Thin Section Characteristics

The well-layered rocks with a laminated or small-scale leucocratic network structure often show features similar to those described and discussed in Section 9.3. In the granular lithology, textural evidence of melting is particularly common. This evidence is provided by the common occurrence of subhedral-euhedral crystals of one phase being partly embedded in interstitial-xenomorphic crystals of another phase. Quartz may form interstitial xenomorphic crystals to both K-feldspar and cordierite (Fig. 9.11a,b), and in many of these rocks a subhedral habit of smallish stubby cordierite crystals is common (Fig. 9.11b). In other cases the quartz forms rounded-subhedral crystals with K-feldspar (Fig. 9.11c) or cordierite (Fig. 9.11d) forming the interstitial phase. Where large rounded-subhedral quartz grains are present, the surrounding matrix may include much finer-grained quartz. In some layers, biotite occurs as large poikilitic crystals enclosing rounded-subhedral quartz grains (Fig. 9.7d). When andalusite is present, it tends to form large, often quite rounded crystals which are associated with a concentration of large biotite crystals and interstitial quartz.

9.5.3 Interpretation

The textural characteristics of the Chaotic Zone rocks as seen in thin section are believed to imply crystallization of melt, as already discussed for similar textures in Sections 9.2.3 and 9.3.3. The main differences in the Chaotic Zone are the greater abundance and wider distribution of such textures, and the much clearer evidence of cordierite with euhedral-subhedral habit.

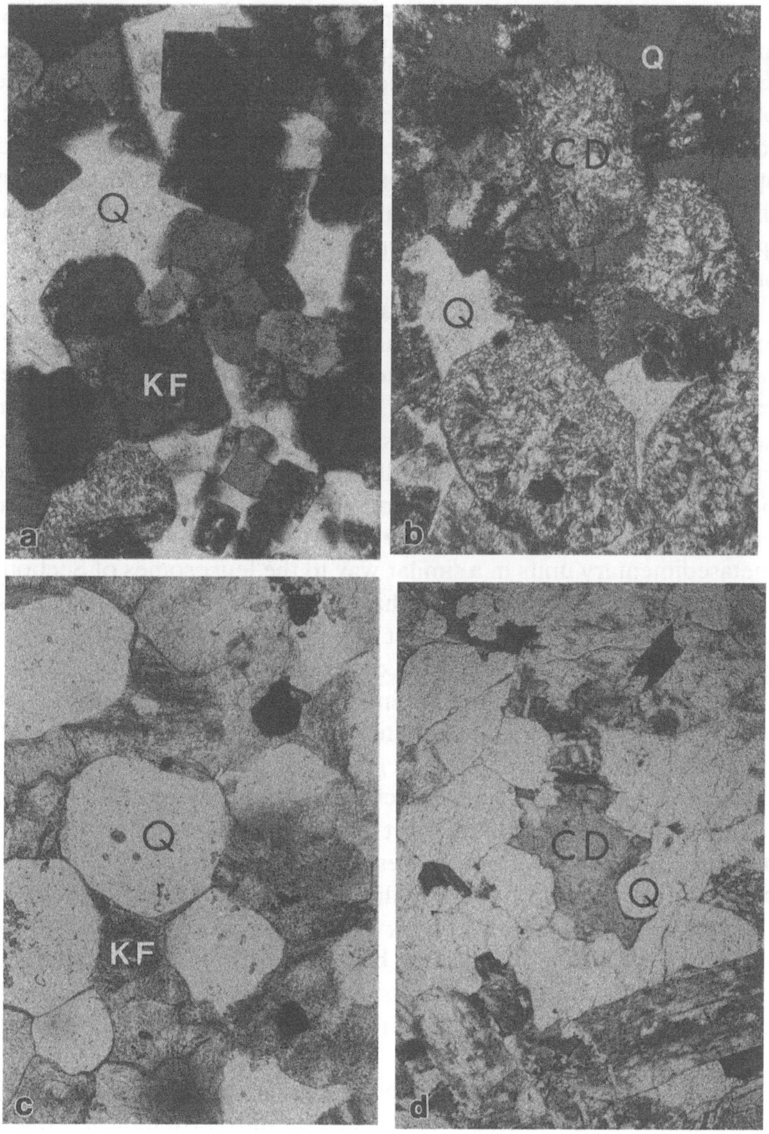

Fig. 9.11. This section characteristics of Chaotic Zone rocks. **a** and **b** Euhedral-subhedral K-feldspar (*KF*) and cordierite (*CD*), respectively, in a matrix of large optically continuous quartz (*Q*) crystals showing interstitial habit. The cordierite crystals show much alteration but their outline shapes are well preserved. **c** Murky K-feldpar (*KF*) crystals show interstitial habit to rounded-subhedral quartz (*Q*) grains. **d** Cordierite (*CD*) in *centre* has interstitial-xenomorphic habit between quartz (*Q*) grains (all figures 1 mm across except **d** which is 2 mm)

Whether a particular phase forms euhedral-subhedral crystals (or in the case of quartz rounded-subhedral crystals), in distinction to interstitial-xenomorphic crystals, is seen to be related to the abundance of the phase in the rocks. Where a particular phase has interstitial-xenomorphic form, it is usually less abundant than where it forms subhedral crystals. Thus the variation in the textures across different layers appears to reflect expected compositional controls on the order of crystallization of different minerals. Specific mineral associations such as andalusite with biotite and quartz, may also be explained by specific melting/crystallization reactions (Pattison and Harte 1988, p. 491).

In the field, the distinctive features of the Chaotic zone particularly relate to the extensive occurrence of the granular textured material. This has a general semi-pelitic compositional appearance, but lacks the well-defined metasedimentary layering expected for such rocks, and has an evident ability to flow around and enclose xenoliths with diverse orientations. These features may be readily explained if the granular rocks represent metasedimentary material which underwent substantial partial melting and acquired magmatic properties (Pattison and Harte 1988). The granular lithology has flowed between and incorporated pieces of more rigid metasedimentary units in a similar way to the leucosomes of Section 9.2, but within the Chaotic Zone this disruption has operated on the scale of metres, in contrast to the centimetre scale seen in most parts of the inner aureole (Sect. 9.2). That the granular lithology has not itself commonly developed leucosomes during these processes, may be related to it having compositions more uniformly similar to the (B) layers of the boudin-type migmatites, and the more extensive melting of this type of material in the Chaotic Zone. The greater degree of melting would have extensively broken up the continuity of contact of the crystalline material, and thereby reduced the rheological contrasts over short distances, so that melt and crystals move together as a suspension in response to stress differences. In many of the boudin-type structures elsewhere in the aureole, as illustrated in Fig. 9.3, bulk flow of melt plus crystals does not appear to have occurred, but rather a process of separation of melt and its infilling of the 'space' between the boudins. In the Chaotic Zone where thin layers of particularly cordierite- or andalusite-rich refractory material occurred, they have been extensively broken up to form disconnected xenoliths floating within moderately uniform granular material.

The apparent ability of the more molten material in the Chaotic Zone to behave as a suspension and undergo bulk (melt + crystals) flow indicates that the critical melt fraction (Van der Molen and Paterson 1979) for such behaviour had been exceeded. This suggests a melt fraction in excess of 30-35% (Van der Molen and Paterson 1979). As we have repeatedly noted elsewhere, assessing the amount of melting from the textures is difficult, because the products of melt crystallization may overgrow unmelted crystalline material and there may be considerable redistribution of melt components during crystallization (Sect. 9.3.3).

Despite the evidence of substantial melting in the Chaotic Zone, the weak foliation of the granular lithology remains dominantly concordant to the relict bedding in the layered rocks, and both show steep dips to the ENE (congruent to the regional lithostratigraphic trend). We have also noted the evidence of limited distances of transport of xenoliths. These features argue against any large-scale diapiric or convective activity in the molten material. On the very small scale, there is also little evidence of crystals sinking through the melt and compacting in response to gravitational forces as found in basic melts (McKenzie 1984; Hunter 1987).

All these characteristics point strongly to the high viscosity expected for a silica-rich melt with limited water content. These observations accord with the behaviour that might be predicted theoretically (Wickham 1987). Thus high viscosities, in combination with a time scale for melting temperatures and therefore melt presence of less than 1.0 million years (Buntebarth, Chap. 18, this Vol.), would render unlikely the possibility of melt segregation by compaction, filter pressing or surface tension processes. However, melt segregation by the extensional fracturing process described in Section 9.2 would be possible on this time scale (Wickham 1987).

The pervasive vein material (Sects. 9.4.1 and 5.1) which cross cuts both the granular and laminated lithologies has clearly not been of limited mobility. The crosscutting relationships indicate a relatively late origin for this material, but the material also shows considerable intimacy with the rocks in the way it may form anastomosing arrays along the contacts between granular and laminated material. It is suggested that this material may represent a relatively hydrous, and therefore mobile, late-stage residue from the crystallization of the melt. It may have segregated to form the veins during the cooling and contraction of the previously extensively molten rocks.

A major cause of the more extensive melting in the Chaotic Zone than other parts of the aureole, is probably the way in which the quartz diorite underlies the metasediments in the Chaotic Zone (Troll and Weiss, Chap. 3, this Vol.). Melting of the quartz-bearing Leven Schists at an H_2O activity of 1.0 would be expected at about 650-680 °C (Pattison and Harte 1988), which is below the estimated temperature of 690-710 ° for devolatilization of the quartz diorite (Weiss 1986). The generation of relatively large melt volumes over a small temperature interval of melting was probably also aided by the greater abundance of quartz within the Leven Schists than other pelitic/semi-pelitic lithologies within the aureole.

An additional factor causing more extensive melting in the Chaotic Zone may have been the liberation of magmatic water into the metasediments. As discussed by Pattison and Harte (1988), the extent of melting within the aureole must have been strongly controlled by the availability of H_2O. Some of this H_2O may have been derived from the metasediments both by dehydration melting of biotite, and by evolution of H_2O from rock layers not forming melt. Additional H_2O from the intrusive complex seems particularly like in the Chaotic Zone because of the underlying disposition of the quartz diorite. Some slight support for this might be taken from the occurrence of two bulk rock ^{18}O values below 11 in the Chaotic Zone. However the overall variation of bulk O^{18} in the aureole makes this evidence weak (Hoernes et al., Chap. 17, this Vol.) and more detailed study is needed.

9.6 Summary

1. In the above descriptions and their associated interpretations, we have provided evidence of the widespread occurrence of localized partial melting in pelitic and semi-pelitic lithologies in the immediate contact zone of the Ballachulish intrusive complex. This evidence is provided by the development of leucosomes as an infilling in boudinage and breccia (agmatitic) structures, and the occurrence of flowage and venous characteristics in small-scale leucocratic segregations. In detail this evidence is strongly supported by textures indicating crystallization from melt.

2. The boudinage and breccia structures are probably the result of hydraulic fracturing, caused by the build-up of relatively high fluid (meaning melt) pressures due to melting, such that effective stresses were reduced and conditions for tensile failure obtained.

3. An area on the western flank of the intrusive complex, the Chaotic Zone, shows more extensive development of melt than elsewhere, leading to wider disruption of pre-existing metasedimentary layering on the scale of metres. This more extensive development of melt is associated with the area being underlain by the intrusive complex, which contributed heat and possibly H_2O to the overlying metasediments.

4. In the Chaotic Zone, some partially molten material appears to have undergone bulk flow as a suspension of crystals within melt, suggesting that the critical melt fraction of 30-35% has been exceeded. Elsewhere in the aureole, in the commoner boudin-type structures, the melt fraction was probably usually less than the critical melt fraction, because under the stress associated with hydraulic fracturing, the melt appears to migrated through the solids into fractures, without transporting the solids themselves.

5. Throughout the innermost aureole, and even in the Chaotic Zone, there was limited flow of the melt and physical segregation of the melt from its associated crystalline elements. Often, flow and segregation of the melt was limited to those regions where there were local stress concentrations, caused by boudinage and brecciation of the more rigid (less melt rich) metasedimentary layers as a result of extensional (hydraulic) fracturing. Given ambient rock pressures of around 3 kbar, rough calculations indicate that less than 10% melting may have been sufficient to raise fluid (melt) pressures the necessary amount to promote such hydraulic fracturing. The lack of occurrence of melt segregation by physical processes other than those associated with extensional fracturing, was probably the result of the high viscosity of the melt in combination with the short (less than 1.0 Ma) time span of melt presence.

6. An exception to (5) above is provided by the occurrence of pervasive and crosscutting veins in some rocks. The material forming these veins appears to have been generated from within the rocks by partial melting, but was clearly sufficiently mobile on the scale of metres to generate thin crosscutting leucosomes. These veins are relatively late in origin and may owe their mobility to the concentration of volatiles during the later stages of melt crystallization. Thus they may be partly hydrothermal in origin.

7. It is exceptionally difficult in most of the partially melted rocks to identify segments of material that represent melt compositions. This is partly due to the limited

segregation of melt and crystalline material noted under (5) above. However it is also related to the connectivity of even very small melt volumes within the solid rock framework, and the redistribution of melt components during crystallization. Thus there is evidence that diffusion of melt components through the largely static melt network caused localized precipitation of different minerals during melt crystallization. This chemical segregation of melt components often appears to have enhanced modal differences in the rocks prior to melting, and led to the formation of selvedges between different modal domains. The occurrence of such redistribution of melt components during melt crystallization, may explain why simple compositional relations between leucosomes, melanosomes and protoliths are rarely observed in migmatites.

10 Decarbonation Reactions in Siliceous Dolomites and Impure Limestones

L. Masch and S. Heuss-Aßbichler

10.1 Introduction

In this chapter the petrography and mineral chemistry of carbonates of the Ballachulish aureole are presented. Carbonates are not widespread in the aureole, so the record of progressive metamorphism is incomplete. The outermost thermometamorphic reaction observed in the aureole leads to the appearance of talc + calcite in siliceous dolomites and impure limestones. In the inner aureole from the point where both forsterite and diopside are formed, the outcrop is continuous up to the formation of periclase.

Mineral analyses of carbonates and silicates (talc, amphibole, clinopyroxene, olivine) were obtained from 15 samples. Carbonate solvus geothermometry was used to construct a thermal profile across the northeastern sector of the aureole. Estimates of pressure during metamorphism and temperature-X_{CO_2} evolution were derived by combining the analyzed data with a calibrated T-X_{CO_2}-plot for the end-member system CaO-MgO-SiO_2-CO_2-H_2O, which was contributed by Gottschalk and Metz using the procedure outlined in their paper of 1984.

These results are the data base for the systematic study of microtextures and the deduction of reaction mechanisms by Heuss-Aßbichler and Masch (Chap. 11, this Vol.).

10.2 Field Relationships and Previous Work

Carbonate rocks, even in the broad sense of carbonate-bearing rocks (i.e., calc-silicates), are minor volumetrically in the Ballachulish aureole. Thin carbonate layers occur to the northeast, east, south and west of the complex. The best outcrops are in the core of a steeply inclined small syncline in the northeastern sector (Allt Giubhsachain). It is here that carbonates (Appin limestone) can be traced towards the contact in four or five sheets (up to several metres thick). These carbonates are interbedded with pelites, semipelites and quartzites. The regional pre-intrusion deformation (Pattison and Voll, Chap. 2, this Vol.) led to folding of the layers. Fold limbs vary in thickness and may pinch out. Minor structures include axial plane cleavage and carbonate veins. Carbonate grains may be flattened and stretched. Mode

G. Voll, J. Töpel, D.R.M. Pattison, F. Seifert
Equilibrium and Kinetics
in Contact Metamorphism
©Springer-Verlag Berlin Heidelberg 1991

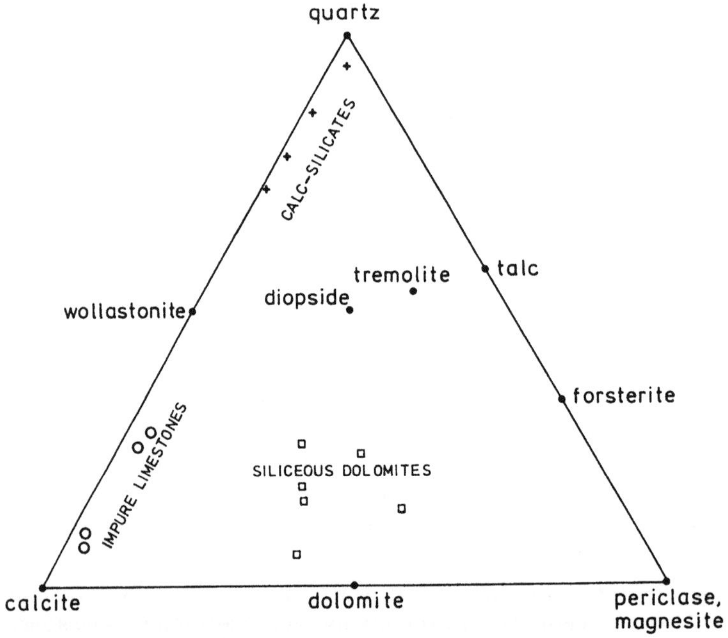

Fig. 10.1. CaO-MgO-SiO_2 diagram showing the position of metamorphic phases and the composition of siliceous dolomites, impure limestones and calc-silicates in the NE sector of the aureole

and chemical composition divide the carbonates in siliceous dolomites and impure limestones. Particularly silica-rich members are referred to as calc-silicates (Fig. 10.1). The assemblage of regional metamorphism (see Pattison and Voll, Chap. 2, this Vol.) is in general very simple: dolomite + calcite + quartz. Slightly siliceous dolomites show additional chlorite and pyrite. Siliceous dolomites have higher silica contents, and contain chlorite, muscovite, biotite and pyrite. Impure limestones have much higher CaO/MgO ratios, and contain the same phyllosilicates as the siliceous dolomites plus albite, microcline, pyrite.

A carbonate rock from Gleann an Fhiodh is exceptional, as here monomineralic reaction rims of phlogopite and calcite are produced between quartzite and dolomite. The basal planes of phlogopite are oriented parallel to the bedding plane. The carbonate solvus temperature of 500 °C fits to the values discussed for regional metamorphism (Pattison and Voll, Chap. 2, this Vol.). A small body of magnesite in the carbonate syncline of the southern sector is unusual.

Mineral symbols taken from Kretz (1983) are listed in Table 8.1.

10.3 Petrography, Isograds and Decarbonation Reactions

Apart from carbonates, the *thermometamorphic* phases are talc, amphibole, clinopyroxene, olivine, wollastonite and periclase. Clinochlore, spinel, phlogopite, muscovite, epidote, garnet, vesuvianite, scapolite, sphene and pyrite are accessory phases.

Seven *isograds*, which are based on specific decarbonation reactions were mapped in the northeastern, eastern and southern sectors of the aureole (Map 2; Figs. 10.2–10.4). Carbonate solvus temperatures are from Section 10.4 (below).

Reactions were numbered using the scheme of Metz and Trommsdorff (1968) for convenience as far as to reaction 8, but using the prefix C. Reaction numbers (C9), (C10), (C11) replace 11, 13 and 14 of this scheme, because of revision of the T-X_{CO_2} topology. The former isobaric invariant point IV is now labelled (CIII). Reaction labels (C14) and (C15) are arbitrary.

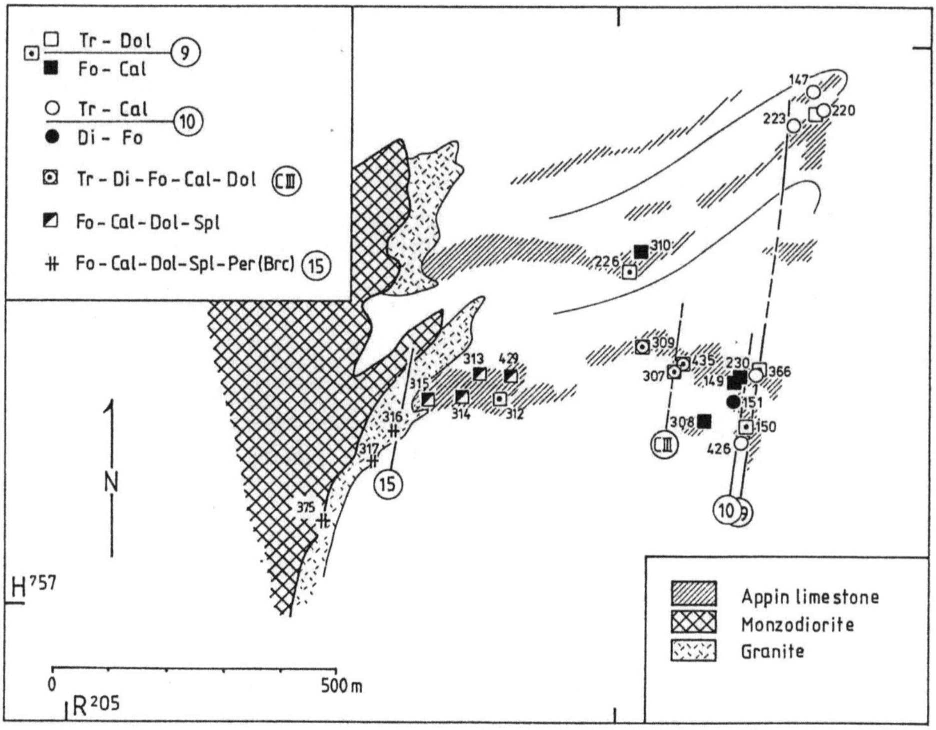

Fig. 10.2. Isograd map of siliceous dolomites in the Allt Giubhsachain area, showing the syncline of Appin limestone carbonates. *Numbers* refer to the relevant decarbonation reaction number

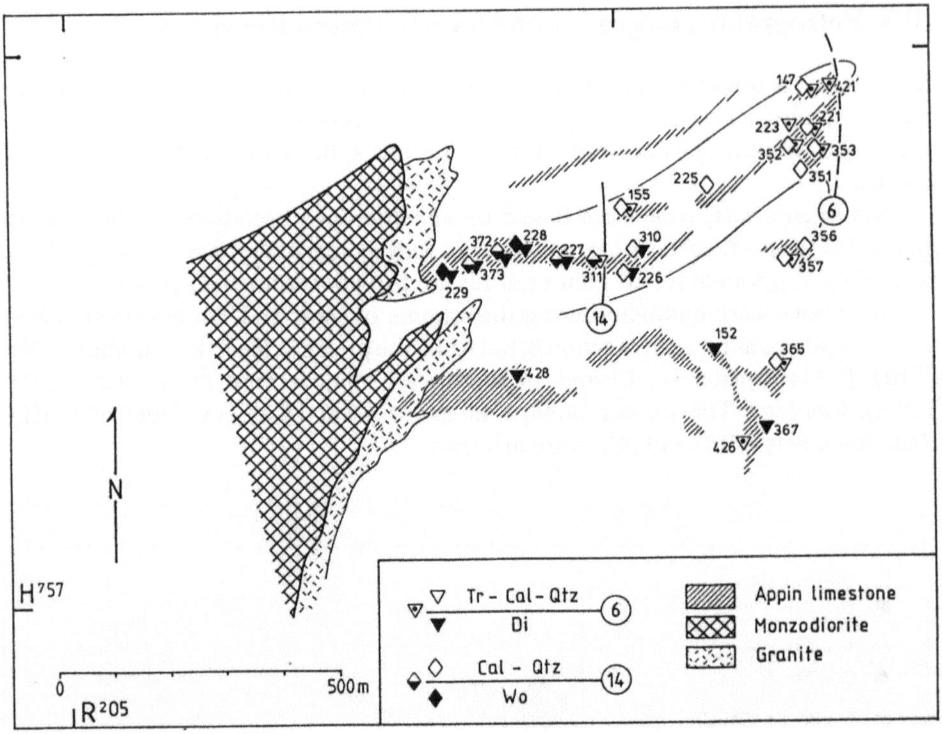

Fig. 10.3. Isograd map of impure limestone in the Allt Guibhsachain area, showing the syncline of Appin limestone carbonates. *Numbers* refer to the relevant decarbonation reactions number

10.3.1 Siliceous Dolomites

The outermost isograd is established by the four-phase assemblage talc + calcite + quartz + dolomite with irregularly distributed talc, replacing the assemblage Cal + Dol + Qtz. Microtextures (see Heuss-Aβbichler and Masch, Chap. 11, this Vol.) show the solid phases of the uncompleted reaction:

$$3\ \mathrm{Dol} + 4\ \mathrm{Qtz} + 1\ H_2O \Leftrightarrow 1\ \mathrm{Tlc} + 3\ \mathrm{Cal} + 3\ CO_2. \qquad \text{(C1)}$$

A *talc + calcite isograd* can be delineated in the northeastern, eastern and southern sectors of the aureole (Map 2; Fig. 10.4). A temperature of 420 °C was determined for a sample from the isograd in the southern sector by applying carbonate solvus geothermometry. The microprobe analysis (Table 10.1A) shows that talc is slightly richer in iron than the other silicates. Fluorine analysis indicates 0.15 wt% F in talc.

In the unique magnesite occurrence in the southern section, coarse flakes of talc are formed at the contact of quartz veins according to a different reaction:

$$3\ \mathrm{Mgs} + 4\ \mathrm{Qtz} + 1\ H_2O \Leftrightarrow 1\ \mathrm{Tlc} + 3\ CO_2. \qquad \text{(C1")}$$

The position of this assemblage downgrade the isograd (C1) is in accordance with the extrapolated data of Johannes (1969) on reaction (C1") and the data of Gottschalk

Table 10.1 The microprobe analyses were obtained using the wavelength dispersive method on an ARL-electron microprobe analyzer at the Mineralogical Institute Tübingen. Natural silicates and carbonates were used as standards. Corrections following Bence and Albee (1968) were applied. (Analysis conditions for carbonates: 8 μm spot size, 10 kV, 30 nA.). For sample locations, see Figs. 10.2–10.4

A

	Talc			Amphibole		Clinopyroxene		Olivine	
	B200		B396	B223b	B220	B309		B307	B375
Wt % sample	–1	–1B	–5	0–1	1	1–5	2–2	2–4	375–3
SiO_2	62.17	62.22	59.17	57.07	51.02	54.83	54.87	41.28	42.48
Al_2O_3	0.44	0.51	4.77	1.76	5.41	0.05	0.20	0.00	0.01
TiO_2	0.00	0.04	0.05	0.28	0.16	0.02	0.01	0.00	0.00
MnO	0.08	0.06	0.03	0.02	0.24	0.19	0.05	0.28	0.07
FeO[a]	3.45	3.47	4.53	0.83	6.02	2.06	0.75	4.87	0.97
MgO	29.20	29.37	26.41	23.54	20.25	17.48	17.86	53.53	55.96
CaO	0.05	0.20	0.15	13.00	10.17	24.46	25.08	0.09	0.05
Na_2O	0.12	0.10	0.13	0.98	3.92	0.14	0.22	0.00	0.00
K_2O	0.01	0.02	0.04	0.12	0.04	0.04	0.00	0.00	0.06
Sum	95.52	96.00	95.28	97.61	97.23	99.24	99.03	100.05	99.60

Calculated formula units

Talc			
Si	7.96	7.94	7.64
Al^{IV}	0.04	0.06	0.36
Al^{VI}	0.03	0.02	0.37
Fe	0.37	0.37	0.49
Mg	5.57	5.58	5.09
Ca	0.01	0.03	0.02
Na	0.03	0.03	0.03
Z	8.00	8.00	8.00
Y	6.01	6.03	6.00
X_{Fe}	0.06	0.06	0.09
X_F[b]	0.015		

Amphibole		
Si	7.79	7.25
Al^{IV}	0.21	0.75
Al^{VI}	0.08	0.16
Ti	0.03	0.02
Mn	0.00	0.03
Fe	0.10	0.72
Mg	4.79	4.07
Mg		0.22
Ca	1.91	1.55
Na	0.09	0.23
Na	0.61	0.85
K	0.02	0.01
X_{Fe}	0.02	0.14
X_F[c]	0.02	

Clinopyroxene		
Si	2.00	2.00
Al^{VI}	0.00	0.01
Ti	0.00	0.00
Mn	0.00	0.00
Fe	0.06	0.02
Mg	0.95	0.97
Ca	0.96	0.98
Na	0.01	0.02
Z	2.00	2.00
Y	1.01	1.00
X	0.97	1.00
X_{Fe}	0.06	0.02

Olivine		
Si	0.99	1.00
Mg	1.91	1.97
Fe	0.10	0.02
Ca	0.01	
X_{Fe}	0.05	0.01

B

	Calcite				Dolomite	
Wt % sample	B375–5	B375–2	B307–0	B307–5	B307–3	B375–5
$CaCO_3$	88.536	88.291	92.376	92.477	55.369	55.450
$MgCO_3$	10.224	11.306	7.073	5.927	43.656	43.975
$FeCO_3$	0.292	0.116	0.282	0.245	1.000	0.677
$MnCO_3$	—	0.023	0.159	0.365	0.243	—
Sum	99.052	99.736	99.890	99.014	100.268	100.102
moleq.						
Xca	0.877	0.867	0.913	0.924	1.023	1.025
Xmg	0.120	0.132	0.083	0.070	0.957	0.965
Xfe	0.002	0.001	0.002	0.002	0.016	0.011
Xmn	—	—	0.001	0.003	0.004	—
Temp (°C)	720	740	630	590		

Chlorite	B375–9
Al_2O_3	70.68
TiO_2	0.22
MnO	0.06
FeO[a]	2.96
MgO	26.09
Cr_2O_3	0.09
Sum	100.11
Al	2.00
Ti	0.00
Mn	0.00
Fe	0.06
Mg	0.93
X_{Fe}	0.06

[a]Total Fe as FeO.
[b]F in talc B2000 0.15 wt.%
[c]F in amphibole B220 0.05 wt%.

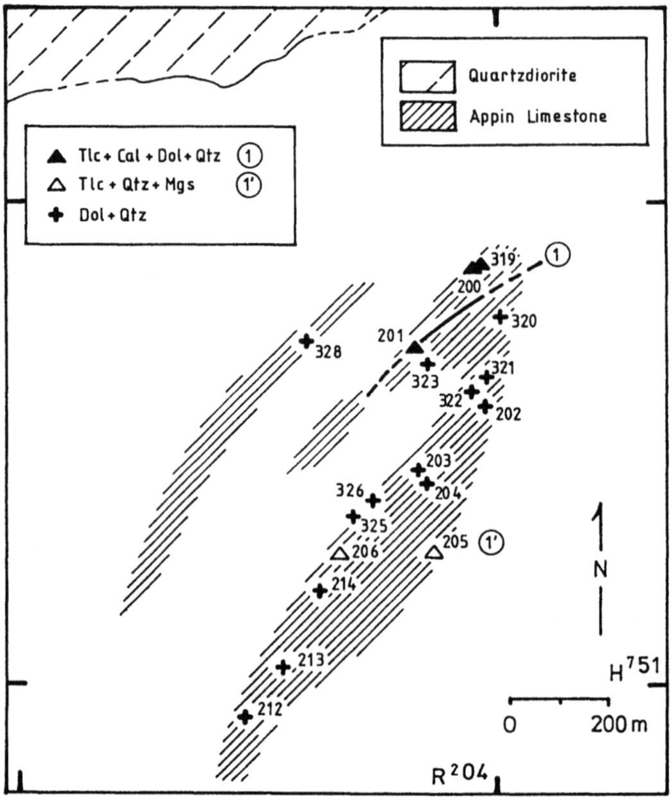

Fig. 10.4. Isograd map. Small syncline of Appin limestone carbonates in southern sector of Ballachulish aureole, east of Fraochaidh (879 m). The *isograd number* refers to the reaction number

and Metz (1984) on reaction (C1). At pressures around 0.3 GPa and X_{CO_2} between 0.1 and 0.5, quartz and magnesite react to produce talc at temperatures 40–60° lower than those at which quartz and dolomite do react to produce talc + calcite.

The sequence of tremolite + dolomite and forsterite + calcite with intercalation of the four-phase assemblage Tr + Dol + Fo + Cal establishes a *forsterite + calcite isograd* and indicates the reaction

$$1\ \mathrm{Tr} + 11\ \mathrm{Dol} \Leftrightarrow 8\ \mathrm{Fo} + 13\ \mathrm{Cal} + 9\ CO_2 + 1\ H_2O. \qquad \text{(C9)}$$

For microtextural evidence for this reaction see Heuss-Aβbichler and Masch (Chap. 11, this Vol.). The Fo + Cal isograd is mappable in the northeastern sector (Allt Giubhsachain) at a distance of 680 m from the contact. The isograd is parallel to the contact surface, when minor irregularities of the latter have been averaged out. A not very well-constrained temperature of 600 °C is determined from the thermal profile (see Fig. 10.11) for this isograd.

The amphiboles are tremolites with Si > 7.0 with the exception of the outermost sample B220. The edenitic composition of amphiboles in the latter, with Na = 0.23, is attributed to an original difference in host rock composition. Whole rock analyses

(Heuss-Aßbichler 1987) show variations in Na_2O and Al_2O_3. The amphiboles have a low F content. Olivines are almost end-member forsterites with X_{Fe} less than 0.01. Clinohumite is not observed.

In the Kentallen area, a single layer of siliceous dolomite with varying Mg/Ca ratios and varying silica content, extends almost parallel to the contact at a distance of approximately 230 m. No isograd can be mapped, as only the product assemblage Fo + Cal is observed.

In silica-rich dolomites, diopside + forsterite appear as the result of the reaction

$$1\,Tr + 5\,Cal \Leftrightarrow 11\,Di + 2\,Fo + 5\,CO_2 + 1\,H_2O. \tag{C10}$$

The reactant assemblage Tr + Cal is not documented. A *diopside + forsterite isograd* was mapped over some distance parallel to the Fo + Cal isograd (Fig. 10.2). Clinopyroxenes associated with the olivine are remarkably pure diopsides (Sample B309 in Table 10.1A).

The five-phase assemblage Tr + Di + Fo + Cal + Dol (CIII, see Fig. 10.2) shows mutual contact of all phases. It occurs in the northeast and in the southeast of the aureole. A *Tr + Di + Fo + Cal + Dol isograd* was mapped only in the northeastern sector and has not been extended to the south, because the single assemblage observed in the southern sector is exposed in a flat thermal trough between the main intrusive and the neighbouring stock (Map 2). In the NE section this assemblage extends over a distance of 40 m (B307 and B309 in Fig. 10.2). In the Kentallen area no five-phase assemblage (CIII) was found, though the products of reactions (C9) and (C10) are well documented.

In sample B307 calcite + tremolite of the five-phase assemblage (Tr + Di + Fo + Cal + Dol) start to be replaced by *diopsite + dolomite*. Dolomite forms a concentric rim around the diopside grains. This microtextural observation is interpreted as evidence of the reaction:

$$1\,Tr + 3\,Cal \Leftrightarrow 4\,Di + 1\,Dol + 1\,CO_2 + 1\,H_2O. \tag{C7}$$

The progressive replacement of diopside, within the forementioned dolomite rim, by *forsterite + calcite* in B309 is taken as evidence for the reaction:

$$1\,Di + 3\,Dol \Leftrightarrow 2\,Fo + 4\,Cal + 2\,CO_2. \tag{C11}$$

The high grade sequence of decarbonation reactions is completed in silica-poor dolomites by the instability of dolomite to produce periclase. All periclases are replaced by pseudomorphs of brucite; in one exceptional case one idioblastic grain of *periclase* is preserved inside a small brucite rim. The periclase-producing reaction is:

$$1\,Dol \Leftrightarrow 1\,Per + 1\,Cal + 1\,CO_2. \tag{C15}$$

In the northeastern sector periclase is only observed in dolomite xenoliths within the granite. The carbonate solvus temperature obtained for a three-phase assemblage (Dol + Cal + Per) of sample B375 from the contact of granite against the quartz diorite is 760 °C (see Fig. 10.2). In the Kentallen area, Per + Dol + Cal are found up to 200 m from the contact, whereas at the immediate contact dolomite is stable. A three-phase assemblage 230 m away from the contact indicates a temperature of 640 °C in this area.

Spinel is restricted to high grade forsterite + calcite + dolomite assemblages, replacing clinochlore + dolomite according to the reaction:

$$1\ \mathrm{Chl} + 2\ \mathrm{Dol} \Leftrightarrow 3\ \mathrm{Fo} + 1\ \mathrm{Spl} + 2\ \mathrm{Cal} + 2\ \mathrm{CO_2}.$$

Near the contact, a metamorphic banding develops; in some places it enhances premetamorphic compositional differences, in others there is no such correlation. The orientations of newly formed carbonate veins are controlled by the minor regional metamorphic structures.

10.3.2 Impure Limestones

Impure limestones contain phases of complex chemistry when metamorphosed to calcic marbles. The outermost impure limestones in the northeastern sector of the aureole (Fig. 10.3) contain a zone of clinopyroxene + amphibole + quartz + calcite. In the six-component model system CaO-MgO-FeO-SiO_2-CO_2-H_2O these are the solid phases of the divariant diopside-forming reaction:

$$1\ \mathrm{Tr} + 3\ \mathrm{Cal} + 2\ \mathrm{Qtz} \Leftrightarrow 5\ \mathrm{Di} + 3\ \mathrm{CO_2} + 1\ \mathrm{H_2O}. \qquad \text{(C6)}$$

The *diopside isograd* cannot be pinpointed to within less than several decametres, as the reactant assemblage Tr + Cal + Qtz is not observable in the small syncline of carbonate rocks. The positions of the other isograds suggest that this particular isograd cannot be any further away.

In one layer only a *wollastonite isograd* based on the reaction:

$$1\ \mathrm{Cal} + \mathrm{Qtz} \Leftrightarrow 1\ \mathrm{Wo} + 1\ \mathrm{CO_2}. \qquad \text{(C14)}$$

could be pinpointed, as rocks of suitable composition are lacking elsewhere (Fig. 10.3). The three-phase assemblage Wo + Cal + Qtz is observed over a distance of 200 m upgrade.

The low-grade reactions (C2) and (C3), which produce tremolite and tremolite + dolomite respectively from talc-bearing assemblages are not exposed in either of the carbonate lithologies:

$$5\ \mathrm{Tlc} + 4\ \mathrm{Qtz} + 6\ \mathrm{Cal} \Leftrightarrow 3\ \mathrm{Tr} + 6\ \mathrm{CO_2} + 2\ \mathrm{H_2O}, \qquad \text{(C2)}$$

$$2\ \mathrm{Tlc} + 3\ \mathrm{Cal} \Leftrightarrow 1\ \mathrm{Tr} + 1\ \mathrm{Dol} + 1\ \mathrm{CO_2} + 1\ \mathrm{H_2O}. \qquad \text{(C3)}$$

In siliceous dolomites neither the potential tremolite-forming reaction (C4) nor the diopside-forming reaction (C8) are observed:

$$8\ \mathrm{Qtz} + 5\ \mathrm{Dol} + 1\ \mathrm{H_2O} \Leftrightarrow 1\ \mathrm{Tr} + 3\ \mathrm{Cal} + 7\ \mathrm{CO_2}, \qquad \text{(C4)}$$

$$2\ \mathrm{Qtz} + 1\ \mathrm{Dol} \Leftrightarrow 1\ \mathrm{Di} + 2\ \mathrm{CO_2}. \qquad \text{(C8)}$$

10.3.3 Calc-Silicates

Calc-silicates are derived from semipelites and carbonaceous quartzites with variable amounts of calcite and clinochlore. In addition to Ca-amphiboles (actinolites and edenitic hornblendes) and clinopyroxenes (salites), they may contain minor amounts of wollastonite, epidote/zoisite, biotite, muscovite, tourmaline, zircon, apatite, sphene, rutile and hematite. Massive sheets of garnet/vesuvianite-clinopyroxene or garnet-epidote-bearing layers are observed between hornfels and calc-silicate horizons, or between calcic-marble and quartzite in the contact zone of the NE section.

10.4 Geothermometry: Mg-Calcite

Seven samples of the Allt Giubhsachain series of progressively thermometamorphosed siliceous dolomites were selected for systematic microprobe analysis of calcites. In each sample several calcites (0.2–2 mm) adjacent to dolomites and product silicates were analyzed (see Fig. 10.5). The iron content of calcite is very low (B375 in Table 10.1B), so the temperatures may be estimated on the basis of subsolidus phase relations in the calcite-dolomite system (Harker and Tuttle 1955; Goldsmith et al. 1961; Graf and Goldsmith 1965; Goldsmith and Newton 1969; Puhan 1979) and no correction for iron (Bickle and Powell 1977) was done.

The Mg-content of calcites, expressed as mole percent $MgCO_3$ was used to evaluate the data. Temperatures up to 600 °C were calculated from the equation

$$\log XMgCO_3 = -1414/T(°K) + 0.4869$$

which was derived by Puhan (1979) from experimental data. Within the range of 600–800 °C, temperatures were read off Puhan's plot of data. Dolomite compositions are not very reliable, due to the steep flank of the asymmetric solvus at Mg-rich compositions, and were therefore not used for temperature evaluation.

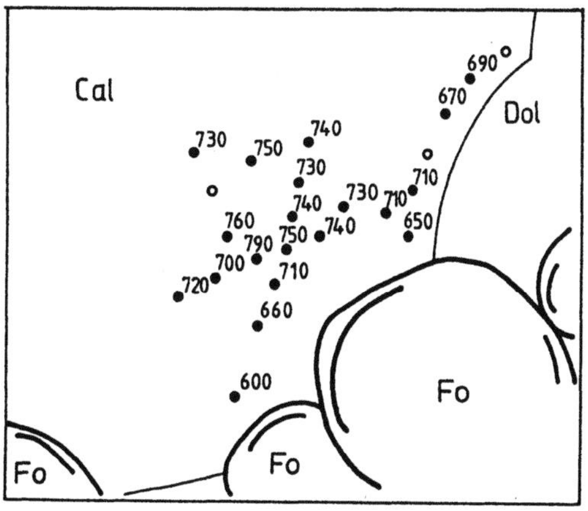

Fig. 10.5. Sketch from a microphotograph of sample B375 (near contact, see Fig. 10.2) showing electron-microprobe spot analysis converted to temperatures (°C) in Mg-calcite

Chemical zoning, with higher Mg contents in the centres and lower Mg contents in the rims, is common to all analyzed grains. The marginal decrease in Mg content occurs in a broad diffuse zone, the centres exhibit a rather irregular plateau of Mg values. A typical grain near from the contact (shown in Fig. 10.5) has an overall range of 16.2–7.3 mol% $MgCO_3$. The zonation is interpreted to be the result of retrogressive resetting, with the cores preserving maximum temperatures.

B317 from the vicinity of the contact produces temperatures as low as 500 °C and is considered to have equilibrated completely during the cooling stage.

In sample B312, relatively high Mg values seem to be the result of submicroscopic blebs of dolomite which remain in calcite in the course of patchy replacement of dolomite by calcite during the decarbonation reaction.

The point analyses of different grains in each sample analyzed are presented in a histogram (Fig. 10.6). At the extremes of the temperature range, five analyses were disregarded as possible artifacts (e.g. representing cut effects). The maximum temperatures establish a *thermal profile* with higher T towards the intrusive contact.

The derived thermal profile is compared in Pattison (Fig. 16.4, this Vol.) with temperature estimates on thermometamorphosed pelites and from within the pluton.

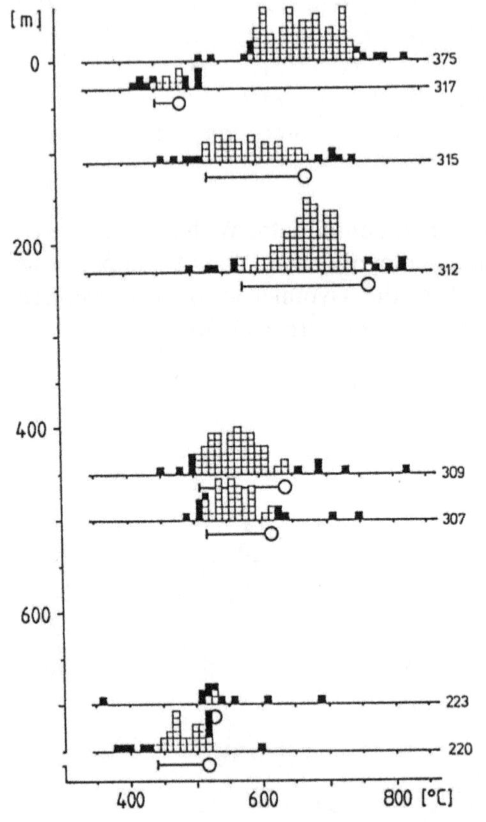

Fig. 10.6. Histogram of electron-microprobe spot analyses (*open boxes*) of Mg-calcites converted to temperatures (°C) vs distance from contact. Maxima values are indicated by *circles*, minima by *dashes*. *Black boxes* = artefacts (see discussion)

There is good agreement with the estimated *temperature at the contact: 760 ± 20 °C*. Similarly the thermal profiles of pelites and carbonates are coincident at distances of 500 m from the contact, with a temperature of 640 ± 20 °C for the pelites and 620 ± 20 °C for the carbonates. The similarity of temperatures derived in different lithologies with calcite-dolomite thermometry supports the assumption that the high Mg contents of calcite cores tend to reflect the *peak temperatures* of thermometamorphism.

The relatively low Mg contents in the samples B220, B223, and B315 could be attributable to more advanced retrograde reequilibration. Curiously, these samples are not from decarbonation isograds, but are out of the reaction zone upgrade of the isograds and have complete reaction assemblages. In contrast, the samples B309, B307 and B375 are from assemblages on or near the isograds. The microtextural investigations (Heuss-Aßbichler and Masch, Chap. 11, this Vol.) suggest that the reactions produce calcite by a dissolution-precipitation process, and these reaction calcites are thought to have the composition corresponding to the temperature at which the decarbonation reaction occurred. Thus carbonate solvus temperatures reflect the maximum temperatures reached in the area near the isograds. In the reaction zones, where the temperatures increased further after completion of reactions, the geothermometry gives lower temperatures than the maximum. The mismatch of samples from the zones with the thermal profile may be the result of sluggish intracrystalline equilibration of carbonate cores after the decarbonation reactions have gone to completion.

10.5 Pressure

The pressure conditions during the thermometamorphism of the Ballachulish aureole can be derived by combining carbonate solvus geothermometry and the quantitative petrogenetic grid for the reactions observed. Several data sets for the pure end-member compositions in the system CaO-MgO-SiO_2-CO_2-H_2O have already been published. We prefer the recent, internally consistent set of Gottschalk and Metz (1984) and Gottschalk (1984), based on new experimental data and calculations, to the earlier data of Metz and Trommsdorff (1968), Skippen (1971, 1974) and Slaughter et al. (1975).

In the P-T diagram (Fig. 10.7) the carbonate-solvus temperature for the five-phase assemblage Tr + Fo + Di + Cal + Dol (B307,B309) indicates a pressure of 0.28 ± 0.04 GPa (2.8 ± 0.4 kbar). The divariant assemblage of reaction (C9) gives a pressure range from 0.16 to 0.32 GPa (Fig. 10.8). Solid solution should modify the position of the equilibria, e.g., fluorine substitution (Abercrombie et al. 1987) will extend the stability of tremolite to higher grade. A shift of some 10 °C will result from even moderate substitution. Similar results are derived for coupled substitution including F and Na_2O (Bowman and Essene 1982) in the tremolites. However, it seems that the shift of equilibria of siliceous dolomites which would occur, given that the fluorine and/or Na_2O concentrations are high enough, is still within the brackets of experimen-

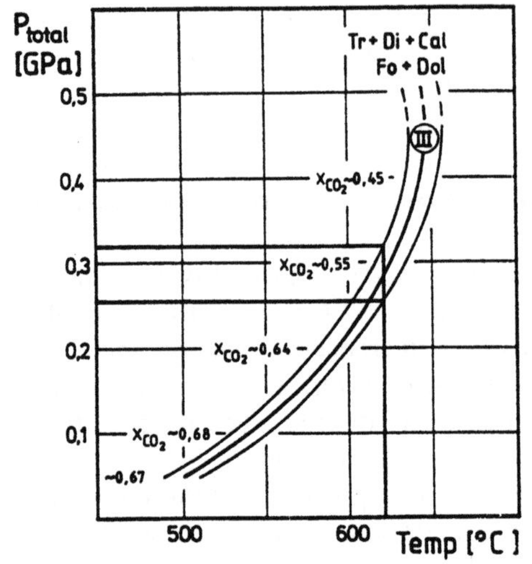

Fig. 10.7. Pressure-temperature diagram of isobaric invariant assemblage CIII Tr-Di-Fo-Cal-Dol (Gottschalk and Metz 1984). *Vertical line* obtained from Mg-calcite geothermometry; *horizontal lines* mark the possible pressure range

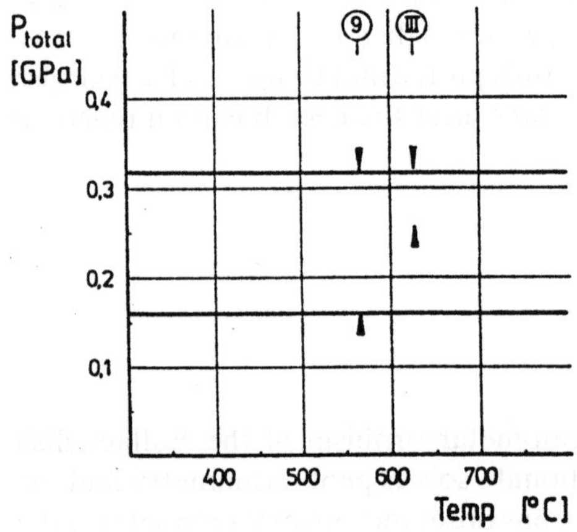

Fig. 10.8. Pressure-temperature diagram with P-T ranges of decarbonation reaction (C9) and the isobaric invariant assemblage CIII. *Arrowheads* are pressure brackets (Gottschalk and Metz 1984); for CIII, see Fig. 10.7

tal uncertainty, as the best estimate for pressure of the Ballachulish pelites (Pattison, Chap. 16, this Vol.) is in excellent agreement with the value derived above.

Fluid pressure is considered to be close or equal to lithostatic pressure during decarbonation reactions. Calcite veinlets (Figs. 10.9a,b, and 10.10), which are associated with reaction sites, indicate fluid phase migration and result from hydraulic fracture mechanisms indicating low effective stresses (Etheridge et al. 1983, 1984). The general good fit of the carbonate solvus temperatures and the temperature brackets for decarbonation reactions for 0.3 GPa (Fig. 10.11) is in favour of this assumption. Isograds would shift to lower temperatures if fluid pressure were lower than lithostatic pressure (Fyfe et al. 1978, p. 136f).

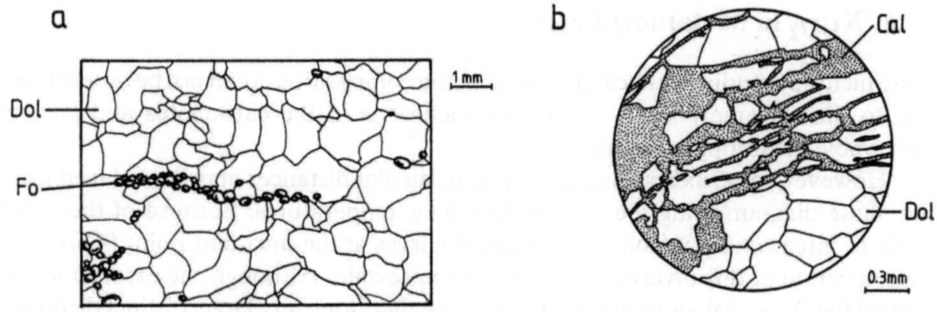

Fig. 10.9. Sketch of forsterite-dolomite marble thin section. **a** String of forsterite grains; **b** calcite veinlets in dolomite matrix. NE sector of Ballachulish aureole

Fig. 10.10. Forsterite-calcite-dolomite marble outcrop, veined with forsterite and calcite. NE sector of Ballachulish aureole

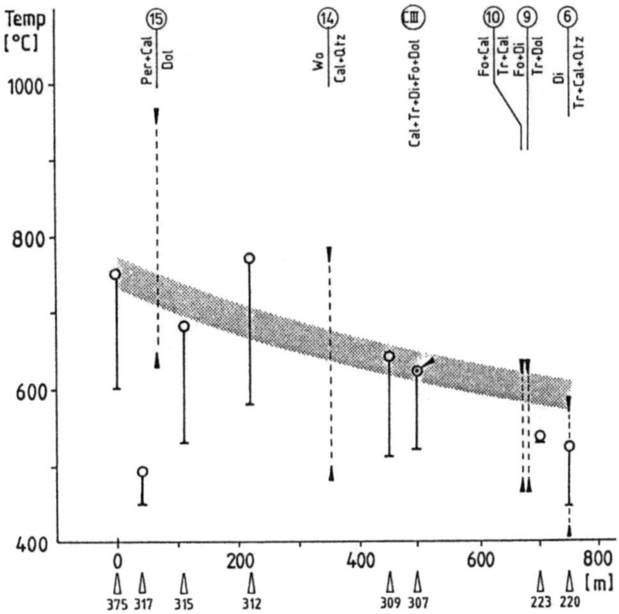

Fig. 10.11. Thermal profile in the NE sector of Ballachulish aureole. T-X_{CO_2} brackets of decarbonation reactions after Gottschalk and Metz (1984) between *black triangles*; Mg-calcite geothermometry between *circles* (max. temperature) and *dashes*. *Numbers* refer to the relevant decarbonation reaction numbers

10.6 X_{CO_2} in Metamorphism

Fluid inclusion studies were carried out, but the composition could not be determined due to the extremely small size of the inclusions in the carbonates and lack of observable inclusions in silicates.

However, some indications of X_{CO_2} at particular distances may be deduced from the phase diagram using the carbonate solvus temperatures. Because of the acute angle of intersection of isobaric univariant curves at the invariant point (CIII), the X_{CO_2} location at this invariant point is very uncertain. Although the error brackets around the X_{CO_2} values read from the grid are therefore very large, distinct differences in various sectors of the aureole are still apparent (see Fig. 10.12).

The 0.3 GPa TX_{CO_2} grid of Fig. 10.12 was calculated by Gottschalk and Metz (for procedure, see their paper of 1984). A stability curve for the Dol + Per + Cal equilibrium is added. This equilibrium was calculated following the procedure of Perkins and Reusser (unpublished). Fugacities calculated with HMRK after Kerrick and Jacobs (1981) and the thermodynamic basis data of Berman (1988) were used.

At the Fo + Cal isograd the first formation of forsterite + calcite is at an X_{CO_2} of about 0.2, as indicated by the intersection of reaction curve (C9) with the isotherm of 580 °C. The five-phase assemblage (CIII) occurs at an X_{CO_2} of 0.55. The stability field of the assemblage Di + Dol (reaction C11) is on the CO_2-rich side of point (CIII)

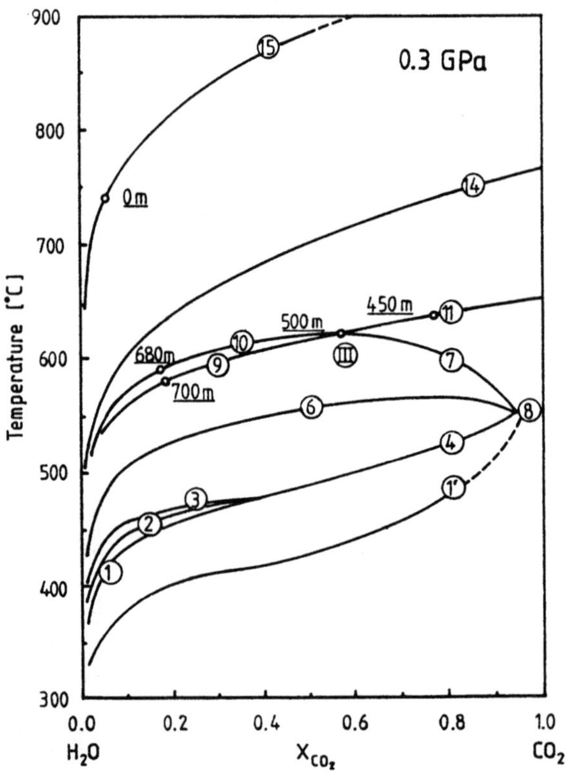

Fig. 10.12. P-T-X_{CO_2} diagram for siliceous dolomite and impure limestone. Data basis from Gottschalk and Metz (1984). Reaction (C15) was calculated after Perkins and Reusser (pers. comm.). Numbering scheme see Section 10.3

in all topologies. Its stability at 640 °C suggests X_{CO_2} of about 0.8. The Per + Cal isograd gives a X_{CO_2} of < 0.02 from the intersection of the 740 °C isotherm with the stability curve. The marked difference between X_{CO_2} of midaureole five-phase assemblage (CIII) and contact area Per + Cal-assemblage would even be greater, if F substitution were to be taken into account. F shifts the isobaric invariant point (CIII) to still higher X_{CO_2} (Abercrombie et al. 1987). Coupled substitution including Na_2O, Al_2O_3, F (Bowman and Essene 1982) will not significantly change the position from the end-member system. In the outer aureole, the apparent lack of tremolite-forming reaction (C4) and diopside-forming reaction (C8) (Sect. 10.3) could be due to relative low X_{CO_2} at the relevant temperatures.

10.7 Discussion

Although quantitative analysis is impossible, there is evidence that X_{CO_2} values attained in the carbonates differ considerably in the region between the Fo-Cal isograd of reaction (C9) and the contact in the NE part of the aureole (Allt Giubhsachain) with X_{CO_2} > 0.5 in the middle part at a distance of 450–500 m from the contact and with X_{CO_2} < 0.1 in the contact area. It is thought that they represent the extremes of fluid development in the system.

The following aspects of fluid evolution are considered common to these portions:

1. The presence of an intergranular fluid at the onset of thermometamorphism as indicated by the formation of the hydrous phase talc from the starting assemblage quartz and dolomite in the outer aureole.
2. The decarbonation reactions liberate a fluid phase which is quite CO_2-rich: 0.9 for reaction (C9), 1.0 for reaction (C11), 0.63 for reaction (C10).
3. The oxygen isotope data (Hoernes et al., Chap. 17, this Vol.) rule out the influx of a magmatic H_2O-rich fluid-phase into the metacarbonate rocks, although there may have been limited exchange with the pelites.

The CO_2-rich fluid phase compositions, which must be assumed for the five-phase assemblage Tr + Di + Fo + Cal + Dol and the assemblage Di + Dol [reactions (C7) and (C11)] in the middle of the Fo + Cal zone, can be explained by the concept of *reaction control* of the fluid phase (Greenwood 1975). The fluid liberated by the decarbonation reactions controls the composition of the fluid phase during metamorphism. The initial X_{CO_2} is shifted towards the X_{CO_2}, which is liberated according to the stoichiometric coefficients of the reaction involved. A T-X_{CO_2} path of small steps following the univariant reaction curve is envisaged for this process (Kerrick 1974). Only a very small proportion of the reactants is consumed on the way up to the isobaric invariant point from which the reaction curve originates. Petrographically, this process should be indicated by isobaric univariant reaction assemblages occurring along the thermal profile and the appearance of isobaric invariant assemblages. The distribution of assemblages in the Ballachulish aureole is not in accordance with these propositions in every respect (Figs. 10.2–10.4). In the vicinity of isograds, there are indeed isobaric univariant assemblages, if the minor substitutions are not taken into account. Furthermore, five-phase assemblages appear which are isobaric invariant in the end-member system. In contrast to Greenwood's proposal, isobaric univariant assemblages do not occur in the inner parts of the zones, but rather divariant product assemblages. The wollastonite-forming reaction (C14) is an exception. Three-phase assemblages with Qtz + Cal + Wo extend over 200 m upgrade in the zone (Fig. 10.3).

Mixing with initial water-rich fluids may change the composition of reaction-produced fluid. This dilution of CO_2-rich fluids will cause a much larger proportion of reactants to be consumed. This is important petrographically, because only then do detectable amounts of product appear. The lack of isobaric univariant assemblages in the major parts of most zones seems to be the effect of more pronounced *mixing of the fluid phases*. As the X_{CO_2} values are uncertain, a discussion can only be conducted for the contact area which has very low X_{CO_2}-values (< 0.1). There is petrographic evidence that within this area high X_{CO_2} values were never attained throughout the entire thermal event. Specimen B312 (see Fig. 10.2) exhibits the isobaric univariant assemblage Tr + Fo + Cal + Dol from reaction (C9). Diopside is missing and therefore the isobaric invariant assemblage (CIII) which indicates elevated X_{CO_2} was not produced. It is concluded that reaction control of fluid phase is less effective than in the middle of the zone and that the influence of initial fluid on the carbonates was more pronounced in this portion of the aureole.

The microtexture systematics (Heuss-Aßbichler and Masch, Chap. 11, this Vol.) provide additional evidence relevant to fluid evolution. The distribution of product phases is indicative of the range of solute transport and relative degree of fluid mobilities. This investigation also rendered propositions on the variations of both in the region between the Fo + Cal isograd and the contact.

In the contact area, the products of decarbonation reactions are either dispersed independently of one another throughout the rocks or form compositional banding. Centimetre-thick veins consisting of calcite and forsterite cut through the carbonates. All these features indicate pronounced solute transport and fluid mobility in this area. This conclusion is in good accordance with the proposal of more pronounced mixing of fluids and reduced efficiency of reaction control of fluid composition in this area, as derived from the discussion of X_{CO_2} above.

In the middle of the Fo + Cal zone, products of decarbonation reactions are not independent of one another, but occur in coronas or intergrowths distributed as isolated reaction sites throughout the rocks. These features correspond to a limited solute transport and hence limited fluid migration. The latter is also thought to favour an effective reaction control of fluid composition (see above) by retarding mixing with the reservoir of initial fluid, thus creating the CO_2-richer fluids deduced in this sector.

The microtextural investigations also indicate the primary nature of these microtextures; this, combined with the observed separate X_{CO_2}-evolution paths in each sector, mean that a different metamorphic path must be considered for each portion of the aureole. Thus the popular concept that the spatial sequence of assemblages inwards reflects the progressive metamorphism of the contact area is not supported by this study. Rather, each group of metamorphic assemblages is the product of the different conditions prevailing at this site during reaction progress. Different transport mechanisms (Walther and Wood 1984) and different states of the interconnected transport system within the carbonate rocks, as indicated by microtexture systematics, are thought to be crucial for the actual path of fluid evolution in each portion of the aureole.

11 Microtextures and Reaction Mechanisms of Carbonate Rocks: A Comparison Between the Thermoaureoles of Ballachulish and Monzoni (N. Italy)

S. Heuss-Aßbichler and L. Masch

11.1 Introduction

The kinetics of metamorphic reactions have become an important aspect of recent investigations. Attempts to determine the step which controls reaction progress are based primarily on experimental results and theoretical calculations (Fisher 1978; Walther and Wood 1984; Rubie and Thompson 1985; Lasaga 1986; Ridley and Thompson 1986). Field-related studies are mainly concerned with the formation of reaction bands (Vidale 1969; Fisher 1970; Joesten 1974, 1977; Sanford 1980; Walther 1983) and are based on detailed textural investigations. The latter treatment is also the basis of our study, which focuses on endothermic decarbonation reactions in the thermoaureoles of the intrusions of Ballachulish and Monzoni. A detailed petrographic and petrological examination was carried out; the results correlated with the temperature-time history in the thermoaureoles, which are characterized by varying heating rates dependent on the distance from the contact. The comparison of the microtextures in these two aureoles, which formed under different pressure conditions, permits the development of a more general theory of the evolution of microtextures during heating. The aim of this chapter is to work out the systematics of the evolution of the microtextures, to determine possible reaction mechanisms and to discuss the effect of changing boundary conditions (T, X_{CO_2} and volume of fluid phase) on their stabilization.

11.1.1 Geology of the Aureoles

The investigation of the microtextures in the carbonates of the Ballachulish thermoaureole was hampered by the fact that the beds are not well exposed and sometimes pinch out. It is possible to follow carbonates from a distance of 700 m right up to the contact in the northeastern section only (see Pattison and Voll, Chap. 2; Masch and Heuss-Aßbichler, Chap. 10, this Vol.). The aureole of Monzoni was selected to study the systematic development of reaction textures because a sequence of thinly bedded sediments of varying composition is exposed there which can be followed through the entire aureole. The intrusion complex of Monzoni, which is located in the western part of the Dolomites, N. Italy, is made up of a sequence of rocks varying in composition from monzodiorites and diorites to pyroxenite and gabbros (Del Monte

G. Voll, J. Töpel, D.R.M. Pattison, F. Seifert
Equilibrium and Kinetics
in Contact Metamorphism

et al. 1967; Spicker 1986; Spicker and Huckenholz 1986, 1988). The Permotriassic sediments in the aureole exhibit prograde metamorphic zonation, and ten different decarbonation reactions have been identified (Masch 1977). The metamorphism of carbonates in the Ballachulish aureole is described in detail by Masch and Heuss-Aβbichler (Chap. 10, this Vol.) and Heuss-Aβbichler (1987). Details of Ballachulish and Monzoni intrusion complex are summarized in Table 11.1. One main difference between these two aureoles is the depth of the intrusion; the pressure deduced for Ballachulish is 0.3 GPa (Pattison, Chap. 16, this Vol.) and for Monzoni 0.05 GPa (Masch 1977). The thermal profile in Monzoni (Fig. 11.1) seems to be much steeper than in Ballachulish (Fig. 10.11), obviously as a result of the smaller volume of the intrusion. This is reflected in the fact that the distances between the isograds are much smaller than at Monzoni. In both areas, the dip of the subsurface contact zone is not precisely known. Both the pattern of the isograds and the thermal profiles deduced from geothermometry support the assumption that the dip is very steep in both cases. In neither case was intrusion accompanied by regional deformation.

The dolomites in the aureole of Ballachulish are very inhomogeneous: the SiO_2 content may vary between 3 and 34 wt%. For the purpose of this study the composition of the minerals can be described in the model system CaO-MgO-SiO_2-CO_2-H_2O (CMS). Tremolites with minor contents of Na_2O, Al_2O_3 and F are the only significant exception (sample B223, Table 10.1). For Monzoni, the model system CMS must be extended by the addition of Na_2O, Al_2O_3 and F, as the composition of

Table 11.1. Comparison of Ballachulish and Monzoni aureoles

	Ballachulish	Monzoni
ø of the complex	5.8 km	2.5 km (E sector)
ø of the thermoaureole	1.9 km	1.2 km
Temp. of intrusion	1050 °C[a]	1000–1150 °C[g]
Temp. of the contact	760 °C ± 40[b]	800–850 °C[h]
Temp of country rock prior to intrusion	250–300 °C[c]	50 °C[i]
Pressure	0.3 GPa[b]	max. 0.05 GPa[i]
Geometry of the intrusion	cylinder model[d]	complex model[g]
SiO_2 content in dolomites	3–34 wt%[e]	1–8 wt%[i]
Distance of the Tlc-Cal isograd from the contact	ca. 700 m (S sector)[f] ca. 1900 m (E sector)[f]	950–1200 m[i]
Distance of the Fo-Cal isograd from the contact	700 m (NE sector)[f]	300–450 m[i]

[a]Weiss and Troll, Chap. 4, this Vol.;
[b]Pattison, Chap. 16, this Vol.;
[c]Pattison and Voll, Chap. 2, this Vol.;
[d]Buntebarth, Chap. 18, this Vol.;
[e]Heuss-Aβbichler (1987);
[f]Masch and Heuss-Aβbichler, Chap. 10, this Vol.;
[g]Spicker and Huckenholz (1986);
[h]Inderst (1987);
[i]Masch (1977).

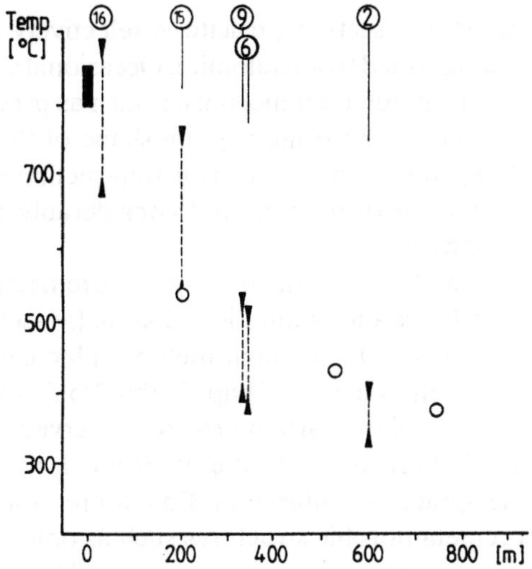

Fig. 11.1. Thermal profile in the eastern aureole of Monzoni intrusion complex, after Masch (1977) and Inderst (1987); T-X_{CO_2} *brackets* of decarbonation reactions between *black triangles*, calcite-dolomite geothermometry (max. temperatures) *circles*, monticellite-forsterite thermometry *black square*. Reaction number, see Chap. 10, this Vol.; (C16) 1 diopside + 1 calcite ⇔ 1 akermanite + 1 CO_2

the amphiboles ranges from tremolite to pargasite, even within one grain. Furthermore, Trommsdorff and Skippen (1986) found NaCl in fluid inclusions. The extended composition system must be taken into account both when interpreting isograds and discussing the T-X_{CO_2} evolution. Nevertheless, we suggest that for this kinetic study we can simplify the phase compositions and describe them in the model system CMS. The simple pattern in the distribution of microtextures which is observed in the extremely inhomogeneous layers supports the assumption that the influence of solid solution of phases may be neglected. In the following text amphiboles are treated as tremolites. (Mineral abbreviations are listed in Table 8.1).

11.2 Microtextures

The assemblages in the siliceous dolomites of the thermoaureoles of the Monzoni and Ballachulish intrusions developed a range of microtextures. The position of the isograd in the aureole is defined by the outermost appearance of the isobaric univariant assemblage. It comprises all solid phases which participate in the reaction. Upgrade of the isograd, the reaction-related assemblages are exposed within a zone; they may be univariant or divariant by consumption of one of the reacting phases. This area is called 'reaction zone' in this study.

Textural investigations presented in this study focus on two decarbonation reactions. Both have dolomite and a silicate as solid reactants and calcite and a silicate as solid products.

4 quartz + 3 dolomite + 1 H_2O ⇔ 1 talc + 3 calcite + 3 CO_2, (C1)
1 tremolite + 11 dolomite ⇔ 8 forsterite + 13 calcite + 9 CO_2 + 1 H_2O. (C9)

In all thin sections, calcite is selectively stained with Alicerin and is thus clearly distinguished from dolomite. Occasional samples may show signs of overprinting due to retrograde metamorphism, such as pseudomorphic serpentinization of forsterite. As long as the original grain shape of the forsterites was unmistakable, they were designated as forsterite in the thin-section drawings. Specimens in which the prograde reaction texture is altered considerably by retrograde metamorphism have been excluded.

In Monzoni, the investigated premetamorphic carbonates are micritic dolomitic limestones and dolomitic oosparite (Masch 1977). In Ballachulish, the dolomites and limestones are regional metamorphic carbonates with granoblastic microtextures (Pattison and Voll, Chap. 2, this Vol.). In both areas the pre-thermometamorphic features of the carbonates are preserved in the samples exhibiting the lowest grade Tlc-Cal reaction. On the other hand, the microtextures in the specimens with the high-grade assemblage Fo-Cal-Dol resulting from reaction (C9) show no evidence of premetamorphic structures such as ooids, as the grain growth of matrix carbonates due to annealing is pronounced. The grain-boundaries of the dolomites in the monomineralic, reaction-free matrix of the high grade samples of both areas are usually straight, though they may be curved, while the grains are generally equi-dimensional. In Ballachulish, the grain size of dolomite increases from about 60 μm (700 m from the contact), though approx. 300 μm (at a distance of 250 m), up to > 1000 μm directly at the contact.

The various 'primary' microtextures can be subdivided into four different types: (1) the term *corona* is used here in the broadest sense. Apart from those textures with concentric polycrystalline reaction rims of the products around a reacting phase, textures are also included which have a silicate product in the core and are surrounded by a mono- or oligocrystalline halo of carbonate. The essential criterion is that one product should enclose the other. (2) In *intergrowth textures*, both product phases are concentrated within domains. The texture may correspond to eutectoid assemblages. (3) The term *granoblastic texture* is used for those textures in which the silicate phase occurs as individual grains distributed throughout the sample, independently of the product carbonate. (4)The term *metamorphic banding* or *layering* indicates that the silicates and carbonates are concentrated in sheets which may be isolated or alternate regularly. It must be emphasized that in this paper, the term is descriptive and does not correspond to the term 'metamorphic layering' in the broadest sense used in some publications. The observed bands or layers are usually parallel to premetamorphic features such as cleavage and there is no evidence that they are caused by anisotropic strain conditions.

The microtextures which result from dissolution-crystallization processes during decarbonation may be overprinted by static annealing and retrogressive processes. As a rule, the grain boundaries are smoothed by grain-boundary migration. The state of the interphase boundary between dolomite, which is reactant in both reactions, and calcite, which is product phase in both reactions, is critical, as it indicates whether the reaction texture is the result of crystallization processes only: this is unequivocal in those samples which exhibit all participating phases and in which the reacting phases show clear signs of dissolution, as may be seen in Fig. 11.9c. The shape and distribution pattern of the product silicate phases suggest that their growth features are not altered by annealing.

The problem of volume change accommodation which accompanies decarbonation reactions cannot be treated in this study, as the textures observed provide no evidence of this important feature of the reactions. In general, the transient permeability in carbonates leaves no imprint on microtextures (Rumble et al. 1982).

11.3 Microtextures and Temperature-Time Path

11.3.1 Temperature-Time Path Diagrams

A thermal model of cooling of the Ballachulish intrusion complex is calculated by Buntebarth (Chap. 18, this Vol.). The following T-t paths are calculated for quartzites. They are considered using a cylindrical intrusion model, an intrusion temperature of 1050 °C (Weiss and Troll, Chap. 4, this Vol.), and a pre-metamorphic temperature of the metasediments of 250 °C (Pattison and Voll, Chap. 2, this Vol.). The influence of the endothermic reactions on the thermal evolution is treated in a general way, as the real reaction times are unknown (see Buntebarth, Chap. 18, this Vol.). A thermal model for Monzoni, however, cannot be calculated yet, as the data which have been obtained indicate complicated conditions for modelling, due to the complex geometry of the intrusion, the unknown intrusion sequence and unknown influence of convection in the aureole. For this reason, the discussion of microtextures in the context of temperature-time history is based on the Ballachulish aureole.

Figure 11.2 shows T-t paths at four different distances in the NE section of the Ballachulish thermoaureole. According to the isograd map (Fig. 10.2) they are all associated with the Fo-Cal-producing reaction (C9). It is not possible to specify at which temperature the decarbonation reaction actually started, as the X_{CO_2} conditions are not known; therefore reaction onset at 500 °C ($X_{CO_2} = 0.05$) was assumed.

The graph for the outermost location, 700 m from the contact, describes the T-t path for a sample which exhibits the onset of the decarbonation reaction, with the first appearance of the isobaric univariant assemblage Tr-Fo-Cal-Dol. The minimum temperature of 500 °C is achieved in about 20 000 years. T-t path at a distance of 300 m shows the thermal evolution for a sample in the reaction zone of reaction (C9). If we assume that at this distance the reaction also starts at 500 °C, this condition is first reached after 3000 years, i.e., 17 000 years earlier than at a distance of 700 m. Thus the term “beginning” of a reaction at the isograd is misleading, if kinetic aspects are considered. A mapped isograd merely specifies the outermost location in the aureole where the conditions for reaction beginning have been achieved. When studying microtextures, it should be considered that near the isograd the reaction progressed at the slowest heating rate possible for this reaction. In specimens which are exposed closer to the contact, in the reaction zone, the reaction began earlier and the heating rate was higher.

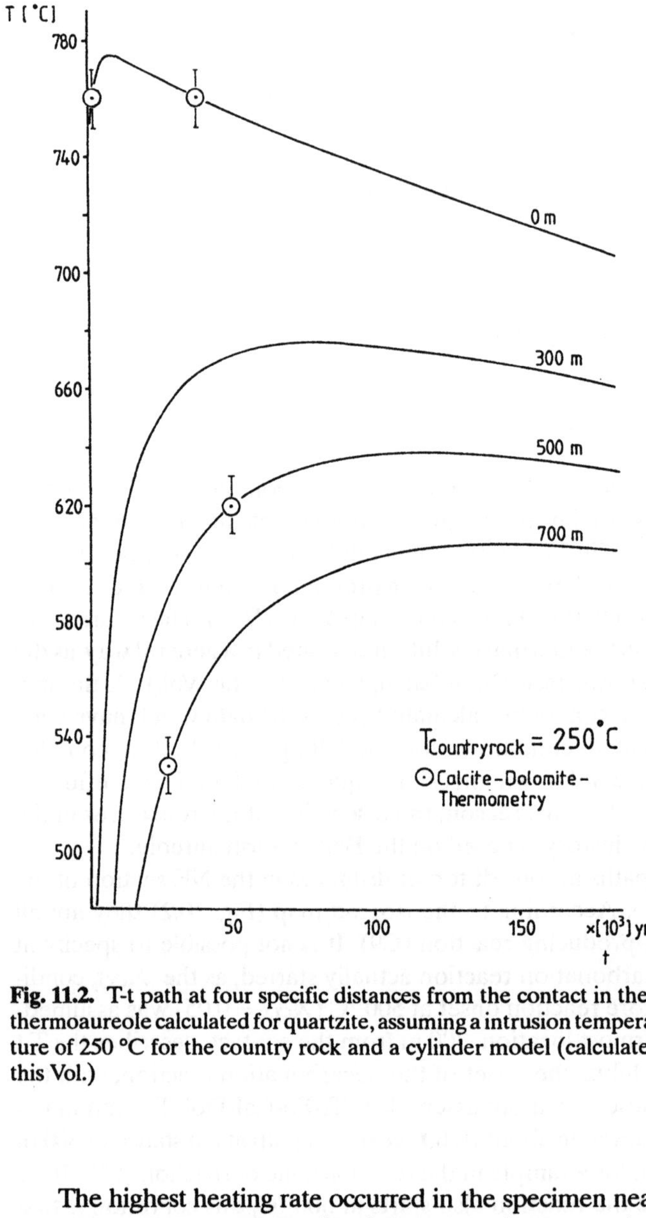

Fig. 11.2. T-t path at four specific distances from the contact in the NE section of the Ballachulish thermoaureole calculated for quartzite, assuming a intrusion temperature of 1050 °C, initial temperature of 250 °C for the country rock and a cylinder model (calculated by Buntebarth, see Chap. 18, this Vol.)

The highest heating rate occurred in the specimen nearest to the contact with the assemblage Fo-Dol-Cal-Per/(Bru) (B375, Om). The maximum temperature of 760 °C, deduced from the Mg-calcite geothermometry, is thought to correspond to the peak temperature achieved during the thermal metamorphism (Masch and Heuss-Aβbichler, Chap. 10, this Vol.). If we assume that the reaction rate was at least as fast as the heating rate, then a time interval of less than 2000 years is necessary in order to achieve this 'maximum-reaction temperature.'

The statements of time deduced from T-t path must be considered critically, since uncertainty concerning any one factor involved in the calculations may change the T-t path considerably, as may be shown in following example. The application of

Mg-calcite geothermometry gives a minimum temperature of 620 °C at 500 m distance from the contact. This temperature corresponds to the maximum temperature possible for the Fo-Cal-producing reaction (C9) according to the equilibrium diagram at 0.3 GPa, with the assemblage Tr-Di-Fo-Cal-Dol (see Fig. 10.12). This temperature is reached after approx. 50 000 years, if an original country-rock temperature of 250 °C is assumed. A higher initial country-rock temperature of 300 °C would shift the maximum temperature at this distance by about 20 °C to a higher value, while the time needed to approach the temperature mark 620 °C would be shortened by half to approx. 30 000 years.

11.3.2 Overstepping

One aspect of studies of kinetics concerns overstepping of the reaction boundary. Because of the variance of (isobaric) decarbonation reactions with regard to T and X_{CO_2}, the only information on this point is provided by the six-phase assemblage Tr-Di-Fo-Cal-Dol-Fluid, which is exposed 500 m from the contact of the Ballachulish intrusion. The temperature of 620 °C deduced by Mg-calcite thermometry corresponds exactly with the calculated temperature for the model system CMS, assuming P = 0.3 GPa, based on reversed experiments and derived from the internally consistent thermodynamic data of Gottschalk and Metz (1984) (see Masch and Heuss-Aβbichler, Chap. 10, this Vol.). The value is within the uncertainty range of the calibrated pelitic reaction (P3) with 640 °C at a distance of 490 m (Pattison and Harte, Chap. 8, this Vol.; see also Fig. 16.4) and it is lower than the maximum temperature calculated by Buntebarth at the distance of 500 m with 640 °C (initial country-rock temperature of 250 °C), or about 660 °C (initial country-rock temperature of 300 °C). According to these data, the overstepping of the reaction boundary seems to be minimal at this location.

The isobaric univariant assemblage Tr-Fo-Cal-Dol (B312), which is found upgrade at a distance of 230 m from the contact, is evidence that overstepping occurred. Tremolites with clear textural signs of dissolution show that the reaction was not completed at the time when the maximum temperature was reached. This occurrence is not an exception, as other specimens near this location exhibit relics of tremolite. The precise amount of overstepping cannot be evaluated, as the data are too uncertain. Nevertheless, according the thermal profile calculated by Buntebarth (Fig. 18.13, curve with T_1 = 1050 °C) a minimum overstepping of 50° must be estimated from the difference between T_{500m} = 640 °C and T_{230m} = 690 °C.

Textural investigations, coupled with considerations of thermal evolution at Ballachulish indicate the importance aspects of overstepping:

1. Consideration of the isobaric invariant assemblage indicates that the ΔT-overstepping before nucleation is small, in accordance with the prediction of Yardley (1977) and Walther and Wood (1984). Thus near an isograd the heating rate seems to be determinative, as reaction progressed readily.
2. The uncompleted reaction (C9) for a sample nearer to the contact in the reaction zone, showing both reacting phases with clear signs of dissolution, is evidence

that overstepping occurred after nucleation. As forsterite and calcite are present in sufficient amounts, this indicates that the overstepping is not caused by suppressed nucleation. These features suggest that overstepping was promoted by reaction rate being more sluggish than heating rate. Thus the ratio between reaction rate and heating rate may change with distance from the contact.

11.4 Microtextures and Reaction Mechanisms

11.4.1 Reaction Mechanisms

The microtextures observed in the aureole show a range of different patterns in the spatial distribution of the product phases within a specimen. The various microtextures developed even for one decarbonation reaction indicate that different reaction paths are possible. Simplified, the progress of a metamorphic reaction can be described by the following distinct reaction stages:

1. Heat flux driving temperature above the equilibrium condition (overstepping);
2. Dissolution of the reacting phases;
3. Transport of the dissolved species;
4. Crystallization, i.e., nucleation and growth.

The result of the interaction of these reaction steps can be described by a reaction mechanism. The most sluggish stage of the sequential process is the rate-controlling step (Fisher 1978). On the other hand, the rate laws can be expressed as the rate of change of the concentration of the species. After a finite time the ratio of concentrations will be the equilibrium constant K at conditions of the reaction. The latter term can be also expressed by the activities of the species. Thus chemical potential diagrams, as used by Fisher (1978) and Ridley and Thompson (1986), can be used to present different types of reaction mechanism. These diagrams illustrate the course of the chemical potential gradient of components between the reactant and product phases. Figure 11.3 shows possible limiting cases.

11.4.1.1 Transport-Controlled Reaction Mechanism

This reaction mechanism requires transport to be the slowest reaction step. It is characterized by saturated or slightly undersaturated solution at the interface of the reacting solid-to-fluid phase, which implies the stabilization of local equilibrium conditions at this position. The chemical potential gradient, the driving force for the reaction, is located *between* the reacting phases within the fluid phase as transport medium. This leads to crystallization of new crystals at those sites along the gradient where the saturation conditions are suitable. Figure 11.3(Ia) represents the establishment of steady state condition with a minimal chemical potential difference

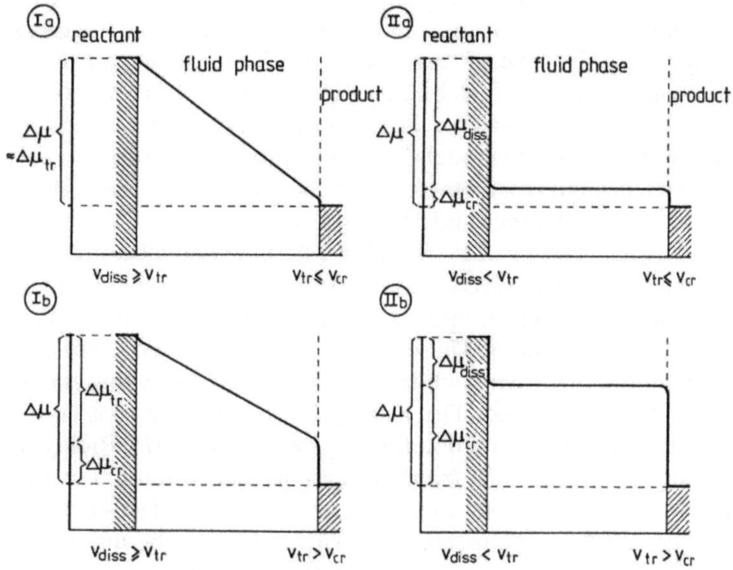

Fig. 11.3. Schematic presentation of possible critical cases in the course of chemical potential (chemical potential vs unit of space) during reaction progress, with the distinct reaction steps dissolution (*diss*), transport (*tr*) and crystallization (*cr*). Two different types of reaction mechanisms may be distinguished: *Ia,b* transport-controlled; *IIa,b* interface-controlled reaction mechanisms

between the fluid phase and solid phases. This reaction condition can be the result of dissolution rates faster than transport rates.

11.4.1.2 Interface-Controlled Reaction Mechanism

Two different interfaces, those of the reacting and of the growing phases, must be considered.

1. *Dissolution* as the rate-limiting step affords a discontinuity in the course of the chemical potential gradient at the interface of the dissolving phase. Interfacial undersaturation conditions in the fluid phase at this position promote preferential dissolution at energetically favoured sites, producing crystallographically oriented ledges or cavities on the surface (Herman and White 1985; Brantley et al. 1986). This reaction condition can be caused by transport rates being faster than dissolution (Fig. 11.3, IIa,b). It will reduce the chemical potential gradient in the fluid phase to a minimum, and the main chemical potential difference (affinity) will be located immediately at the interface of the dissolving phase. The discontinuity of the chemical potential at this position means that an exact evaluation of the course of chemical potential gradient in the reaction system is not possible. Two possible situations are demonstrated in Fig. 11.3 (IIa,b).

2. *Crystallization* as the slowest step within the reaction sequence will induce high saturation conditions with respect to the product phases, and thus affect the

nucleation rates and growth conditions of the product phases (Sunagawa 1981, 1984). It is unlikely that nucleation and/or growth, as the most sluggish step, will give rise to a change in the type of reaction mechanism. We suggest that the type of reaction mechanism which may stabilize is determined by the course of the chemical potential gradient at the interface of the dissolving phase, as shown in Fig. 11.3.

Several reaction models have been proposed in order to determine the reaction mechanisms valid for metamorphism, and to form a concept of the rate laws in general (Fisher 1978; Walther and Wood 1984; Rubie and Thompson 1985; Lasaga 1986; Ridley and Thompson 1986; Rubie 1986). It should be stressed that the discussions are based primarily on conclusions drawn from experimental results and on theoretical calculations which apply to heterogeneous devolatilization and, in particular, dehydration reactions. These models may vary in detail, but they have some aspects in common. In most models the interface of the dissolving and that of growing phase are not differentiated. The assumption is common that crystallization, especially growth, is the usual rate-determining step (Fisher 1978; Ridley and Thompson 1986). Fisher (1978) concludes from calculations that with *reaction progress* ($\Delta\xi$) there will be a change in reaction mechanism from initial interface-control, to transport-control, and finally to heat flux control. With regard to the amount of *overstepping*, it is estimated that in the case of low overstepping of the equilibrium conditions, processes at the interface such as dissolution and/or crystallization will be the rate-determining step, and high overstepping will cause transport to be the controlling step (Fisher 1978; Walther and Wood 1984; Lasaga 1986; Ridley and Thompson 1986). The catalytic effect of H_2O in the *fluid phase* and its function as a transport medium is thought to be a further factor which may influence the establishment of a particular reaction mechanism (Walther and Wood 1984; Rubie 1986). According to Rubie (1986), a change in reaction mechanism from interface-controlled to transport-controlled occurs as diffusion coefficients become smaller. A large volume of fluid phase can favour a reaction-controlled mechanism, whereas under 'dry' reaction conditions, transport is thought to be the rate-determining step.

There have been a few attempts to explain natural reaction textures by specific reaction mechanisms (Fisher 1973; Fisher and Elliott 1974; Joesten 1974, 1977; Yardley 1977; Heuss-Aßbichler 1987). The concept of transport-controlled reaction mechanism has been applied to explain reaction rims or coronas, which are characterized by well-developed mineral zones between the reacting phases, with the individual zones arranged in a sequence of increasing or decreasing chemical potential (Korzhinskii 1959; Vidale 1969; Fisher 1970, 1973; Fisher and Elliott 1974; Joesten 1974, 1977; Heuss-Aßbichler 1987). However, correlation of microtextures with interface-controlled or heat-flux controlled reaction mechanisms is still in its infancy.

11.4.2 Reaction Mechanism and Changing Boundary Conditions

The above studies concentrate on the determination of the rate laws in metamorphic reactions. The effect of a continued change in boundary conditions on the reaction is generally not taken into account. Reflection on the thermal evolution in the aureole and on overstepping (see above, Sect. 11.3.3), however, suggests a change in the ratio of reaction rate to heating rate with distance from the contact. A change in boundary conditions (T, X_{CO_2}) during reaction progress becomes more likely when reaction rate is slower than heating rate. Changing reaction conditions not only influence the rates of the different reaction steps, but also the equilibrium constant by affecting the activities of the participating species. If this change is faster than the reaction system can react with change in concentration, it is likely that the prevailing saturation condition in the fluid phase will deviate from that needed at equilibrium.

A decrease in the chemical potential means that equilibrium is attained at lower activities. A change in the reaction conditions means that the concentration of the species dissolved in the fluid phase is higher than is the case at equilibrium. Thus dissolution of the reacting phase will be retarded as long as interfacial saturation conditions prevail. The resulting course of the chemical potential gradient corresponds with that of transport-controlled reaction mechanism. On the other hand, a shift of the chemical potential to higher values corresponds to a shift to higher activities in the fluid phase. Thus a sudden change leads to undersaturation in the fluid phase. This reaction condition will promote dissolution of the reacting phase in order to compensate the non-equilibrium condition at the interface. The discontinuity in the chemical potential gradient at the interface of the dissolving phase corresponds with that associated with interface-controlled reaction mechanism.

A general trend in the evolution of the reaction system can be indicated using chemical potential diagrams or activity-diagrams. The diagrams in Fig. 11.4 are taken from the tabularized logarithmic activity diagrams of the system CaO-MgO-(SiO_2)-H_2O-CO_2-HCl as functions of P, T and X_{CO_2}, published by Bowers et al. (1984). They show the stability fields of the silicate phases at SiO_2-saturation, intercepted by the stability fields of calcite (horizontal dashed line), dolomite (inclined dashed line) and magnesite (vertical dashed line). Within the enclosed sector the minerals are stable, outside of the sector metastable. These diagrams imply that within the stability fields of the minerals, the fluid phase is saturated with regard to the solid phase (Helgeson 1968). In the reaction system CMS, the non-ideal behaviour of the fugacity of H_2O and CO_2 (Flowers 1979; Kerrick and Jacobs 1981) affects the stability fields of the silicates and carbonates (Walther and Helgeson 1980; Bowers et al. 1984). A decrease in the chemical potential of both silicates and carbonates, which implies saturation conditions at the interface of *all* reacting phases, is only attained by increase in temperature at low prevailing pressures (see Fig. 11.4, $P = 0.05$ GPa). At higher pressures ($P \geq 0.2$ GPa) the chemical potential of the silicates decreases also with rising temperature, but the stability field of dolomite and calcite shifts to higher values. Change in X_{CO_2} may shift the activities of the reacting phases to higher values independently of prevailing pressure. The resulting undersaturation condition at the interface of one of the dissolving phases is thought to be critical for the stabilization of interface-controlled reaction mechanism.

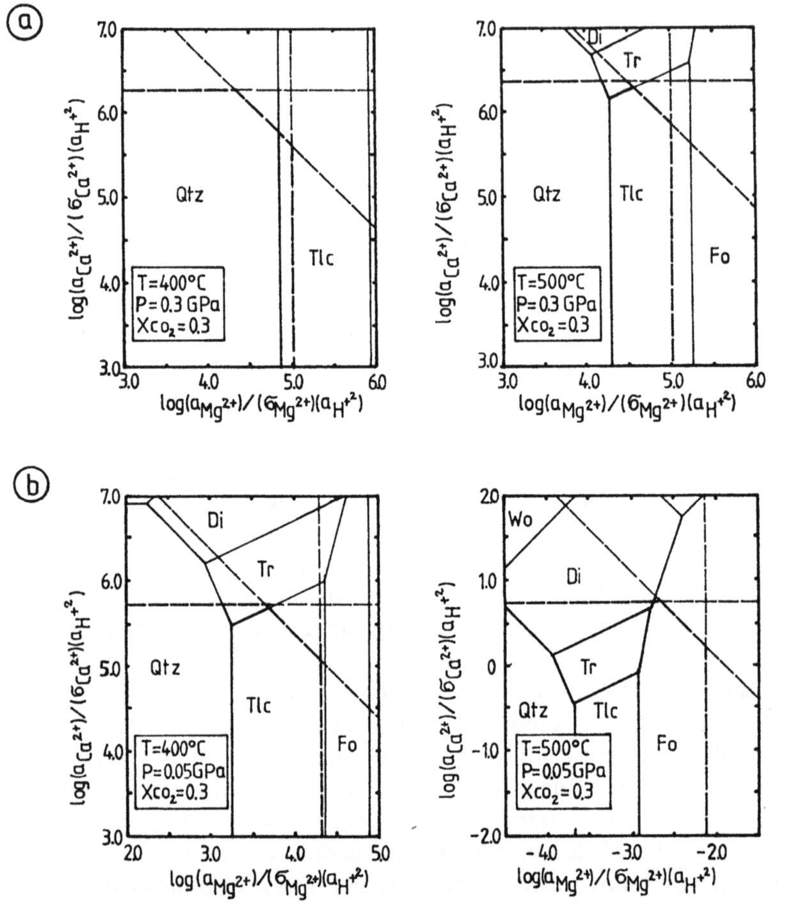

Fig. 11.4. Logarithmic activity diagrams for the system CaO-MgO-SiO_2-HCl-CO_2-H_2O at X_{CO_2} = 0.3 for 400 and 500 °C at a pressure of **a** 0.3 GPa and **b** 0.05 GPa. The *dashed lines* correspond to the saturation of the fluid phase with respect to calcite (*horizontal dashed line*), dolomite (*inclined dashed line*), and magnesite (*vertical dashed line*). $\sigma_i = a_{H_2O}^{\Delta n_i}$, $\Delta n_i = n_i - Z_i n_{H^+}$ with Z_i: charge of the *i*th aqueous species. Diagrams after Bowers et al. (1984)

11.4.4 The Correlation of Microtextures with Reaction Mechanism

The comparison of the textures developed in the siliceous dolomites of Ballachulish with those exposed in Monzoni indicates a common trend in the evolution of microtextures. The main difference observed concerns the textures developed near the isograd, as in Monzoni, in contrast to Ballachulish, three different decarbonation reactions result in corona textures (Heuss-Aβbichler 1987). The *low-grade Tlc-Cal-producing reaction* (C1) is remarkable, as only one type of microtexture developed in both aureoles. It is the outermost reaction observed in the dolomite beds of the two aureoles. In both cases premetamorphic features are preserved, all phases are present, and with the onset of the reaction, two completely different types of microtexture are observed. These microtextures deviate both in the pattern of nucleation site and

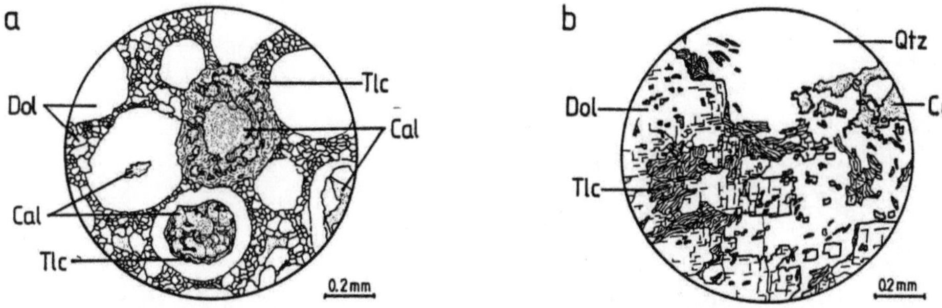

Fig. 11.5 a. Corona texture of monomineralic talc and calcite layers embedded within a dolomite matrix with pre-metamorphic ooids (Monzoni). For clarity the very fine-grained dolomite in the ooliths are not drawn. **b** Talc and calcite distributed independently within quartz and dolomite. Dolomite and quartz may be in direct contact without reacting (Ballachulish)

growth features of the products, and in the dissolution structures of the reacting phases.

In *Monzoni* only type (1) *corona* textures occur (Fig. 11.5a), which are characterized by isolated reaction sites, on the scale of grain size, within a reaction-free dolomite matrix. Rims of talc and calcite are located between the reacting phases quartz and dolomite. The zones are arranged according the chemical potential gradient, and are consistent with steady state diffusion at local equilibrium conditions. The order of the monomineralic layers alters in a systematic way with distance from the contact (Heuss-Aβbichler 1987). The formation of this texture can be described perfectly by *transport-controlled reaction mechanism.*

The samples from the south and the east of the *Ballachulish* intrusion (see Fig. 10.4) exhibit a *granoblastic texture* [type (3); (Fig. 11.5b]. In contrast to above, the reaction site cannot be localized, as the product phases talc and calcite occur independently within quartz and/or dolomite. It is therefore on the scale of thin section or specimen. The clear signs of dissolution in dolomite and quartz are evidence for undersaturation conditions in the fluid phase immediately at the interface of the reacting solid. Quartz and dolomite frequently coexist. Obviously, at this site the fluid was undersaturated with regard to the products. All these features suggest that the development of the granoblastic texture is the result of *interface-controlled reaction mechanism.*

The *high-grade Fo-Cal-producing reaction* (C9) exhibits a great variety of microtextures; textural investigations confirm the establishment of a systematic pattern of types of microtextures from the isograd to the contact (Fig. 11.6). The spatial sequence of reaction textures for this one decarbonation reaction is: coronas, intergrowths, granoblastic textures and metamorphic banding. In comparison to Monzoni, the Ballachulish samples show a higher variability in the development of microtextures. In particular, those near the contact exhibit a variety of features, even within a single thin section. Except for two cases, the completed reaction assemblage Fo-Cal-Dol is exposed.

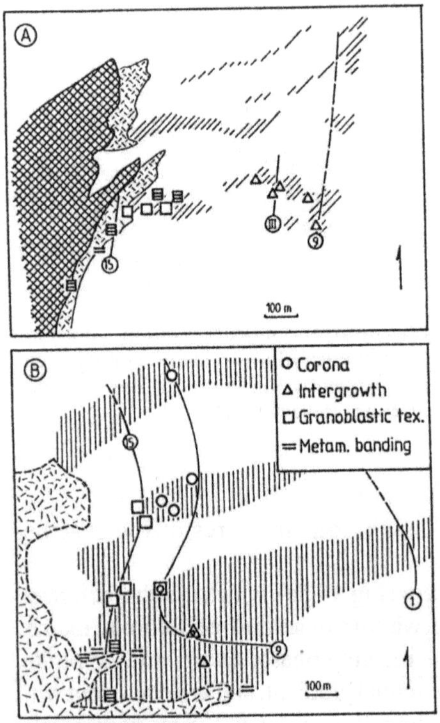

Fig. 11.6. The systematic change in microtextures along a profile in the thermoaureole of the **A** Ballachulish (see Fig. 10.2) and **B** Monzoni intrusion complexes (*vertical line*: siliceous dolomites of the southern Alpine series (Perm-Trias); *irregular dashes*: eastern contact of Monzoni intrusion complex) for the reaction (C9) 1 TR + 11 Dol ⇔ 8 Fo + 13 Cal + 9 CO_2 + 1 H_2O

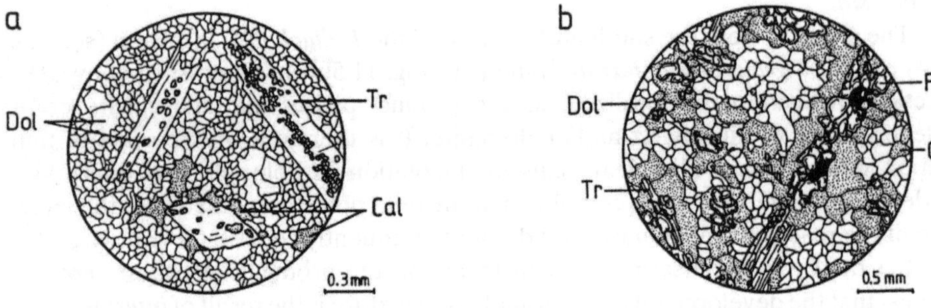

Fig. 11.7 a. Starting assemblage Tr-Dol with prismatic tremolites distributed homogeneously throughout the dolomite matrix; **b** development of corona texture near the isograd: cores of forsterite are surrounded by calcite and embedded in a fine-grained dolomite matrix (Monzoni)

11.4.4.1 Texture Type (1): Corona

Only in Monzoni are corona textures developed in the specimens nearest to the isograd with the isobaric univariant assemblage Tr-Cal-Fo-Dol. In the sample with the reacting assemblage Tr-Dol, prismatic subhedral tremolites are nearly homogeneously distributed throughout an equidimensional fine-grained dolomite matrix (Fig. 11.7a). Tremolites usually have very fine-grained inclusions of dolomitein the core. Upgrade near the isograd, forsterite partially or totally replaces tremolite, and is surrounded by calcite showing corona texture (Fig. 11.7b). At higher grades, the nucleation sites of forsterite seem to be restricted, as at distinct reaction sites forsterite grows to form large rounded or euhedral crystals. They all have a halo of calcite, whereas residual tremolite distributed throughout the matrix dolomite shows features of dissolution.

11.4.4.2 Texture Type (2): Intergrowth Texture

In Monzoni, this microtextural type is restricted to the southern section of the carbonates, near the isograd (Fig. 11.6b). Coarser calcitic domains are scattered throughout fine-grained dolomite-rich regions of the sample (Fig. 11.8a). Just adjacent to these domains, dolomite is more coarse-grained than in the surrounding matrix. Within the segregations, many fine-grained forsterites with embayed forms are irregularly disseminated. They show preferred crystallographic orientation in distinct patches, which suggest that they are the result of fast growth conditions and restricted independent nucleation. The intergrowths of forsterite and calcite correspond to eutectoidal textures.

In Ballachulish, in the specimens from 700 m up to 450 m from the contact, the product phases forsterite and calcite are limited to isolated sites which stand out against the more fine-grained dolomite matrix. The intergrowths may vary in size and they show high variability in detail (Fig. 11.8b,c). It is striking that the forsterites

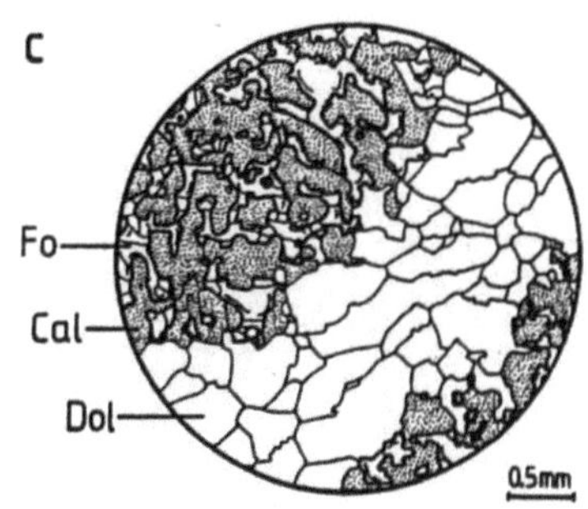

Fig. 11.8 a. Intergrowth of forsterite and calcite within a domain surrounded by a fine-grained dolomite matrix (Monzoni); **b** intergrowth of forsterite and calcite with many fine-grained forsterites disseminated throughout the calcite segregation (Ballachulish); **c** intergrowth of forsterite and calcite; embayed forsterites with preferred crystallographic orientation fingering into the calcite segregation (Ballachulish)

exhibit preferred crystallographic orientations within the domains, similarly to those in Monzoni.

11.4.4.3 Texture Type (3): Granoblastic Texture

In the adjacent higher grade area forsterite occurs irregularly throughout the dolomite-calcite aggregate, at triple junctions or along the grain boundaries (Fig. 11.9a,b). Obviously the nucleation sites of the phases forsterite and calcite are independent of one another. They are likely to have no immediate local relationship with the reacting phases. Dissolution features of dolomite are still preserved, which indicates that the textures are primary. Further upgrade in this section, the microtextures tend to exhibit smooth, embayed interphase contacts of dolomite with calcite. They are thought to be caused by grain-boundary migration during annealing processes. The distribution of forsterite does not seem to be affected by this process.

In Ballachulish, the change in microtexture from intergrowth to granoblastic type is not exposed. The granoblastic textures occur within approx. 300 m of the contact (Fig. 11.9b). The grain size of forsterite within one sample is highly variable (between 40 and approx. 400 μm); similarly, the grain shape varies from rounded to embayed forms, with inclusions of carbonate which correspond to atoll structures. One specimen (B312, 230 m from the contact) exhibits unequivocal primary textures (Fig. 11.9c). This sample contains large amounts of tremolite, which shows dissolution features resembling those produced in the experimental studies on the mechanisms

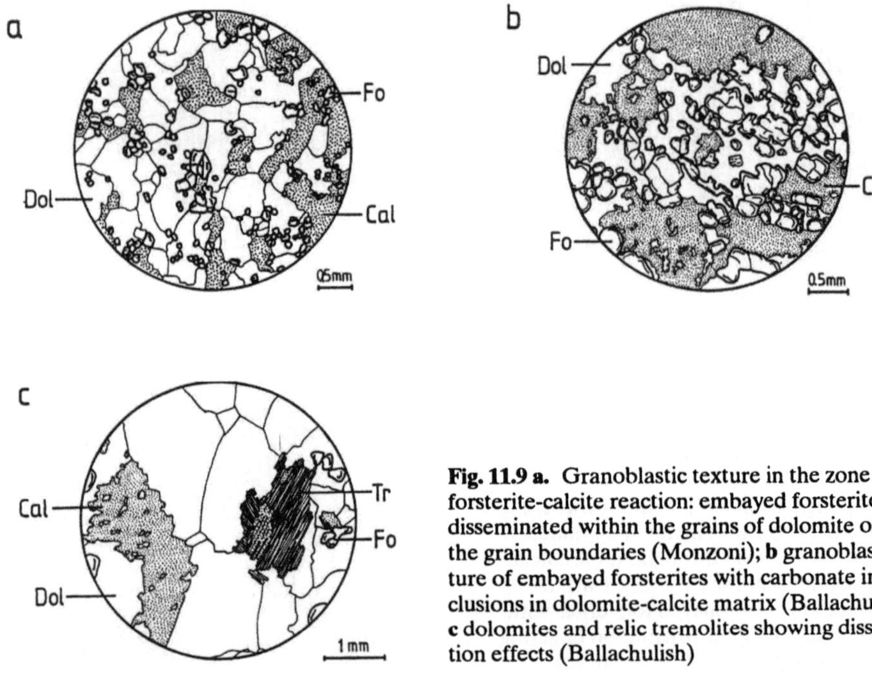

Fig. 11.9 a. Granoblastic texture in the zone of the forsterite-calcite reaction: embayed forsterites are disseminated within the grains of dolomite or along the grain boundaries (Monzoni); **b** granoblastic texture of embayed forsterites with carbonate inclusions in dolomite-calcite matrix (Ballachulish); **c** dolomites and relic tremolites showing dissolution effects (Ballachulish)

of two decarbonation reactions by Heinrich et al. (1986) and Dachs and Metz (1988). Dolomite also shows clear signs of dissolution and is replaced by calcite. At higher grades, the primary texture is overprinted by recrystallization processes due to annealing.

11.4.4.4 Texture Type (4): Metamorphic Banding or Layering

Closest to the contact of the intrusion, the samples with the assemblages Fo-Cal-Dol exhibit a distinctive metamorphic banding or layering. The modal composition and bulk chemistry cannot be attributed to the primary composition of the sedimentary protolith. In Monzoni, forsterite-rich layers alternate with coarse-grained, calcite-rich layers containing few forsterites (Fig. 11.10a). Groups of forsterite grains, which are much larger than in the lower grade samples, may also grow along the grain boundaries of dolomite and/or calcite to produce a kind of garland texture.

In Ballachulish, large-scaled domains of calcite accompanied by a ring-shaped arrangement of forsterite grains (Fig. 11.10b) may occur in close spatial association with metamorphic banding. Furthermore, calcite-rich layers often contain a concentration of single forsterite grains which defines a band of variable thickness (Fig. 11.10c). As in Monzoni, grains of forsterite may form a garland within dolomite as well (see Fig. 10.9a).

To summarize, in both areas, zoning of microtextures parallel to the isograds is observed. The textures may have been modified by static annealing (grain-boundary

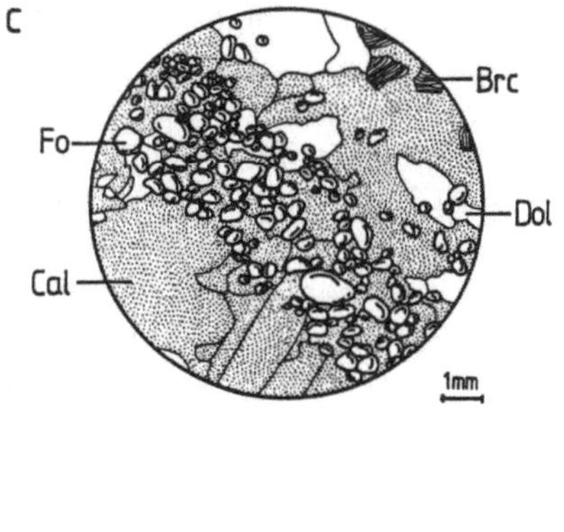

Fig. 11.10 a. Metamorphic layering of forsterite and calcite in the zone of the forsterite-calcite reaction (Monzoni); **b** fine-grained forsterites, showing a ring-like distribution within calcite segregations in a dolomite matrix (Ballachulish); **c** metamorphic banding of a forsterite-rich layer (Ballachulish)

migration, grain growth) and grain-boundary migration induced by thermal stress, but in general the reaction-induced distribution of the phases in the microtexture is preserved. This is verified by the granoblastic texture of the isobaric univariant assemblage Tr-Fo-Cal-Dol (B312) observed near the contact of the Ballachulish aureole, which did not form a precursory texture such as intergrowth. The change in types of microtextures within a specimen is restricted to where the margins of texture-type zones in the aureole overlap. In Monzoni in one sample, single coarse forsterite grains, which are surrounded by a halo of calcite similar to corona texture, may occur within segregations showing intergrowth texture (see Fig. 11.8a). In another specimen, forsterites appear more frequently in the core of the calcites, as in corona textures, as well as being distributed independently, as in granoblastic textures. Metamorphic banding may be superimposed on the granoblastic texture in both Monzoni and in Ballachulish.

The systematic change in microtextures in both areas correlates with a systematic change in the dimension of the reaction site. Coronas and intergrowth textures occur near the isograd. In both cases, the distribution of product phases is restricted to domains within a reaction-free matrix, and the reaction sites are on the scale of the grain size. These textures suggest that the transport of the fluid phase was limited. Nearer the contact, in the reaction zone, granoblastic textures and metamorphic banding predominate. Due to the independent nucleation of the products, the reaction site cannot be determined exactly; it varies on the scale of specimen or even outcrop. Transport distances are definitely larger in this section of the aureole.

This subdivision of microtextures according to the reaction site indicates the influence of the fluid phase on the development of reaction textures. The transport distances decrease definitely with increasing distance from the contact, which can be explained by decrease in transport rate. This correlates with the deduced decrease in overstepping (Sect. 11.3.2). The differing behaviour of fluid phase with distance from the contact is reflected in the X_{CO_2} values determined for the Ballachulish aureole (Masch and Heuss-Aβbichler, Chap. 10, this Vol.). The high X_{CO_2} values determined for the specimens with intergrowth textures in the middle of the reaction zone (X_{CO_2} > 0.5, 500 m distance from the contact) suggest that buffering was effective. This evolution can be explained by reaction rates being faster than transport rates. A limited exchange of the fluid generated by the reaction with the initial fluid present along the interconnected grain boundary system causes a change in composition of the fluid phase to higher values within these locally restricted reaction sites. There are, however, no indications that high X_{CO_2} values were ever reached in the specimens near the contact. The low values which have been deduced (X_{CO_2} < 0.2) mean that dilution by H_2O-rich fluids must have occurred during reaction progress. Oxygen isotope data show that these fluids could not have had a magmatic source, and that the volume contributed by the neighbouring pelites was minimal (Hoernes et al., Chap. 17, this Vol.). This suggests that the H_2O-rich fluids must therefore have originated to a great extent within the carbonates themselves. This may be explained by high transport rates relative to reaction rate. The components dissolved are transported faster than the product phases may crystallize at that site. Thus the CO_2

released by the reaction may be distributed over a large area along the interconnected grain boundary system, and total change in composition is therefore low.

Several features point to the effect of change in fluid volume. This is reflected by coarsening of matrix dolomite adjacent to calcite intergrowth textures in Monzoni (Fig. 11.8a), which seems to be promoted by local increase of fluid volume due to reaction progress (Tullis and Yund 1982). A striking feature in the samples with intergrowth textures from the Ballachulish aureole is that the separate segregations are interconnected by a network of fine sub-parallel veinlets of calcite, which may even pierce through dolomite (Fig. 10.9b). Centimetre-thick veins in the high-grade carbonates of Ballachulish are characteristic (Fig. 10.10), whose orientation corresponds with the main pre-intrusion deformation structures. Both the veins apparent in outcrop and microfractures within the dolomite matrix are thought to be caused by hydraulic fracture mechanism at $P_{fl} = P_{load}$ (Etheridge et al. 1984), due to the volume of fluid phase liberated during progress of the decarbonation reaction.

11.5 Conclusions

The examination of textural evidence in two different aureoles with different thermal profiles shows that usually the textures preserved in the specimens developed during the last period of the heating phase; this corresponds with the peak of the metamorphic event. Thus any discussion of the evolution of microtextures must be restricted to this interval of time. The range of different microtextures exposed indicates that there is no texture typical of either of the examined decarbonation reactions. The systematic change in the spatial distribution of the microtextures observed in both aureoles suggests that the T-t path is of great importance for the evolution of microtextures. According to the T-t path, the assemblages of a specific reaction developed first are exposed closer to the contact, i.e., in the reaction zone, and those produced last are exposed at the isograd. Consideration of the concept of overstepping indicates that it may occur by heating rate faster than reaction rate, even if overstepping before nucleation is minimal. This leads to a simple metamorphic model for a specific reaction: the increasing distance from the contact may be correlated with decreasing overstepping of the reaction boundary.

The development of the textures seems to be the result of a complex interplay of reaction constraints such as composition, volume and mobility of the fluid phase, and overstepping. The study of reaction kinetics, and thus of the rate laws, is hampered by the fact that many variables are unquantified. Nevertheless, several conclusions may be drawn, based on the textural investigations. The textures developed for the low-grade Tlc-Cal-producing reaction enabled the correlation of corona textures with transport-controlled reaction mechanism and granoblastic textures with interface-controlled reaction mechanism. According to the spatial distribution of microtextures of the Fo-Cal-producing reaction (C9) in Monzoni, coronas occur near the isograd and granoblastic textures nearer the contact, in the reaction zone. Thus, the combination of deductions with regard to T-t-evolution and the correlation of textures with reaction mechanisms suggests that transport-controlled reaction mechanism occurs at low overstepping, and interface-controlled reaction mechanism at higher overstepping. This result does not correspond with the assumptions made in other kinetic studies (Fisher 1978; Walther and Wood 1984; Ridley and Thompson 1986).

In both thermoaureoles, field relations suggest that fluid phase was present during reaction progress. Two different textures, granoblastic textures and metamorphic banding, which are exposed in the reaction zone mean that extensive fluid migration must have occurred. The corona and the intergrowth textures exposed near the isograds seem to be the result of restricted mass transport. This evidence suggests that the rate of mass transport is important for the development of particular microtextures. The different textures possible for the two different ranges of transport rates give no clear indication that transport rate alone may cause a change in reaction mechanism, as has been assumed by Walther and Wood (1984) and Rubie (1986). The conceptual approach proposed in this paper suggests that the stabilization of various reaction mechanisms, or a change in them, may be the result of overstepping due to heating rate being faster than reaction rate. A 'sudden' change in prevailing boundary conditions such as T and X_{CO_2} may cause under- or oversaturation in the fluid phase, if the rates of dissolution and/or transport are not fast enough to equalize the deviation from the equilibrium constant in the reaction system. The dynamic concept of operative processes would explain the development of different types of microtextures at the isograd for Tlc-Cal-producing reaction (C1) in the aureole of the high-level intrusion Monzoni and the deeper intrusion of Ballachulish, without variation in the rates of the different reaction steps; with increase in temperature, the stability fields of the minerals may change differently at different pressures. The concept of dynamic change of boundary conditions, coupled with the effects of the fluid phase, may explain the development of the different microtextures in the reaction zone for the Fo-Cal-producing reaction (C9).

Field relations and textural evidence suggest that at a specific distance from the contact, the specimen follows a defined thermal path, which is accompanied by a definite change in boundary conditions. For a distinct T-t-range, only one kind of reaction mechanism is stable, causing a specific type of microtexture as the result of the prevailing thermodynamic properties and reaction kinetics. Specimens with different types of microtexture are thought to be borderline cases, in which the reaction system reacts sensitively to change in boundary conditions during reaction progress.

11.6 Summary

In two thermoaureoles, Ballachulish and Monzoni, the variation of the microtextures related to two different decarbonation reactions with distance from the mapped isograd was investigated. The endothermic Fo-Cal-producing reaction (C9) has been selected for detailed discussion, as the assemblage shows the greatest range of variation. Four types of microtexture can be distinguished. Analysis of the observed pattern of microtextures indicate a corresponding systematic evolution in both thermoaureoles.

According to the thermal cooling model, the assemblage exposed nearest to the contact in the reaction zone was produced first and that near the isograd developed last. Theoretical consideration of heating rate supports high overstepping conditions in the reaction zone and minimal overstepping at the isograd.

The position of the chemical potential gradient in the reacting system and the resulting equilibrium conditions at the interface of dissolving phases are thought to be decisive for the stabilization of a particular reaction mechanism. Local equilibrium conditions at the interface of the dissolving phase support transport-controlled reaction mechanism. Undersaturation with regard to one of the reacting phases may cause interfacial nonequilibrium and promotes interface-controlled reaction mechanisms. Textural investigations enabled the correlation of microtextures with reaction mechanism: coronas are thought to be transport-controlled, and granoblastic textures the result of interface-controlled reaction mechanisms. The development of different types of microtextures at the isograd of the two aureoles can be explained by the effect of overstepping. A change in equilibrium constant due to changing boundary conditions (T, X_{CO_2}) may change the saturation conditions, and thus the course of the chemical potential in the reaction system.

In a different approach, the systematic change in both type of microtexture and the size of reaction site with distance from the contact may be correlated with the change in reaction mechanism by increase in transport rate, as predicted in kinetic studies. The formation of various microtextures and their systematic variation, however, cannot be explained by this concept alone. They seem to be the result of interaction of various reaction constraints such as composition of the fluid phase, the volume of the released fluid and its mobility, and to be a function of temperature and time.

12 Quartz Grain Coarsening by Collective Crystallization in Contact Quartzites

G. Buntebarth and G. Voll

12.1 The Appin Quartzite: Sedimentary Features

Clastic minerals of the Appin Quartzite are: ca. 85 vol % quartz, ca. 15% alkali feldspars, heavy minerals (zircone, rutile — partly formed from titanomagnetite during diagenesis, and apatite and tourmaline as very rare exceptions.

The Appin Quartzite is a current-bedded orthoquartzite. Currents came from the W. They formed spoon-shaped units, often intersecting lower units, rarely parallel-sided. Foreset laminae are centimetre-thick, often graded, enriched in feldspar and in heavy minerals at the lower side of the slope. Zircon and rutile are enriched as millimetre to centimetre layers, with up to more than 50% heavy minerals in some top set beds. Clastic quartzes, feldspars and the rare rock-fragments appear as pebbles, transported by saltation, up to 3–4 cm in diameter. Approximately 15% of the quartz pebbles were polycrystalline, often with preferred c-axis orientation. The other grains are much smaller, transported in suspension. Locally, pebbles may dominate and form pebble beds. Approximately 5% of all clastic quartzes contain rutile needles, mostly oriented, as they are typical for quartzes from granulite facies terrains. During diagenesis, quartzes continued to grow into the pores and K-feldspars developed thin adularia rims.

12.2 The Appin Quartzite: Deformation and Regional Metamorphism

The Appin Quartzite was deformed by the Caledonian orogeny, together with the other members of the Ballachulish Dalradian. It was folded round the big first folds, Appin Synform and Ballachulish Synclinorium. Quartzes were flattened in the first steep cleavage, and stretched in the steep stretching lineation. Flattening and stretching are inhomogeneous, but mostly weak. Deviations of the mean strain ellipsoid with x:y:z = 1.6:1:0.6 are usually small and of no importance for our study. Shear zones, up to 100 m thick, extend the fold limbs. Within them, flattening and stretching are strongly increased, all quartzes recrystallized and preferred orientation (cross girdles) are pronounced. Voll (1969) has shown that this orientation has no significant influence on grain coarsening and that the

G. Voll, J. Töpel, D.R.M. Pattison, F. Seifert
Equilibrium and Kinetics
in Contact Metamorphism

orientation is blurred very little during coarsening. Shear zones were excluded from this study.

During and after deformation, the quartzites were heated to ca. 400 °C near Port Appin and 450 ± 20 °C in the range of the later aureole. Clastic quartz was plastically deformed by Caledonian deformation, mainly by basal-a-glide. Included rutile needles were bent and broken accordingly. One hundred percent of the clastic quartzes in shear zones, and 50–90% in the quartzites with normal deformation, recrystallized. Preserved clastic grains, mainly pebbles, developed subgrains, elongate parallel c, and bounded by subgrain boundaries, mainly parallel to prism faces, subordinately by others subparallel to basal planes. Their size is ca. 2/3 of that of recrystallized grains. In larger quartzes, gas-liquid inclusions are usually enriched on subgrain boundaries. From Port Appin to the area near the Ballachulish Aureole, the size of recrystallized grains increases little, but increases steadily by increasing T during regional metamorphism. Recrystallized grains form at grain boundaries and replace the clastic grains entirely or partially by new grains seeding and growing inward. Recrystallized grains develop from parent grain, i.e., subgrains with greater misorientation and less internal strain. The amount of recrystallization increases with deformation; the size of recrystallized grains is smallest within the shear zones, but varies little at normal deformation.

Recrystallized grains are also deformed, but show less undulatory strain and no or few subgrains compared to larger clastic relics. Strain-induced suturing of grain boundaries is common between deformed grains. Straight parts in the sutures often are near rhombohedral faces of one or both neighbours. Their length is bound to the subgrain size and subgrain boundaries usually adhere to reentrant angles in the sutured boundaries. Features of post-tectonic annealing are rare, e.g., polygonal grains without strain and equilibrium angles at common edges and corners; deformation continued as long as quartz boundaries were mobile. Relics of clastic grains are built into this pattern of recrystallized grains. They are recognizable where they are much larger pebbles or where they still contain rutile needles related to the quartz lattice. During recrystallization, these needles are overrun by grains of different orientation. Their clustered occurrence still marks the circumference of an original clastic grain, even after the strongest contact coarsening.

Therefore the system of reference for contact heating is a quartzite with predominantly recrystallized grains with weak strain, subgrain-subdivision and sutured boundaries and an average grain size of 140 μm for the recrystallized grains. More or less pebble relics and clastic feldspars are built into this pattern.

12.3. Contact Heating of Quartz and Grain Coarsening

Grain growth of quartz was first studied in a preliminary study by Voll (1969) at the same aureole. So far, coarsening has been studied from four profiles normal to the contact: (1) from Beinn Bhan towards the NE contact; (2) at the E contact, N of Sgorr Dhearg; (3) at the W contact, S of Kentallen; and (4) at the SE contact, S of Sgorr Dhearg.

Of these profiles, the third, at the W contact, is unusual in that it reaches larger grain sizes near the contact, where the country rocks are underlain by granite. From profiles (1) and (2), many quartzites contained sufficiently feldspar to hinder quartz grain growth and reduce the grain sizes accordingly. If one uses specimens yielding maximal grain sizes only, however, they are identical to the fourth profile. We chose this fourth profile, at the SE contact, as a reference profile, since it can be based on a large collection of deformed and annealed quartz veins with quartz only, and to a small part on feldspar containing quartzites with very coarse quartz pebbles. Both allow the exclusion of the influence of a second phase, i.e., feldspar.

12.3.1 Methods of Measuring

Eighty-four specimens were selected over a distance of 1000 m towards the contact. Sixty-eight specimens are from the zone of coarsening extending to 750 m from the contact. In nearly all specimens, 300–700 grains were measured; only six points are based on 250 or more measurements. From 20 specimens, 2 thin sections were measured and the mean value was taken. A circle of equal area was fitted upon each grain and its diameter taken as the grain diameter. The mean and quartile values were determined (Table 12.1). The mean values and their contact distance are shown in Fig. 12.1.

Grain growth of the phase (in our case: quartz) is usually hindered by presence of grains of a second phase (feldspar). Therefore, quartz becomes much coarser in layers with little or no feldspar compared to those layers with much small and finely divided feldspar. Or: quartz replacing quartz pebbles is much larger than quartz in the feldspar-rich matrix between them. The amount of feldspar and its subdivision between the quartz grains are both of influence. Therefore, great care was taken to use quartzes only from either quartz veins or from the inside of the largest quartz pebbles, in order to avoid grain growth-locking effects.

The quartz veins formed during diagenesis or orogenic deformation. They were filled with coarse quartz grains, usually one to several cm thick. Many of them were deformed sufficiently to recrystallize (nearly) entirely. Veins in which a larger amount of parent grains were preserved develop very unequal grain sizes and were rejected.

12.3.2 The Morphology of Grain Coarsening

Grain coarsening of quartz sets in very sharply, at distinctly higher T than K-feldspar disordering and little higher T than cordierite formation in neighbouring metapelites. Up to the T of beginning coarsening, the quartz is not changed microscopically with respect to the reference quartzes outside. Study with electron microscopy may, however, show a first change of dislocation type, distribution and density towards slightly lower T of maximum contact heating.

Table 12.1. Mean (M) and quartile (Q) values of the diameter of quartz grains (mm), depending on the distance (x) to the contact; n = number of measurements

x(m)	n	Q_1	M	Q_2	x(m)	n	Q_1	M	Q_2	x(m)	n	Q_1	M	Q_2
16.5	389	0.7402	1.0173	1.4288	435	635	0.3717	0.6160	0.8456	645	391	0.2743	0.4149	0.5790
23	243	0.6839	1.0094	1.4320	450	402	0.5371	0.7198	0.9141	655	418	0.2013	0.3189	0.4366
44	281	0.6948	0.9473	1.3021	465	371	0.3388	0.5242	0.6886	660	441	0.1858	0.3158	0.4552
57	297	0.7085	0.9708	1.3681	465	384	0.3521	0.5282	0.7120	665	450	0.1802	0.2917	0.4224
93	317	0.6752	0.9682	1.6427	485	328	0.3646	0.5947	0.8295	673	502	0.1872	0.2924	0.3944
115	316	0.6182	0.9123	1.2285	495	312	0.3834	0.5712	0.7590	685	462	0.1922	0.2843	0.3787
125	328	0.6701	0.9549	1.4008	505	348	0.3560	0.4892	0.6614	690	498	0.1944	0.3028	0.4128
165	478	0.5822	0.8592	1.1127	510	685	0.3406	0.4923	0.6688	695	613	0.1380	0.2284	0.3252
185	600	0.5744	0.8412	1.0916	517	356	0.3654	0.5047	0.6664	707	526	0.1232	0.1948	0.2649
192.5	248	0.5775	0.7825	0.9922	535	340	0.3185	0.4664	0.6690	725	633	0.1072	0.1628	0.2312
192.5	314	0.5853	0.8138	1.0392	535	309	0.3333	0.4898	0.6808	727	413	0.1601	0.2143	0.2812
200	471	0.4992	0.8357	1.1190	542	523	0.3542	0.4862	0.6224	735	270	0.0922	0.1548	0.2342
200	606	0.4632	0.7215	1.0861	550	437	0.3506	0.5282	0.7356	745	545	0.0881	0.1563	0.2396
245	329	0.6013	0.8643	1.1901	550	561	0.2895	0.4773	0.6964	747	508	0.0807	0.1652	0.2572
265	365	0.7254	0.8060	1.0407	555	387	0.3560	0.4892	0.6614	760	337	0.0991	0.1503	0.2130
275	503	0.4382	0.7215	1.0329	557	505	0.3902	0.5109	0.6360	770	378	0.0921	0.1600	0.2428
275	663	0.5086	0.8012	1.0955	575	418	0.3013	0.4789	0.7043	800	297	0.0876	0.1472	0.2204
300	463	0.4945	0.7684	1.0673	577	579	0.2892	0.3994	0.5345	800	387	0.0931	0.1492	0.2240
335	279	0.5165	0.7669	1.0392	580	541	0.2812	0.4273	0.5685	825	403	0.0974	0.1556	0.2192
345	458	0.5086	0.7356	0.9547	585	468	0.2706	0.4044	0.5357	825	383	0.0952	0.1349	0.1989
345	518	0.4539	0.7152	0.9515	593	511	0.3239	0.4521	0.6069	835	395	0.0849	0.1365	0.1982
350	266	0.4539	0.6964	0.9194	593	489	0.3066	0.4490	0.6100	885	255	0.0447	0.0783	0.1208
350	442	0.5125	0.7551	0.9820	600	498	0.3685	0.4985	0.6639	890	479	0.0766	0.1269	0.2000
370	564	0.4812	0.7316	0.9640	600	572	0.3443	0.4707	0.6224	890	568	0.0801	0.1361	0.2074
390	298	0.4699	0.7748	1.1807	610	501	0.2961	0.4131	0.5503	928	407	0.0916	0.1525	0.2227
407	600	0.4898	0.7246	0.8889	615	754	0.2960	0.4211	0.5527	930	444	0.0852	0.1521	0.2223
425	327	0.4695	0.6612	0.8529	623	330	0.2849	0.4100	0.5264	930	420	0.0852	0.1392	0.1968
425	318	0.4147	0.6025	0.7708	633	422	0.2774	0.3877	0.4923	940	445	0.0921	0.1501	0.2114
										995	384	0.1152	0.1810	0.2491

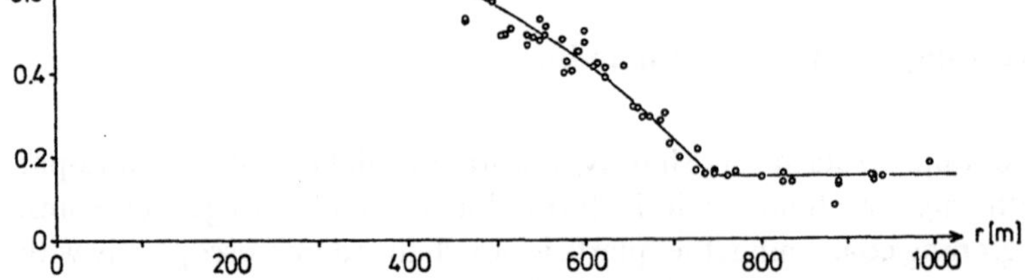

Fig. 12.1. The mean diameter of quartz grains in Appin Quartzite versus the distance from the contact of the pluton

The first effect of contact heating seen microscopically is a slight reduction in grain size, caused by subgrains healing, developing high-angle boundaries and thus becoming independant. In our reference profile this slight effect is at a contact distance of ca. 850 m. We nevertheless consider it real, as a similar effect appears at the E and NE profile at the same distance. As this affects only a smaller part of subgrains, and as the subgrains are not much smaller than the recrystallized grains, this effect would be small in any case. A formation of new independant grains by subgrain coalescence was observed by Voll (1969) during static annealing after deformation of quartz pebbles in Dalradian greywackes near the Highland border. Here we did not observe indications for subgrain coalescence during regional or contact metamorphism.

During the next stage these healed subgrains or clastic and recrystallized grains, start to grow at the expense of more deformed parent grains. This can be considered as a continuation of recrystallization under static conditions. Where these extending grains (or old grains sufficiently healed not to be replaced) meet, they immediately develop polygonal grains with equilibrium angles at common edges and corners. Such angles at common edges, here and at more developed stages of coarsening, show a sharp maximum of angle-frequency at 120°. This indicates that by far the most interfaces are high-angle boundaries with similar and high interfacial free energy.

During further coarsening, relics of older grains, especially pebbles, are continuously replaced by the growing grains. They are, however, internally stabilized by further development of subgrain structure. Up to the highest contact T, gas-liquid inclusions are still found in such pebble relics. There they are enriched along the subgrain boundaries and corners and gradually liberated while the pebbles are replaced. Water which is set free is not included in the replacing grains and is not found on the grain boundaries between them, or at boundaries between the replaced parent and the extending grain replacing it. There are sufficient quartzites with larger patches of only newly formed grains, therefore, such relics do not interfere with grain size determination. No enhanced coarsening has been found around such replaced and dewatering pebbles. However, their presence, even if decreasing, into the innermost contact zones proves that at least the small amount of water stored in these pebbles must have been available on grain boundaries. There is no quartzite of quartz vein where the distance to such dewatering parent grains is larger than a few decimetres. As this water is not found on the grain boundaries of the coarsening grains, they must be viable to water. At the first stage of parent grain replacement, greater local differences in the development of growing grains are observed. They are determined by differences in deformation, strain stored in and size of parent grains. Though most of the quartzites behave similarly, some show a more or less patchy distribution of the new grains. Once the new grains meet and form more than 75% of all grains, such differences are evened out. The grains grow now amongst themselves, following the laws of collective crystallization (Voll 1960). Sutured boundaries are eliminated, equilibrium angles are introduced at common edges and corners, where necessary by introduction of grain boundary curvatures. Larger grains are concave and grow at the expense of their smaller, convex neighbours.

Kroll et al. (Chap. 13, this Vol.) found that within the contact aureole, the Na content of clastic feldspars is increased and homogenized. Pyrite, later changed to pyrrhotite, settles preferentially within feldspars of the contact zone. Rare muscovite dehydrates under formation of sanidine. All this points to the presence of a fluid with dissolved Na, K and S. There is no indication that local variation in amount or composition of this fluid enhances or slows down grain growth locally. Only in the W profile do quartz grains become distinctly coarser within that part near the contact which is underlain by granite. There, partial melting within Leven Schists becomes strongest. However, the curve of grain-size increase from W, where heat is supplied from the E sideways, towards the E, where the quartzite is underlain by granite, does not show steepening in the eastern part. Therefore, we interpret the stronger coarsening there as caused by higher or longer heating, and not by a surplus of water.

Once the coarsening grains dominate, the population is extremely uniform and coarsening proceeds smoothly. Within the last 3 m at the contact, the quartzites may be feldspathized. K-Na-feldspar penetrates along grain edges and boundaries and extends from these, corroding the quartzes to rounded hexagonal bipyramids. This feldspathization is combined with formation of a partial melt, as Voll (1969) could show at the Glen Coe intrusion: once feldspar content becomes more than 40%, preferred orientation of the coarsened quartzes is suddenly lost. Biotite and plagioclase appear where this migmatization proceeds, very close to the granite. No further grain growth is observed within these migmatites.

Where grain diameters of coarsened grains are ca. 0.5 mm or more, the grains show signs of cooling strain and develop subgrains. These are very large; even the largest grains contain only two to a few of these subgrains, and the misfit-angles are very low, usually less than 1°. However, the subgrain boundaries are now normal to c, and not parallel to it as they are during regional metamorphic polygonization. The grain boundaries are very coarsely sutured; subgrain boundaries usually adhere to reentrant angles within these boundaries. Certainly these quartzes have reached the high-quartz field on coarsening. However, it cannot be told whether this plastic strain and following polygonization have been acquired above or below the inversion T. The attitude of subgrain boundaries points toward a prism-c glide system as having been used. As no tectonic deformation accompanies the contact heating, this strain must be set up by cooling tensions. Pebbel-relics, having preserved a subgrain structure with liquid inclusions, may clearly be distinguished from coarsened grains with these late strain features. This strain does not influence grain size. Where coarsening was less, i.e., in the outer zone of coarsening, such strains are missing. Probably strain accumulated within these smaller grains could not accumulate to values necessary to produce plastic deformation. Figure 12.2 shows stages of original quartzite and progressive coarsening, including the late cooling strain.

12.4 Theoretical Approaches to Grain Coarsening

The growth of the grains is assumed to be a function only of the temperature and the time interval at which the grains are exposed to a given temperature. Since the

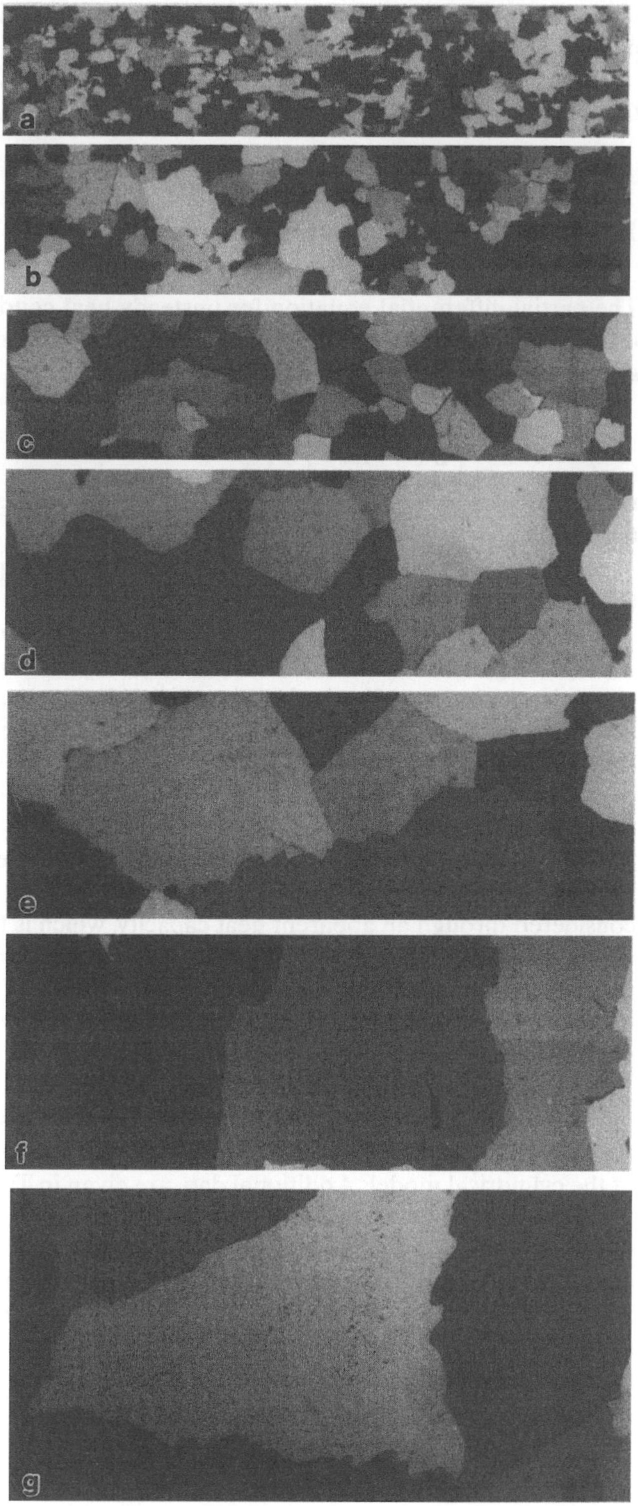

Fig. 12.2. Appin Quartzite at different stages of coarsening by contact metamorphism at the SE contact of the Ballachulish Igneous Complex. Distances from the contact in m in parentheses: **a** (1040) no contact effect. Fabric produced by regional metamorphism; **b** (805) first annealing effects. Many original grains left; **c** (708) all grains coarsened. Here and in next picture: polygonal grain shapes and simple boundaries. **d** (385); **e** (197); **f** (65); **g** (12). The last three stages show strain-induced boundary migration due to cooling strains. For all pictures long side = 4 mm

temperature-time-path for each distance to the contact is different, a simple Arrhenius plot cannot be applied to estimate the activation energy.

The non-stationary conditions need another approach for estimating the activation energy as well as the pre-exponential factor of the Arrhenius equation.

12.4.1 Temperature History of Coarsening

The partial differential equation for unsteady heat conduction is approximated by using constant thermal conductivities in concentric rings and by applying them to one space coordinate only. Thus simplification leads to

$$c\,\rho\,\frac{\partial T}{\partial t} = K\left[\frac{\partial^2 T}{\partial r^2} + \frac{1}{r}\frac{\partial T}{\partial r}\right], \qquad (1)$$

where c = specific heat, ρ = density, K = thermal conductivity, T = temperature, t = time and r = radial distance.

The finite difference equation which approximates Eq. (1) is given by Buntebarth (Chap. 18, this Vol.).

The solution of the finite difference equation is found by the explicit method. The spatial grid elements used here is $\Delta r = 100$ m and the time step $\Delta t = 1$ year.

A large amount of latent heat is released by the magma during crystallization, and two approaches are applied here to take the latent heat (L) into account. For the ring model, the phase boundary liquid/solid gradually moves from both contact zones of the ring towards the centre, and at the interface between liquid and solid, the heat of fusion is released.

The symmetrical solidification model from both sides of the ring is not suitable for the cylindrical model. Another approach has been applied: the latent heat is considered through an apparent heat capacity, which is the sum of the specific heat of the magma and the ratio of the latent heat of fusion (L) and the temperature interval between the liquidus and solidus temperature of the magma. This approach introduces some simplification because solidification proceeds gradually from the contact zone to the centre instead of an spatially equally released heat.

The boundary conditions and the thermal properties which are applied are given by Buntebarth (Chap. 18, this Vol.). The temperature distribution dependent on the time after intrusion is demonstrated in Fig. 12.3 for several distances from the contact of the cylindrical model. Additional data are given in Table 18.1 and in Fig. 18.8.

The good thermal conductivity of Appin quartzite yields high maximum temperatures in the adjacent rocks. Comparing the ring model with the cylindrical model, the contact temperature is approximately 40 °C below the value for the cylindrical model. At distances between 200 and 700 m, however, this difference exceeds even 120 °C (Table 12.2). The much larger amount of heat which can be supplied by a cylindrical intrusion is due to this significant difference.

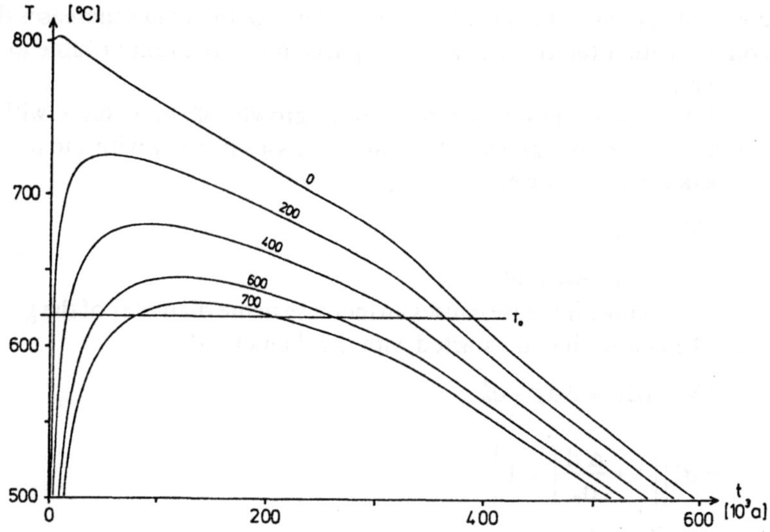

Fig. 12.3. Temperature of Appin Quartzite for several distances from the contact versus the time t after setting of the first intrusion for a cylindrical intrusion model

Table 12.2. Quartz grain coarsening $(d/d_0)^3$ depending on the distance from the contact (x) and the maximum temperature as well as the sum of Eq. (11) for the cylindrical model with E = 8.5 kcal/mol and the ring model with E = 8 kcal/mol

x (m)	$\left(\frac{d}{d_0}\right)^3$	Cylindrical		Ring model	
		T_{max} (°C)	Sum of Eq. (11) (years)	T_{max} (°C)	Sum of Eq. (11) (years)
0	362	803	2050	761	927
100	284	756	1600	662	733
200	216	726	1212	609	558
300	154	701	866	575	399
400	102	680	582	555	279
500	58.4	662	350	538	172
600	26.2	645	155	524	86
700	4.7	629	34	511	28

12.4.2 Grain Growth and Activation Energy

The Appin quartzite exhibits a grain coarsening between the contact to the intrusive body and a distance of about 750 m from the contact. Within this interval, the medium grain diameter drops continuously from 1.019 mm at the contact to the constant value of 0.143 mm at distances greater than 750 m, as demonstrated with Fig. 12.1.

A measure of the grain coarsening is obviously the volumetric growth of the grains in comparison to the volume of the non-affected grains. This volumetric grain coarsening is given in Table 12.2. The volume of the grains increases thereafter at the contact with a factor of 362 in comparison to the grain volume at a distance of r > 750 m.

A basic assumption is a continuous growth of the volume with reaction time. Therefore, the volume (V) of a grain increases linearly with the reaction time (t) at a constant temperature:

$$\Delta V = A\ t, \tag{2}$$

with A = the growth rate.

The volume increase causes an increase in the diameter of the growing grain (Δd) after the coarsening has started with the diameter d_0:

$$\Delta V \sim (d_0 + \Delta d)^3 - d_o^3 \tag{3a}$$

$$= d_o^3\left[\left(1 + \frac{\Delta d}{d_o}\right)^3 - 1\right] \tag{3b}$$

$$= d_o^3\left[\left(1 + \frac{d - d_o}{d_o}\right)^3 - 1\right] \tag{3c}$$

$$= d_o^3\left[\left(\frac{d}{d_o}\right)^3 - 1\right].$$

The grain coarsening Q is defined as

$$Q = (d/d_o)^3. \tag{4}$$

From Eqs. (2), (3d), and (4) follows

$$Q - 1 = d_o^{-3}\ A\ t. \tag{5}$$

The reaction constant k is identified with

$$k = A/d_o^3. \tag{6}$$

The growth rate A, respective the reaction constant k, is dependent on temperature, and the temperature dependence is formulated with the Arrhenius equation:

$$k = k_{max} \exp(-E/RT), \tag{7}$$

where k_{max} = pre-exponential factor, E = activation energy, R = molar gas constant, T = temperature (°K).

Equations (5), (6), and (7) yield then

$$Q - 1 = t\ k_{max} \exp(-E/RT). \tag{8}$$

Since the grain coarsening begins at temperatures above the threshold temperature T_o, Q must be equal to 1 at $T = T_o$. This demand is satisfied by

$$Q - 1 = k_{max} t\left[\exp\left(-\frac{E}{RT}\right) - \exp\left(-\frac{E}{RT_o}\right)\right] \tag{9}$$

for $T \geqq T_o$. Because Q is much greater than unity, according to Table 12.2, this value may be neglected on the left side of Eq. (9).

The temperature T is not constant during the time t. The temperature varies with time, so that $T = T(t)$. In this case, the grain growth is assumed as

$$Q = k_{max} \int_0^t \left[\exp\left(-\frac{E}{RT(t)} \right) - \exp\left(-\frac{E}{RT_o} \right) \right] dt. \tag{10}$$

The integrated value can be approximated with a sum. A stepwise increase of the temperature causes a proceeding grain growth during the duration of Δt at the temperature step $T = T_o + \Delta T$. After the time t = sum of Δt_i, n temperature steps T_i yield a grain coarsening of

$$Q = k_{max} \sum_{i=1}^{n} \left[\exp\left(-\frac{E}{R(T_o+T_i)} \right) - \exp\left(-\frac{E}{RT_o} \right) \right] \Delta t_i . \tag{11}$$

The grain coarsening, according to Eq. (11), agrees with the observed coarsening, if both constants, i.e., the activation energy and the pre-exponential factor, are known.

Both values can be estimated because several Q values with different temperature histories are known. A linear relation between Q and the sum in Eq. (11) exists for the correct activation energy only. In this case, the slope of the relation yields the pre-exponential factor.

In order to get the mean grain coarsening being representative at discrete steps, a curve matches the data to give a mean diameter in steps of 100 m from the contact. For these discrete points, the cooling history of the intrusion is calculated (Buntebarth, Chap. 18, this Vol.) and the course in temperature and time during cooling can be related to the grain coarsening according to Eq. (11).

At the time when the temperature exceeds the threshold value, the grain coarsening starts, and it stops when the temperature falls below this value. The duration of grain growth is subdivided into equal time intervals of $\Delta t = 2.10^4$a. Each interval represents a mean temperature which is used to calculate the contribution to the grain growth during each time step according to Eq. (11). The sum of Eq. (11) is calculated at all discrete distances. This approximation to the integrated value of Eq. (10) is given in Table 12.2 for an activation energy of E = 8.5 kcal/mol.

If the observed grain coarsening is plotted versus the parameterized integrated value, the linear relation of Eq. (10) is satisfied for the correct activation energy only. The highest integrated values are most sensitive to activation energy variations and any deviation from the linear behaviour of Eq. (10) can be realized at first at the contact as demonstrated with Fig. 12.4. It is demonstrated that an activation energy of E = 8.5 kcal/mol matches the observed grain growth in the best way. The value of E = 10 kcal/mol would yield coarser grains at the contact than the observed ones. On the other hand, the value of E = 7 kcal/mol underestimates slightly the grain coarsening. This trial and error method provides not only the most suitable activation energy but also controls the limits of the method. The data set coincides with an activation energy of E = 8.5 ± 1 kcal/mol.

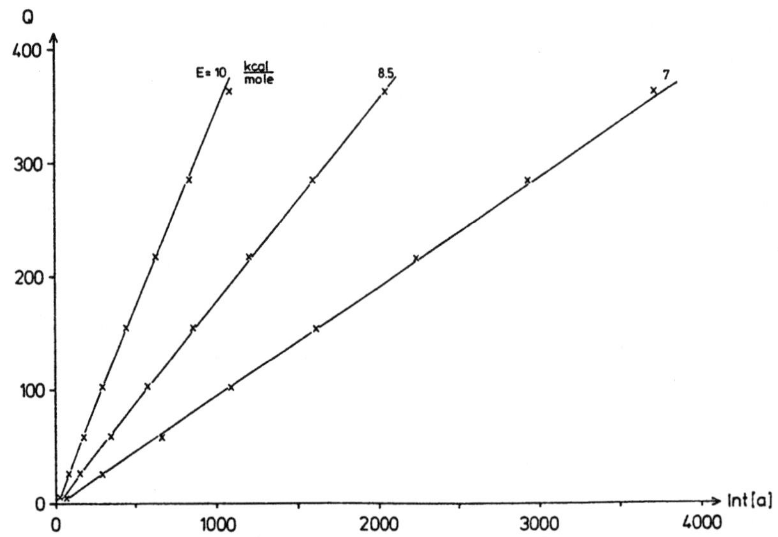

Fig. 12.4. Quartz grain coarsening versus the integrated value of Eq. (10) applying different activation energies

Using the temperature distribution which results from the ring intrusion model of Chapter 18 (this Vol.), where some temperature distributions are presented in Fig. 18.8, the integrated values according to Eq. (11) are given in Table 12.2. Since the maximum temperatures, reached in the Appin quartzite, are lower than in the first case, the threshold temperature for grain coarsening is different. At the distance of 750 m from the contact where the grain growth ceased, the temperature does not exceed 500 °C which is taken as the threshold value in this case. The heat deriving from the ring intrusion dissipates faster than that from the cylindrical intrusion on account of which the time step Δt is reduced to $\Delta t = 10^4$a. The best fit of the data provides an activation energy of E = 8 ± 1 kcal/mol.

The slope of the relation of Eq. (11) yields the pre-exponential factor k_{max}. This factor is estimated with $k_{max} = 0.18\ a^{-1}$ for the cylindrical intrusion model, resp. 0.39 a^{-1} for the ring intrusion model.

The following mean values are deduced from both models:

1. Activation energy: E = 8.2 ± 1 kcal/mol;
2. Pre-exponential factor: $k_{max} = 0.28\ a^{-1}$;
3. Threshold temperature: $T_o = 560$ °C.

12.5 Interpretation

From model calculations, it can be said that grain coarsening of quartz starts at 500 °C for the ring and at 620 °C for the cylinder model. This latter model seems to agree better with T-data derived from isograds. This threshold T is a special one for our case. Voll (1960, 1969, 1976) has shown that the true threshold T for quartz-polygonization, -recrystallization and -collective crystallization is near, or just under 300 °C, i.e., no stronger deformation or longer heating can shift this T to lower values. This statement is derived for the tens of millions of years given for processes of the named kind, as they are available during orogenies and regional metamorphism. Xenoliths ejected from the Eifel crust in NW Germany by subrecent volcanoes consist of metamorphic rocks, deformed and regionally metamorphosed ca. 300 Ma ago. Part of these xenoliths has spent this time at a T just under 300 °C and lower after the Variscan orogeny, and yet they show no sign of quartz recrystallization after this orogeny.

Voll has found that heating of quartz micaschists with quartzes similar to our reference quartzes at basalt dykes in Ardnamurchan, Scotland to more than 600 °C for days, weeks and months has not been able to produce the slightest microscopic indication of beginning healing, recrystallization or collective growth, though these rocks had developed partial melts at quartz/alkali-feldspar grain boundaries at ca. 1 kbar pH_2O. In this case, apparently the time of heating has been too short, though higher temperatures were reached than for our quartzites, to initiate grain growth.

For determining at what T quartz grain growth starts for different time-T histories, other cases would have to be studied with longer and shorter heating.

Voll (1969) has, in a preliminary study of our quartzites, suggested that possibly the beginning of grain coarsening in the Ballachulish aureole may be triggered by the low/high-quartz inversion and the lattice changes and vibrations connected to it. At surface pressure, this inversion T is 573 °C and for increase of p for 100 MPa it is shifted to higher T for 26 °C for single crystals.

Kern (1978) determined a higher shift of T for polycrystalline rocks and van der Molen (1981) has shown that it may amount to 60–70 °C for pressure increase of 100 MPa for quartz in granite. As our quartzites contain ca. 15 vol % of feldspar such an additional effect might shift the inversion-T even above the one for the 300 MPa in our aureole, 650 °C for single crystals of quartz. In each case the threshold value of 620 °C observed in our case would be somewhat too low to be caused by the low/high-quartz inversion.

Fielitz (1971) and Kern (1978) have observed a pronounced decrease of the Poisson-ratio below the inversion T. At about 100 °C beneath it, the Poisson ratio of quartzite, calculated from ultrasonic velocities, becomes negative and is much more sensitive to T variations than at lower T. This elastic softening of the structure may contribute to factors enhancing grain growth and thus fix it not at but at least to a T interval near and below it for intermediate times of heating.

Tullis and Yund (1982) studied quartz grain coarsening experimentally at 800–1100 °C, at confining p of 2–15 kb for 2–790 h. As they did not achieve dry grain growth, they added 1–2% of water and estimated the pH_2O as close to p_{total}. They found linear coarsening when plotting log increase of grain diameter against log time.

Increase of T and of p enhanced coarsening. They determined the activation energy for coarsening from their plots as 11 kcal/mol.

They explained the acceleration up of grain growth by water and by increased p by the diffusion within a water-film at grain boundaries, which involved solution and precipitation for the transfer across the boundaries.

Under these conditions, they found an activation energy of 11 kcal/mol, which they think corresponds to ionic diffusion through an intergranular fluid.

Our study derives an activation energy of 8.2 ± 1 kcal/mol for the grain growth of quartz. However, if the grain coarsening is calculated while taking into account an activation energy of 11 kcal/mol, a slightly increased mean grain diameter of 1.028 mm instead of 1.019 mm should be found at the contact. The scatter of the data measured (Fig. 12.1) does not exclude such scarcely elevated diameters, so that in principle the results are in accord with the activation energy reported by Tullis and Yund (1982). In order to support our value, we have argued that, in our case, water was present during grain growth and Na, K and S were dissolved in it. On the other hand, grain growth is strictly confined to grains surrounded by high-angle boundaries. Subgrains surrounded by boundaries with a misfit angle below ca. 15–18° do not grow essentially and liquid inclusions on them are not drawn out to form thin films. From this we conclude that the viability of quartz grain boundaries depends on the misfit angle and thus on grain boundary structure. Migration of water in the quartzites must have occurred early in the heating history as homogenization and enrichment of feldspars in albite is achieved early, at T little more than 500 °C. At this T, no grain growth of quartz can be detected. Dehydration of relic pebbles goes on to higher T but we cannot say whether grain boundaries behaved in a similar way at these higher T. The amount of hydrous fluid must, however, be small and it seems likely that it forms a very thin, possibly even discontinuous, film on the predominant high angle boundaries. Hydration of feldspars on cooling occurs rarely and is only locally found. Therefore, water on the grain boundaries may not have the thermodynamic state of a solution and it seems doubtful whether solubility of quartz in a solution is an adequate model. Using the argument of Tullis and Yund (1982), water deficiency or water in a state similar to that of ice should decrease grain growth rate and increase the activation energy, while we arrive at a lower activation energy. Dissolved species in the water film on grain boundaries should have an influence on the SiO_2 solubility. They found no effect caused by traces of calcite in the experimentally coarsened quartz-aggregates. But it could be that the presence of alkalies increases the solubility and thus enhances grain growth, lowers the activation energy. Certainly alkali-feldspars do not dissolve (quickly) enough to redistribute feldspars in the aggregate of coarsening quartzes. At the beginning, all feldspars have connection to the network of high-angle grain boundaries. However, at advanced stages of coarsening, the feldspars may be included by coarse quartz grains without essential change of their size and distribution. This means that the feldspars do not dissolve readily and are not reprecipitated elsewhere. Therefore their presence provides an effective locking mechanism.

The size of activation energy and the rate of grain growth are, however, at least of a similar order in Tullis and Yund's and our study. Both values are achieved on the basis of observations and model assumptions. Only further experimental and field studies can reconcile the small differences existing so far.

High/low-quartz inversion is combined with Dauphinee-inversion twinning. We polished and etched quartzites in order to obtain the inversion isograd by differences in inversion twinning. However, the number of twin seeds is determined by the cooling rate. Very slow cooling may even lead only one right- or left-handed quartz single crystal to form. At the Ballachulish aureole, the Dauphinee twins are uncharacteristic; ca. 15–20% of the grains of similar size show twins. They are mostly restricted to grain boundaries, corners of inclusions or continue grain boundaries into a neighbouring grain, i.e., they occur where stress peaks resulted. We cannot decide which of these were formed on inversion and which are deformation twins acquired in the low quartz field. Here again it seems necessary to extend the study to aureoles with higher cooling rate in order to assess the mode of inversion-twin formation, and thus to pin down the inversion isograd.

13 Disordering, Re-Ordering and Unmixing in Alkali Feldspars from Contact-Metamorphosed Quartzites

H. Kroll, C. Krause and G. Voll

13.1 Introduction

For kinetic reasons it is difficult to study the sanidine ⇄ microcline transition in the laboratory. A contact aureole, however, in which K-feldspars have passed the transition at rising and falling temperatures provides the opportunity to examine the disordering and re-ordering processes in some detail.

In a series of papers, Hart (1964), Steiger and Hart (1967) and Wright (1967) investigated K-feldspars from the contact aureole of a Tertiary quartz-monzonite stock (3 km in diameter) at Eldora, near Boulder, Colorado. The quartz-monzonite intrudes into Precambrian gneisses and schists. These contain numerous, usually concordant pegmatites, which too are thought to be of Precambrian age.

The authors distinguished three zones within the aureole: (1) the zone unaffected by contact metamorphism; (2) the transition zone; and (3) the orthoclase zone. In the first zone, the pegmatitic K-feldspars mostly occur as microperthitic low microcline, with some additional intermediate microcline and orthoclase. The relative amounts of orthoclase increase in the transition zone toward the contact, although many K-feldspars still show the original sharp M-twinning pattern, but may contain untwinned areas. The width of the transition zone and its distance from the intrusive contact vary greatly with each of the seven traverses investigated, the width ranging from ≈100 to ≈300 m. The orthoclase zone was also found to be of variable width. Here orthoclase was the predominant or exclusive variety of K-feldspar. M-twinned microcline had not (or almost not) survived, but intermediate microcline was occasionally observed in six of the traverses, and occurred as the major variety in the seventh.

The occurrence of orthoclase in the transition and orthoclase zones was interpreted by the authors as the result of a prograde microcline-orthoclase transition. We now know that this was a misinterpretation: with increasing temperature, microcline transforms to sanidine, and sanidine may transform during subsequent cooling to orthoclase or microcline. In fact, based on the work by Laves and Goldsmith (e.g., Laves 1952, 1960; Goldsmith and Laves 1954), Voll (1969) gave this interpretation to the occurrence of the feldspar varieties he observed in the contact aureole of the Ballachulish intrusion, NW Scottish Highlands. In this chapter, we will extend his early investigation to contribute to the behaviour of feldspar minerals in contact aureoles.

G. Voll, J. Töpel, D.R.M. Pattison, F. Seifert
Equilibrium and Kinetics
in Contact Metamorphism

13.2 The Object of Investigation

The Ballachulish igneous complex is one of the numerous Caledonian intrusions found in the Scottish Highlands. It postdates the orogenesis and subsequent uplift and erosion. The magma crystallized at a depth of ≈10 km corresponding to a pressure of ≈0.3 GPa (Masch and Heuss-Aβbichler, Chap. 10, this Vol.). The intrusion heated the metasediments of the Ballachulish Dalradian, creating a well-defined contact aureole.

Part of the metasediments is formed by the Appin Quartzite, which is coarse-grained and contains quartz pebbles, layers enriched in zircon and rutile and clastic feldspars up to 15% by volume. More than 95% of them are perthitic K-rich feldspars, up to 2 cm in diameter. They are derived from granitic magmatites, pegmatites and gneisses of the amphibolite and granulite facies. During the time of their deposition they had different degrees of Al,Si order, different (sub)microscopic textures and different contents of Na-feldspar in solid solution or as exsolution products. Spindle-perthites, mesoperthites, film-, band-, patchy perthites, with and without plate perthite, all occur.

During diagenesis, the K-feldspars became surrounded by thin, fragmentary rims of adularia. Sometimes perthitic albite or even the K-feldspars themselves were partially converted to adularia. Clastic plagioclases were transformed to low albite.

During the cycle of Caledonian regional metamorphism, these sediments were heated from surface temperature up to ≈480 °C, i.e., just below the temperature of the microcline-sanidine transition (see below), and were then uplifted and cooled back to surface temperature. During the long time span of Caledonian heating from ≈300 °C to ≈480 °C and back, ordering and unmixing proceeded, whereas very Na-poor K-feldspars incorporated additional Na. The whole varied suite of structurally different K-feldspars was turned into fully ordered low microcline with $2V_x$ = 80–88°. In this status the crystals entered the contact aureole.

It is important to point out that all changes of K-feldspars observed within the aureole are changes on low microclines which all had the same completely ordered state to start with. This is a favourable factor and certainly a rare event, since usually K-feldspars show different structural states in a rock or in a hand specimen or even in one and the same crystal.

13.3 Methods of Investigation

The K-feldspars were investigated by light optical, transmission electron microscopic (TEM) and X-ray powder methods.

To ensure high resolution in the X-ray work we used a Guinier-Jagodzinski type camera ($CuK_{\alpha 1}$ radiation) rather than a Bragg-Brentano diffractometer. Lattice parameters were determined for the smaller part of the specimens following standard procedures (e.g., Kroll et al. 1980, 1986). For the major part, only those reflections were measured which were necessary to determine composition and structural state (e.g., Kroll and Ribbe 1983, 1987). Both procedures where checked against each other to ensure consistency. In most specimens, two K-feldspar varieties plus exsolved albite occurred. Special care was taken to resolve overlapping peaks to avoid undesirable averaging of states. The relative amounts of coexisting orthoclase and microcline were estimated from the intensities of the 131 and $1\bar{3}1$ reflections. Carefully measured mechanical mixtures were used for calibration.

In the TEM work the K-feldspar varieties were identified by their contrast, their diffraction pattern and their γ^* angles measured in a^*b^* patterns. To calibrate the TEM we used a Si specimen, so that γ^* angles could be measured to ±0.3°. Care was taken to get the exact relation between dark field image and diffraction pattern which was taken with a small (ca. 1 µm) selected area diffraction aperture. The specimens were prepared from petrographic thin sections: a K-feldspar oriented approximately parallel to (001) was identified in the polarizing microscope and photographed to relate its appearance in the microscope to its TEM contrast. The crystal was then prepared for the TEM by ion-beam thinning. To find the exact orientation of the diffraction pattern a tilt rotation holder was used. All patterns were taken in the a^*b^* orientation. The dark-field images were taken approximately under two-beam conditions with a reflection of type (hk0).

13.4 The Contact Aureole

13.4.1 The Low Microcline Zone

To facilitate comparison with the work by Steiger and Hart (1967), we will subdivide the aureole into three zones and denote them similarly. Figure 13.1 shows the NE part of the Ballachulish intrusion and its aureole. The sample locations are indicated; they all occur within the Appin Quartzite and the so-called Transition Series between Appin Quartzite and Ballachulish Slates. Two straight lines have been drawn in Fig. 13.1, the right one separates the 'low microcline zone', i.e., the 'zone of no contact metamorphism' from the 'transition zone', and the left line is the boundary between the 'transition zone' and the 'sanidine zone'.

In the low microcline zone, the K-feldspars exclusively occur as low microcline, mostly showing a well-developed, sharp M-twin pattern. The fragmentary rims of adularia which they acquired during diagenesis are still morphologically visible. However, their optical properties are identical to low microcline. These adularia rims clearly demonstrate the ordering effect of Caledonian regional metamorphism on the structural state.

Due to long-term heating up to ≈480 °C, the twin domains started to coarsen in some microcline crystals. At points where albite and pericline twins meet, the domains become rounded. Planar twin composition planes become sutured. Some grains with completely irregular, ragged twin boundaries apparently are stranded on their way to becoming untwinned single crystals.

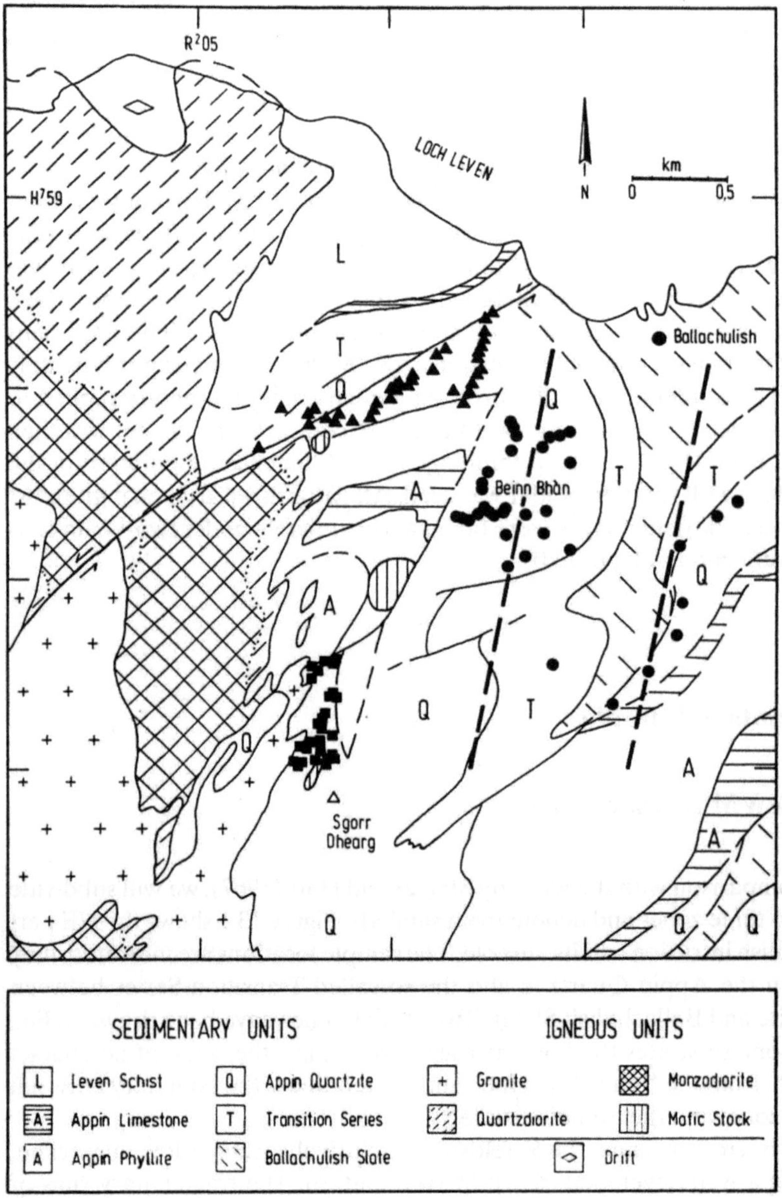

Fig. 13.1. Geological map of the northeastern area of the Ballachulish contact aureole (after Troll and Weiss, Chap. 3, this Vol.). Two *dashed lines* subdivide the aureole into three sections: (1) the low microcline zone extends to the east of the eastern line, (2) the sanidine zone lies between the intrusive contact and the western line, and (3) a transition zone occurs between the two lines. The eastern line approximately represents the microcline-sanidine transition isograd

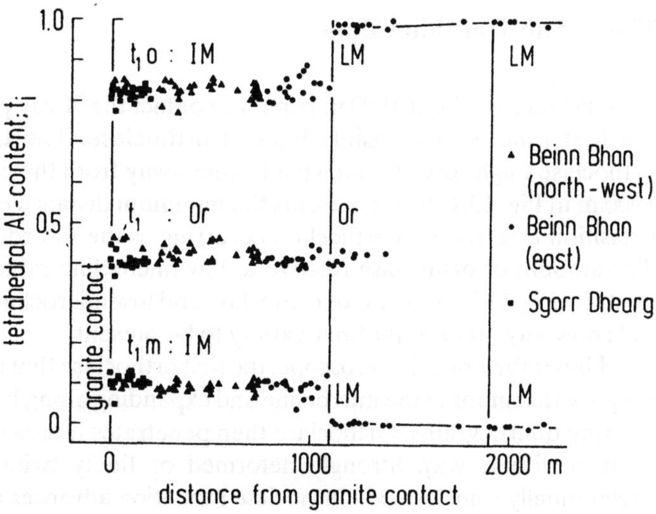

Fig. 13.2. Structural state of K-feldspars (given as tetrahedral Al content) as a function of their distance from the contact ($t_1 = X_{Al}^{T1}$ of orthoclase; $t_1o = X_{Al}^{T1o}$, $t_1m = X_{Al}^{T1m}$ of microcline). *Vertical lines* at 1900 and 1100 m subdivide the aureole into the low microcline zone (*LM*), the transition zone (*LM* + *Or*), and the sanidine zone (*IM* + *Or*). Orthoclase (*Or*) and intermediate microcline (*IM*) are re-ordering products of sanidine

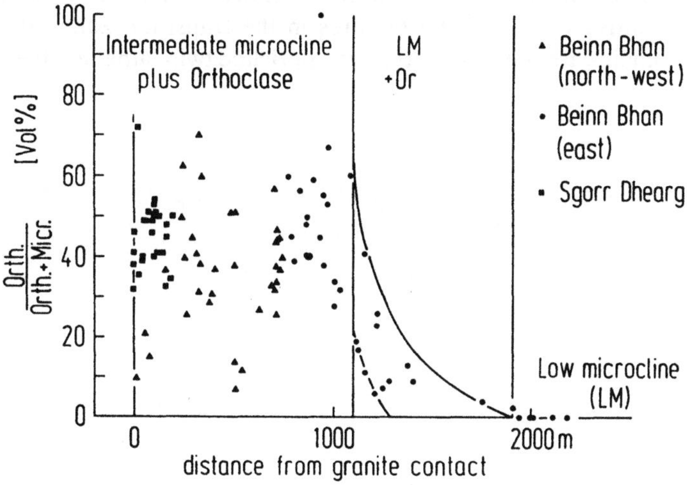

Fig. 13.3. Volumetric ratio orthoclase/microcline as a function of distance from contact. See Fig. 13.2 for further comments

All K-feldspars give large optic axial angles ($2V_x$ = 80–88°) and large extinction angles on (001) typical of fully ordered low microcline.

X-ray powder patterns of K-feldspar fractions separated from quartzite consist of sharp diffraction lines of low microcline, with no other K-feldspar variety being present (Figs. 13.2, 13.3).

13.4.2 The Transition Zone

At a distance of about 1900 m from the contact the X-ray powder pattern shows, for the first time, faintly visible lines of orthoclase. The true, first appearance of orthoclase might occur somewhat further away from the contact. The line drawn at 1900 m in Fig. 13.1 thus represents the minimum distance of the microcline-sanidine transition isograd, with orthoclase occurring as the re-ordering product of sanidine. The amount of orthoclase relative to low microcline rapidly increases towards the contact (Fig. 13.3). Except for orthoclase and low microcline, the X-ray diagram does not show any other K-feldspar variety to be present.

Under the optical microscope, the first orthoclase that appears forms incomplete outgrowths rimming the microcline and expanding along boundaries between neighbouring quartz grains. Orthoclase then penetrates microcline as veinlets or patches or in a diffuse way. Strongly deformed or finely twinned parts of a grain are preferentially transformed. Orthoclase formation advances along cracks and twin and exsolution boundaries. The orthoclase is always turbid, with a reddish-brown tint. This all indicates the influence of 'water' and/or the release of strain. If two microcline grains are in contact, orthoclase is seeded by each of the neighbouring crystals and advances into the other neighbour. Clear microcline single crystal domains between turbid orthoclases show a transition in $2V_x$ from larger to smaller angles.

Corresponding to the X-ray and light optical results, mainly (tweed-) orthoclase and (regular) microcline were observed in the TEM work. Figure 13.4 is representative of microcline textures in the transition zone. Sharp, narrow albite twin domains are seen that intersect a pericline twin lamella. Due to its texture we call this

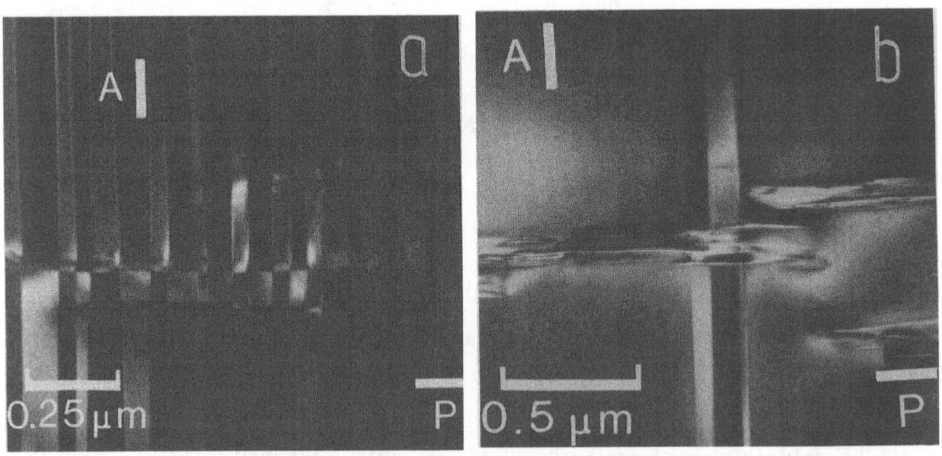

Fig. 13.4. a Regular microcline with polysynthetic albite twin domains crossing a pericline twin domain. **b** Almost homogeneous regular microcline with small pericline twin domains which resemble a planar distortion. A single albite twin domain is crossing (DF image). *A* and *P* symbolize the orientation of albite and pericline twin domains, respectively. Specimen No. K83/6, 1753 m east of the contact

type of microcline 'regular microcline' to distinguish it from a differently textured microcline, named 'irregular microcline' which is described below. The coexistence of regular microcline with orthoclase, denoted 'tweed orthoclase' due to its texture, is documented in Fig. 13.5. This is a characteristic feature found in the transition zone. The matrix in Fig. 13.5 is tweed orthoclase, the lamellae are 'regular microcline' lamellae.

Energetically, upon cooling pericline lamellae become apparently less favourable than albite lamellae. This is seen from Fig. 13.6, where broad pericline lamellae are replaced by fine albite lamellae, which grow from the pericline composition plane into both the pericline lamellae and the tweed areas.

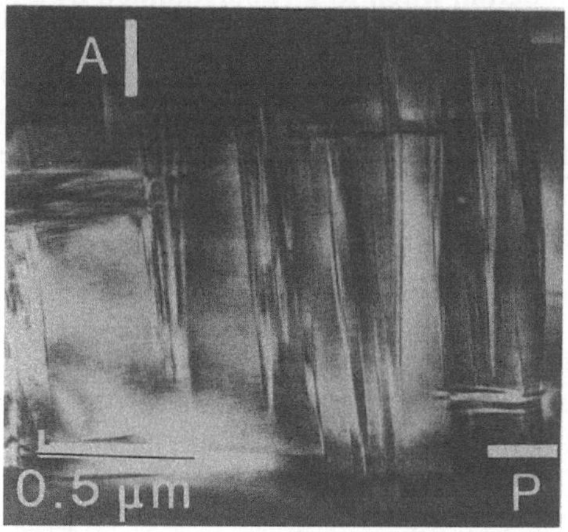

Fig. 13.5. Characteristic microtexture in the transition zone: Tweed orthoclase and lamellar domains with and without tweed texture (DF image). *A* and *P* symbolize the orientation of albite and pericline twin domains, respectively. Specimen No. 717, 1222 m east of the contact

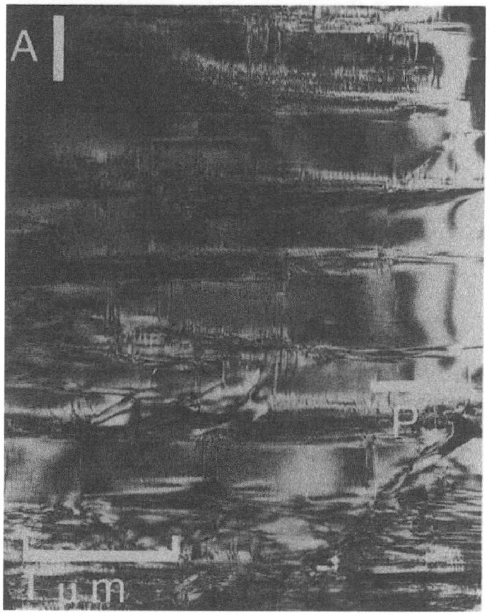

Fig. 13.6. Tweed orthoclase and regular microcline with pericline twin domains decomposing into fine albite twin domains (DF image). Specimen No. 721, 1250 m east of the contact

In his investigation of the Eldora contact aureole, Wright (1967) reports the occurrence of microcline in diagonal association. Figure 13.7 is a photograph of a disturbed area of a microcline grain exhibiting subgrain boundaries. The bright subgrain near the scale bar is especially interesting. When putting an aperture onto the subgrain boundary, it turns out that grain and subgrain are related by the diagonal association (Fig. 13.8). It is seen from Fig. 13.8 that the name is based on the appearance of the a^*b^* diffraction pattern. The geometry of the diagonal association is contrasted to the albite and pericline twin laws in Figs. 13.9 and 13.10. In Fig. 13.9, unit cells are arranged according to the diagonal association and the albite and pericline twin laws. The relative orientation of their reciprocal lattices is given in Fig. 13.10. The diagonal association might be the position of minimum strain for a triclinic subgrain when totally surrounded by a matrix with reversed Al,Si distribution and thus reversed triclinicity.

Besides regular microcline and orthoclase, irregular microcline is also observed near the inner boundary of the transition zone. Figure 13.11 represents a specimen from virtually the same position as the regular microcline of Fig. 13.4. The left part

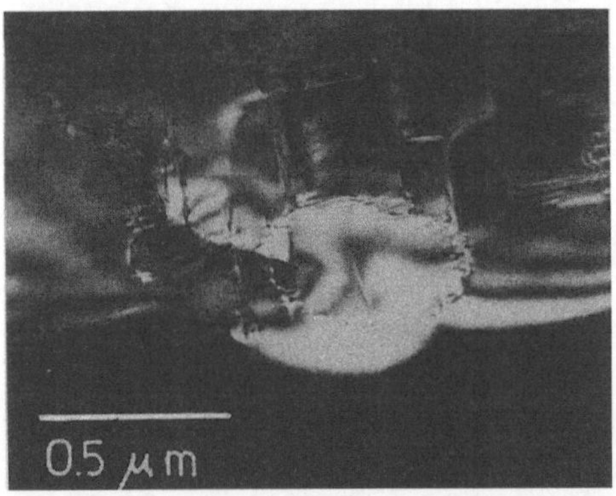

Fig. 13.7. Irregular microcline showing tweed areas, ill-defined twin boundaries and subgrain boundaries (DF image). Specimen No. 717, 1222 m east of the contact

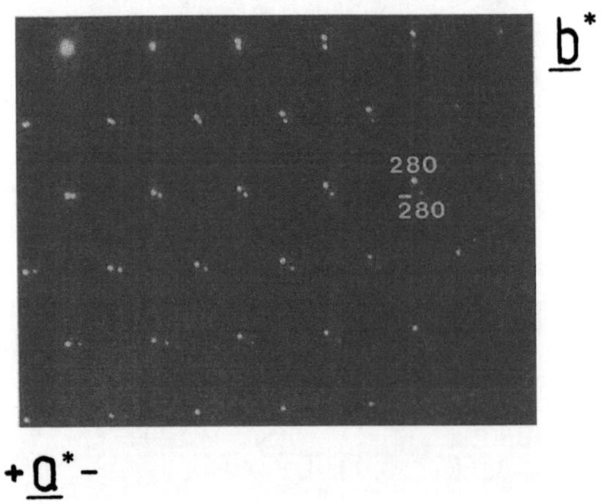

Fig. 13.8. Diffraction pattern of the bright subgrain near the scale bar of Fig. 13.7 (see text). The subgrain and the matrix are in diagonal association

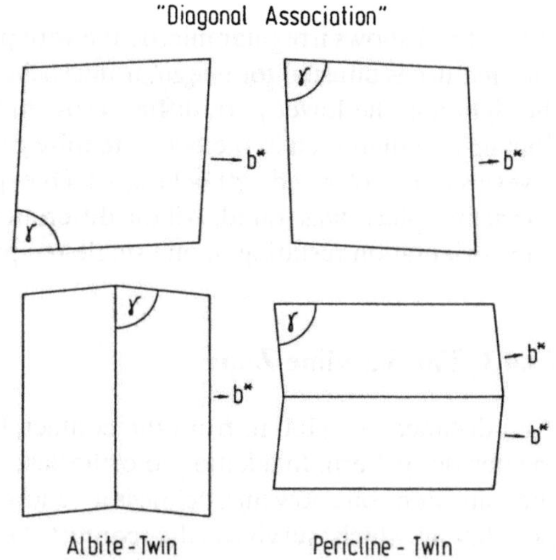

Fig. 13.9. Relative orientation of unit cells twinned after the albite and pericline laws or arranged according to the diagonal association (see text)

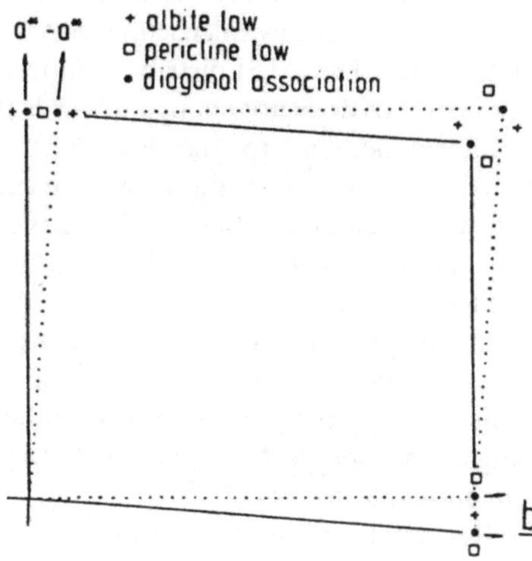

Fig. 13.10 Relative orientation of reciprocal lattices for albite and pericline twinning, and the diagonal association (see text)

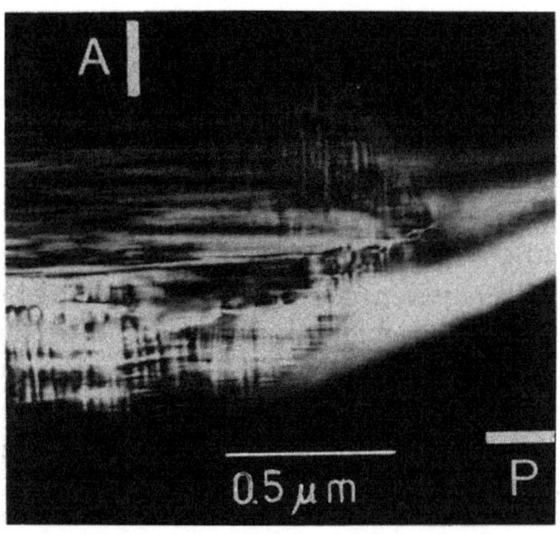

Fig. 13.11. Irregular microcline with ill-defined pericline twin area, albite plus pericline twin areas, and tweed-textured area (DF image). *A* and *P* symbolize the orientation of albite and pericline twin domains, respectively. Specimen No. 717, 1222 m east of the contact

of Fig. 13.11 shows irregular microcline with predominant ill-defined pericline twins. This feature is unusual for irregular microcline, which mostly shows albite twins. To the right and the lower part, diffuse cross-hatching sets in which has a much larger domain size than seen in the tweed texture on the right side. The diffraction pattern is very similar to tweed orthoclase, which explains the fact that in powder patterns no triclinic phase was found. All the different parts of the grain in Fig. 13.11 have the same orientation resulting in one single reciprocal lattice with $\gamma^* \approx 90°$.

13.4.3 The Sanidine Zone

At a distance of ≈1100 m from the contact, low microcline totally disappears from the powder pattern. In addition to orthoclase, intermediate microcline is observed. In the transition zone, low microcline is a 'relic' of the regional-metamorphic K-feldspar population which survived the contact heating. In the sanidine zone, all low microcline was converted to sanidine, which on cooling re-ordered to orthoclase and intermediate microcline. Intermediate microcline has a powder diagram which is less sharp than that of low microcline, but it is still clearly defined with a single pair of $131/1\bar{3}1$ lines. The Al,Si distribution derived from lattice parameters or diffraction peak positions does not change towards the contact. This is true for both orthoclase and intermediate microcline (Fig. 13.2).

Whereas in the transition zone the amount of orthoclase is increasing toward the contact, the relation in the sanidine zone is less clear. On the whole, there seems to be a slight decrease of orthoclase relative to intermediate microcline, at least very close to the contact.

In the optical microscope, most grains show a diffuse and patchy extinction. They have a reddish-brown colour which is inhomogeneously distributed and usually consist of tiny, fibrous or patchy domains of variable optics. $2V_x$ values vary within a single grain: less turbid crystals show a general increase of $2V_x$ towards the contact from about 55° to 75–80°.

Recrystallization of deformed grains is observed, and recrystallized grains become coarser towards the contact. They show a sharp and homogeneous extinction. The brown turbidness is partly older than recrystallization, being eliminated by this process, but partly may also affect grains being already recrystallized.

At the boundary between transition and sanidine zones, the type of unmixing changes. Lamellar exsolution is rarely preserved, the lamellae are converted into perthitic flames and patches. Towards the contact they steadily get coarser. Their maximum thickness increases from 10 to 100 μm (Fig. 13.12), their average thickness increases from 1 to 9 μm (Fig. 13.13), and their volume proportion rises from 15 to 40% (Fig. 13.14). Often rims free of exsolution are developed which become broader with increasing temperature. Near the contact sometimes a second generation of fine lamellae originates between coarse perthitic albite grains.

The change in exsolution texture and the general increase in Na content indicates an influx of Na from the intrusion supplied by a fluid phase.

Another observation concerning unmixing is noteworthy. Outside the contact aureole the clastic K-feldspar grains contain various amounts of exsolved albite (0–65

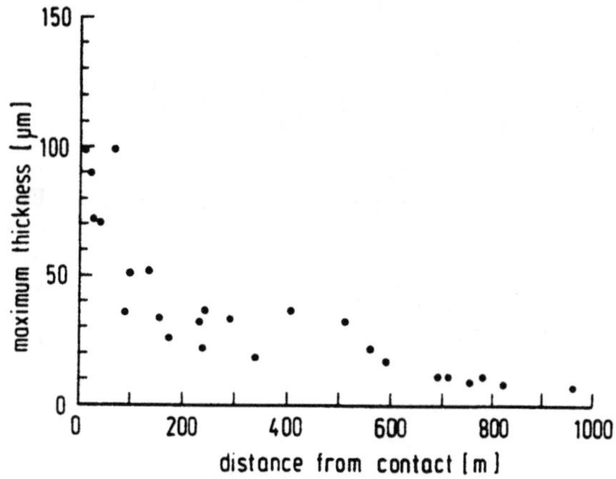

Fig. 13.12. Variation with distance from contact of the maximum size of perthitic exsolution products in K-rich feldspars

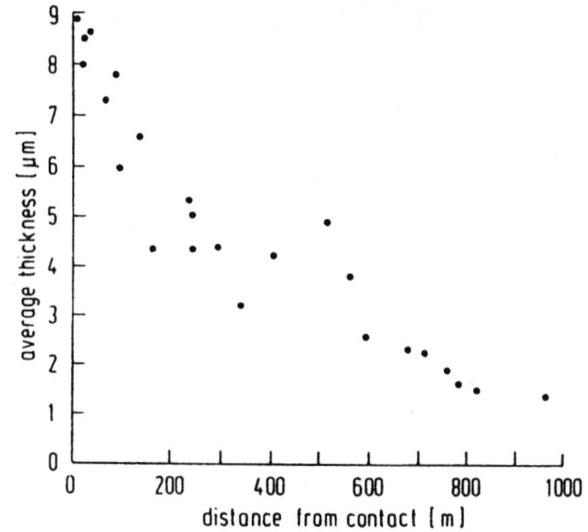

Fig. 13.13. Variation with distance from contact of the average size of perthitic exsolution products in K-rich feldspars

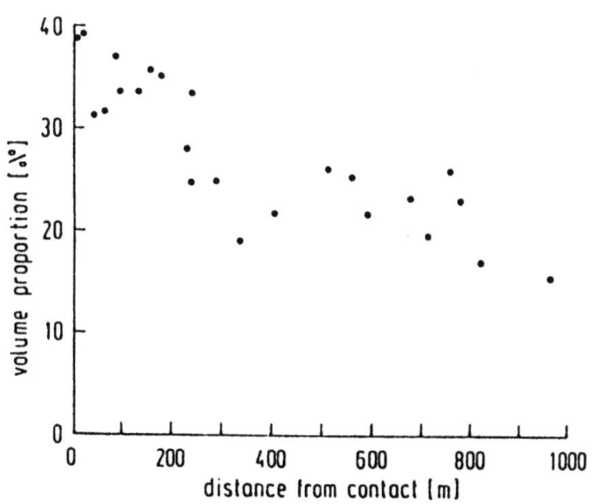

Fig. 13.14. Variation with distance from contact of the amount of perthitic exsolution in K-rich feldspars, given as volume percentage

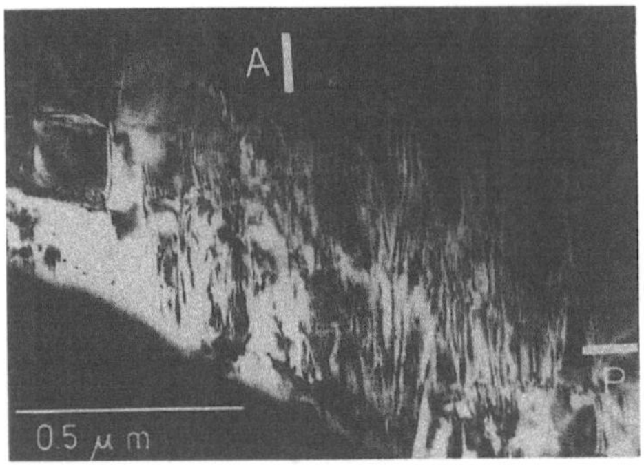

Fig. 13.15. Irregular microcline showing ill-defined (albite) twin boundaries, tweed-textured areas, and subgrain boundaries (DF image). *A* and *P* symbolize the orientation of albite and pericline twin domains, respectively. Specimen No. 790, 15 m east of the contact

Fig. 13.16. Irregular microcline showing ill-defined, microcline-like twin domains with tweed-textured 'windows', and some subgrain boundaries (DF image). *A* and *P* symbolize the orientation of albite and pericline twin domains, respectively. Specimen No. 792, 77 m east of the contact

vol %), the shape and distribution of albite varies greatly from grain to grain. In the sanidine zone all these differences are lost. All grains within a sample contain equal amounts of Na, equal portions of perthitic unmixing and the same exsolution textures. It thus appears that the influx of Na was combined with a chemical homogenization between grains.

The typical feldspar in the sanidine zone is irregular microcline associated with tweed-orthoclase. Figures 13.15 and 13.16 represent a mixture of twin-like features (mainly in albite twin orientation), subgrain boundaries with dislocations (especially clear in Fig. 13.16) and window-like tweed areas. Figures 13.15 and 13.16 were taken from specimens occurring close to the contact, but similar contrast is found as well in specimens throughout the sanidine zone. The a^*b^* diffraction pattern corresponding to this disturbed and complicated contrast usually consists of streaked, albite twin oriented reflections corresponding to a variety of γ^* angles in the irregular microcline areas with a centering reflection representing tweed-orthoclase (Fig. 13.17). Even more frequent than the window-like occurrence of tweed-orthoclase is a patchy arrangement side by side with irregular microcline, both with the same orientation, as seen in diffraction patterns.

In X-ray work, the boundary between the transition and sanidine zones was found to be sharp. This is not true when looked at by TEM. We already mentioned the

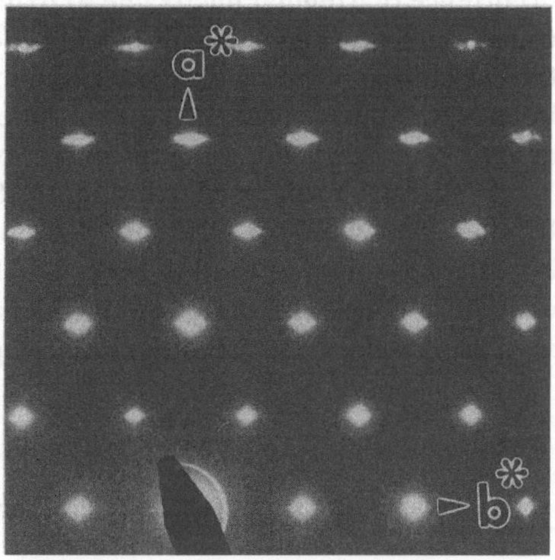

Fig. 13.17. Diffraction pattern of irregular microcline interwoven with tweed orthoclase similar to Fig. 13.18 (Specimen No. 729, 955 m east of the contact). The diffraction pattern was taken with a small diffraction aperture put onto the boundary between irregular microcline and tweed orthoclase areas. It is seen that both varieties have the same orientation. The albite twin reflections and the streaks parallel to b^* are due to predominantly albite twinned irregular microcline, whereas the centering reflection is caused by tweed orthoclase

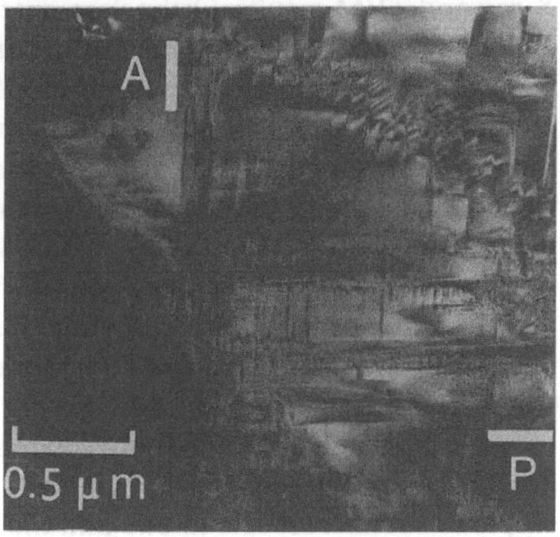

Fig. 13.18. Regular microcline with tweed orthoclase showing a band of a dislocation network extending NW-SE (DF image). *A* and *P* symbolize the orientation of albite and pericline twin domains, respectively. Specimen No. 960, 1009 m east of the contact

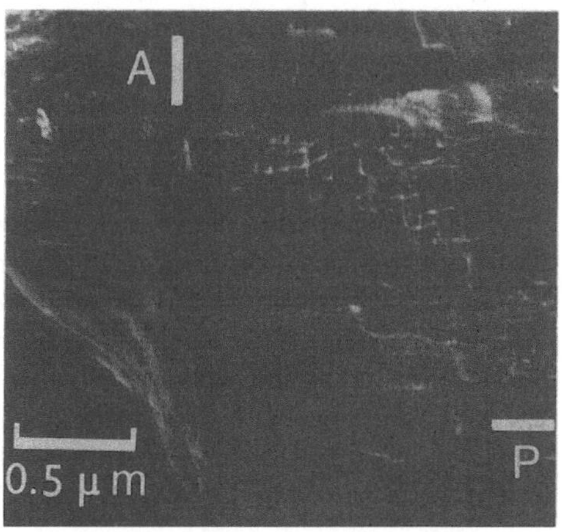

Fig. 13.19. Weak-beam image of the same specimen as in Fig. 13.15. Only the distorted areas are in a reflecting position and the dislocation network can clearly be seen. *A* and *P* symbolize the orientation of albite and pericline twin domains, respectively

occurrence of irregular microcline ≈120 m *east* of the inner boundary, i.e., within the transition zone (Figs. 13.7, 13.11). On the other hand, Fig. 13.18 shows a contrast typical for regular microcline occurring in a sample taken ≈100 m *west* of the inner boundary. It thus appears that the transition/sanidine zone boundary is smeared over at least ≈200 m, judging from the microtextures seen by TEM.

In the regular microcline of Fig. 13.18, pericline lamellae are partly substituted by albite twin lamellae, as in Fig. 18.6 from a sample taken 150 m east of the inner boundary. Compared to most other K-feldspars, the specimen shown in Fig. 13.18 contains a dense concentration of dislocations. The dislocation network is clearly seen using weak-beam conditions as in Fig. 13.19.

13.5 The K-feldspar Varieties in TEM and X-Ray Work

In our description of K-feldspars, we have used different terms when dealing with X-ray and TEM results: Orthoclase, intermediate and low microcline were used to define structural states identified from lattice parameters; tweed-orthoclase, irregular and regular microcline refer to different types of contrast seen in the TEM. Both methods give different kinds of information and there is no a priori linkage between them. We tried to provide this linkage in two ways: (1) TEM, X-ray and optical work were done on the same specimen; (2) γ^* angles were measured on electron diffraction patterns to provide a direct comparison with X-ray results, although the precision is much smaller, i.e., 0.3° compared to 0.01°.

Broadly speaking, we can state that regular microcline corresponds to low microcline, irregular microcline to intermediate microcline or (X-ray) orthoclase, and tweed-orthoclase generally to (X-ray) orthoclase.

An important difference between regular and irregular microcline concerns their domain sizes. Whereas the domain size in regular microcline may vary from microscopic to submicroscopic in one and the same grain, it seems to be restricted to the submicroscopic scale in irregular microcline. Therefore, diffuse cross-hatching under the light microscope is not suited as a means to distinguish between regular and irregular microcline. In optical orthoclase, however, when showing a fairly homogeneous extinction, the tweed texture usually dominates. Diffuse optical behaviour may be caused by the submicroscopic size of domains, or by the presence of various types of submicroscopic intergrowths, or by strain induced by an outer anisotropic stress field.

The chaotic texture of irregular microcline and the submicroscopic tweed texture of orthoclase raises the question: what is meant by the term structural state described by the Al,Si distribution and determined by structure refinement or from lattice parameters? When we are dealing with a macroscopic or microscopic feldspar phase, like sanidine or low microcline, the distribution of Al and Si over the tetrahedral sites (the long range order) is homogeneous and can thus be determined unequivocally. However, in submicroscopic textures with domain sizes smaller than the coherency area of the X-ray beam like tweed orthoclase and irregular microcline, some kind of

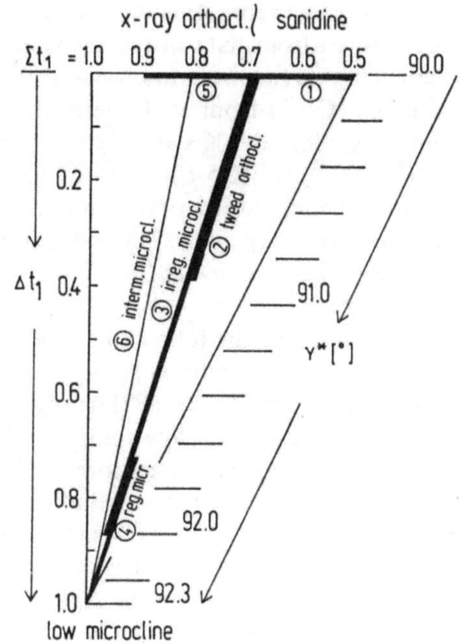

Fig. 13.20. Two different K-feldspar Al,Si-ordering paths in the Σt_1 (= $2t_1$ or $t_1o + t_1m$) vs Δt_1 (= $t_1o - t_1m$) diagram: line segments *1,5,* and *6* constitute the average ordering path found from X-ray methods; similar paths are shown in Fig. 13 of Kroll and Ribbe (1983), and Fig. 7 of Kroll and Ribbe (1987). Line segments *1,2,3,* and *4* constitute a possible ordering path derived from TEM results. γ^* contouring is valid for K-rich compositions (Or > 80), cf. Fig. 4 of Kroll and Ribbe (1987); from Bambauer et al. (1989)

averaging must occur. Therefore, the site occupancies we derive from lattice parameters are to some degree determined by domain size and type of intergrowth.

We have seen that the Al,Si distribution of orthoclase and intermediate microcline does not vary with distance from the contact (Fig. 13.3). This may be related to the observation that the relevant textures seen in the TEM do not show a distinct variation in type or coarseness.

In an attempt to compare the results on structural state obtained from X-ray and TEM work, we have prepared Fig. 13.20 (also see Bambauer et al. 1989, and their discussion). In this diagram, values $\Sigma t_1 = 2t_1$ (monoclinic feldspars) or $t_1o + t_1m$ (triclinic feldspars) have been plotted versus $\Delta t_1 = t_1o\text{-}t_1m$. The t values denote the Al content of the respective tetrahedral sites (Kroll 1973). Between monoclinic K-feldspars and low microcline, equidistant contours of γ^* have been drawn because this angle is linearly related to $t_1o\text{-}t_1m$ and can be measured in a^*b^* electron diffraction patterns. γ^* is relatively insensitive to Or content in K-rich alkali feldspars (Kroll and Ribbe 1987, Fig. 4).

Two straight lines starting at low microcline have been drawn into Fig. 13.20. The first one intersects the abscissa at $\Sigma t_1 = 0.81$. It represents a *mean* straight line drawn through numerous data points of (X-ray) intermediate microclines. Their Σt_1 and Δt_1 values were derived from either single crystal structure refinements (Kroll and Ribbe 1987, Fig. 7) or lattice parameters (Bambauer and Bernotat 1982, Fig. 10). The second straight line intersects the abscissa at $\Sigma t_1 =$ 0.68.

This value represents the lower stability limit of Na-containing sanidine. It was determined from data obtained by Knitter (1985), Kroll and Knitter (1985) and Papageorgiou (in preparation) who investigated the Al,Si equilibrium distribution in sanidine (Or85) from Volkesfeld, Eifel, between 850 °C and 650 °C. From their results, a plot of $\ln K_D$ vs $1/T$ was found to be linear, where $K_D = (X_{Al}^{T2} \cdot X_{Si}^{T1})/(X_{Al}^{T1} \cdot X_{Si}^{T2}) = [(0.5\text{-}t_1)(1\text{-}t_1)]/[t_1(0.5 + t_1)]$. Extrapolation to T_{diff} = 480 °C, the presumed temperature of the sanidine/microcline transformation for Na-containing sanidine, gives $\Sigma t_1 = 0.68$. This value may be compared with $\Sigma t_1 = 0.74$ given by Su et al. (1984, Fig. 4) and with $\Sigma t_1 = 0.67$ found by McLaren and Fitz Gerald (1987) in a tweed orthoclase.

We may give the following tentative interpretation to Fig. 13.20:

When Na-sanidine is slowly cooled down to $T_{diff} \approx$ 480 °C, its Al,Si order increases to reach $\Sigma t_1 = 0.68$ (line segment '1'). At T_{diff}, sanidine should transform to microcline. At such a low temperature, however, it is difficult for the crystal to produce long-range ordered domains as are found in low microcline. However, as an alternative, local order in small alternately oriented domains can be achieved, i.e., tweed orthoclase originates. As found from TEM diffraction photographs γ^* angles are usually different from 90°: locally tweed orthoclase is *not* monoclinic! Thus line segment '2' may represent the Al,Si distribution as averaged over the small volumes from which information is derived in the TEM diffraction pattern ('TEM diffraction average'). When sanidine is cooled so slowly that its kinetic cut-off temperature is lower than $T_{diff} \approx$ 480 °C, some transfer of Al from one T_1 site (T_1m) to the other (T_1O) must occur below this temperature, i.e., the crystal becomes locally triclinic. Therefore, line segment '2' must leave the Σt_1-axis at 0.68 and extend into the triclinic region. We have chosen a straight line just for simplicity. Any non-linear curve may be possible, since it merely represents an *average* structural state.

When X-rayed, tweed orthoclases usually, but not always, appear monoclinic, because averaging occurs over a relatively large volume ('X-ray diffraction average'). We may denote such orthoclases as 'X-ray orthoclases' and plot them onto line segment '5', because their average Al,Si distribution derived from structure refinement or lattice parameters appears to be monoclinic.

Irregular microcline and tweed orthoclase occur side by side. We could not find evidence that irregular microcline had developed from tweed orthoclase by a continuous growth of domains. They appear to occur independently of each other. γ^* angles of irregular microcline range from 90–92° (line segment '3'). When investigated by X-rays some of them will appear monoclinic (X-ray orthoclase on line segment '5') and some will appear triclinic (line segment '6').

Regular microcline has γ^* angles between 91.7° and 92.5° (line segment '4'). Apparently it does not originate from irregular microcline. We regard regular microcline as a stable triclinic phase of K-feldspar. Irregular microcline may represent a nucleation event that failed to produce regular microcline.

Similarly, tweed orthoclase is a stranded structure. It consists of transverse waves in a strained, locally triclinic, but usually overall monoclinic configuration and thus does not represent the state of minimum free energy.

Therefore, line segments '2', '3', and '4' do *not* imply a continuous transition sanidine-microcline, but merely represent possible Al,Si distributions — *averaged*

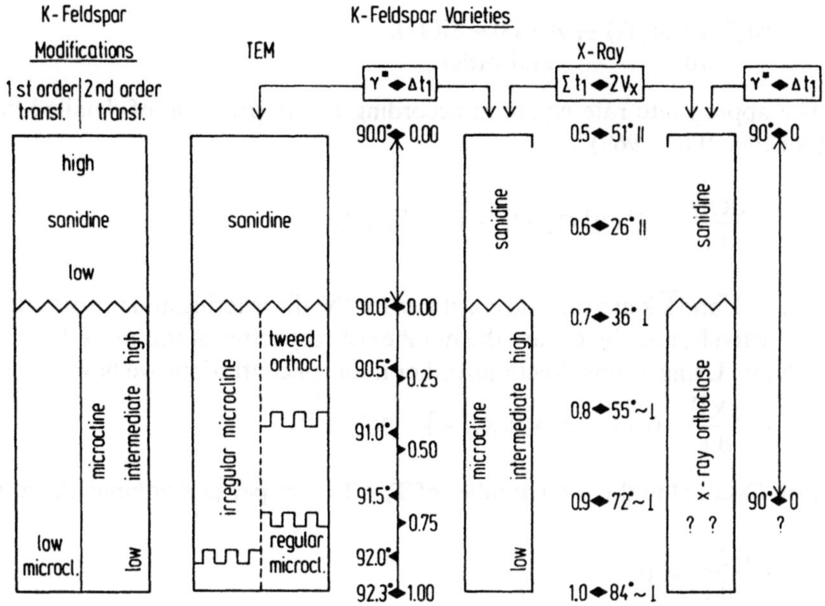

Fig. 13.21. Comparison of K-feldspar modifications and varieties on the basis of Fig. 13.20. The 'TEM-varieties' are distinguished on the basis of their contrast and diffraction patterns, whereas the 'X-ray varieties' are characterized by X-ray diffraction methods only. Σt_1, Δt_1, and γ^* values are taken from Fig. 13.20; the optical data which hold for a composition of Or85 are taken from Su et al. (1986). The orientation of the optic axial plane relative to (010) is specified by | |, ⊥, ~⊥, respectively; from Bambauer et al. (1989)

over small regions by TEM methods — which characterize metastable, stranded structures like tweed orthoclase and irregular microcline as well as stable structures like regular microcline.

Figure 13.21 schematically summarizes the mutual relations between K-feldspar modifications and K-feldspar varieties on one hand and γ^* angle, optic axial angle and Al,Si distribution on the other hand. The figure is taken from Bambauer et al. (1989).

13.6 Kinetics of the Al,Si Ordering Process in Sanidine

Knitter (1985), Kroll and Knitter (1985) and Papageorgiou and Kroll (1986) have investigated the kinetics of the Al,Si order/disorder process in sanidine. We will summarize their method and results as far as necessary for discussing the K-feldspar behaviour in the Ballachulish aureole. A full account will be published elsewhere.

The Al,Si order/disorder process is considered as a homogeneous exchange process:

$$Al(T_1) + Si(T_2) \leftrightarrows Al(T_2) + Si(T_1). \quad (1)$$

order anti-order

The appropriate rate equation according to the principle of detailed balancing is (Mueller 1967, 1969):

$$-\frac{dC_{Al}^{T1}}{dt} = \overrightarrow{k}\, C_{Al}^{T1}\, C_{Si}^{T2}\, C_{Si}^{T1} - \overleftarrow{k}\ \ C_{Al}^{T2}\, C_{Si}^{T1} \quad (2)$$

where $\overrightarrow{k}$ and $\overleftarrow{k}$ k are the rate constants for the disordering and the ordering reaction, respectively, and the C's are the number of respective atoms per sublattice and unit volume. Using atomic fractions rather than concentrations we have

$$-\frac{dX_{Al}^{T1}}{dt} = 0.5\, C_o\, (\overrightarrow{k}\, X_{Al}^{T1}\, X_{Si}^{T2} - \overleftarrow{k}\ \ X_{Al}^{T2}\, X_{Si}^{T1}) \quad (3)$$

with C_0 equal to the total number of $T_1 + T_2$ sites per unit volume. At equilibrium

$$-\frac{d\,X_{Al}^{T1}}{dt} = 0\,, \quad (4)$$

and

$$\frac{\overrightarrow{k}}{\overleftarrow{k}} = \frac{X_{Al}^{T2}\, X_{Si}^{T1}}{X_{Al}^{T1}\, X_{Si}^{T2}} = K_D\,. \quad (5)$$

Because in sanidine

$$X_{Al}^{T2} = 0.5 - X_{Al}^{T1},\ X_{Si}^{T1} = 1.0 - X_{Al}^{T1},\ \text{and}\ X_{Si}^{T2} = 0.5 + X_{Al}^{T1}\,,$$

the rate equation can be written

$$-\frac{d\,X_{Al}^{T1}}{dt} = C_o\overrightarrow{k}\,[\,a\,(\,X_{Al}^{T1}\,)\,2 + bX_{Al}^{T1} + c],\quad (6)$$

where $a = 0.5\,(1 - K_D^{-1}$
$b = 0.25 + 0.75\,K_D^{-1}$
$c = -0.25\,K_D^{-1}$.

Integrating and solving for X_{Al}^{T1} gives

$$X_{Al}^{T1}(t) = -\frac{b}{2a} + \frac{f}{2a} \cdot \frac{1 + g \cdot \exp\,(-\,C_o\overrightarrow{k} \cdot f \cdot \Delta t)}{1 - g \cdot \exp\,(-\,C_o\overrightarrow{k} \cdot f \cdot \Delta t)}\,, \quad (7)$$

where $f = (b^2 - 4ac)^{\frac{1}{2}}$.

$$g = \frac{2aX_{Al}^{T1}(t_o) + b - f}{2aX_{Al}^{T1}(t_o) + b + f}\,.$$

Equation (7) can be used to calculate the isothermal ordering of Al into the T1 tetrahedral site. This requires that we know the initial Al content in T1 ($= X_{Al}^{T1}(t_o)$), the equilibrium and rate constants at the temperature and pressure of interest, and the run duration. The process of interest is continuous cooling. Such a process can be approximated to an arbitrary degree by a discontinuous process, consisting of small steps of quenching and isothermal annealing. To follow the increase of order, we have to predetermine the temperature and pressure dependences of the equilibrium and rate constants.

The equilibrium constant K_D does not depend on $P(H_2O)$ to a measurable degree (Papageorgiou in preparation). Its temperature dependence is given by Kroll and Knitter (in preparation) as

$$\ln K_D = \frac{\Delta S^\circ}{R} - \frac{\Delta H^\circ}{R} \cdot \frac{1}{T}, \tag{8}$$

$$= \frac{1.94}{R} - \frac{2979}{R} \cdot \frac{1}{T} \left(R = 1.98586 \frac{\text{cal}}{\text{mol} \cdot \text{K}} \right), \tag{9}$$

where ΔS° and ΔH° are the molar standard state entropy and enthalpy differences between the ordered and anti-ordered states.

By contrast, the influence of $P(H_2O)$ on rate constants is considerable. Papageorgiou (in preparation) has performed experiments between 0.01 and 0.1 GPa. The amount of water added to the samples was 1 wt% and 10 wt% with no difference in result. Two Arrhenius type equations were fitted for $P(H_2O) = 0.01$ GPa (data of Papageorgiou) and $P(H_2O) = 0.1$ GPa (data of Kroll and Knitter):

$$C_o \vec{K} = \vec{A} \exp(-\vec{E}_a/RT) \tag{10}$$
$$\vec{A}\,(0.01\ \text{GPa}) = 7.71 \cdot 10^{16}\,\text{d}^{-1};\ \vec{A}\,(0.1\ \text{GPa}) = 1.625 \cdot 10^{10}\,\text{d}^{-1}$$
$$\vec{E}_a(0.01\ \text{GPa}) = 88.3\ \text{kcal/mol};\ \vec{E}_a(0.1\ \text{GPa}) = 53.3\ \text{kcal/mol}.$$

Strictly spoken, Arrhenius equations are valid only within the range of experiments. To some degree, extrapolations are always suspicious. In our case this is certainly true for the high temperature range, where the extrapolated straight lines would intersect, which would mean that the rate constants would decrease with increasing $P(H_2O)$, a highly unlikely prediction. Probably, the straight lines will become curves flattening off at high temperatures. However, we are interested in the low temperature region, into which we dare to extrapolate the straight lines.

In Fig. 13.22, the time-temperature variation is given at the contact and at 1100 m, as calculated by Buntebarth (Chap. 18, this Vol.) for the cylinder intrusion model. For decreasing temperatures these curves can be nicely approximated by the equation

$$1/T = 1/T_o + \eta \cdot \Delta t, \tag{11}$$

where η is a cooling parameter, ranging between

$$1.0 \cdot 10^{-12}\,\text{K}^{-1}\text{d}^{-1} \text{ and } 2.0 \cdot 10^{-12}\,\text{K}^{-1}\text{d}^{-1}.$$

Using these cooling rates, we have calculated the increase of order in sanidine during continuous cooling, starting from 773 and 555 °C. These temperatures are maximum temperatures reached at distances of 0 m and 1100 m (Fig. 13.22). Calculations were done for $P(H_2O) = 0.01$ GPa and 0.1 GPa. For each case we made two assumptions concerning the state of order at the beginning of cooling: (1) X_{Al}^{T1} corresponds to the equilibrium value at the respective starting temperature; (2) X_{Al}^{T1} equals 0.342 for each

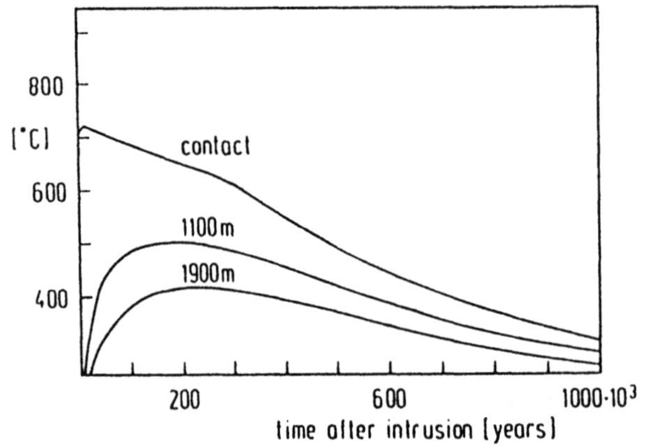

Fig. 13.22. Variation of temperature with time, calculated by Buntebarth (Chap. 18, this Vol.) for Appin Quartzite at distances of 0, 1100, 1900 m from the contact. A cylinder-type intrusion was assumed. The intrusion temperature was taken as 1050 °C, the temperature of the country rocks was 250 °C

of the two cooling curves, corresponding to the equilibrium Al,Si distribution at 480 °C (see Fig. 13.25).

When we start cooling at equilibrium Al,Si distributions using $P(H_2O) = 0.01$ GPa [points (2) in Fig. 13.23] we find that the ordering curve (a) with starting temperature = 773 °C virtually coincides at higher temperatures with the equilibrium distribution curve, calculated from Eq. (9). The ordering curves (a) and (b) leave this curve at about 550 °C. From this temperature on, ordering is slower than cooling; it is left behind and effectively freezes in at a kinetic cut-off temperature $T_{co} \approx 450$ °C. The state of order finally reached ($X_{Al}^{T1} = 0.334$) is smaller than corresponds to the lower limit of sanidine ($X_{Al}^{T1} = 0.342$).

When we use as the initial Al,Si distribution $X_{Al}^{T1} = 0.342$ corresponding to T_{equil} = 480 °C [points (1) in Fig. 13.23], then on starting at 773 and 555 °C the Al,Si distribution has first to disorder and then to order. This disordering process is very rapid, as can be seen from Fig. 13.23, where the disordering parts of the curves extending from right to left are virtually horizontal lines. The Al,Si distribution disorders toward the equilibrium curve in a time equivalent for the temperature to decrease over a few degrees only, corresponding to 2000–20 000 years. This clearly demonstrates the drastic effect which a large difference between starting and equilibrium conditions has on kinetic processes.

The result of these calculations is: when ordering in the Ballachulish aureole had proceeded under a small (partial) water vapour pressure, the sanidines in the transition and sanidine zones had not been inverted into orthoclases, but their structural states were frozen in at sanidine states.

We find a totally different result when we perform the same calculations assuming $P(H_2O) = 0.1$ GPa. Then, all curves follow the equilibrium curve to much lower temperatures than before (Fig. 13.24). Because ordering continuous below 480 °C,

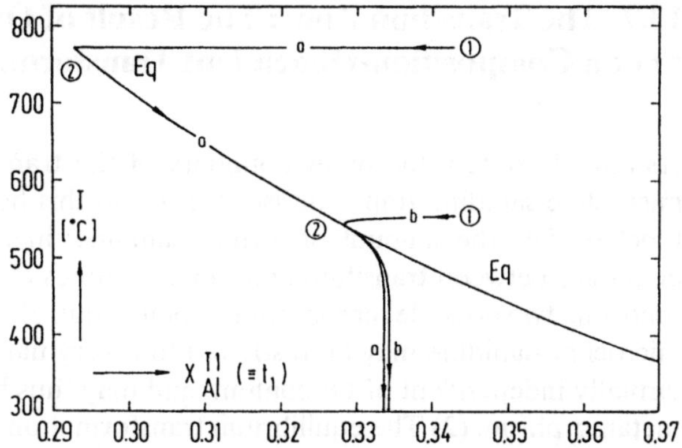

Fig. 13.23. Calculated increase of Al,Si order in snidine during continuous cooling under P(H_2O) = 0.01 GPa, starting at T_o = 773 °C (**a**) and 555 °C (**b**), respectively. The temperature was assumed to decrease according to $1/T = 1/T_o + \eta \bullet \Delta t$, with η approximately chosen to model the descending branches of the curves in Fig. 13.22. Two assumptions have been made for the Al,Si order at the beginning of cooling: *1* X_{Al}^{T1} = 0.343 corresponding to T_{equil} = 480 °C; *2* X_{Al}^{T1} = 0.294 and 0.328 for T_{equil} = 773 and 555 °C, respectively. The equilibrium Al,Si distribution follows the curve denoted by *Eq*. Extension beyond T = 480 °C corresponds to the fictitious situation that there was no sanidine to microcline transition boundary

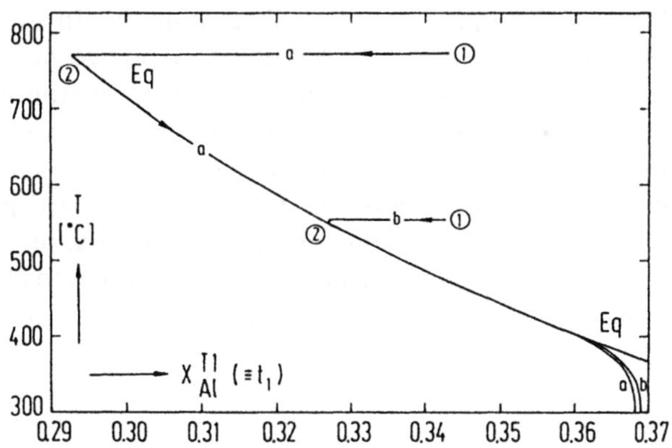

Fig. 13.24. Calculated increase of Al,Si order in sanidine during continuous cooling under P(H_2O) = 0.1 GPa. Further comments in Fig. 13.23

these sanidines could transform to low temperature varieties. The extension of the ordering and equilibrium curves below 480 °C into the microcline stability field is metastable, of course.

This result demonstrates the drastic effect P(H_2O) has on ordering rates and shows that a (partial) water vapour pressure of some hundred bars would have been sufficient to invert sanidine into orthoclase.

13.7 The Transition Zone: The Result of Overheating or of a Composition-Dependent Transformation Temperature?

As already stated, the outer boundary of the transition zone corresponds to the microcline-sanidine transition isograd. From this boundary on, the amount of orthoclase, i.e., the amount of former sanidine, increases nonlinearly towards the boundary between transition and sanidine zones (Fig. 13.3). We will discuss in the following two possible reasons for this behaviour: (1) the equilibrium transformation microcline-sanidine may be restricted to a very narrow temperature region, being virtually independent of Or content, and may thus be overheated during prograde metamorphism. (2) The equilibrium transformation microline-sanidine may extend over a considerable temperature range, being strongly dependent on Or content, and therefore no overheating is necessary to produce the transition zone.

We will first ask the question whether or not the microcline and sanidine regions are separated by a two-phase loop. Existence of a two-phase loop means that the transformation is first order. Do we have any experimental evidence for a first-order character?

Goldsmith (1988) performed heating experiments on low microcline at temperatures between 750 and 900 °C and pressures between 1.1 and 2.8 GPa to transform it to sanidine. He reports that "the development of monoclinic material took place progressively, coexisting with triclinic microcline until all the microcline had transformed". The process is thus discontinuous. In similar experiments on low albite, Goldsmith and Jenkins (1985) observed no two-phase behaviour; the order/disorder process was continuous. This contrast in disordering behaviour suggests that the microcline-sanidine transition is first order. However, the disordering reactions of low microcline and low albite are not strictly comparable: whereas K-feldspar changes its space group from $C\bar{1}$ to $C2/m$ Na-feldspar retains its $C\bar{1}$ symmetry. It is conceivable that due to overheating, a nucleation event could mask the true, continuous character of the microcline-sanidine transition and make it appear discontinuous. This possibility should be kept in mind when interpreting Goldsmith's (1988) results in favour of a first-order transition.

Salje (1985) and Salje et al. (1985) investigated the low-to-high albite transition using Landau theory. In agreement with Goldsmith and Jenkins (1985), they concluded that the transition is continuous in character. To describe the equilibrium states of order, it was necessary to employ two order parameters, one describing the displacive transition (Q), the other one describing Al,Si-order (Q_{od}). Coupling between Q and Q_{od} changes the character of the low-to-high albite transition from discontinuous to continuous. The transition would be first order if only Q_{od} were operative. Whereas in Na-feldspar Q dominates at high temperatures and Q_{od} at low temperatures, the reverse is true for K-feldspar, if Q plays a significant rôle at all. Openshaw et al. (1979) suggested that a 'smeared' displacive-like transition occurs in low microcline near room temperature. In that case, the order parameter Q in addition to Q_{od} would be necessary to describe the equilibrium order in microcline. Like in Na-feldspar, coupling between Q and Q_{od} could change the otherwise discontinuous transition to a continuous one. We repeated Openshaw's experiments carefully, but could not find any indication of a 'smeared' transition. We therefore assume that Q is not operative and treat the microcline-sanidine transformation as first order.

Considering only the effect of Q_{od} in Na-feldspar it is seen that the first order character is only weakly developed, but is close to tricritical (Carpenter 1988). Assuming the same to be true for pure K-feldspar means that the two-phase microcline-sanidine field is narrow.

The temperature T_{diff} of the phase change is poorly known, and estimates have been made both from field observations and experimental work. Voll (1969) placed it at ≈500 °C from a comparison with the oligoclase boundary in the Barrow-type metamorphic zones in NW Scotland. In the Ballachulish aureole he assumed T_{diff} ≈430 °C because the transition occurs close to the cordierite-producing reaction

$$\mathrm{Ms + Chl + Qtz \Leftrightarrow Crd + Bt + H_2O}.$$

However, Pattison (Chap. 16, Tables 16.3, 16.4, this Vol.) places this reaction at 550 ± 20 °C. Thus 430 °C is too low an estimate. Bambauer and Bernotat (1982) choose 450 °C. They found that the transition isograd in the Central Swiss alps occurs in the middle of a zone for which Frey et al. (1980) estimate a temperature of ≈450 °C. They concede, however, that this estimate "may be rather low". Pattison and Voll (Chap. 2, Table 2.2, Fig. 2.6, this Vol.), using garnet-biotite thermometry, report a temperature range of 467–506 °C in garnet-bearing pelites next to the Appin Quartzite.

Experimental work dates back to Goldsmith and Laves (1954). They produced sanidine from microcline by hydrothermal treatment at 525 °C, whereas no such change was observed at 500 °C. The reaction occurred as a solution-reprecipitation process, and in principle could therefore occur at $T < T_{diff}$ (Carpenter and Putnis 1985).

The work of Senderov and Yas'kin (1975, 1976) is especially important. They converted microcline into sanidine hydrothermally at $T \geq 550$ °C, whereas sanidine, prepared by previous heating of microcline, turned back to microcline at $T \leq 450$ °C, but did not transform at $T \geq 550$ °C. These experiments bracket the transition between 450 and 550 °C.

Combining the results of field work and experiments we choose 480 ± 20 °C as the temperature of the microcline-sanidine transformation in Na-containing K-feldspars.

Figure 13.25 shows two possible topologies of the alkali feldspar phase diagram at K-rich compositions. (1) The upper part of the solvus represents sanidine solid solutions; (2) the lower part represents microcline solid solutions. The solvi were schematically modified from P. Smith and Parsons (1974) and Bachinski and Müller (1971) to account for the fact that increasing Al,Si order in sanidine and microcline causes the solvus to broaden. Considering the sanidine-microcline transition to be discontinuous, we have to separate their stability fields by a transition loop. We have chosen two extreme cases: in Fig. 13.25a the loop extends between 480 and 550 °C, in Fig. 13.25b it extends between 480 and 490 °C.

The first choice is based on Buntebarth's calculation of the temperature distribution in the contact aureole (Chap. 18, this Vol.). Maximum temperatures are plotted versus distance in Fig. 13.26 assuming a cylindrical intrusion model. Also plotted are data points derived from contact-metamorphic reactions occurring in pelites (Pat-

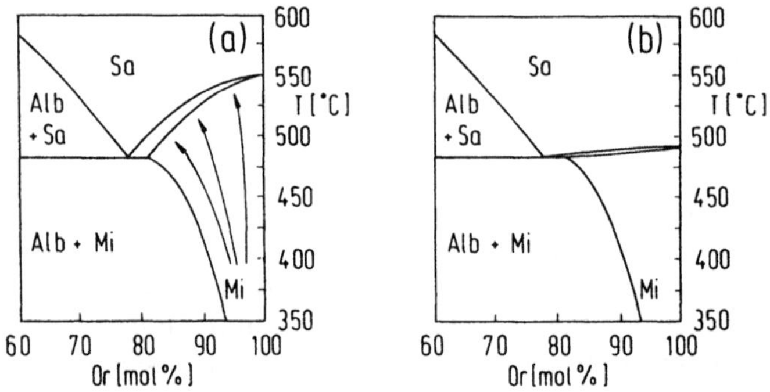

Fig. 13.25. Two possible phase relations for K-rich alkali feldspars. The high temperature solvus is based on P. Smith and Parsons' (1974) solvus, the low temperature solvus is based on Bachinski and Müller's (1971) low albite-low microcline solvus. The modifications schematically account for the dependence of the width of the solvi on Al,Si order. A first-order phase transformation (close to tricritical) is assumed. **a** The lower and upper transition temperatures (≈480 and 550 °C, respectively) were chosen to accord with the temperatures at the outer and inner transition zone boundaries, respectively (see Fig. 13.26). **b** The transition temperature is assumed to be virtually independent of composition. *Alb* Albite; *Sa* sanidine; *Mi* microcline

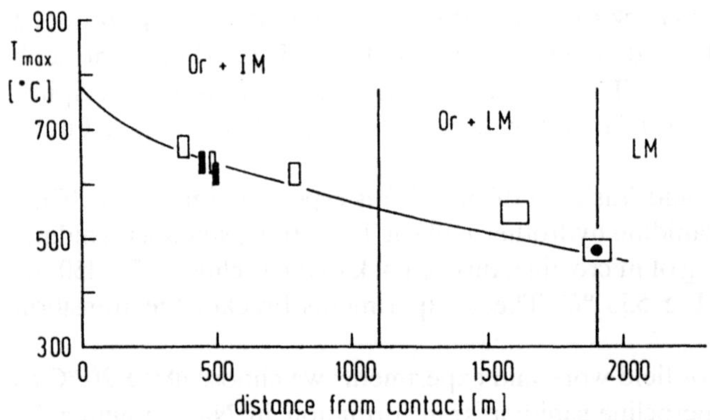

Fig. 13.26. Variation of the maximum temperatures in Appin Quartzite with distance from contact. The curve was calculated by Buntebarth (Chap. 18, this Vol.) for a cylinder-type intrusion (temperature of country rocks = 250 °C, intrusion temperature = 1050 °C). *Open rectangles* indicate isotherms for contact-metamorphic reactions in pelites (Pattison, Chap. 16, this Vol.), *filled rectangles* give temperatures derived from the calcite-dolomite thermometer (Masch and Heuss-Aβbichler,Chap. 10, this Vol.), the *square* with a *dot* indicates the microcline-sanidine isograd

tison, Chap. 16, Table 16.4, this Vol.) and from the Mg content of calcites coexisting with dolomites in thermometamorphosed impure limestones and siliceous dolomites, both from the northeastern part of the aureole (Masch and Heuss-Aβbichler, Chap. 10, this Vol.). According to these authors their temperature data at 450 and 500 m are most reliable. Buntebarth's curve nicely fits the data points. It gives T ≈ 550 °C at

1100 m, corresponding to the microcline-sanidine transformation at Or100 in Fig. 13.25a.

The Or content of low microcline in the zone of no contact metamorphism is ≈96 mol%. If a grain of this composition is heated, the transition loop is crossed at ≈550 °C. If re-solution of Na occurs, this temperature is lowered. Three re-solution paths have schematically been indicated in Fig. 13.25a. Re-solution is observed in the optical microscope: fine film-perthite lamellae disappear, whereas coarse patchy perthites are preserved. The original low microcline grains contain various amounts of exsolved Na-feldspar, so that a range of compositions may be covered due to re-solution. If at a given temperature (e.g., 480 °C) the microcline compositions are more or less equally distributed between ≈Or80 and Or100, the nonlinear increase of the amount of orthoclase in the transition zone (Fig. 13.3) is explained by the chosen curvature of the transition loop.

Steiger and Hart (1967) also report a volumetric increase of the orthoclase content in the transition zones of the various traverses which they investigated around the Eldora stock. They found that the width of the transition zone and its distance from the contact varied with the shape of the intrusive. Near planar or concave contacts, the transition zone not only was further from the contact, but it appeared to be wider, whereas narrow transition zones were found adjacent to convex contacts. Steiger and Hart (1967) related these observations to their heat-flow calculations. Near convex contacts, the temperature gradient was steeper and the time span during which the feldspars were held above the transformation temperature was shorter than near planar and concave contacts.

The Eldora stock is smaller than the Ballachulish intrusion, 3 km in diameter compared to 4 × 7 km, and the transition zone in the Ballachulish aureole is wider and more distant from the contact.

The inverse correlation between width and distance of the transition zone on one hand and temperature gradient on the other hand, as observed in the Eldora and Ballachulish traverses (Fig. 13.27), is consistent with the range of transition temperatures drawn in Fig. 13.25a. The temperatures T1 and T2 in Fig. 13.27a correspond to ≈480 and ≈550 °C.

However, there is a serious objection to a transition range of about 70 °C. Salje et al. (1985) have shown that in Na-feldspar, the continuous order/disorder transition, if it were not modified by the displacive influence, would occur as a discontinuous C2/m-C$\bar{1}$ transformation at ≈600 °C. The analogous (Na-)sanidine-(Na)microcline inversion is lower by only ≈120 °C, and it is unlikely that it should increase by ≈70 °C between ≈Or80 and Or100. Figure 13.25b accounts for this objection. To explain the transition zone, it is then necessary, however, to assume that the microcline-sanidine transformation has been overheated. Figure 13.28 is a schematic TTT diagram for a reaction taking place with rising temperature. The temperature variation at 1100 m was calculated by Buntebarth (Chap. 18, this Vol.) and corresponds to the cylindrical intrusion model. Some overheating is necessary to get the microcline-sanidine transformation started. An even greater overheating is necessary to get the transformation finished, at rates increasing with temperature. The difference between the start and finish temperatures is certainly strongly dependent on the amount of 'water' present. With a restricted

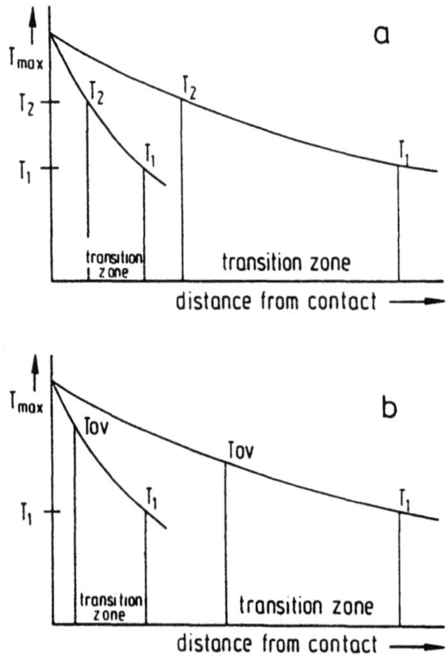

Fig. 13.27. Schematic drawing to illustrate the inverse correlation between the gradients of maximum temperatures in the Eldora and Ballachulish contact aureoles and the widths and distances from contact of their transition zones. **a** T_1 is the temperature of the microcline-sanidine isograd (≈480°C), T_2 corresponds to the transition temperature of pure microcline (≈550 °C) as assumed in Fig. 13.25a. **b** T_{ov} schematically indicates the temperatures at which microcline is finally inverted into sanidine when overheating of the reaction occurs. In a wide contact aureole T_{ov} is lower than in a narrow one

amount of water, overheating could be considerable. Thus this model alone could explain the non-linear increase of the orthoclase/low microcline ratio in the transition zone.

In addition, it could still be consistent with the inverse relation between widths and distances of transition zones and temperature gradients. The temperature difference between the end and beginning of the transformation process would be smaller in a wide aureole than in a narrow one (Fig. 13.27b), and yet the inverse relation between width and temperature gradient could still be preserved.

The 'overheating theory', however, also faces a major difficulty. When discussing the Al,Si exchange kinetics, we have seen that only some hundred bars of partial water vapour pressure are sufficient to allow sanidine to convert to orthoclase during cooling of the aureole. Under such a water pressure the disordering process in sanidine is very fast, as seen from the virtually horizontal lines in Figs. 13.23 and 13.24 [numbered (1)]. Only about 25 000 years are necessary to disorder sanidine at T = 550 °C, $P(H_2O)$ = 0.01 GPa. The heating and cooling cycle in Fig. 13.28 lasts approximately 20 times as long. We have therefore to assume that heating occurred under essentially dry conditions, but cooling did not. This would have to apply to the Ballachulish as well as the Eldora traverses. Furthermore, even in traverses through the regional metamorphic terrain of the Central Swiss Alps transition zones are found which are 5–10 km wide (Bambauer and Bernotat 1982, Fig. 7, and Bambauer, pers. comm.). To explain them by overheating, one would again have to assume dry heating and 'wet' cooling, which is not a likely assumption. It will be argued in the next section that internal and external non-hydrostatic stresses favour the development of

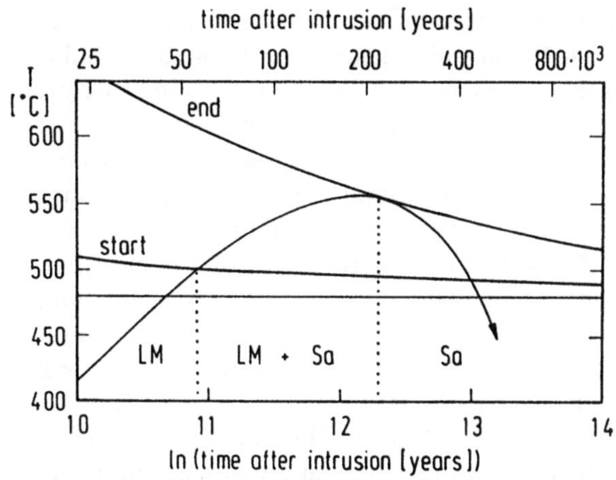

Fig. 13.28. Schematic time-temperature-transformation (TTT) diagram for the microcline (*LM*) to sanidine (*Sa*) transformation occurring with rising temperature. The time-temperature (TT) curve is given for a distance of 1100 m from the contact. It was calculated by Buntebarth (Chap. 18, this Vol.) assuming a cylinder-type intrusion. Low microcline would continue to exist as an overheated phase until the TT curve intersects the start curve of the transformation. In the interval between the *start* and *end* curves, sanidine would continuously grow at the expense of low microcline until the transformation is completed. The time interval in which microcline transforms to sanidine is enclosed by the *dotted lines*

triclinic K-feldspar during cooling. Vice versa, non-hydrostatic stresses exerted by quartz grains surrounding microcline possibly hinder the attainment of monoclinic geometry during heating. This speculation would solve the discrepancy between the 'overheating theory' and the results of the experiments on Al,Si exchange kinetics.

13.8 The Re-Ordering Products

In Fig. 13.3, the volumetric ratio of orthoclase to low microcline in the transition zone steeply increases toward the contact. We cannot decide, however, whether the increase smoothly reaches 100% at ≈1100 m, or whether it is cut off by some process at this distance, as it may seem.

In any case, at this very same boundary intermediate microcline appears in the X-ray diagram. No intermediate microcline is found in the transition zone, i.e., in this traverse low microcline and intermediate microcline exclude each other. This statement has to be modified regarding the TEM results, as we have seen, but it is true for the X-ray results.

In the Eldora traverses, orthoclase is the dominant re-ordering product of both the transition and sanidine zones. In one traverse, however, intermediate microcline occurred to more than 50%. It was this traverse, where the temperature was brought up faster than in the other ones and had a much slower cooling rate. On extrapolation to the Ballachulish heating conditions one would expect that intermediate microcline would be by far the dominating re-ordering product. Yet it does not occur in the transition zone at all. Apparently, besides slow cooling, additional conditions must be fulfilled for intermediate microcline to occur.

We have seen that a certain water vapour pressure is necessary to avoid that ordering is frozen in at the sanidine state during cooling. When considering the drastic difference in the microtexture of tweed-orthoclase and irregular microcline, it is conceivable that a certain amount of water, not just partial pressure, is necessary to produce irregular microcline. Maybe this water was part of the fluid that transported Na outward from the intrusion to produce the large amount of perthite exsolution mentioned earlier (cf. Figs. 13.12, 13.13, 13.14). Its radius of action would then have been limited to 1100 m. The water migration front might have occurred at 1100 m at a time when the temperature was not dropped below 400 °C, so that textural re-organization to irregular microcline was still possible.

Another speculative source of water (and Na?) are fluid inclusions in quartz crystals. These might be released to an increased rate when recrystallization of the quartz grains sets in at ≈620 °C, which is close to the low-to-high quartz inversion temperature at 0.3 GPa. This temperature is reached to the west of the inner transition zone boundary. However, the sanidine-to-microcline transition occurs on cooling, and by the time the transition temperature is reached, the fluid might have migrated outward.

A totally different mechanism which drives the sanidine-to-microcline transition is stress. We know from light optical and electron optical work, that microcline lamellae preferentially grow at points of increased stress, e.g., inclusions, exsolution lamellae or grain boundaries. Often a direct relation is observed between the size of microcline lamellae so induced and the size of the inclusion or exsolution lamella causing the stress field. Stress exerted from the outside should have a similar effect. During cooling, the quartz grains produce an inhomogeneous stress field around a feldspar grain. If a certain threshold is exceeded, twinning may set in. The stress produced would depend on the maximum temperature reached, and thus the difference between this maximum temperature and the transformation temperature must exceed a minimum value. If this mechanism is responsible for the formation of irregular microcline, this temperature difference was insufficient at distances > 1100 m to build up the stress necessary to transform sanidine to microcline.

Besides stress from the outside, we have to consider another factor. During regional metamorphism all low microcline grains have been deformed to some degree. In response to deformation twinning in low microcline often is developed in a non-regular way. Within the inner aureole, recrystallization is observed that eliminates plastic strain. In non-recrystallized grains, the deformation features are preserved. Even after transformation to sanidine, the grains must have consisted of somewhat differently oriented and deformed domains. Also, remnants of twin and exsolution boundaries may have been present.

If such a crystal cools down, inhomogeneous stress is not only built up by contraction of its surrounding, by irregularities of grain boundaries or by inclusions, but, in addition, the internal irregular domain structure causes inhomogeneously distributed stress fields. It appears unlikely that in such a grain a texture should develop that is uniform over large distances. Tweed areas may be replaced by twin areas starting at misoriented subgrain boundaries, whereas in 'windows' between twinned areas, where stress is released, the tweed texture may be preserved.

A clear indication of the relation between microtexture and deformation is provided by the comparison of recrystallized and non-recrystallized grains. In contrast to the optically inhomogeneous non-recrystallized parts of a grain, the recrystallized ones are homogeneous. Their exsolution characteristics are also different. They contain fine film-perthite lamellae rather than patchy and irregular perthitic unmixing.

In conclusion, although it is impossible to decide which mechanism or combination of mechanisms discussed has controlled the development and distribution of irregular microcline, we believe that stress release played a major rôle.

13.9 Conclusion

Investigation of the K-feldspars in the Ballachulish aureole has not answered all questions regarding the microcline-to-sanidine and sanidine-to-microcline transition, but at least it has brought some of them closer to an answer.

1. TEM work has revealed three main types of microtextures: Regular microcline, irregular microcline and tweed orthoclase. When X-rayed, these correspond to low microcline, intermediate microcline or X-ray-orthoclase, and (mostly) X-ray-orthoclase, respectively.

2. Tweed orthoclase has a modulated structure. Such a structure can originate when at a certain temperature below a first-order transition the nucleation barrier vanishes and continuous (spinodal) ordering can commence (Carpenter 1985). That is, the tweed texture may start to develop some degrees below the transition temperature (also see Kroll, Chap. 20, this Vol.).

3. Kinetic studies of the Al,Si order/disorder process show, that at least some hundred bars of partial water vapour pressure must have been present in the aureole during the cooling process. Otherwise, sanidine could not have been converted to orthoclase and intermediate microcline, but would have been frozen in as a partially ordered sanidine.

4. Transition zones found in the Eldora and Ballachulish aureoles exhibit an inverse correlation between width and distance from contact on one hand and gradient of maximum temperatures on the other hand. Two possible explanations are discussed: (a) overheating of a virtually composition-*independent* microcline-sanidine transition, and (b) transformation at a strongly composition-*dependent* boundary. Each explanation faces its own problems, either kinetically or thermodynamically. The 'overheating theory' would be acceptable if one could support the speculation that the microcline-to-sanidine transformation is hindered by non-hydrostatic stress exerted by the surrounding of the microcline crystals, which prevents the attainment of monoclinic geometry.

5. Within the transition zone, sanidine re-orders mainly into orthoclase (considering X-ray *and* TEM results), whereas intermediate microcline is abundant in the sanidine zone. We have discussed various possibilities to explain this observation, including a fluid phase as well as stress, but to arrive at a conclusion one would have to know, for example, the fluid regime in the aureole in space and time. However, comparison of the different optical appearances of recrystallized and non-recrystallized parts of a former sanidine grain shows that internal stress-modification by the deformation texture favours the formation of (irregular) microcline.

14 A Search for Variations in the Structural States of Cordierite in Contact-Metamorphosed Pelites

W.V. Maresch, P. Blümel and W. Schreyer

14.1 Introduction

The mineral cordierite is an important rock constituent of pelites that has developed throughout the Ballachulish aureole as a consequence of contact metamorphism. Cordierite is a framework silicate with distinct similarities to ring silicates, whereby rings of six (Al,Si)-tetrahedra are linked by additional (Al,Si)-tetrahedra to form a three-dimensional framework. The rings are stacked in the crystallographic *c*-direction to form the large open channels that are permeable to rock fluids and cationic species of various compositions and types, leading to the unusual crystal chemical properties and thermodynamic behaviour observed for this mineral. The general structural formula has been summarized by Schreyer (1986) as follows:

$$(Na,K)_{0-1}{}^{[CHANNEL]}\,(\mathit{Mg},\mathit{Fe}^{2+},Mn^{2+},Li)_2{}^{[6]}$$
$$[(\mathit{Si},\mathit{Al},Be,Fe^{3+})_9{}^{[4]}O_{18}] \bullet (\mathit{H_2O},\mathit{CO_2},Ar...)^{[CHANNEL]},$$

where the most important site occupants found in nature are italicized.

Of particular importance is the fact that the Al,Si ordering process within the tetrahedral framework of the cordierite structure is a potential 'geospeedometer'. Although disordered, hexagonal, pure Mg-cordierite is thermodynamically stable only at temperatures in excess of approximately 1450 °C (e.g., Schreyer and Schairer 1961; Smart and Glasser 1977; Putnis 1980a), there are numerous indications in nature and in experimental studies that the stable, ordered, orthorhombic cordierite now found in rocks or observed at the end of an experimental run initially nucleated and grew metastably as the disordered form (Schreyer and Yoder 1964; Schreyer 1966; Armbruster and Bloss 1981; Putnis and Bish 1983; Putnis and Holland 1986). In a contact metamorphic aureole such as that surrounding the Ballachulish intrusive, the Al,Si ordering state observed in any hand specimen must be a consequence of the interplay between the particular time-dependent heating-and-cooling path of that sample with the time-temperature-transformation kinetics of the initial cordierite-forming reaction(s), followed by the kinetics of the Al,Si ordering process itself. Schreyer (1966) has predicted that the most ordered structural states should be found in cordierite that nucleated and grew in rocks close to the intrusive contact. Harwood and Larson (1969) have reported observations on cordierite from the contact metamorphic aureole of the Cupsuptic quartz-monzonite pluton in Maine, USA, which appear to support this prediction, although the differences in the distortion index determined throughout the Cupsuptic aureole are not large. Kamineni (1974)

G. Voll, J. Töpel, D.R.M. Pattison, F. Seifert
Equilibrium and Kinetics
in Contact Metamorphism

has attributed distortion index variations of cordierite in metagreywackes and argillites near the Sparrow Lake granite pluton (NWT, Canada) to thermal effects, but the small number of samples and the compositional variability of the cordierites hamper unequivocal interpretation.

The dimensions of the Cupsuptic pluton and those of the Ballachulish intrusion are comparable, so that this study was undertaken to determine whether similar trends in the structural state of cordierite exist in the latter. If so, it was anticipated that the detailed models of the thermal history of the Ballachulish intrusion and its aureole as well as the detailed analyses of the phase petrology of the aureole rocks being developed by other workers in this cooperative project would allow kinetic analyses of these trends.

14.2 The Determination of the Structural State of Cordierite

Crystallographic studies indicate that the disordered cordierite polymorph is hexagonal, while ordered cordierite is orthorhombic (Gibbs 1966; Meagher and Gibbs 1977), so that the determination of metric variations by X-ray methods appears to offer a simple method for monitoring the progress of the hexagonal-to-orthorhombic transformation. Miyashiro (1957) first introduced the commonly employed distortion index Δ, which is based on the splitting of the (211) reflection for hexagonal symmetry into the corresponding triplet in the orthorhombic case. Δ is defined as $2\Theta_{131} - [(2\Theta_{511} + 2\Theta_{421})/2]$ for CuK_α radiation in X-ray powder diffractograms. Miyashiro (1957) discovered that observed values of Δ for cordierite varied continuously between 0° (hexagonal) and 0.29°, and related this observation to the existence of structures with all possible states of intermediate Al,Si order. For nomenclature purposes, Schreyer and Schairer (1961) defined as high cordierite all those yielding distortion indices of 0°, as structurally intermediate cordierites all those with $0° < \Delta < 0.20°$, and as low cordierite all those with $\Delta > 0.20°$.

Although Δ is certainly a relatively simple parameter to determine, it is nonetheless not a straightforward measure of Al,Si disorder, because all factors that influence cordierite lattice metrics will affect the measured distortion index as well. Of primary importance are compositional variations. As Schreyer et al. (1979) have discussed, the combined presence of significant amounts of Na^+ and H_2O in the cordierite channels leads to a drastic lowering of the distortion index. However, highly sodian cordierites appear to develop almost exclusively through the coupled substitution $NaBeAl_{-1}$ and are therefore limited to pegmatitic and granitic parageneses. More relevant to the Ballachulish aureole rocks are the statistical analyses of Selkregg and Bloss (1980), who show that Δ is a direct function of the average radius of the cations in the octahedral sites. A typical distortion index of ≈ 0.24° for a magnesian low-cordierite may increase to 0.31° when Mg is completely substituted by Fe^{2+}.

In addition to this chemical influence, the early infrared work of Langer and Schreyer (1969) on pure Mg-cordierite of various structural states showed that Δ can obviously not always be correlated linearly with increasing short-range order. These authors emphasized that nearly perfect short-range order can already have been achieved even when the Mg-cordierite still exhibits low Δ values of intermediate structural states. Nevertheless, the distortion index persisted as a practical tool for rapidly estimating structural states of cordierites.

Since the initiation of the Ballachulish cooperative project, the true physical significance of the distortion index and the mechanisms of the hexagonal-to-orthorhombic transformation have been the subject of several further papers (e.g., Putnis 1980a,b; Putnis and Bish 1983; Carpenter et al. 1983; Putnis and Angel 1985), based on experimental results obtained on synthetic, anhydrous, end-member Mg-cordierite. The experimental techniques involved NMR and IR spectroscopy, synchrotron X-ray powder diffraction, transmission electron microscopy, and heat of solution calorimetry. These studies have culminated in two reports by Putnis et al. (1987) and Salje (1987), incorporating Landau theory, in which a physical interpretation of the transformation process is proposed. Two parameters are considered necessary to understand the nature of the hexagonal-to-orthorhombic transition. One is the state of Al,Si ordering, which varies continuously from highly disordered to entirely ordered and can be measured by NMR spectroscopy. The other monitors the tendency of the structure to distort from hexagonal to orthorhombic symmetry and may be considered analogous to the classical distortion index obtained by X-ray diffractometry.

According to the original interpretation of Δ, already discounted by Langer and Schreyer (1969), these two parameters are linearly coupled and should plot on the dashed line indicated in Fig. 14.1. Putnis et al. (1987) now clearly show that Δ remains at 0° until a very high degree of order is achieved, and that at a given, specific state of Al,Si order, the structure adjusts abruptly from a Δ of 0° to a value on the dashed line (Fig. 14.1). Only from this point on in the hexagonal-to-orthorhombic transformation may there be a linear correlation between the distortion index and Al,Si order. The abrupt, non-linear behaviour of Δ is explained by the formation of a thermodynamically poorly defined, intermediate, modulated hexagonal phase (Fig. 14.1), which allows local strain and order modulations in the structure, while retaining long-range hexagonal symmetry.

On the basis of these new experimental results and their interpretation by Putnis et al. (1987) and Salje (1987), the many intermediate distortion indices between O° and ≈ 0.20° reported in the literature for both natural and synthetic cordierites (e.g., Miyashiro 1957; Schreyer and Schairer 1961; Schreyer 1966; Putnis and Bish 1983; and many more) are difficult to explain. Some are undoubtedly due to special chemical compositions (Schreyer et al. 1979). For pure Mg-cordierite, Putnis et al. (1987) contend that such determinations are artefacts caused by insufficient resolution in routine X-ray diffractograms. Conceivably, various intermediate Δ's could reflect summation curves of the superposition of the single broadened peak of the modulated phase with the orthorhombic triplet in different intensity ratios, if some inter-grain heterogeneity in Al,Si order is assumed to exist in the powders investigated. In our opinion, such intermediate distortion indices could then still be an at least qualitative measure of the hexagonal-to-orthorhombic reaction progress, because they reflect decreasing hexagonal/orthorhombic volume ratios. The physical interpretation of the empirical, but experimentally verifiable and reproducible in-

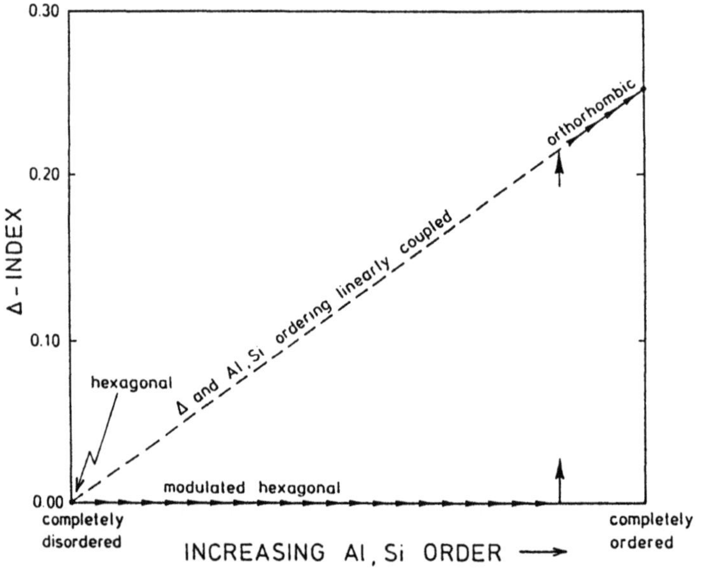

Fig. 14.1. Schematic diagram to illustrate the coupling between the distortion index Δ and the state of Al,Si order as the hexagonal-to-orthorhombic transformation proceeds, according to Putnis et al. (1987) and Salje (1987). The particular state of order at which the abrupt increase in Δ occurs depends critically on temperature and also on cordierite composition (E. Salje, pers. comm. 1987)

crease in the classical Δ index from essentially 0° to maximum values of 0.20° during the hexagonal-to-orthorhombic transition still requires further clarification.

Studies comparable to those of Putnis et al. (1987) on end-member Mg-cordierite have so far been carried out only on K-substituted varieties (Redfern et al. 1989). Here the transformation behaviour was found to be essentially similar to that in end-member Mg-cordierite, except that the maximum values of Δ obtained are lower and the collapse of the hexagonal modulated structure also occurs at lower values of Δ. It is not known, however, if the region where Δ and state of order vary linearly might expand relative to the region of non-linearity for other compositions more relevant to natural occurrences. It is theoretically not inconceivable that for certain cordierites far from end-member composition, the modulated intermediate phase might even be completely suppressed, and linear coupling could hold over the *entire* range of observed values.

At the present state of knowledge, we must conclude from many experimental studies that, although clearly not in a linear manner, the transformation from highly disordered hexagonal cordierite to the ordered orthorhombic phase can be monitored on the basis of the distortion index as determined on standard X-ray diffractometers, as long as compositional effects can be ruled out. Intermediate values of Δ must be regarded as purely empirical and may in part be instrumental artefacts. However, because natural cordierites always contain some iron, there is practically no other choice. NMR studies not only require sophisticated apparatus and several grams of pure cordierite concentrate, but they are also impossible on Fe-bearing samples.

On the basis of the above considerations, we feel that the classical determinations of structural state of cordierite by virtue of the distortion index, as carried out in this study, are still an important and justifiable *first step* in searching for variations in Al,Si ordering states in the metapelites of the Ballachulish aureole. In fact, the Δ values measured were found to be consistently high, so that the problem of intermediate states was actually never encountered. Thus any trends observed should only apply to structural variations of the orthorhombic phase (long-range order of Langer and Schreyer 1969), for which according to Fig. 14.1 there is indeed a linear relationship between Δ and Al,Si order.

14.3 Sampling Procedures and Experimental Techniques

Samples for the present investigation were obtained from contact metamorphosed rocks of the Creran Succession at the southern tip of the Ballachulish intrusion (Fig. 14.2), where the aureole rocks are well exposed on the steep northern flank of a crescent-shaped topographic high. The main sampling profile was located parallel to the NNE/SSW regional strike of the schists in order to minimize any cryptic bulk compositional effects arising from bed-to-bed inhomogeneity in the original sedimentary sequence. The distance here between the intrusive contact and the cordierite-in isograd is approximately 575 m. A subparallel, auxiliary profile was sampled 300–400 m to the NW. In addition, two broad sampling lines, each approximately 1000 m in length and crossing the strike of the schists at a high angle, follow the intrusive contact and the cordierite-in isograd, respectively. Several samples were obtained from metapelitic schollen within the intrusive itself. In all, 36 samples yielded acceptable data on the cordierite distortion index (Fig. 14.2), a major problem being the strong secondary pinitization observed in a number of cases, such that not enough unaltered cordierite could be obtained for X-ray work. Strong pinitization was found to be particularly prevalent in the auxiliary profile of the western part of the study area. All samples are deposited at the Institut für Mineralogie, Bochum, and are identified by their Bochum catalogue number.

Cordierite grains are typically highly poikiloblastic and intimately intergrown with other phases such as muscovite, biotite, quartz and feldspar, so that pure cordierite concentrates could not be obtained. For routine determinations of the distortion index, pieces of whole rock were crushed, ground and sieved to obtain the 45–63-μm grain-size fraction. Several passes in a Frantz isodynamic separator sufficed to concentrate cordierite to an appropriate extent. For lattice parameter determinations, 10–15 mg of cordierite fragments were hand-picked from these concentrates.

X-ray diffraction work was carried out on an automated Siemens D500 powder diffractometer using Ni-filtered CuK_{α} radiation and a secondary graphite monochromator. $K(\alpha_2)$ reflections were eliminated mathematically during data reduction. Raw data for distortion indices were collected by step-scanning the interval between 28 and 31°2Θ CuK_{α} with steps of 0.004° and 8 s counting time. The

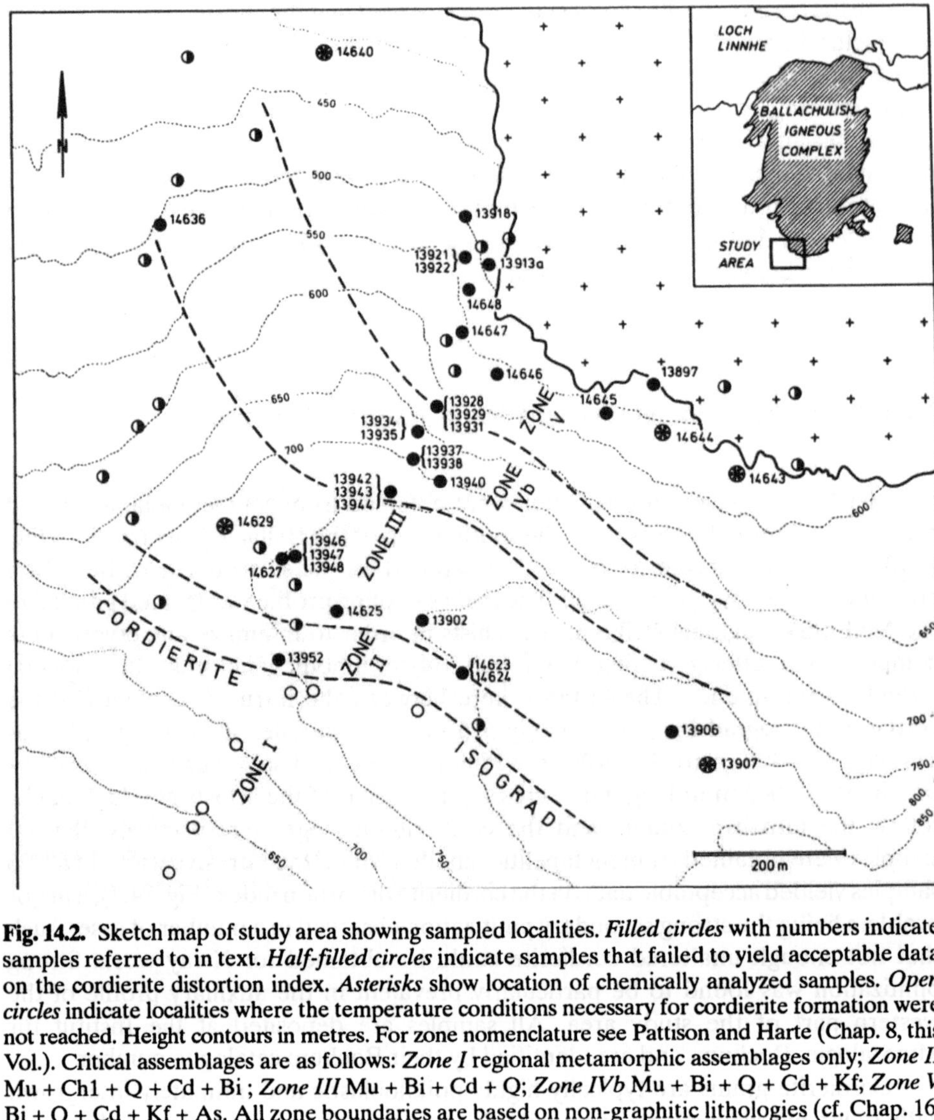

Fig. 14.2. Sketch map of study area showing sampled localities. *Filled circles* with numbers indicate samples referred to in text. *Half-filled circles* indicate samples that failed to yield acceptable data on the cordierite distortion index. *Asterisks* show location of chemically analyzed samples. *Open circles* indicate localities where the temperature conditions necessary for cordierite formation were not reached. Height contours in metres. For zone nomenclature see Pattison and Harte (Chap. 8, this Vol.). Critical assemblages are as follows: *Zone I* regional metamorphic assemblages only; *Zone II* Mu + Chl + Q + Cd + Bi ; *Zone III* Mu + Bi + Cd + Q; *Zone IVb* Mu + Bi + Q + Cd + Kf; *Zone V* Bi + Q + Cd + Kf + As. All zone boundaries are based on non-graphitic lithologies (cf. Chap. 16, this Vol.). Their location outside the profile area is based on samples described by Pattison (1985). Trace of the contact between the igneous complex and the Creran Succession as mapped by S. Weiss (pers. comm. 1988)

2Θ-positions of the three overlapping peaks of the orthorhombic triplet were determined using a least-squares profile-fitting program provided by the D500 system, which is based on an algorithm by Marquardt (1963). Both Cauchy and pseudo-Voigt functions were employed in fitting. Because only *relative* peak positions are of interest, no calibration procedures for instrumental shifts in *absolute* 2Θ-positions were carried out. The errors quoted for each distortion index are obtained from the standard deviations (1σ) of the refined position of each peak, propagated according to the definition of Δ as given by Miyashiro (1957, see discussion in previous section). For the determination of lattice parameters, the interval 5–80°2Θ was measured in steps of 0.006°2Θ, 6 s counting time, using synthetic F-phlogopite as an internal standard (National Bureau of Standards SRM 675; Hubbard 1983). Data reduction and refinement procedures are those described by Maresch and Czank (1985). At least 30–40 unequivocally indexed reflections were available for each refinement.

Major and trace element analyses of whole-rock powders were obtained by XRF and routine wet chemical techniques, respectively (Analyst: G. Werding, Bochum). Wave-length electron-microprobe spot analyses were carried out on a fully-automated Cameca (model Camebax) instrument, with the on-line correction procedures provided by the system. The standards used are discussed by Abraham and Schreyer (1973). Standard operating conditions were: focussed beam, 15 kV, 14 nA sample current, and 10 s counting time.

Attempts to carry out single-crystal optical measurements using spindle-stage techniques (Medenbach 1985), in order to estimate H_2O/CO_2 ratios in the cordierite channels, proved unsuccessful, mainly due to the many inclusions and the development of undulatory extinction and subgrains in the cordierite porphyroblasts.

14.4 Petrography and Aureole Zonation

The salient petrographic features of the metamorphosed pelitic rocks within the Ballachulish aureole have been described by Pattison (1985) and Pattison and Harte (1985; see also Chap. 8, this Vol.). The zone terminology used here follows that defined by these authors. In the present contribution, emphasis is placed upon the microscopic appearance of the cordierite found from the peripheral zone of spotted phyllites to the hornfels of the inner aureole.

According to Pattison and Harte (1985 and Chap. 8, this Vol.) there are essentially two cordierite-forming reactions in the aureole. At lower grades, cordierite grows at the expense of muscovite and chlorite (Zone II), while at higher grades, muscovite and biotite are the reactants (Zone IVb). Both reactions are trivariant (Pattison 1987), so that the production of cordierite is not abrupt along specific isograds, but continues towards higher grades in Zones II and IVb.

The first cordierites found in spotted phyllites along the Zone I/II boundary in the outer aureole are isolated rounded or ellipsoidal poikiloblasts up to 2 mm in diameter, containing a multitude of minute matrix inclusions. They are generally elongated parallel to the old regional schistosity, but show no twinning. In the metapelites closer to the contact, these porphyroblasts increase in number and gradually coalesce, thus leading to poikilitic, interlocking cordierite grains in the dense hornfels of the inner aureole. These are also untwinned.

Only in two exceptional cases (samples 13906 and 13944) were typical cordierite trillings (e.g., Venkatesh 1954; Schreyer and Yoder 1964; Putnis and Holland 1986) with circular to hexagonal cross sections found. In sample 13944, these twinned crystals lie within a fabric of almost contiguous, untwinned, irregularly shaped cordierite porphyroblasts. To all appearances, the twinned crystals belong to an older generation of cordierite that has been overgrown in amoeboid fashion by the untwinned cordierite representing the bulk of the cordierite volume in the rock. The trillings do not exhibit the enrichment of matrix minerals along the sector boundaries observed by Kitamura and Yamada (1987, Fig. 2) for growth twins ('a-sectors'). Therefore these Ballachulish cordierite trillings could well be transformation twins formed during Al,Si ordering of initially hexagonal cordierite. A similar history of transformation from initially disordered cordierite appears likely for the trillings of sample 13906. However, this rock represents one of the subordinate graphitic interbeds found in the rocks of the Creran Succession. As Pattison and Harte (1985, and Chap. 8, this Vol.) have pointed out, the metamorphic histories of graphitic and non-graphitic pelites exhibit distinct differences, so that caution should be exercised with direct comparisons between the two lithological types.

In rocks of the inner aureole a third type of cordierite appears, together with the usual poikiloblasts, which contains much fewer inclusions, shows clearly developed prism faces and is characterized by strongly undulatory extinction and complicated subgrain formation. The latter is suggestive of a relatively high defect density acquired during rapid growth. Significantly, these very high-grade cordierites again do not exhibit any kind of twinning.

It is clear from the very nature of the progressive metamorphism in the Ballachulish contact aureole that the rocks of the innermost aureole must have experienced the most complicated history of mineral reactions, but it was found impossible to reconstruct all the stages of cordierite growth in these rocks on textural grounds. Any cordierites generated during the lower-grade reaction comparable to Zone II should have endured the longest periods of heating and probably grain growth. They may even have fully recrystallized. Certainly no cordierite trillings, if ever present during earlier stages of contact metamorphism, were found in these highest-grade hornfels.

14.5 Observed Distortion Index Variations

The results of the X-ray power investigations are listed in Table 14.1 and shown in terms of regional distribution in Fig. 14.3. Without exception, all determinations point to low-cordierites according to the nomenclature of Schreyer and Schairer (1961). The range of observed values with $0.228° \leq \Delta \leq 0.295°$ is small but significant, when the maximum 2σ-uncertainties of approximately 0.02° are considered. Samples near the intrusive contact yield the highest distortion indices, as predicted by Schreyer (1966), but also the largest observed range of Δ at a given distance from the contact. This fact is best appreciated when the distortion indices are plotted as function of distance to the contact, as shown in Fig. 14.4. Values near the contact are quite variable between ≈ 0.24° and ≈ 0.30°, whereas at distances >300 m from the intrusive, relatively constant distortion indices of ≈ 0.23°–0.25° are observed. In Fig. 14.4, no

Table 14.1. Summary of measured distortion indices[a]

Sample No.	Δ	Sample No.	Δ
13897	0.258(3)	13944	0.239(4)
13902	0.249(9)	13946	0.248(5)
13906	0.251(9)	13947	0.241(10)
13907	0.227(5)	13948	0.242(6)
13913a	0.239(5)	13952	0.237(5)
13918	0.249(5)	14623	0.248(5)
13921	0.247(8)	14624	0.235(9)
13922	0.249(3)	14625	0.251(11)
13928	0.245(6)	14627	0.244(7)
13929	0.250(3)	14629	0.234(7)
13931	0.245(4)	14636	0.252(5)
13934	0.261(6)	14640	0.268(11)
13935	0.265(4)	14643	0.265(4)
13937	0.255(3)	14644	0.295(8)
13938	0.257(3)	14645	0.257(2)
13940	0.253(3)	14646	0.250(9)
13942	0.252(3)	14647	0.267(4)
13943	0.253(4)	14648	0.262(5)

[a]Numbers in parentheses represent one standard deviation. For sample locations, see Fig. 14.2.

correction for the topography has been made because the dip of the intrusive contact is unknown, so that the actual values of the distances plotted may not reflect the true lengths of the perpendiculars to the contact surface. However, almost all the sampled outcrops lie on an erosion surface that dips with constant 20–25° towards the exposed contact, so that the *relative* positions of the samples along the abscissa in Fig. 14.4 should be unaffected. In absolute terms, the distance scale should not be greatly distorted, because the surface expression of the intrusive contact suggests that the contact surface should be very steep if not vertical.

Based on the results of this Δ-index survey, five samples were chosen for more detailed investigation. These are indicated by starred symbols in Figs. 14.2–14.4. Three are located in Zone V and exhibit comparatively high distortion indices. Two lie in Zone III near the cordierite-in isograd and have relatively low Δ's. Chemical analyses are given in the next section. From handpicked cordierite concentrates, lattice constants have been determined (Table 14.2), which allow calculation of the distortion index according to the relationship $\Delta = 1.094\,(a\text{-}b \cdot \sqrt{3})$ (e.g., Selkregg and Bloss 1980). The agreement with the distortion indices obtained by the method of Miyashiro is quite good, with the exception of sample 14644, which had yielded the highest value observed in this study. These values have been substantiated by repeated measurement. The reason for the discrepancy is unclear.

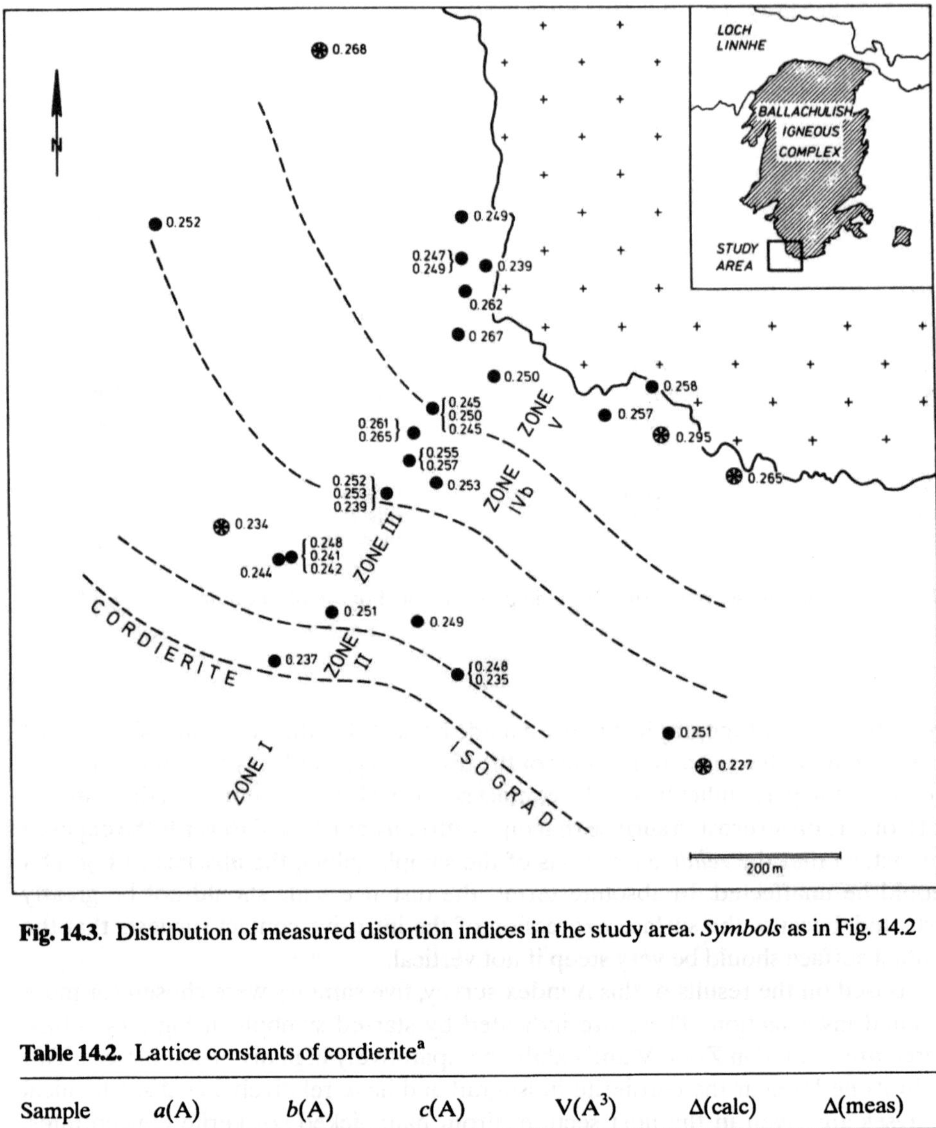

Fig. 14.3. Distribution of measured distortion indices in the study area. *Symbols* as in Fig. 14.2

Table 14.2. Lattice constants of cordierite[a]

Sample	*a*(A)	*b*(A)	*c*(A)	V(A^3)	Δ(calc)	Δ(meas)
13907	17.108(3)	9.759(3)	9.336(2)	1558.7(4)	0.224(8)°	0.227(5)°
14629	17.159(2)	9.772(1)	9.318(1)	1562.4(2)	0.255(4)°	0.234(7)°
14640	17.176(1)	9.7737(5)	9.3119(5)	1563.3(1)	0.271(2)°	0.268(11)°
14643	17.164(3)	9.768(2)	9.305(2)	1560.1(3)	0.268(7)°	0.265(4)°
14644	17.171(1)	9.775(1)	9.309(1)	1562.4(2)	0.263(3)°	0.295(8)°

[a]Distortion indices calculated according to $\Delta = 1.094\ (a\text{-}b \bullet \sqrt{3})$, compared to directly measured distortion indices from Table 14.1. Numbers in parentheses are the standard deviations (1σ) of the refinement in terms of the last digit(s) quoted. Space group *Cccm*.

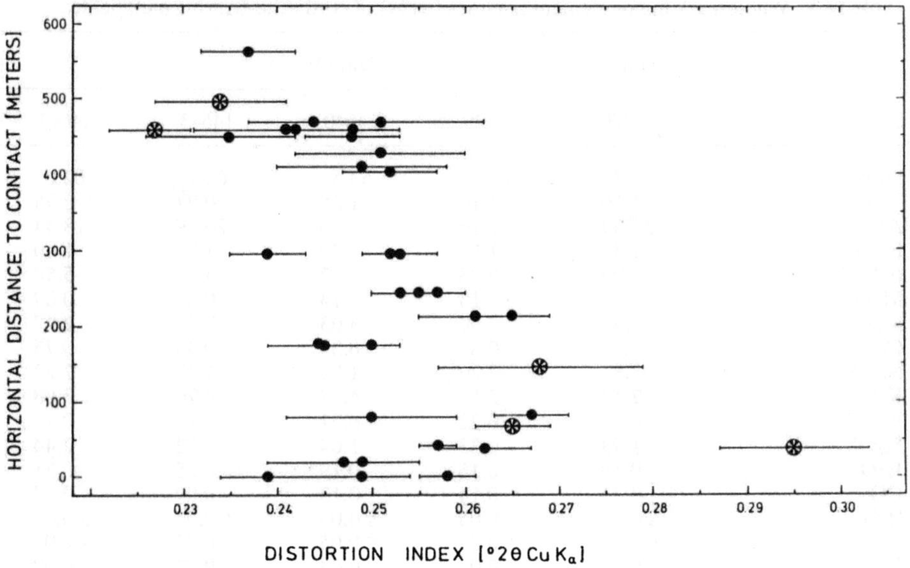

Fig. 14.4. Variation of cordierite distortion indices as a function of horizontal distance to contact with the pluton. *Error bars* represent 1σ uncertainty. *Symbols* as in Fig. 14.2

14.6 Cordierite and Whole-Rock Chemistry

Although chemical data on the metamorphic rocks and rock-forming minerals of the Ballachulish aureole are presented in detail in Chapters 8–10, this Vol. (see also Pattison 1987), whole-rock data and microprobe analyses of cordierite were obtained for the five critical samples indicated in Figs. 14.2–14.4 to allow direct correlation with the distortion indices measured on concentrates from the same rock.

The major and minor element contents (cf. Tables 14.3, 14.4) of the five whole rock samples are comparable in all respects except for Na_2O, K_2O, Rb and Ba. The three samples nearest the intrusive contact, which also yield cordierite with the highest distortion indices, clearly have higher K/(K + Na) ratios, higher Rb contents, and in part also higher Ba contents. The bulk $Mg/(Mg + Fe^{2+})$ ratios, which Pattison and Harte (1985, and Chap. 8, this Vol.) and Pattison (1987) have shown to be of critical importance in the development of assemblage zonations in the aureole, are extremely uniform and are close to the average of 0.38 (15 samples, range 0.28–0.47) cited by Pattison (1987) for non-graphitic rocks of the Leven Schists or Creran Succession as a whole. The Be contents are uniformly low. Because of the high modal abundances of cordierite in these rocks, it is unlikely that this mineral will contain significant proportions of beryllium. The oxidation state is low and comparable in all samples.

The compositions of cordierite from the Zone V samples are strikingly uniform both within samples and between localities (Table 14.5). X^{Mg} varies from 0.46 to 0.50, exactly the range determined by Pattison (1987; Pattison and Harte, Chap. 8, this Vol.) for cordierites from quartz-bearing metapelites from these zones. The measured

Table 14.3. Whole-rock major element analyses (wt%) of cordierite-bearing metapelites

	Zone III		Zone V		
Sample	13907	14629	14640	14643	14644
SiO_2	62.4	62.0	53.5	60.7	62.0
TiO_2	1.16	1.10	1.25	0.97	0.95
Al_2O_3	17.39	17.16	24.20	20.19	18.44
Fe_2O_3	1.32	1.64	1.30	1.21	1.40
FeO	5.90	6.85	7.27	4.85	5.54
MnO	0.12	0.13	0.14	0.07	0.09
MgO	2.52	2.90	3.03	2.27	2.27
CaO	0.82	0.85	0.39	0.34	0.75
Na_2O	3.83	2.20	1.85	1.77	2.22
K_2O	2.17	2.12	4.42	4.76	3.08
P_2O_5	0.22	0.22	0.21	0.16	0.22
H_2O^+	1.99	1.83	1.64	1.82	2.44
H_2O^-	0.19	0.16	0.19	0.25	0.21
CO_2	0.23	0.14	0.13	0.09	0.17
B_2O_3	≤ 0.05	≤ 0.05	≤ 0.05	≤ 0.05	≤ 0.05
F	≤ 0.05	≤ 0.05	≤ 0.05	≤ 0.05	≤ 0.05
S	0.04	0.07	0.04	0.12	0.09
Cl	0.01	0.01	0.01	0.01	0.01
Σ	100.41	99.48	99.67	99.68	99.98
$Mg/(Mg + Fe^{2+})$	0.43	0.43	0.43	0.45	0.42
K/(K + Na)	0.27	0.39	0.61	0.64	0.48
$\frac{200\ Fe_2O_3}{(2Fe_2O_3 + FeO)}$	16.8	17.7	13.9	18.3	18.5

Table 14.4. Whole-rock trace element analyses (ppm) of cordierite-bearing metapelites

	Zone III		Zone V		
Sample	13907	14629	14640	14643	14644
Li	79	95	86	68	59
V	88	83	100	84	81
Cr	74	76	95	84	75
Co	17	23	19	12	15
Ni	46	59	41	47	55
Cu	10	32	4	17	21
Zn	133	161	184	120	129
Rb	98	99	186	171	135
Sr	230	147	169	158	153
Y	47	39	66	48	44
Zr	570	437	689	493	318
Nb	33	25	81	22	22
Ba	582	404	897	801	506
Be	<2	<2	4	<2	<2

Table 14.5. Representative electron microprobe analyses of cordierite in pelitic hornfels

	14640		14643		14644	
	1.	2.	1.	2.	1.	2.
SiO_2	46.89	46.64	46.62	46.94	47.00	46.93
Al_2O_3	32.17	32.37	32.34	32.22	32.03	32.27
FeO	12.24	11.70	11.84	11.14	12.08	12.04
MnO	0.33	0.32	0.13	0.22	0.27	0.25
MgO	5.75	5.98	6.09	6.30	5.88	5.84
CaO	—	—	—	—	—	—
Na_2O	0.21	0.22	0.21	0.22	0.22	0.18
K_2O	—	—	—	—	—	—
	97.59	97.23	97.25	97.04	97.48	97.51
Cation proportions on the basis of 18 oxygens						
Si	4.97	4.95	4.95	4.97	4.98	4.97
Al	4.02	4.05	4.04	4.02	4.00	4.02
Fe	1.08	1.04	1.05	0.99	1.07	1.07
Mn	0.03	0.03	0.01	0.02	0.02	0.02
Mg	0.91	0.95	0.96	0.99	0.93	0.93
	2.02	2.02	2.02	2.00	2.02	2.02
Na	0.04	0.04	0.04	0.02	0.05	0.04
Mg/(Mg + Fe)	0.46	0.48	0.48	0.50	0.47	0.47

values of Na are low throughout. These are generally linked with a comparable small deficiency of Si and an excess of Al, so that Na is probably incorporated into the cordierite structural channels according to the substitution $NaAl\square_{-1}Si_{-1}$. Other potential cordierite channel constituents such as K and Ca remain below the detection limit. The oxide sums vary between 97.0 and 97.7 wt% and can be taken to indicate high and relatively uniform contents of volatiles in the channels.

The fine-grained aspect of the rock and the strongly poikiloblastic nature of the cordierites in the spotted phyllites of Zone III unfortunately precluded accurate cordierite analyses. Pattison (1987) indicated similar difficulties, but presented two analyses of Zone III cordierite. The X^{Mg} values of 0.46 and 0.49 are indistinguishable from the Zone V results.

14.7 Discussion

The Δ-index survey of cordierite in the Ballachulish thermal aureole has shown that there is a recognizable, but only minor trend in the Δ values as a function of distance to the intrusive contact (Fig. 14.4). All values are indicative of low-cordierite exclusively. A comparison with the results of Harwood and Larson (1969) in Fig. 14.5 shows that the Ballachulish trend is strikingly similar in both shape and range of values observed to the trend determined around the Cupsuptic pluton in Maine (USA). Although the Cupsuptic aureole is approximately 30% wider, both aureoles indicate a high Δ-index variability near the contact, as well as generally lower values with less scatter in the outer parts of the aureole. The results thus appear generally applicable to intrusive bodies of this size.

During the course of the Ballachulish cooperative project, experimental data on the kinetics of the hexagonal-to-orthorhombic transformation in Mg-cordierite

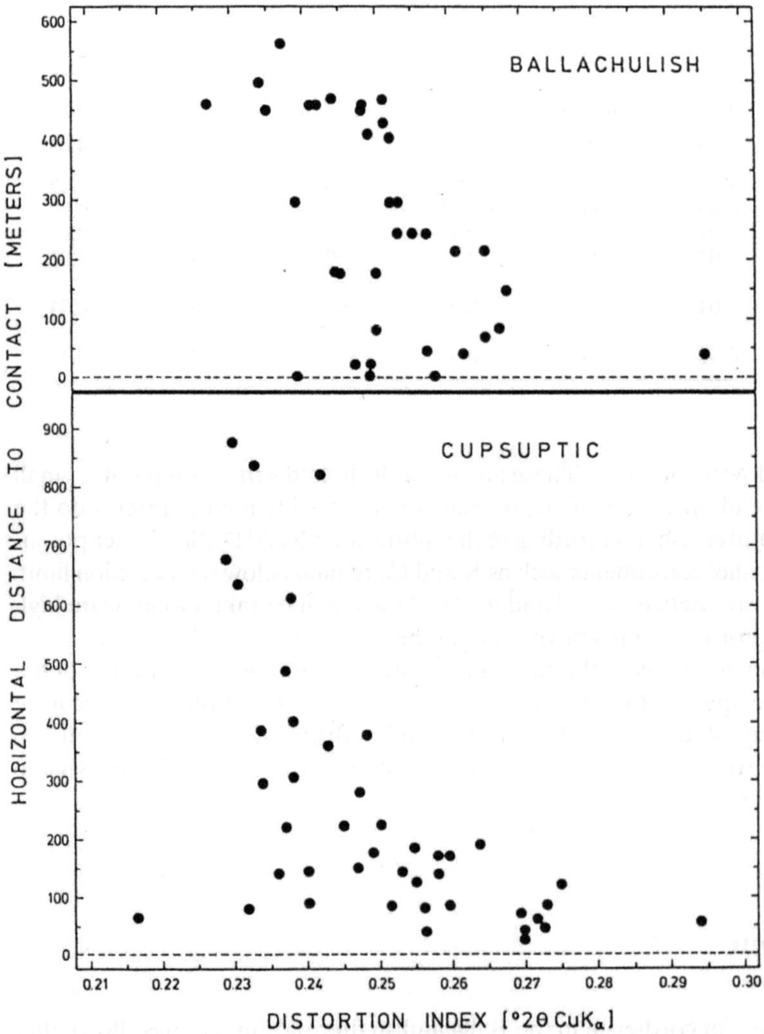

Fig. 14.5. Comparison of the observed trends in Δ as a function of distance to intrusive contact between the present study and the Cupsuptic aureole (Harwood and Larson 1969)

have become available (Putnis and Bish 1983; Langer and Stürzebecher 1983), which, in conjunction with the thermal models for the Ballachulish aureole developed by Buntebarth (Chap. 18, this Vol.), allow theoretical prediction of the possible structural states to be encountered in the aureole. All pertinent time-temperature-transformation (ttt) and heating/cooling curves are shown in Fig. 14.6. The most complete ttt data set (Putnis and Bish 1983) is available for pure, end-member Mg-cordierite under anhydrous conditions only. However, the point data of Langer and Stürzebecher (1983) at $P(H_2O) = 0.5$ GPa show an important and dramatic shift of the ttt-curves to lower temperatures when H_2O is available. Experimental runs by these authors under anhydrous conditions and 0.5 GPa total pressure show that the difference in pressure in itself is not responsible for the shift. The most important additional compositional difference with respect to the natural cordierite of the Ballachulish aureole is the FeO-content of the latter. The data of Kitamura and Hiroi (1982) indicate that the ttt-curves for the hexagonal-to-orthorhombic transformation should again be shifted to lower temperatures with the introduction of Fe^{2+}, but the contention of these authors, that disordered high cordierite (indialite) of intermediate Mg/Fe^{2+} ratios is stable to below 700 °C appears unrealistic. No experimental evidence for such low transformation temperatures was found in a recent study of the system Mg-cordierite-Fe^{2+}-cordierite (Boberski and Schreyer, 1990).

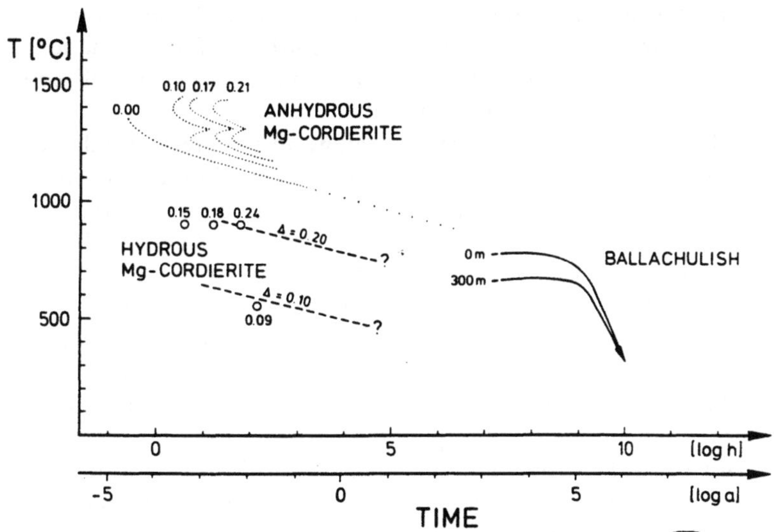

Fig. 14.6. Comparison of experimental time-temperature-transformation data for the hexagonal-to-orthorhombic cordierite transformation (in terms of the distortion index) with calculated heating/cooling paths at 0 and 300 m from the intrusive contact in the Ballachulish thermal aureole. Data for anhydrous Mg-cordierite at atmospheric pressure from Putnis and Bish (1983); data on hydrous Mg-cordierite at 0.5 GPa $P(H_2O)$ from Langer and Stürzebecher (1983). The heating/cooling paths have been calculated from the data of Buntebarth (Chap. 18, this Vol.), assuming that all cordierite crystallizes as soon as the first cordierite isograd is crossed (see text for further details)

From the heating/cooling curves of the aureole rocks developed by Buntebarth (cf. Chap. 18, this Vol.), the time-temperature path for any cordierite crystal may be ascertained (Fig. 14.6), once the time of nucleation has been established. The thermal model chosen is that for a cylindrical intrusion with conductivities applicable to Leven Schist (assumed to equal those of the Creran Succession) parallel to schistosity, and a 3 wt%-H_2O loss due to dehydration reactions during contact metamorphism. Furthermore, it has been assumed that all cordierite nucleates and grows metastably as the disordered polymorph, according to the reaction from muscovite and chlorite, as represented by the Zone II assemblage. The temperature at which this metastable reaction begins is assumed ≈ 100 °C higher (Putnis and Holland 1986) than the corresponding reaction involving low-cordierite, as derived by Pattison and Harte (cf. Chap. 8, this Vol.) and has been set at 650 °C. It is immediately clear from the Buntebarth data that the assumption of high-cordierite nucleation must lead to a much narrower cordierite-bearing aureole than that observed in the field. Maximum temperatures of 650 °C or more are reached only within ≈ 300 m of the contact. Due to the logarithmic nature of the plot in Fig. 14.6, all cordierite time-temperature curves converge to an essentially similar cooling path. It is therefore immaterial for a comparison with the ttt-curves whether cordierite nucleates immediately and begins to grow at 650 °C, or at any time until peak temperatures are reached at any point in the aureole.

Although the experimental ttt-data for cordierite ordering in the presence of H_2O and the thermal models of the natural rocks are orders of magnitude apart in time, it seems inconceivable on the basis of Fig. 14.6 that any disordered high-cordierite could have survived during the course of contact metamorphism. Available experimental and model data thus predict that all cordierite in the Ballachulish aureole should now be in the fully-ordered, low-cordierite form. Variations in the Δ-index in the aureole, and by induction in the Cupsuptic aureole as well, should not be due to variations in Al,Si ordering states.

The petrographic data at hand allow an even more fundamental conclusion to be reached, namely that most of the cordierite in the Ballachulish aureole never went through the hexagonal, disordered stage at all. There can be no doubt that, with the two exceptions mentioned in Section 14.4 above, all cordierites along the investigated profile lack any form of obvious twinning. We interpret this fact as evidence that cordierite has crystallized *ab initio* in its orthorhombic, low-temperature form. This interpretation is at variance with earlier experience and predictions (e.g., Schreyer and Schairer 1961; Schreyer 1966), namely that cordierite initially tends to nucleate and grow as metastable, hexagonal high-cordierite (indialite), and subsequently transforms to orthorhombic low cordierite as a result of Al,Si ordering. The lack of cordierite trillings in the Ballachulish aureole is typical of practically all areas of regional metamorphism as well. We suggest that both observations can be traced back to a common cause. For regional metamorphic rocks, Kitamura and Yamada (1987) had explained the lack of trillings by suggesting that "indialite is unstable under regional metamorphic conditions". This statement contrasts with the earlier conclusion of Kitamura and Hiroi (1982), that the hexagonal-to-orthorhombic transformation of cordierites with intermediate Mg/Fe^{2+} ratios lies below 700 °C. Obviously many regional metamorphic cordierites formed above this temperature. Although this contradiction appears clarified by the recent study of Boberski and Schreyer (1990), who found no experimental evidence for such low transition temperatures in the system Mg-cordierite — Fe-cordierite, we suggest instead that the lack of cordierite trillings in both the Ballachulish aureole and regional metamorphic rocks is due to a relatively low rate of heating, thus precluding significant overstepping of stable reactions. The metastable nucleation and growth of hexagonal cordierite, with subsequent transformation and twinning to orthorhombic cordierite, should then be confined to environments with a critically high rate of heating, such as in some narrow, buchite-bearing contact aureoles around very hot intrusives, or in rapidly chilled igneous rocks themselves.

As discussed above, the calculations of Putnis and Holland (1986) have shown that for cordierite-producing reactions involving metastable high-cordierite, an overstepping of ≈ 100 °C is required with respect to the corresponding reaction involving stable low-cordierite. This would entail dramatic shifts in the relative positions of the metamorphic zones mapped around the Ballachulish pluton. For instance, if the requisite overstepping required for the formation of high-cordierite from muscovite and chlorite has occurred, then the Zone I/II and the Zone IVb/V boundaries should be practically superposed. Pattison (1989, and Chap. 16, this Vol.) report no significant shifts of this type in the entire Ballachulish aureole and argue against any significant overstepping. The logical consequence of this line of reasoning is that no overstepping should have occurred during regional metamorphism either.

The rare occurrence of cordierite trillings in two samples must mean that only locally were conditions conducive to nucleation and initial growth of high-cordierite realized. We can at present only speculate on the possible causes. There are a number of small intrusive stocks and bodies of more basic composition in the Ballachulish area. Pattison (1989, and Chap. 16, this Vol.) report some evidence of disequilibrium and overstepping in the narrow aureole surrounding such a body approximately 4 km east of the study area. The textures of sample 13944 could be explained by the formation of initially hexagonal cordierite in a cryptic aureole around an early small stock not now exposed. During subsequent thermal metamorphism by the Ballachulish pluton, these crystals were overgrown by orthorhombic cordierite. Alternatively, it is not inconceivable that local kinetic barriers could lead to local overstepping during the main thermal event. However, the formation of such hexagonal cordierites would then have to *post-date* that of the bulk of the orthorhombic cordierite.

If variations in Al,Si ordering state can be ruled out as a cause of the Δ-index trend in Fig. 14.4, then only variations in cordierite composition remain as a viable alternative explanation. However, the available data on cordierite chemistry offer no ready solution to the problem. Although cordierite should theoretically become somewhat more iron-rich toward the contact (cf. Fig. 8.7, this Vol.), and this should lead to somewhat higher Δ-indices (Selkregg and Bloss 1980), actual microprobe analyses document no palpable changes in Mg/Fe ratio between Zone II and Zone V (cf. Fig. 8.22, this Vol.). There remains the possibility of systematic variations in the volatile content of the cordierite channels, but Hoernes et al. (Chap. 17, this Vol.) have analyzed oxygen isotope ratios in samples from our profile and have found that only Zone V rocks show some evidence of isotopic equilibration with magmatic fluids. Harte et al. (Chap. 9, this Vol.) document migmatization processes and melt mobility in the metapelites as far as 100 m from the contact. The striking differences in K/(K + Na) bulk rock ratios for schists from Zone III and Zone V (Table 14.3) must surely be an expression of such processes. Although the cordierite channels may hold the key to an explanation for the Δ-index variations observed, we have found no ready analytical technique that would allow us to document such subtle differences with sufficient accuracy.

15 Detrital Quartz and K-Feldspar in Quartzites as Indicators of Oxygen Isotope Exchange Kinetics

S. Hoernes and G. Voll

15.1 Introduction

Oxygen isotope fractionation between coexisting minerals may indicate equilibrium or disequilibrium, provided the temperature of formation is known. Depending on the time of blocking of specific $^{18}O/^{16}O$ ratios, the fractionations contain information on:

1. Pre-peak-metamorphic conditions as assumed by Wenner and Taylor (1973) for serpentine-magnetite fractionations or by Hoernes and Hoffer (1979) for metamorphic parageneses in a more general sense. In this case oxygen isotope temperatures represent formation temperatures.
2. Peak-metamorphic conditions; equilibrium is thought to be reached by exchange reactions via a fluid phase, by recrystallization (Hoernes and Friedrichsen 1978), or by mineral reactions.
3. Post-peak-metamorphic conditions; based on different rates of oxygen diffusion in different minerals, it is thought that $^{18}O/^{16}O$ ratios are fixed at different temperatures during cooling, as demonstrated by Deines (1977) for specific minerals and recently discussed by Giletti (1986). The postmagmatic exchange reactions discussed in Hoernes et al. (Chap. 17, this Vol.) belong in this category. For further discussion see Hoefs (1987), O'Neil (1986), Valley (1986).

The oxygen isotope fractionation between quartz and K-feldspar, though not a precise geothermometer because of the similar slope of the respective mineral-water fractionation curves, may be used to obtain some information concerning the state of isotopic equilibrium.

At temperatures close to the solidus of water-saturated granitic systems (650–700 °C at 0.3 GPa) quartz-K-feldspar equilibrium fractionations will be between 1.0 and 0.9‰ (Richter and Hoernes 1988). Usually fractionations of 1 to 1.5‰ are taken as 'normally magmatic' (c.f. Taylor 1977). Deviations from this fractionation indicate an isotopic disequilibrium, in hydrothermally altered rock sequences frequently observed as 'reversals' (Taylor 1977).

Diffusion coefficients for oxygen have been determined in hydrothermal experiments using ^{18}O enriched water. Diffusion profiles have been measured using an ion microprobe (Giletti and Yund 1984). For a compilation of diffusion data see Freer and Dennis (1982) and Cole and Ohmoto (1986).

G. Voll, J. Töpel, D.R.M. Pattison, F. Seifert
Equilibrium and Kinetics
in Contact Metamorphism

Based on these diffusion data and some additional data concerning particle sizes, a mineral specific geometrical factor and an arbitrarily chosen cooling rate Giletti (1986) calculated closure temperatures (T_c) for different minerals. The equation he used is that of Dodson (1973). Combining the T_c values with the known whole-rock isotopic composition and the modal composition, he was able to estimate the cooling rate of the rock. It has been shown by Hoernes et al. (Chap. 17, this Vol.) that, provided the cooling rate is known, this method can yield additional information on the mechanism and the kinetics of oxygen isotope exchange. It has been demonstrated that only surface reactions are fast enough to achieve a near-isotopic equilibrium at the cooling rates of the Ballachulish complex (Hoernes et al., Chap. 17, this Vol).

15.2 Geological Situation and Sample Description

The Appin Quartzite forms a significant part of the wall rocks W and E of the Ballachulish Igneous Complex. Its clastic quartz- and K-feldspar grains show bimodal grain size: 0.3-mm-sized grains were transported in suspension, 5–50-mm-sized pebbles by saltation.

Prior to contact metamorphism, the quartzite suffered moderate Caledonian deformation and regional metamorphism at greenshist facies temperatures. Quartz grains were partly recrystallized, and K-feldspars changed to maximum microcline, including adularia rims acquired during diagenesis. Within shear zones, which thin the limbs of first folds, the quartzites are very strongly deformed. The quartz is recrystallized completely, and K-feldspar converted to chess board albite.

Approximately 5–10% of the clastic quartz grains contain fine, oriented rutile needles, acquired in granulite-facies parent rocks. During recrystallization these needles are again incorporated in quartz.

Heating by the Ballachulish intrusion causes K-feldspar to disorder and to become sanidine (Kroll et al., Chap. 13, this Vol.). While outside the contact aureole KNa-feldspars contain variable, but generally low amounts of unmixed albite, disordering causes them to become homogeneous and generally enriched in sodium. Spindle-, film-, band- and patch-perthites, sedimented with different types and amounts of unmixed albite, are all remixed and homogenized. After this contact homogenization, they all contain equal amounts of admixed albite, which during cooling is unmixed again, forming film-, band-, and patch-perthites.

Relics of clastic quartz grains and recrystallized quartz grains left behind from Caledonian deformation and metamorphism were coarsened within the contact aureole and increasingly, towards the contact (Buntebarth and Voll, Chap. 12, this Vol.).

This study tries to correlate the state of Al, Si order in K-feldspar, and the degree of quartz recrystallization and coarsening with changes or preservation of oxygen isotopes. Comparison with data on oxygen diffusion should yield information about mechanism and kinetics of oxygen isotope exchange.

15.3 Experimental Procedures

During initial stages of this investigation, quartz and K-feldspar were separated in the usual manner: samples of appropriate size were crushed, and sieve fractions prepared. From the fraction, > 500 μm quartz and K-feldspar was separated by hand picking. These integral separates do not allow conclusions about the degree of equilibration within one mineral population. We therefore decided to change the technique of separation: from slices 1–3 mm thick, single grains were broken out. These grains were either non-recrystallized pebbles or aggregates of recrystallized grains, each derived from a single pebble. It is highly unlikely that all quartz pebbles sedimented at the same isotopic composition, since they were derived from totally different source rocks. The predominance of relatively heavy $\delta^{18}O$ values suggests a strong component of metasedimentary rocks, whereas quartz values close to 11‰ could be related to igneous sources. Without homogenization this should yield an according spread in the isotopic composition. Uniform composition, on the other hand, indicates homogenization by processes after deposition.

O-isotope analyses has been performed by extraction of O_2 using F_2 as reagents (purified according to Asprey 1976). The liberated O_2 was converted to CO_2 and isotope ratios measured using a VG SIRA-9 gas mass spectrometer.

15.4 Results

15.4.1 Integrated Separates

The results of oxygen isotope analyses of mineral separates prepared in the classical manner are summarized in Table 15.1. Equilibrium fractionations between quartz and K-feldspar for the respective metamorphic temperatures are given for comparison (fractionations according to Richter and Hoernes 1988; metamorphic temperatures based on the work of Pattison and Harte, Chap. 8, this Vol.). It is obvious that none of the measured fractionations corresponds to the equilibrium value. The lowest measured fractionations correspond to two samples from Salachan Glen which were affected by only the Caledonian regional metamorphism. These small fractionations demonstrate that the coarse (< 0.5 mm) detrital quartz and K-feldspar grains did not equilibrate during regional metamorphism, although all K-feldspar was ordered to maximum microcline. The large and unsystematic scatter of O-isotope fractionations obtained by this preparation technique was one reason to change the separation procedure.

Table 15.1. Oxygen isotopic composition of integral separates of quartz and K-feldspar from Appin quartzites

Sample No.	Locality Distance from cont. Metam. temp	Phase	$\delta^{18}O_{SMOW}$	Measured fractionation	Equilibrium[a] fractionation
B 20	Salachan Glen 3 km $T_{reg.\ met.}$ 480 °C	Qtz Kfsp	+12.8 +12.3	Qtz-Kfsp 0.5	1.65
B 22	Salachan Glen 3 km $T_{reg.\ met.}$ 480 °C	Qtz Kfsp	+11.9 +11.0	Qtz-Kfsp 0.9	1.65
B 2	Loch Leven beach Hotel Creag Dhu 350–500m $T_{cont.\ met.}$ 550 °C	Qtz (fine) Qtz (coarse) Kfsp	+12.5 + 9.9 +11.6	Qtz-Kfsp 0.9 Qtz-Kfsp rev.	1.35
B 9	Quarry (diss.) above Kentallen 150 m $T_{cont.\ met.}$ 670 °	Qtz Kfsp	+11.5 + 9.1	Qtz-Kfsp 2.4	1.0
B 41	Above Kentallen 100 m $T_{cont.\ met.}$ 700 °C	Qtz Kfsp	+12.6 +11.4	Qtz-Kfsp 1.2	0.9
B 44	Above Kentallen 25 m $T_{cont.\ met.}$ 720 °C	Qtz Kfsp	+13.3 +11.7	Qtz-Kfsp 1.6	0.85

[a]Calculated using fractionation factors given by Richter and Hoernes (1988).

15.4.2 Single Grain Separates

The isotope data obtained from single grain analyses are presented in Table 15.2. The samples are listed in the order of their distance from the contact. Where duplicate analyses were performed, the 1σ error is given; in some cases, only one analysis could be obtained due to the small mass of the single grains. The variation of the single grain analyses is illustrated in two histograms. As can be seen in Fig. 15.1, quartz-$\delta^{18}O$ values range from +11 to +14‰ with a bimodal distribution: a clear maximum at +12 ± 0.2‰ and a second population with $\delta^{18}O$ values heavier than 12.6‰. It is important to note that there is no correlation of oxygen isotope composition with distance from the contact. In some samples, the coarse quartz fraction consists mainly of the second, isotopically heavy population, i.e., V 795 and V 790; others display only the lighter group as V 640, and others again show a mixed population as V 802. The rutile-bearing quartz pebbles, most likely derived from granulite facies terranes, mainly belong to the population with low $\delta^{18}O$ values.

The $\delta^{18}O$ variation of the K-feldspar grains is illustrated in Fig. 15.1. Again a bimodal distribution of $\delta^{18}O$ values is evident, one group with values less than +11‰, the other group with values heavier than +11.4‰. The isotopically light feldspar-grains are exclusively from samples collected at a contact-distance > 1000 m. Their Na content, estimated from the amount of unmixed albite, is less than in the samples collected closer to the contact.

Table 15.2. Oxygen isotopic composition of coarse clastic grains from Appin quartzites (eastern part of the aureole)

Sample	Distance from contact in m	Phase description	$\delta^{18}O_{SMOW}$	Standard deviation 1 σ
V 716	1209	Quartz	+12.0	0.05
		Quartz with rutile	+11.33	0.05
		Quartz	+12.0	0.05
		Quartz	+11.62	0.05
		Quartz with rutile	+11.80	0.02
		Quartz with rutile	+12.11	0.02
		Quartz with rutile	+11.91	
		Quartz with rutile	+11.51	
		Quartz with rutile	+11.71	0.05
		Quartz	+11.45	
V 720	1204	Quartz with rutile	+11.22	
		K-fsp	+10.26	0.02
		K-fsp	+ 9.9	0.1
V 723	1070	Quartz	+11.8	0.2
		Quartz with rutile	+11.78	0.05
		Quartz with rutile	+11.51	0.05
		Quartz	+11.80	0.02
		Microcline	+10.2	0.2
		K-fsp	+10.4	0.15
		K-fsp	+10.28	0.05
		K-fsp	+10.85	*
		K-fsp	+10.7	0.1
V 726	958	Quartz	+12.6	0.1
		Quartz	+13.95	0.1
		Quartz	+12.03	0.05
V 802	270	Quartz pebble milky-blue d = 4 mm	+12.7	0.2
		Quartz pebble glassy-grey d = 3 mm	+13.1	0.1
		Quartz pebble glassy, bluish	+12.8	0.1
		Quartz recrystallized and coarsened from matrix	+12.7	0.05
		Quartz pebble grey-bluish	+11.9	0.1
		Quartz recrystallized and coarsened from matrix	+12.5	0.15
		K-fsp, single crystal	+12.5	0.1
V 800	241	Quartz pebble grey d = 8 mm	+12.6	0.1
		Quartz pebble grey d = 8 mm	+12.3	0.1
		Quartz pebble with rutile	+12.7	0.05
		K,Na-fsp single crystal	+12.4	0.1
		K,Na-fsp single crystal	+11.9	0.1
		K,Na-fsp single crystal	+12.7	0.1
		K,Na-fsp single crystal	+12.1	
V 795	159	Quartz pebble glassy-grey	+13.2	0.1
		Quartz pebble slightly bluish	+13.2	
		Quartz pebble bluish	+13.0	0.05
		K,Na-fsp single crystal	+12.5	0.1
V 635	105	Quartz pebble recrystallized coarsened	+12.2	0.1
		K-fsp single crystal	+12.1	0.1
		K-fsp single crystal	+11.7	0.05

Table 15.2. (continued)

Sample	Distance from contact in m	Phase description	$\delta^{18}O_{SMOW}$	Standard deviation 1 σ
V743	67	Quarts recrystallized and coarsened from matrix	+12.9	0.2
		Quartz pebble recrystallized and coarsened	+12.5	0.1
		K,Na-fsp single crystal	+12.5	0.1
		K,Na-fsp single crystal	+12.2	0.1
		K,Na-fsp single crystal	+12.4	0.05
		K,Na-fsp single crystal	+12.3	0.1
V 640	40	Quartz recrytallized and coarsened from matrix	+12.0	0.02
		Quartz pebble recrystallized and coarsened	+12.3	*
		Quartz pebble recrystallized and coarsened	+12.3	0.1
		Quartz, recrystallized and coarsened from matrix	+12.2	0.1
		K,Na-fsp single crystal	+11.5	
		K,Na-fsp single crystal	+11.5	0.2
		K,Na-fsp single crystal	+11.8	0.05
		K,Na-fsp single crystal	+12.0	0.1
		K,Na-fsp single crystal	+11.8	0.1
V 790	24	Quartz recrystallized and coarsened from matrix	+13.2	0.1
		Quartz pebble recrystallized and coarsened	+13.2	0.05
		Quartz pebble recrystallized and coarsened	+13.2	0.1
		Quartz from discordant vein, recrystallized and coarsened	+13.2	0.05
		K,Na-fsp single crystal	+12.7	0.1
ED 6 quartzite		Quartz pebble	+11.75	
Damara Orogenetic Belt		Quartz pebble	+11.63	0.1
Namibia		Quartz pebble	+11.50	0.02
T_{met} 600 °C		Quartz pebble	+11.56	0.02
	Mean	Quartz	+11.61	0.11
		K-fsp single crystal	+ 9.9	0.1
		K-fsp single crystal	+10.1	0.05
		K-fsp single crystal	+ 9.96	0.05
		K-fsp single crystal	+10.0	0.1
	Mean	K-fsp	+ 9.99	0.08

For comparison, Table 15.2 includes data from a regionally metamorphosed quartzite of similar grain size where single grain separation and analysis were possible. The sample was collected from the Damara Orogen in Namibia. At this sample locality, regional metamorphism reached approximately 600 °C, as indicated from mineral parageneses in pelitic rocks (Hoffer 1983) and from O-isotope data (Hoernes and Hoffer 1979). Here, the spread in the $\delta^{18}O$ values of the single grains is less than in the Appin quartzites of the Ballachulish aureole.

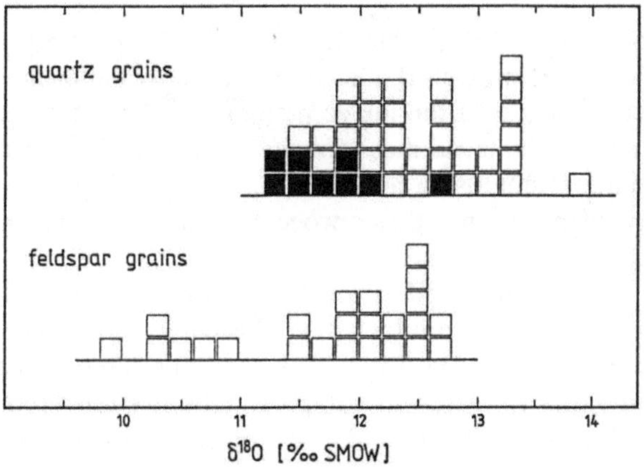

Fig. 15.1. Histograms of quartz- and K-feldspar single grain analyses: *black*, granulite facies-quartz pebbles containing needles of rutile. The low $\delta^{18}O$ population of K-feldspar grains has been obtained on samples collected from contact distances > 1000 m

15.5 Discussion and Conclusions

Before drawing conclusions about kinetics of the isotope exchange, we have to prove whether or not variation of isotopes is due to contact heating. The variation of quartz $\delta^{18}O$ values as illustrated above clearly contradicts contact induced homogenization by a pervasively streaming fluid phase. This can be proved by using a simple model calculation:

The oxygen isotopic composition of the magmatic water released during crystallization of the Ballachulish intrusive complex is close to +8‰ throughout the entire period of cooling (Hoernes et al., Chap. 17, this Vol.). The fractionation between quartz and water at 700 °C is 0.7‰ (Richter and Hoernes 1988). Thus, the effect of a significant contribution of water of this composition would be to lower the $\delta^{18}O$ values of quartz grains, at least in samples close to the contact (D < 100 m).

Based on O-isotope data integrated W/R ratios can be calculated. For the following model calculation the equation (20b) of Criss and Taylor (1986) has been used: to lower quartz from +13‰ to an assumed value of +12.5‰, taking +8‰ for the infiltrating magmatic water and the above-mentioned quartz-water fractionation, the W/R ratio in weight units is 0.07. At the other extreme, to lower it to 11‰, the W/R ratio would have to be 0.49. This ratio would be even larger, if the composition of the invading fluid was heavier than +8‰. At such high W/R ratios a homogenization of the quartz $\delta^{18}O$ values had to be expected.

This model calculation demonstrates that if any water-rich fluid derived from the pluton had invaded the quartzites at the contact, its amount cannot have been very large.

Even the recrystallized and coarsened matrix quartzes display a similar isotopic composition as the coarse quartz pebbles within the matrix. The grain size of the matrix quartzes is much smaller than that of the large-pebble single crystals. Therefore one would expect the fine matrix quartzes to exchange more rapidly if an externally derived fluid phase was present during recrystallization and coarsening (collective crystallization). The lack of equilibration between matrix quartz and K-feldspar suggests that even the amount of internally present fluids must have been very small, even though some amount of a fluid phase must have been present to enable volumetric mass transfer during collective crystallization of recrystallized quartz (Buntebarth and Voll, Chap. 12, this Vol.). It is a surprising result of this investigation, that these processes can take place without concomitant O-isotope equilibration.

$^{18}O/^{16}O$ data differ for K-feldspar single grains from close to the contact, compared to those from far away. This could suggest contact-induced rise of $\delta^{18}O$ close to the contact (Fig. 15.1). However, this is more likely to reflect different feldspar individua in different samples. This is supported by different $\delta^{18}O$ values for feldspars from B 20 and B 22, both 3 km from the contact. B 9 (150 m from the contact) and V 795 (159 m) also yield widely different $\delta^{18}O$ values, this time close to the contact. Although these B-labelled samples represent integrated separates, their values are comparable to the tendency of the single grain analyses, especially as the separation has been performed from the fraction > 500 μm where a few grains are sufficient for isotopic analysis. The variation of $\delta^{18}O$ values of single grains within one sample even close to the contact (V640–40 m) clearly demonstrates the lack of O-isotope homogenization.

The data discussed so far justify to neglect fluid-controlled exchange reactions like surface reactions or recrystallization as the mechanism of isotope exchange in these rocks. Therefore the only alternative mechanism is diffusion. We can try to answer the following questions: was the time-span at a specific elevated temperature long enough, and is oxygen diffusion fast enough to equilibrate quartz and K-feldspar?

For simplicity we restrict this discussion to the temperature of 700 °C. The first question can be answered based on the model calculations of Buntebarth (Chap. 18, this Vol.). In Fig. 15.2 the time-dependent temperature evolution of a typical Appin quartzite is shown for two distances. It can be seen that quartzites at a distance of ≥ 200 m stayed at least 27×10^3 years at ≥ 700 °C, but not more than 40×10^3 years for samples immediately adjacent to the contact.

For both minerals, O-diffusion data are available. For K-feldspar we can take the data of Yund and Anderson (1974) valid for microcline. The diffusion coefficient, D, at 700 °C is 6.2×10^{-13} cm^2/s. For quartz the data of Giletti and Yund (1984) for the same temperature yield $D = 9.078 \times 10^{-15}$ cm^2/s. (For a comprehensive compilation of O-diffusion data for rock-forming minerals see Freer 1981; and Cole and Ohmoto 1986.)

These data are valid for diffusional exchange of oxygen between the solid phase and an aqueous fluid at 0.1 GPa water pressure. Little is known on the amounts of water necessary to promote diffusion rates. At absolutely dry conditions, even in millions of years at this temperature, no measurable exchange would occur (Cole and Ohmoto 1986; Elphick et al. 1988 and references therein). The above discussion on the presence or absence of externally derived fluids does not exclude the presence of internally present, small amounts of water, possibly entrapped as fluid inclusions or adsorbed at grain boundaries.

Using a relation given by Freer (1981), i.e., $x^2 \approx Dt$, where x means the radius of a spherical mineral, we can approximate the time necessary to equilibrate a mineral

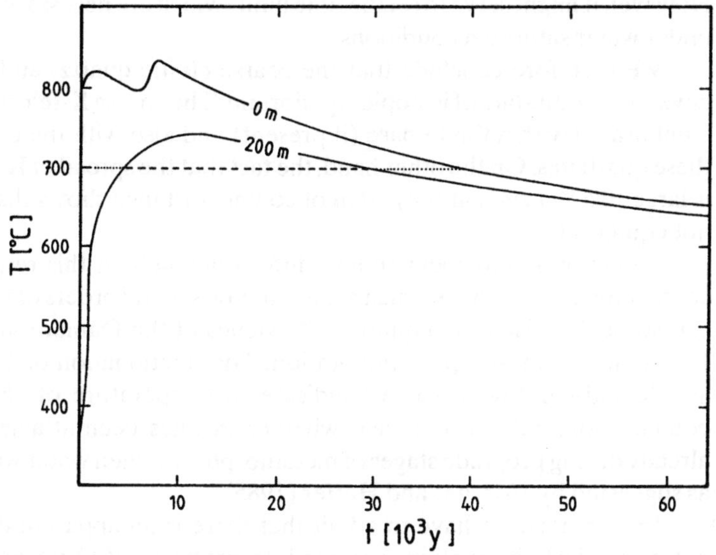

Fig. 15.2. Temperature-time diagram for quartzites immediately at the contact and in 200 m distance according to Buntebarth, pers. comm.; (cf. also Buntebarth, Chap. 18, this Vol.). Assumed conditions: initial temperature of the quartzite 350 °C; cylindrical model of intrusions

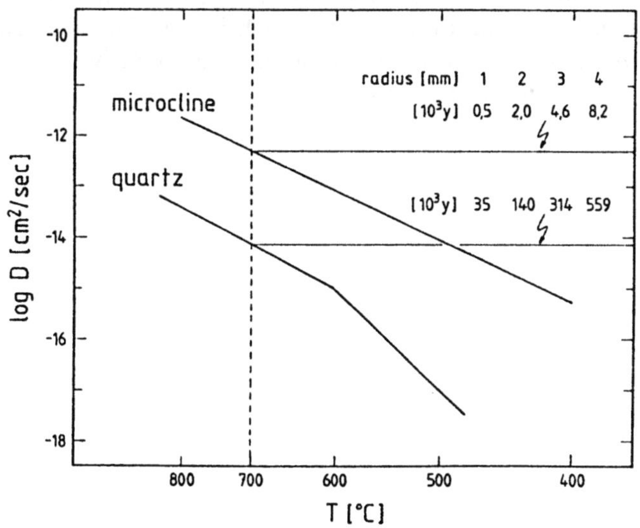

Fig. 15.3. Oxygen diffusion under wet conditions. Equilibration times for grains of different radius are calculated according to Freer (1981)

of a given radius with a fluid phase. Figure 15.3 summarizes the results of this calculation. It is obvious that sufficient time was available to equilibrate K-feldspar even of 4-mm-radius grains, but that the heating time was too short to equilibrate quartz except for grain sizes < 2 mm. This calculation assumes an infinite reservoir of fluids, which is not a realistic assumption for these quartzitzes as discussed above.

However, it might serve to set the time limits which are necessary to reach equilibrium under water saturated conditions.

We therefore conclude that the coarse clastic quartz- and K-feldspar pebbles never reached a state of isotopic equilibrium. The coarse K-feldspar grains could have equilibrated with a fluid phase (if present) and also with the fine grained matrix of these quartzites. On the other hand, the fact that the spread of K-feldspar $\delta^{18}O$ values is large (0.5 ‰) even at only 40 m of contact distance shows that these minerals did not equilibrate.

Metamorphosed rocks should differ regionally in this respect, because of the contrasting cooling times which are greater by several orders of magnitude. The small spread of K-feldspar and quartz-$\delta^{18}O$ values of the Damara sample listed in Table 15.2 indicates an isotopic equilibration. The fractionation of 1.6‰ between quartz and K-feldspar (mean values) indicates a temperature of about 490 °C for this equilibration, but it is not clear whether this has been attained during cooling or already during prograde stages of metamorphism, when water was available in excess as suggested by Hoernes and Hoffer (1985).

In summary, we have to admit that there is an apparent discrepancy between petrographical observations and the interpretation of O-isotope data. Thin section observations indicate a sodium input at some stage of contact heating. If this influx occurred via a water-rich fluid, this could not have happened at peak metamorphic conditions; otherwise we would observe granitic melting. The O-isotope data indicate essentially dry conditions at peak metamorphic conditions. Massive influx of an externally derived fluid at any time of the contact metamorphism should have an imprint on the isotope data which is not observed. The liberation of briny fluid inclusions during recrystallization and collective crystallization is a possible, yet hypothetical explanation.

Part IV
Interactions Between the Intrusion and the Contact Aureole

16 P-T-a(H_2O) Conditions in the Thermal Aureole

D.R.M. Pattison

16.1 Introduction

In this chapter estimates are made of the pressure of emplacement of the Ballachulish Igneous Complex and the range of maximum temperatures in the surrounding thermal aureole. Using these P-T estimates, quantitative estimates of a(H_2O) are made for pelites at different grades in the aureole. Activity of H_2O is shown to vary between pelites above and below the onset of partial melting, and between graphitic and non-graphitic pelites at the same grade.

The first part of the chapter summarizes a quantitative calibration by Pattison (1989) of the well-defined sequence of pelitic reactions in the aureole (Pattison and Harte, Chap. 8, this Vol.). P-T estimates from calibrated geothermometers and geobarometers (Pattison 1989), carbonate reactions (Masch and Heuss-Aßbichler, Chap. 10, this Vol.), microcline-sanidine transformation (Kroll et al., Chap. 13, this Vol.) and late-stage haplogranites in the igneous complex (Weiss and Troll, Chap. 4, this Vol.) are used to constrain further P-T conditions in the aureole. Four thermal profiles, including ones across the widest and narrowest parts of the aureole, are constructed to constrain the thermal modelling of Buntebarth (Chap. 18, this Vol.).

Quantitative estimates of a(H_2O) show that dehydration reactions occur at lower temperatures and melting reactions occur at higher temperatures in graphitic pelites compared to non-graphitic units. Lowered a(H_2O) accounts for several conspicuous features of assemblages in the graphitic Ballachulish Slate, such as the development of incipient cordierite farther from the igneous contacts in the Ballachulish Slate than in the other non-graphitic pelitic units, the exclusive development of andalusite-bearing assemblage IVa in the Ballachulish Slate, and the anomalously magnesian compositions of cordierite and biotite in assemblage IVa compared to assemblages in non-graphitic units (Pattison and Harte, Chap. 8, this Vol.).

In the zone of partial melting, variation in a(H_2O) is assessed by the construction of a semi-quantitative T-a(H_2O) diagram. It is argued that melting reactions internally buffered a(H_2O) to low values as melting proceeded. Prograde paths for the most common pelitic and semipelitic assemblages are discussed with reference to melting reactions in the T-a(H_2O) diagram.

The chapter closes with a short discussion of the importance of equilibrium and kinetics in controlling the distribution and chemical composition of pelitic assemblages in the aureole. Most of the evidence suggests that gross equilibrium prevailed during contact metamorphism, although kinetic effects are in evidence in

G. Voll, J. Töpel, D.R.M. Pattison, F. Seifert
Equilibrium and Kinetics
in Contact Metamorphism

a number of cases. The discussion focuses on the stability of cordierite-producing reactions, and the problem of the andalusite-sillimanite transition.

16.2 Calibration of the Pelitic Petrogenetic Grid

The method of calibration of the petrogenetic grid of Fig. 8.20 is outlined by Pattison (1989). The minerals involved in the subsolidus portion of the grid include Ms, Chl, Bt, Crd, Kfs, And, Sil, Crn and Qtz in the model pelitic system KFMASH (Pattison and Harte, Chap. 8, this Vol.). Mineral abbreviations and model mineral compositions are listed in Table 8.1, and end-member reaction stoichiometries are listed in Table 8.2. Pattison (1989) attempted to derive an internally consistent thermodynamic data set for these minerals in order to reproduce the experimentally determined positions of the end member reactions. Using this approach, the reaction positions could then be shifted for non-end member fluid compositions and measured mineral compositions, thereby giving more realistic pressure and temperature estimates in the aureole. Although considerable difficulties were encountered in the derivation of the data, a best-compromise grid was calculated that satisfied the large majority of the constraining experimental data.

Figure 16.1 illustrates a calibrated petrogenetic grid for typical mineral compositions in the Ballachulish aureole (Pattison 1989). The mineral compositions and activity formulations used in the calculation of Fig. 16.1 are found in Tables 3 and 4 of Pattison (1989). Isopleths of Mg/(Mg + Fe) in cordierite, derived from Holdaway and Lee (1977), cover the majority of measured cordierite compositions at Ballachulish (see Fig. 8.22). In Fig. 16.1, the andalusite-sillimanite boundary is located to be consistent with the first appearance of sillimanite at reaction (P3), Ms + Qtz = Al_2SiO_5 + Kfs + H_2O, for a pressure of 0.3 GPa (for a more detailed discussion of andalusite and sillimanite, see Pattison 1991 and Sect. 16.7.1). For the calculation of reaction stabilities above reaction (P3), sillimanite is used as the Al_2SiO_5 polymorph, even though andalusite and sillimanite occur together in many assemblages; this choice does not substantially affect the positions of the reactions concerned.

Two vapour-consuming melting reactions are included in Fig. 16.1; these are representative of the most important initial melting reactions in the aureole (Pattison and Harte 1988). High-grade dehydration-melting reactions (P4a), (P10), (P11) and (P12) in the zone of partial melting are indicated by dashed lines; these are only approximately constrained in P-T space (Holdaway and Lee 1977; Thompson 1982; Grant 1985, Le Breton and Thompson 1988). Isopleths from subsolidus reactions (P2a) and (P4a) have been extended through the high grade reactions, assuming negligible variation in K_D(Fe-Mg,Crd-Bt) between 650–750 °C (see below).

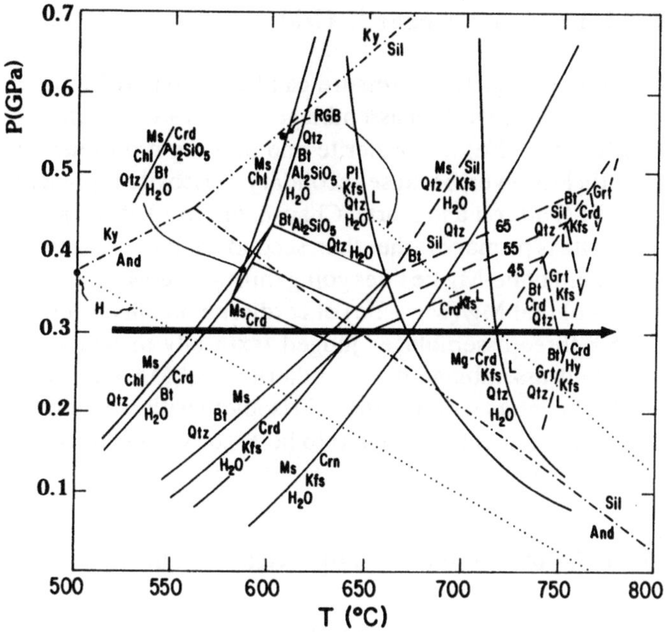

Fig. 16.1. Calibrated pelitic petrogenetic grid for typical mineral compositions in the Ballachulish aureole (for details of the calibration, see Pattison 1989). Isopleths of Mg/(Mg + Fe) in cordierite cover the majority of measured compositions in the aureole (0.45–0.65). The two vapour-consuming melting reactions are An_{30} + Kfs + Qtz + H_2O = L, from Thompson and Tracy (1979), and Mg-Crd + Kfs + Qtz + H_2O = L, from Seifert (1976). For dehydration reactions (*solid lines*), a(H_2O) is assumed to be 1. High-grade Grt- and Hy-producing reactions in the migmatitic zone (*dashed lines*) are located approximately from Holdaway and Lee (1977), Thompson (1982), Grant (1985) and Le Breton and Thompson (1988); in these reactions a(H_2O) is assumed to be internally buffered (see text and Fig. 16.11). The And = Sil boundary is calculated from the data set of Pattison (1989), and is consistent with the first appearance of sillimanite at reaction (P3), Ms + Qtz = Al_2SiO_5 + Kfs + H_2O (Pattison 1991); the And = Sil boundaries of Holdaway (1971) (*H*) and Richardson et al. (1969; *RGB*) are shown for comparison. The *solid arrow* represents the prograde isobaric trajectory in the Ballachulish aureole

16.3 Pressure

Independent estimates of pressure in the Ballachulish Igneous Complex and thermal aureole have been made by a variety of methods. It is assumed that the metamorphic field gradient in the aureole is sensibly isobaric, owing to the lack of evidence for downward or upward drag of country rock adjacent to the igneous contacts (Pattison 1989; Troll and Weiss, Chap. 3, this Vol.), and the lack of tilting during uplift following emplacement (Pattison and Voll, Chap. 2, this Vol.). It is further assumed that the prograde P-T trajectory in the thermal aureole is also isobaric, owing to the rapidity of heating in the aureole (< 0.2 Ma for rocks heated above 500 °C, Buntebarth, Chap. 18, this Vol.) relative to regional uplift rates at the time of emplacement (ca. 0.5 mm/a between 410 and 390 Ma, Dempster 1985).

16.3.1 Pelitic Petrogenetic Grid

Referring to Fig. 16.1, pressure can be estimated from pelitic rocks containing the reaction (P3) univariant assemblage Ms + Bt + Qtz + Crd + Al_2SiO_5 + Kfs by analyzing the Mg/(Mg + Fe) in cordierite and plotting it on reaction (P3) at the appropriate isopleth. However, because secondary muscovite is present in all assemblages above Zone III (Pattison and Harte, Chap. 8, this Vol.), it is essential to demonstrate that the muscovite is primary rather than secondary. Pattison and Harte demonstrated by the movement of Fe-Mg tie lines going through Fe-Mg continuous reactions (P2b), (P4a) and (P4b) that Mg/(Mg + Fe) of cordierite on univariant reaction (P3) must be close to 0.50; three assemblages judged texturally to contain primary muscovite in the univariant assemblage have cordierite Mg/(Mg + Fe) ratios of 0.49–0.50 (Pattison 1987). The intersection of the 0.50 isopleth with reaction (P3) gives a pressure of 0.31 GPa. Uncertainty is estimated to be ± 0.05 GPa (Pattison 1989).

16.3.2 Geobarometry of High-Grade Pelites

In Tables 16.1 and 16.2, mineral chemical data and derived pressures and temperatures from a variety of published geothermometers and geobarometers are listed for high-grade quartz-bearing and quartz-absent assemblages, respectively. For pressure calculations, a temperature of 775 °C has been chosen for Grt + Qtz + Hy-bearing assemblages from pelitic screens within the igneous complex, and 750 ° for Spl-bearing, Qtz-absent assemblages from the main part of the aureole (Fig. 8.1).

The Grt + Pl + Opx + Qtz barometer is in principle the best of the methods listed in Tables 16.1 and 16.2, because all the minerals in the equilibrium are present in the natural assemblage, allowing the calculation of a unique pressure. The calibrations of Perkins and Chipera (1985; Fe reaction) and Bohlen et al. (1983) are favoured over Perkins and Chipera (1985; Mg reaction) and Newton and Perkins (1982) because they are based on the Fe rather than Mg end-member reaction: from Tables 16.1 and 16.2, it can be seen that the natural minerals are relatively Fe-rich, which means that activity-composition models do not have to be extrapolated as far using the Fe-based calibrations as for the Mg-based calibrations. Pressures from the Fe-based calibrations are 0.27 and 0.32 GPa, respectively; pressures from the Mg-based calibrations are 0.48 and 0.41 GPa, respectively.

Pressure estimates from the Crd + Grt + Sil + Qtz and Crd + Spl + Qtz barometers provide upper limits, because sillimanite is absent in quartz-bearing assemblages and quartz is absent in spinel-bearing assemblages. Water contents of cordierite must also be assumed. Droop and Charnley (1985) argued that cordierite water contents in a variety of comparable assemblages in contact aureoles in NE Scotland ranged between 0.0 and 0.25 mol. Using these as bracketing values, upper pressure limits of 0.36 and 0.40 GPa, averaging all specimens and calibrations, are obtained from Crd + Grt + Sil + Qtz; for Crd + Spl + Qtz (Seifert and Schumacher 1986), averaging all specimens, upper pressure limits of 0.37 and 0.43 GPa are obtained.

Table 16.1. Geothermometry and geobarometry of high-grade quartz-bearing metapelites

		DP567	DP568	DP568b	SW2a	SW2b
Garnet	X_{Fe}[a]	0.782	0.761	0.761	0.780	0.772
	X_{Mg}	0.158	0.169	0.165	0.140	0.153
	X_{Ca}	0.034	0.035	0.036	0.025	0.023
	X_{Mn}	0.026	0.035	0.039	0.054	0.052
Biotite	X_{Fe}[a]	0.619	0.628	0.621	0.679	0.668
	X_{Mg}	0.381	0.372	0.379	0.321	0.332
	X_{Al}^{VI}	0.103	0.053	0.061	0.106	0.118
	X_{Ti}^{VI}	0.087	0.107	0.112	0.078	0.082
Cordierite	X_{Fe}[a]	P	0.482	P	0.499	0.476
	X_{Mg}	P	0.518	P	0.501	0.524
Orthopyroxene	X_{Fe}[a]	—	0.652	—	—	—
	X_{Mg}	—	0.348	—	—	—
	X_{Al}^{VI}	—	0.055	—	—	—
Plagioclase	X_{Ca}	0.404	0.357	0.370	0.255	0.264
	X_{Na}	0.587	0.620	0.619	0.727	0.724
T(Grt-Bt)[b]	T	777	839	810	841	860
	HL	740	791	766	793	808
	FS	844	938	892	942	972
	HS	858	953	908	952	981
	IM1	692	741	691	784	804
	IM2	676	745	696	790	802
T(Grt-Crd)[b]	T	—	848	—	791	780
	HL	—	799	—	753	743
T(Grt-Opx)[b]	H	—	768	—	—	—
	HG	—	787	—	—	—
P(Grt-Pl-Opx-Qtz)[c]	NP(Mg)	—	0.41	—	—	—
	BWB(Fe)	—	0.32	—	—	—
	PC(Mg)	—	0.48	—	—	—
	PC(Fe)	—	0.27	—	—	—
P(Grt-Crd-Sil-Qtz)[c]	NW(Fe-Mg)	—	<0.38(0.41)[d]	—	<0.36(0.39)	<0.37(0.40)
	L1(Fe-Mg)	—	<0.29(0.33)	—	<0.29(0.33)	<0.30(0.34)
	L2(Fe-Mg)	—	<0.36(0.40)	—	<0.36(0.40)	<0.37(0.41)
	MS(Fe-	—	<0.42(0.48)	—	<0.40(0.46)	<0.41(0.47)

[a]All Fe assumed to be Fe^{2+}.
[b]T(°C) calculated for P = 0.3 GPa.
[c]P(GPa) calculated for T = 775 °C
[d]P(GPa) calculated for $X^{Crd}_{Hyd\,Crd}$ = 0.0 and 0.25 (the latter in parentheses)
P = pinitized cordierite. (Fe), (Mg) and (Fe-Mg) mean that the barometric expression is based, respectively, on the Fe-end member reaction, Mg-end member reaction, and both.
Garnet X_i = i/Fe + Mg + Ca + Mn. Cordierite: X_i = i/Fe + Mg. Orthopyroxene: X_i = i/Fe + Mg; X_{Al}^{VI} = (Al calculated for 6 oxygens)/2. Plagioclase: X_i = i/Na + K + Ca. T = Thompson 1976; Table 1. HL = Holdaway and Lee 1977; Table 7. FS = Ferry and Spear 1978; Eq. (7). HS = Hodges and Spear 1982; Eq. (9). IM1, IM2 = Indares and Martingnole 1985; Eqs. (18) and (19). H = Harley 1984; Eq. (15). HG = Harley and Green 1982; Eq. (5). NP = Newton and Perkins 1982; Eq. (12). BWB = Bohlen et al. 1983; Fig. 5. PC(Fe), PC(Mg) = Perkins and Chipera 1985; Eqs. (16) and (17). NW = Newton and Wood 1979, Figs. 4 and 5. Ll, L2 = Lonker 1981; Appendix 4, Reaction (11), Eqs. (1) and (2). MS = Martingnole and Sisi 1981; Figs. 5 and 6.

Table 16.2. Geothermometry and geobarometry of high-grade quartz-absent metapelites

		DP83a	DP608–1	DP433	DP441
Cordierite	X_{Fe}[a]	0.494	0.543	0.512	0.527
	X_{Mg}	0.506	0.457	0.488	0.473
Spinel	X_{Fe}	0.89	0.88	0.77	0.89
	X_{Mg}	0.10	0.10	0.10	0.10
	X_{Mn}	0.01	0.01	Nd	Nd
	X_{Zn}	Nd	Nd	0.13	0.01
	X_{Al}^{IV}	0.94	0.95	0.99	0.99
T(Crd-Spl)[b]	V	689	812	819	766
P(Crd-Spl-Qtz)[c,d]	SS(Fe)	0.39(0.45)	0.35(0.41)	(0.41)	0.40(0.46)

[a] All Fe assumed to be Fe^{2+}.
[b] T(°C) calculated for P = 0.3 GPa.
[c] P(GPa) calculated for T = 750 °C.
[d] P(GPa) calculated for $X^{Crd}_{Hyd\ Crd}$ = 0.0 and 0.25 (the latter in parentheses). Nd, not detected. Cordierite: $X_i = i/Fe + Mg$. Spinel: $X_i = i/Fe^{2+} + Mg + Mn + Zn$; X_{Al}^{IV} = (Al calculated for 4 oxygens)/2. V = Vielzeuf 1983; Eq. in abstract. SS = Seifert and Schumacher 1986; Eqs. (9) and (10). (Fe) means that the barometric expression is based on the Fe-end member reaction.

16.3.3 Carbonate Reactions

Masch and Heuss-Aßbichler (Chap. 10, this Vol.) obtained a pressure estimate of 0.28 ± 0.04 GPa from the intersection of a calcite-dolomite temperature of 620 °C with the P-T trace of the isobaric univariant assemblage Tr + Fo + Di + Cal + Dol (CIII). The P-T trace of isobaric invariant point CIII is based on the reversed experiments and derived internally consistent thermodynamic data of Gottschalk and Metz (1984).

Upper and lower pressure limits of 0.32 and 0.26 GPa are obtained from the intersection of calcite-dolomite temperatures with isobaric univariant reactions C6 (Qtz + Tr + Cal = Di + H_2O + CO_2), C9 (Tr + Dol = Fo + Cal + H_2O + CO_2) and CII (Di + Dol = Fo + Cal + CO_2) (Heuss-Aßbichler 1987).

16.3.4 Crystallization of Granites

Weiss and Troll (Fig. 4.7, this Vol.) projected C.I.P.W.-normative albite-orthoclase-quartz components from their analyzed suite of samples of the igneous complex into the P(H_2O) = 0.3 GPa haplogranite system of Tuttle and Bowen (1958). Quartz diorites, granodiorites and granites followed the thermal valley towards the ternary minimum; late-crystallizing, H_2O-saturated, Ca-poor leucogranites plotted on or near the ternary minimum. From the close coincidence of the plotted leucogranites with the ternary minimum, Weiss and Troll estimated a water pressure of 0.25–0.30 GPa.

16.3.5 Summary of Pressure Estimates

The above four independent methods of estimating pressure in the igneous complex and aureole are in excellent agreement with each other; the preferred estimates from each method range between 0.28 and 0.32 GPa. It is concluded that the igneous complex was emplaced at a pressure of 0.30 GPa. Uncertainty is conservatively estimated as ± 0.05 GPa, although consideration of all of the uncertainty brackets gives a tighter constraint of ± 0.03 GPa. 0.30 ± 0.03 GPa corresponds to a depth of 10 ± 1 km, assuming a mean density of 2.7 g/cm^3.

16.4 Temperature Variation in the Aureole

Several independent methods of estimating temperature within the aureole are discussed below. At the end of the section, four composite thermal profiles through different parts of the aureole are constructed to constrain the thermal modelling of Buntebarth (Chap. 18, this Vol.).

16.4.1 Calibrated Pelitic Petrogenetic Grid

Taking a 0.30-GPa isobaric trajectory through Fig. 16.1, the schematic subsolidus T-X(Fe-Mg) diagram of Fig. 8.21 has been calibrated in Fig. 16.2. From Figs. 16.1 and

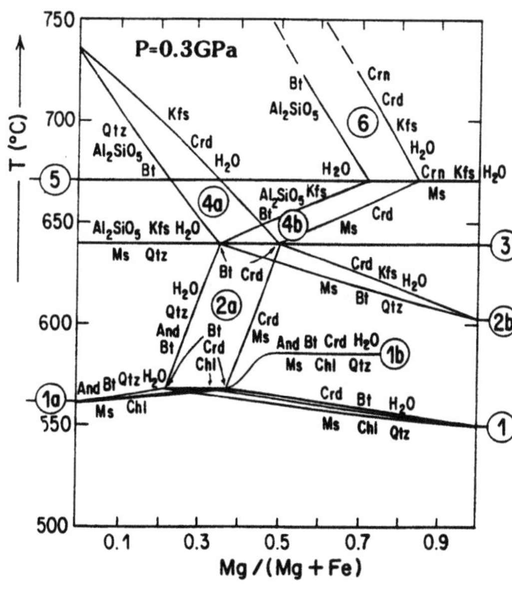

Fig. 16.2. Calibrated 3-kbar isobaric T-X(Fe-Mg) diagram for the subsolidus reactions of Fig. 16.1. Activity of water is assumed to be 1

Table 16.3. Temperatures (°C) of major mineral assemblage boundaries in the Ballachulish aureole, for P = 0.3 GPa[a]

Non-graphitic pelites	Prograde reaction	Graphitic pelites
560	Ms Chl Qtz / Crd Bt H_2O	550
	Ms Crd / And Bt Qtz H_2O	600
620	Ms Bt Qtz / Crd Kfs H_2O	
640	Ms Qtz / Kfs $Al_2SiO_5H_2O$	625
670	Ms / Crn Kfs H_2O	655
670	Kfs Pl Qtz H_2O / L	715
720	Kfs Crd Qtz H_2O / L	775
750	Bt Crd Qtz / Grt Kfs L	N/A
760	Bt Grt Qtz / Hy Crd Kfs L	N/A

[a]Uncertainties are discussed in the text.

16.2, temperatures may be assigned to the pelitic reactions. These are listed in Table 16.3. For Fe-Mg continuous reactions, which may commence over a range of temperatures depending on Mg/(Mg + Fe) of the minerals, temperatures in Table 16.3 represent the lowest possible for the measured mineral compositions in the appropriate zone (see Fig. 8.22). For non-graphitic pelites, selected temperatures include: 560 °C at reaction (P1), first growth of cordierite; 640 °C at reaction (P3), Ms + Qtz breakdown; 670–720 °C at the onset of partial melting (depending upon the presence and composition of plagioclase); and 750 °C for high-grade garnet-bearing pelitic assemblages in the zone of partial melting. In graphitic pelites, temperatures of subsolidus reactions will have been somewhat lower, whilst temperatures of vapour-consuming melting reactions will have been somewhat higher; this effect is discussed in more detail below. Temperature uncertainty for subsolidus reactions (P1), (P2b), (P3) and (P5) is estimated as ± 20 °C, whilst for reaction (P2a), which has a shallow P-T slope, it will be higher (Pattison 1989).

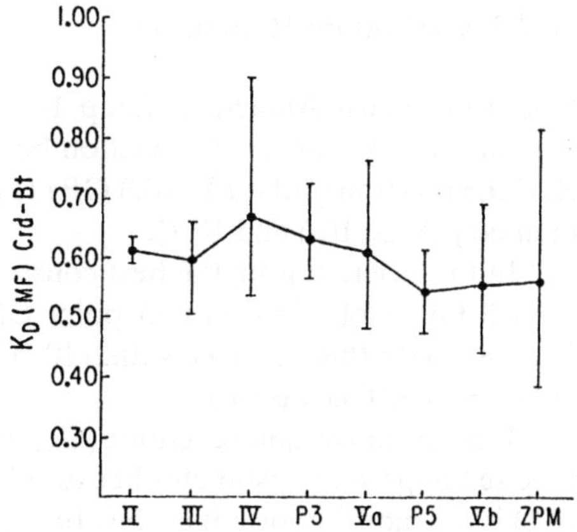

Fig. 16.3. Variation in Kd(Fe-Mg) between cordierite and biotite going through the metamorphic zones. In each zone, the *brackets* show the absolute range in values, and the *dot* is the mean. Numerical data from Pattison (1987)

16.4.2 Geothermometry

With the exception of rare high-grade Grt + Crd + Bt- and Grt + Hy + Crd ± Bt-bearing pelites in the zone of partial melting, assemblages in most of the thermal aureole do not contain useful mineral pairs for exchange thermometry. Figure 16.3 shows the variation in Kd(Fe-Mg) between cordierite and biotite going through the subsolidus contact metamorphic zones. It can be seen that there is huge scatter within individual zones, and that this scatter is far greater than the slight overall decrease in Kd(Fe-Mg) with grade; consequently, this is not a good method of geothermometry, regardless of calibration. The K-feldspar-plagioclase thermometer gives equally unsatisfactory results (Pattison 1985).

Tables 16.1 and 16.2 list geothermometric estimates for high-grade quartz-bearing and quartz-absent assemblages in the zone of partial melting, assuming a pressure of 0.3 GPa. There is also considerable scatter in these results, with temperatures ranging from 676 to 981 °C. This scatter is probably due to a combination of calibration problems, activity-composition formulations and different closure temperatures (Lasaga 1983) of the various participating minerals.

It is difficult to judge if any of the methods are reliable. The popular Hodges and Spear (1982) and Ferry and Spear (1978) Grt-Bt calibrations give very high estimates in the range 850–980 °C, which seem too high compared with other thermometric constraints (see Fig. 16.5). The Grt-Opx calibrations of Harley (1984) and Harley and Green (1982) range between 760 and 785 °C for DP568, the single Opx-bearing specimen. It is felt that, of the exchange thermometers involving garnet, the Grt-Opx calibrations are the most reliable: Opx has a relatively simple anhydrous composition, resulting in less uncertainty in activity-composition corrections for non-end member components than for more complex minerals such as Bt and Crd. Additionally, Grt and Opx have among the slowest Fe-Mg diffusion rates amongst the minerals in Tables 16.1 and 16.2 (Lasaga 1983), which means that they are the most likely to preserve their cation ratios from the peak of metamorphism. These estimates also fit best with the approximate location of reactions (P11), ca. 750 °C, and (P12), ca. 760 °C, in the calibrated pelitic grid (Fig. 16.1).

16.4.3 Carbonate Reactions

Masch and Heuss-Aβbichler (Chap. 10, this Vol.) use two different methods to assign temperatures to carbonate reaction boundaries in the aureole: (1) a calibrated T-X(CO_2) reaction grid for P = 0.3 GPa (Fig. 10.12); and (2) calcite-dolomite geothermometry (Figs. 10.6 and 10.7).

In the reaction grid, the best constrained temperature is 620 °C obtained from the 0.3 GPa isobaric invariant point CIII (Dol + Cal + Tr + Di + Fo). In Coire Guibhsachain, this is located within 10 m of pelitic univariant reaction (P3), estimated at 640 ± 20 °C (see above).

Isobaric univariant reactions (C1), (C6), (C9), (C10), (C14) and (C15) give less precise temperature estimates because X(CO_2) can vary along the curves. Reaction (C1) is the most important of the above, because the development of talc outside of the first development of cordierite in the pelites extends the range of temperature estimates at the margins of the aureole. From Fig. 10.12, this reaction occurs in the range 370–480 °C (Masch and Heuss-Aβbichler, Chap. 10, this Vol.). The appearance of periclase by reaction (C15) in the highest grade marbles suggest very high temperatures of 750–800 °C (Masch and Heuss-Aβbichler, Chap. 10, this Vol.).

Calcite-dolomite geothermometry has been applied by Masch and Heuss-Aβ-bichler to a range of samples in the aureole (see Fig. 10.7). This geothermometer typically gives a range of temperatures for a given sample. Down-temperature resetting is thought to be a major problem with this method, so that for a given sample, the highest temperatures of a coherent population have been favoured by Masch and Heuss-Aβbichler as the best estimate of peak temperature (see Fig. 10.6). Using this approach, it can be see that temperatures rise toward the contact, and are generally consistent with the temperature ranges of the univariant reactions and of invariant point CIII. However, Masch and Heuss-Aβbichler consider that in several specimens the thermometric temperature estimates may be too low.

16.4.4 Microcline-Sanidine Transformation

Kroll et al. (Chap. 13, this Vol.) note that as one approaches the igneous complex in the Appin Quartzite, there is a transition zone between 1900 and 1100 m in which the volumetric proportion of low microcline decreases relative to orthoclase. Outside 1900 m, only microcline is present, whilst inside 1100 m, only orthoclase is present. Kroll et al. take the first occurrence of orthoclase at 1900 m to represent the onset of the microcline to sanidine transition.

Kroll et al. estimate a temperature range of 470–490 °C for the transformation, based on the relative location of the equivalent transformation in a series of isograds from the Central Swiss Alps (Frey et al. 1980). Experimental data of Senderov and Yas'kin (1975, 1976) suggest an upper temperature limit for pure Na-free microcline of 550 °C. Kroll et al. (Chap. 13, this Vol.) note that the transformation temperature is dependent on the Na content of the K-feldspar.

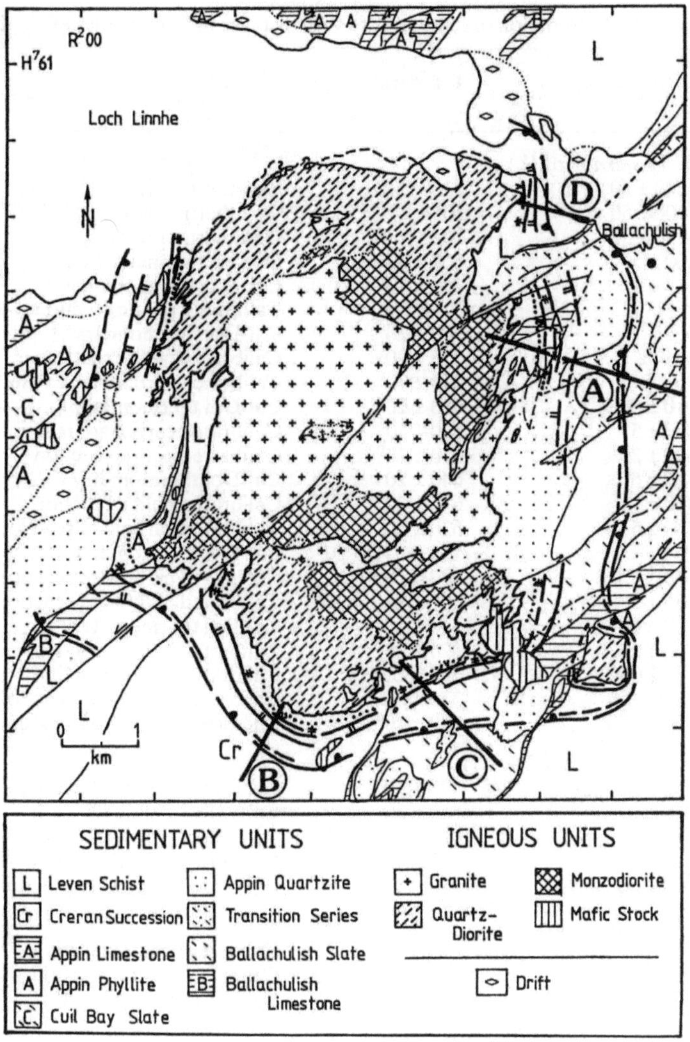

Fig. 16.4. Location of the four thermal profiles illustrated in Figs. 16.5–16.8

16.4.5 Temperature Profiles in the Aureole

Using all of the temperature estimates from the above equilibria, four thermal profiles were constructed across different parts of the thermal aureole (see Figs. 16.4–16.8). The most widely applicable method of constraining the thermal profiles is the series of calibrated subsolidus pelitic reactions, which have been located around the whole igneous complex. Table 16.4 lists the distances, temperatures and methods used to construct the thermal profiles. Uncertainties in temperature and distance from the contact are from the authors of the different methods.

Table 16.4. Variation of temperature with distance in four profiles through the Ballachulish aureole

Distance (m)[a]	Temperature (°C)	Method	Reaction number or sample
A. Eastern profile			
−500 ± 50	1100 ± 70	Opx thermometry	M1[c]
−400 ± 50	1050 ± 70	Opx thermometry	M2
−250 ± 50	1020 ± 70	Opx thermometry	M3
100 ± 50	800 ± 60	Opx thermometry M4	
−15 ± 5	760 ± 20	Grt-Opx thermometry	DP568[d]
		Calibrated pelitic reaction	P12[e]
0 ± 5	760 ± 20	Cal-Dol thermometry	C(375)[f]
10 ± 5	750 ± 20	Calibrated pelitic reaction	P11
70 ± 5	650–970	Calibrated carbonate reaction	C15[g]
110 ± 5	680 ± 20	Cal-Dol thermometry	C(315)
350 ± 5	500–750	Calibrated carbonate reaction	C14
380 ± 20	670 ± 20[b]	Calibrated pelitic reaction	P5
450 ± 5	640 ± 20	Cal-Dol thermometry	C(309)
490 ± 10	640 ± 20[b]	Calibrated pelitic reaction	P3
500 ± 5	620 ± 20[b]	Calibrated carbonate reaction	CIII[g]
		Cal-Dol thermometry	C(307)
665 ± 5	500–620	Calibrated carbonate reaction	C10
685 ± 5	500–620	Calibrated carbonate reaction	C9
700 ± 5	530 ± 20	Cal-Dol thermometry	C(223)
790 ± 20	620 ± 20[b]	Calibrated pelitic reaction	P2b
1600 ± 50	550 ± 20[b]	Calibrated pelitic reaction	P1
1800 ± 50	370–480	Calibrated carbonate reaction	C1
1900 ± 50	480 ± 20[b]	Microcline-sanidine transition	—
B. Southern profile			
150 ± 10	670 ± 20	Calibrated pelitic reaction	P5
210 ± 10	640 ± 20	Calibrated pelitic reaction	P3
320 ± 20	620 ± 20	Calibrated pelitic reaction	P2b
620 ± 50	560 ± 20	Calibrated pelitic reaction	P1
900 ± 50	370–480	Calibrated carbonate reaction	C1
C. Southeast profile			
350 ± 30	625 ± 20	Calibrated pelitic reaction	P3
500 ± 20	600 ± 30	Calibrated pelitic reaction	P2a
1050 ± 50	550 ± 20	Calibrated pelitic reaction	P1
D. Northeast profile			
60 ± 5	670 ± 20	Calibrated pelitic reaction	P5
80 ± 5	640 ± 20	Calibrated pelitic reaction	P3
140 ± 10	620 ± 20	Calibrated pelitic reaction	P2b
340 ± 30	560 ± 20	Calibrated pelitic reaction	P1

[a]Negative distances are distances from contact *in the igneous complex*; positive distances are distances from the contact *in the country rocks*.
[b]Most reliable estimates in the eastern profile (see text for discussion of methods). Pressure in all profiles is assumed to be 0.3 GPa.
[c]Ml, etc. Magmatite sample from Weiss and Troll (Chap. 4, this Vol.).
[d]DP568, Metapelite sample from Pattison and Harte (Chap. 8, this Vol.).
[e]P12, etc., Pelitic reaction number (Pattison and Harte, Chap. 8, this Vol.).
[f]C(375), etc., Carbonate sample from Masch and Heuss-Aβbichler (Chap. 10, this Vol.).
[g]CIII and C15 etc., Isobaric invariant point and univariant reactions from Masch and Heuss-Aβbichler (Chap. 10, this Vol.).

Included in Table 16.4 and Fig. 16.5 are some temperatures within the igneous complex (Weiss and Troll, Chap. 4, this Vol.), based on Opx-Cpx geothermometry of Lindsley (1983) modified by Lindsley and Andersen (1983). The combined use of temperatures within the igneous complex and the thermal aureole gives a very complete thermal profile across the intrusive/country rock boundary. From these estimates, the contact temperature on the east flank is estimated to be in the range 750–800 °C.

The profile in Fig. 16.5 is taken across the east flank, one of the widest parts of the aureole (see Fig. 16.4); it is also the best constrained of all the profiles. This profile contains a broad expanse of Appin Quartzite, a feature common to all the widest parts of the aureole. The presence of quartzite is shown by Buntebarth (Chap. 18, this Vol.) to be of great importance in the transfer of heat away from the igneous complex. Figures 16.6 to 16.8 are from: the southern profile, in which Creran Succession pelites comprise the country rocks to the igneous complex, with D_1 schistosity striking into the igneous complex (normal to the contact); the southeastern profile, in which graphitic Ballachulish Slate comprises most of the country rocks, with D_1 schistosity striking parallel to the contact; and the northeastern profile, in which pelitic Leven Schist comprises the country rocks, with D_1 schistosity also striking parallel to the contact. The northeastern profile is the narrowest in the aureole.

Buntebarth (Chap. 18, this Vol.) discusses the profiles in more detail with respect to thermal modelling.

16.5 The Effect of Graphite on Pelitic Phase Relations

Using the thermodynamically-based calibration of the pelitic reactions in the aureole, it is possible to assess variations in a(H_2O) and its effects on phase relations in different pelitic units. The discussion focuses on the importance of graphite, which interacts with H_2O to dilute the hydrous fluid phase with C-bearing fluid species, thereby lowering a(H_2O). This effect is shown to influence the temperature at which pelitic reactions commence (see Table 16.3), and accounts for many of the unusual features of pelitic assemblages in the graphitic Ballachulish Slate noted in Pattison and Harte (Chap. 8, this Vol.).

16.5.1 Stability of Pelitic Reactions in Graphitic Rocks

The interaction of graphite with hydrous vapour introduces C-bearing fluid species such as CO_2, CH_4 and CO to the hydrous vapour (e.g., Ohmoto and Kerrick 1977). At fixed P and T, the abundances of all C-O-H fluid species may be determined if one variable, such as f(O_2), is fixed. In the large majority of the pelites in the aureole, however, ilmenite is the only oxide phase present, which means that an independent

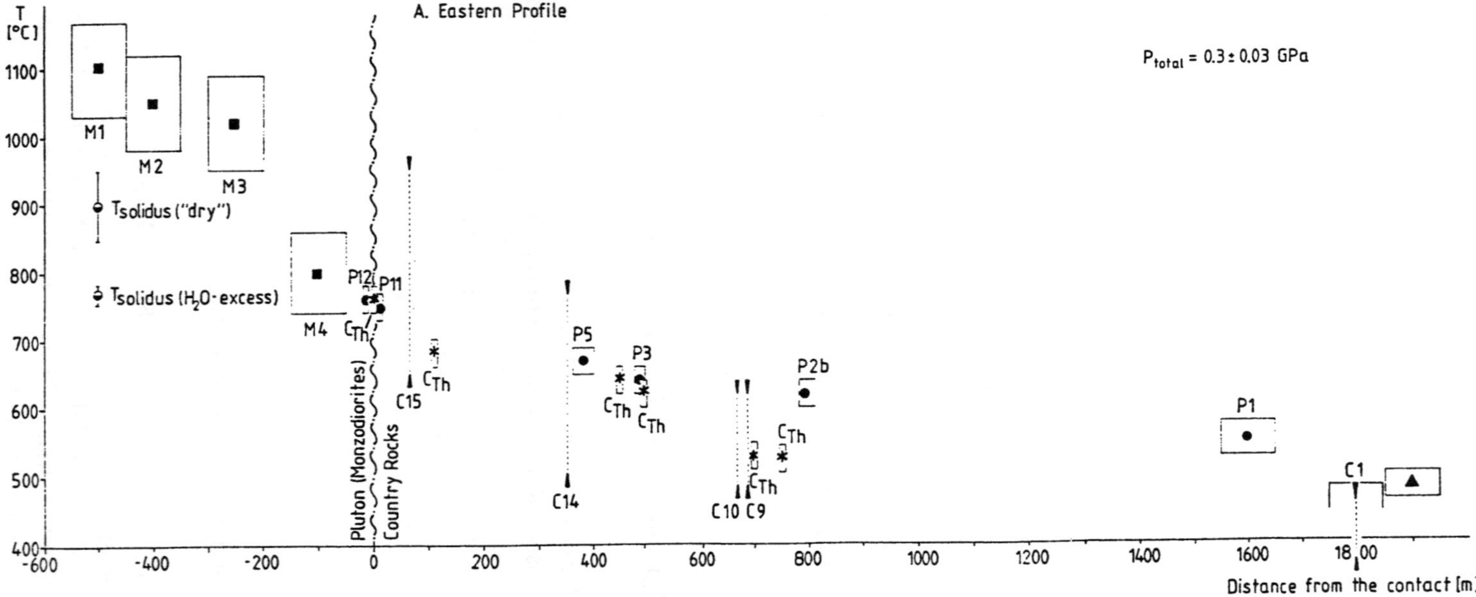

Fig. 16.5. East profile, taken across the widest part of the aureole. *Black dots* represent best estimate temperatures and locations for pelitic reactions (P1), (P2b), etc., with *boxes* representing uncertainties in temperature and location; the same convention is adopted for *filled squares* and surrounding *boxes* (magmatic temperatures, Weiss and Troll, Chap. 4, this Vol.), *stars* and *surrounding boxes* (calcite-dolomite geothermometry, Masch and Heuss-Aßbichler, Chap. 10, this Vol.) and *triangle*and *surrounding box* (microcline-sanidine transformation, Kroll et al. Chap. 13, this Vol.). *Inverted pointed triangles* connected by *dotted lines* represent the temperature limits of X(CO_2)-dependent carbonate reactions (C1), (C6), etc. (Masch and Heuss-Aßbichler, Chap. 10, this Vol.).

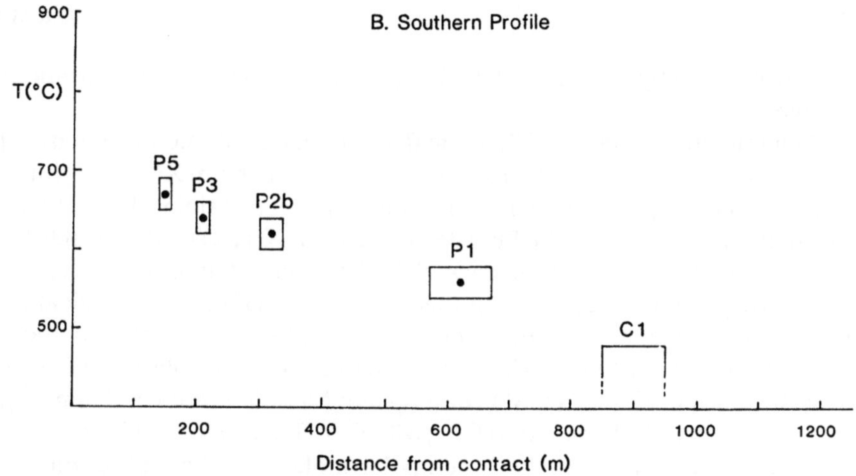

Fig. 16.6. Southern profile. All *symbols* as in Fig. 16.4

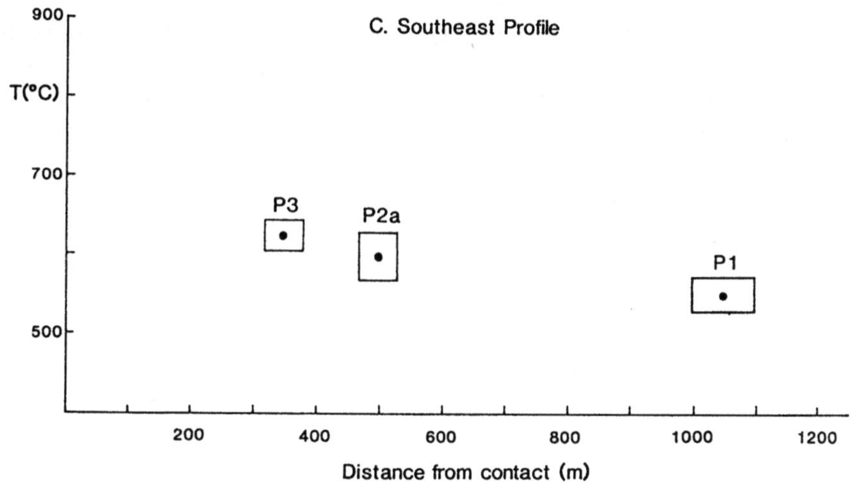

Fig. 16.7. Southeast profile. all *symbols* as in Fig. 16.4

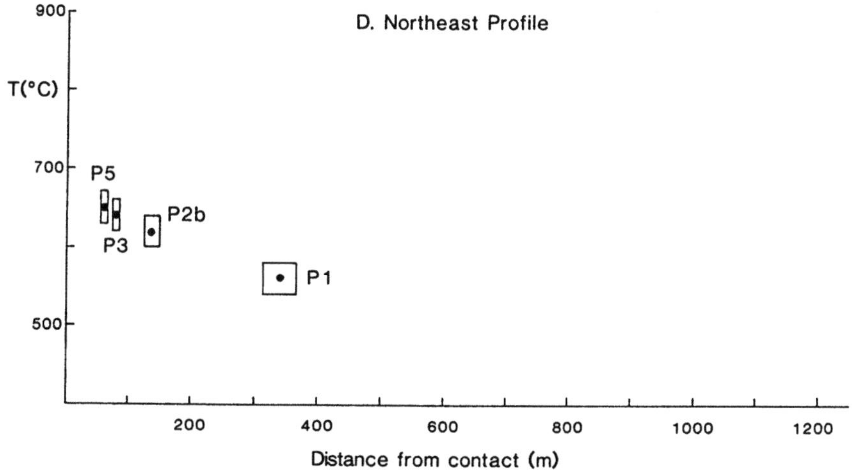

Fig. 16.8. Northeast profile, taken across the narrowest part of the aureole. All *symbols* as in Fig. 16.4

estimate of $f(O_2)$ by magnetite-ilmenite or magnetite-graphite oxybarometry is not possible.

Ohmoto and Kerrick (1977) argue that during dehydration reactions in pelites, the fluid is internally buffered to a composition of maximum $X(H_2O)$, with CH_4 and CO_2 in equal proportions making up most of the remaining fluid. At 0.3 GPa, maximum H_2O content of the fluid decreases from $X(H_2O) = 0.90$ at 500 °C to 0.67 at 800 °C (Fig. 3, Ohmoto and Kerrick 1977). These calculations were made assuming ideal mixing in the fluid, meaning that $X(H_2O) = a(H_2O)$. If non-ideal mixing is assumed, then $X(H_2O)$ will be somewhat lower, because activity coefficients are positive for H_2O in H_2O-CO_2-CH_4 mixtures (Kerrick and Jacobs 1981). For example, using Kerrick and Jacobs' (1981) Hard Sphere Modified Redlich Kwong equations for binary H_2O-CO_2 fluids at 0.30 GPa, 640 °C, $X(H_2O) = 0.79$ whilst $a(H_2O) = 0.84$. Because CH_4 and CO_2 are both non-polar molecules, their effect on $a(H_2O)$ is comparable; consequently, activity coefficients for H_2O derived from the H_2O-CO_2 binary are not thought to be significantly different from those for H_2O-CO_2-CH_4 mixtures (Kerrick and Jacobs 1981).

Figure 16.9 illustrates in a T-$a(H_2O)$ diagram (e.g., Tyler and Ashworth 1982) the limit of maximum $a(H_2O)$ in equilibrium with graphite. Reactions (P1) (Mg-end member), (P3) and (P5) have been calculated for lowered $a(H_2O)$, using the thermodynamic data of Pattison (1989) and fluid fugacity equations of Kerrick and Jacobs (1981). Prograde paths for graphitic and non-graphitic pelites are significantly different. Reaction (P1) is predicted to occur 10–15 °C lower in graphitic pelites than in

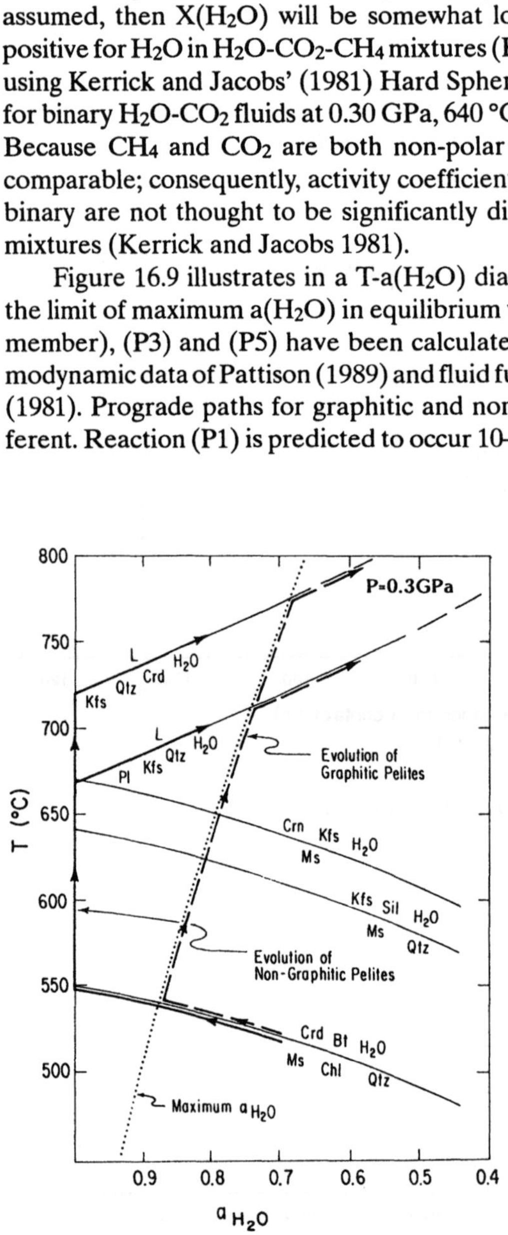

Fig. 16.9. Calculated 0.3 GPa isobaric T-$a(H_2O)$ diagram for reactions (P1) (Mg end member reaction), (P3) and (P5). The melting curves correspond to those in Fig. 16.1. The shape of the melting curves in T-$a(H_2O)$ space are obtained from a 0.3-GPa section through Fig. 16–6 of Burnham (1979). The line of maximum $a(H_2O)$ is derived from a 0.3-GPa section through Fig. 3 of Ohmoto and Kerrick (1977) using H_2O-CO_2 parameters from Kerrick and Jacobs (1981). The *dashed line* shows the prograde path of graphitic pelites, whilst the *solid line* shows the prograde path of non-graphitic pelites

non-graphitic pelites. Although this is quite a small temperature difference, its effect appears to be manifested in the onset of reaction (P1) farther from the contact in graphitic Ballachulish Slate than in adjacent non-graphitic Appin Phyllite on the east flank (see Map 2). Reactions (P3) and (P5) are expected to occur 15–20 °C lower in graphitic pelites than in non-graphitic pelites.

In contrast to dehydration reactions, vapour-consuming melting reactions are predicted by Fig. 16.9 to occur at higher temperatures in graphitic pelites than in non-graphitic pelites. The difficulty in reliably identifying incipient partial melting in the aureole (Harte et al., Chap. 9, this Vol.) has meant that this effect has not been documented. Table 16.3 summarizes contrasting temperatures for pelitic reactions in graphitic and non-graphitic rocks.

16.5.2 Assemblage IVa in the Graphitic Ballachulish Slate

Two conspicuous features of andalusite-bearing assemblage IVa are that (1) it occurs exclusively in the Ballachulish Slate, and (2) that the Mg/(Mg + Fe) ratios of cordierite and biotite in this assemblage contradict the orderly sequence measured in the non-graphitic units (Pattison and Harte, Chap. 8, this Vol.). In assemblage IVa, cordierite and biotite are more Mg-rich by Mg/(Mg + Fe) = 0.04–0.07 than cordierite and biotite on univariant reaction (P3), rather than being more Fe-rich as predicted by Fig. 16.2 (see Fig. 16.10).

The T-X(Fe-Mg) locations of reactions (P1), (P2a), (P2b), (P3), (P4a) and (P4b) were displaced by reducing a(H_2O) according to Fig. 3 of Ohmoto and Kerrick (1977) and using the H_2O-CO_2 mixing parameters of Kerrick and Jacobs (1981). Because reaction (P4a) only releases 0.5 moles of H_2O, the displacement of this reaction due

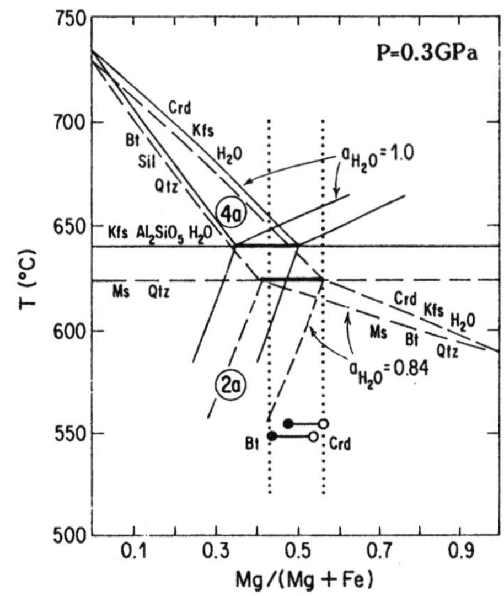

Fig. 16.10. Illustration of the displacement to more magnesian compositions of cordierite and biotite on reaction (P3) due to lowering of a(H_2O). Reaction loops for graphitic pelites are represented by *dashed lines*, whilst reaction loops for non-graphitic pelites are represented by *solid lines*. Lowered a(H_2O) expands the stability field of reaction (P2a) (assemblage IVa) relative to reaction (P2b) (assemblage IVb). The two *tie lines* are for measured compositions of cordierite and biotite in And + Bt + Qtz + Crd + Ms (IVa) pelites from the graphitic Ballachulish Slate (DP266 and DP272; see Fig. 8.22)

to lowered a(H_2O) is very small. In contrast, reactions (P2a), (P3) and (P4b) release more water and are therefore more substantially displaced [e.g., ~15 °C shift for reaction (P3)]. The results are shown in Fig. 16.10. Relative to non-graphitic pelites, the Mg/(Mg + Fe) compositions of cordierite and biotite on reaction P3 are displaced to more magnesian compositions by 0.06–0.07, thereby eliminating the apparent inconsistency with respect to assemblage IVa.

It is likely that the displacement of reaction (P2a) to more magnesian compositions, due to the presence of graphite, accounts for the exclusive development of assemblage IVa in the graphitic Ballachulish Slate. If graphite were not present in the slates, the measured range of Mg/(Mg + Fe) in cordierite and biotite would result in the development of Kfs-bearing assemblage IVb rather than And-bearing assemblage IVa (see Fig. 8.22).

16.6 Variation in a(H_2O) in the Zone of Partial Melting

Applying principles from Greenwood (1975), Powell (1983) used schematic isobaric T-a(H_2O) diagrams to present a theoretical treatment of the variation of a(H_2O) going through a sequence vapour-absent melting reactions. His main conclusions were that the participating minerals and melt phase would internally buffer a(H_2O) to reduced values as melting proceeded, and that melting would proceed in a series of steps when isobaric invariant points were reached.

Figure 16.11 illustrates a semi-quantitative 0.3 GPa isobaric T-a(H_2O) diagram constructed for melting reactions in the Ballachulish aureole. The location of dehydration reactions (P3), Ms + Qtz = Sil + Kfs + H_2O, and (P4a), Bt + Sil + Qtz = Crd + Kfs + H_2O, were calculated using the thermodynamic data of Pattison (1989) and the Kerrick and Jacobs (1981) fugacity equations (see above). Isopleths of Mg/(Mg + Fe) in cordierite are taken from Fig.4 of Pattison (1989). Vapour-consuming melting, for a(H_2O) = 1.0, is represented by the reaction Kfs + Qtz + Bt + Sil ± Pl + H_2O = L + Crd (PM1); the exact stoichiometry is uncertain owing to the variable presence or absence of plagioclase, and uncertainty in the exact melt composition (e.g., Grant 1985). The location of this reaction at 670 °C on the a(H_2O) = 1.0 axis corresponds to that of the reaction Kfs + Qtz + Pl + H_2O = L in Fig. 16.1. In details, the stability of reaction (PM1) will depend on the presence and composition of Bt, Crd, Sil and especially Pl.

The other anchoring point for Fig. 16.11 is the calculated T-a(H_2O) condition for the hypersthene-bearing assemblage Grt + Hy + Crd + Bt + Kfs + Pl + Qtz (DP568) (recognizing the possibility that biotite in this assemblage may be of late stage origin). The preferred geothermometric estimate for this assemblage is 760 °C at 0.3 GPa (see above). Activity of water was calculated from the equilibrium:

$$KMg_3AlSi_3O_{10}(OH)_2 + 3SiO_2 = 3MgSiO_3 + KAlSi_3O_8 + H_2O.$$

Enstatite is not included in the data set of Pattison (1989), so that the data set of Berman (1988, updated in 1989) was used for the following calculation. Measured

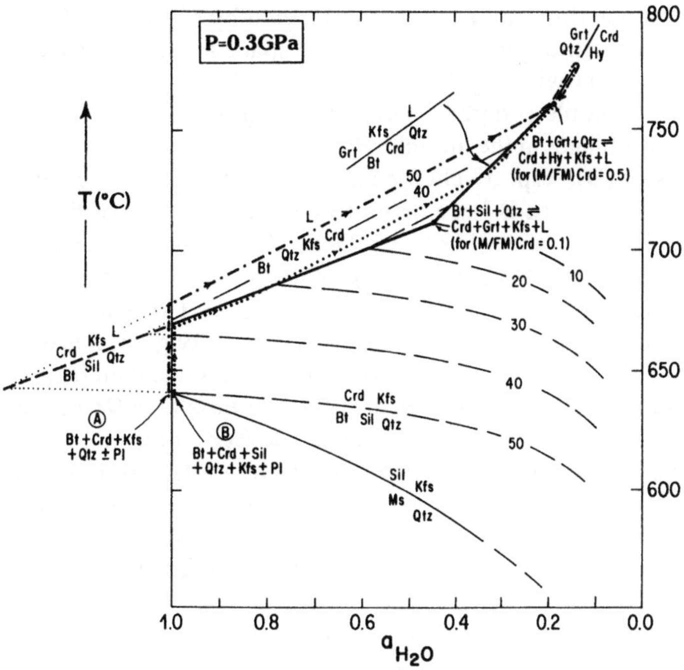

Fig. 16.11. Semi-quantitative 0.3-GPa isobaric T-a(H_2O) diagram for pelites in the zone of partial melting. Reactions (P3) (Ms + Qtz + Kfs + Sil + H_2O) and (P4a) (Bt + Sil + Qtz = Crd + Kfs + H_2O) are calculated from the thermodynamic data of Pattison (1989). *Dashed lines* with *numbers* are isopleths of Mg/(Mg + Fe) in cordierite for reaction (P4a). See text for discussion

compositions of biotite, hypersthene and K-feldspar are $K_{0.87}Na_{0.04}Mg_{0.87}Fe_{1.49}Ti_{0.30}Al_{1.44}Si_{2.71}O_{10}(OH)_2$, $Mg_{0.33}Fe_{0.63}Mn_{0.01}Al_{0.06}Si_{0.96}O_3$, and $K_{0.80}Na_{0.19}Ca_{0.01}Al_{1.02}Si_{2.98}O_8$. Activity models and resultant activities are:

$$a^{Hy}_{MgSiO_3} = \left(a^{Hy}_{Mg_2Si_2O_6}\right)^{1/2} = \left(X^{M_2}_{Mg}\, M^{M1}_{Mg}\right)^{1/2} = 0.345$$

$$a^{Bt}_{KMg_3AlSi_3O_{10}(OH)_2} = 9.48\left[X^{A}_{K} \bullet \left(X^{M}_{Mg}\right)^3 \bullet \left(X^{IV}_{Al}\right) \bullet \left(X^{IV}_{Si}\right) \bullet \left(X_{OH}\right)^2\right]$$

$$= 0.021 \quad (1)$$

or $$= (X^{M}_{Mg})^3 = 0.030 \quad (2)$$

$$a^{Kfs}_{KAlSi_3O_8} = X^{A}_{K} = 0.80$$

Activity of SiO_2 in quartz is assumed to be 1.0. The resultant values of a(H_2O) are 0.18 (first biotite activity model) and 0.25 (second activity model); an a(H_2O) of 0.2 has been used in Fig. 16.11.

Between the 670 °C point on the a(H_2O) = 1.0 axis and the 760 °C, a(H_2O) = 0.2 point, the other reactions have been drawn in to be consistent with Fig. 16.1. Although plagioclase is present in many assemblages, for simplicity the reactions below just consider KFMASH phases. The melt phase is assumed to have a composition similar to that described by Grant (1985, p. 416). The relationships between the reactions in T-a(H_2O) space and P-T space can be seen by comparing Fig. 16.11 with Fig. 16.1.

Possible prograde paths for the most common KFMASH pelitic and pelitic assemblages in the aureole, Bt + Crd + Qtz + Kfz (A) and Bt + Al_2SiO_5 + Crd + Kfs + Qtz (B), respectively, are illustrated in Fig. 16.11. Above reaction (P3), Crd and Bt in assemblage B become more Fe-rich as they pass through dehydration reaction (P4a) along the a(H_2O) = 1.0 axis. Assemblage B commences melting by reaction (PM1) on the a(H_2O) = 1.0 axis. With the consumption of free vapour, further melting proceeds by reaction (P4a), Bt + Sil + Qtz = Crd + Kfs + L, the dehydration-melting analogue of dehydration reaction (P4a). Along this reaction, minor melting (Powell 1983) is fluxed by H_2O released directly into the melt from the breakdown of biotite. As temperature increases, a(H_2O) is internally buffered along the reaction to lower values, while at the same time Crd and Bt become more Fe-rich. No specimens in the aureole contain the sub-assemblage Grt + Sil, so at some point along this reaction Sil is consumed and melting proceeds by the divariant reaction (PM2), Bt + Qtz + Kfs + Crd = L. Garnet appears when isobaric univariant reaction (P11), Bt + Crd + Qtz = Grt + Kfs + L, is intersected. Melting proceeds along reaction (P11) with further decrease in a(H_2O), but in contrast to reaction (P4a), coexisting Crd, Grt and Bt become more Mg-rich as melting proceeds. At 760 °C, a(H_2O) = 0.2, hypersthene is produced at the invariant point by reaction (P12), Bt + Grt + Qtz = Crd + Opx + Kfs + L. If biotite is consumed at the invariant point by reaction (P12), further temperature increase may result in the progress of degenerate reaction Opx + Crd = Grt + Qtz. The late, poikiloblastic texture of biotite in DP568 suggests that it may have formed by 'retrograde reaction' at the invariant point on the down-temperature path.

Above reaction (P3), assemblage A passes unaffected through dehydration reaction (P4a) because it does not contain the reactant sub-assemblage Bt + Sil + Qtz. For the same reason, melting does not commence by reaction (PM1); instead, melting commences at higher temperatures by reaction (PM2), Bt + Qtz + Kfs + Crd + H_2O = L. With the consumption of free vapour, melting proceeds along reaction (PM2) to progressively lower values of a(H_2O). Depending upon Mg/(Mg + Fe), univariant reaction (P11), Bt + Crd + Qtz = Grt + Kfs + L, or invariant point reaction (P12), Bt + Grt + Qtz = Crd + Opx + Kfs + L, may be intersected, resulting in Grt- or Grt + Hy-bearing assemblages.

Figure 16.11 demonstrates that as temperature rises and melting proceeds, a(H_2O) is internally buffered by the melt and solid phases to lower values. Accompanying the lowering of a(H_2O), Mg/(Mg + Fe) ratios of coexisting ferromagnesian phases also change, most particularly along isobaric univariant reactions.

16.7 Equilibrium vs Kinetics in the Metamorphism of Pelites

All of the above discussion implicitly assumes that equilibrium prevailed during the contact metamorphism of the pelites. Although detailed examination of reaction textures were not made in the pelites, such as were made by Heuss-Aßbichler and Masch (Chap. 11, this Vol.) to document kinetic effects in carbonate reactions, a number of lines of evidence suggest that gross equilibrium was attained during contact metamorphism:

1. The mineral assemblages fit consistently and with no overlap into the various trivariant and divariant fields of Fig. 8.20. If there had been significant reaction overstepping due to sluggish kinetics, one might expect to see a higher proportion of low variance assemblages in the aureole. The main exception is the overlap in development of andalusite and sillimanite (see separate discussion below).

A contrast to the orderly behaviour of assemblages in the main aureole may be provided by the narrow aureole surrounding the small stock in the southeast of the area (see Map 2). In the narrow aureole, there is a disproportionately large number of low variance assemblages in Zone V (Pattison 1985). Some Ms + Qtz-rich pelites contain both andalusite and corundum, which may indicate reaction overstepping of reaction (P3). These observations suggest that kinetics (e.g., rate of heating exceeding rate of reaction) may have exerted some control in the development of pelitic assemblages in parts of the narrow aureole, but probably not in the main aureole.

2. The variations in Mg/(Mg + Fe) and Al(Al + Si) between coexisting minerals going through the contact metamorphic zones are fully consistent with an equilibrium model involving continuous and discontinuous reactions (see Figs. 8.21 and 8.22, and detailed discussion in Pattison and Harte, Chap. 8, this Vol.).

3. The sequence and spacing of pelitic mineral zones in the field corresponds to that of the calibrated reaction grid (cf. Figs. 8.1 and 16.1). This correspondence provides strong evidence against significant overstepping of experimentally determined cordierite-producing reactions, in contrast to the suggestions of Putnis and Holland (1986). Putnis and Holland argued that natural cordierite nucleates in a metastable disordered form, in contrast to cordierite used in experiments (e.g., Seifert 1976) which is ordered. According to their thermodynamic estimates, the growth of disordered cordierite would require an up-temperature shift, relative to the experimentally-determined reactions, of about 85 °C for reactions (P1) and (P2b). Referring to Fig. 16.1, this would place reaction (P1) on top of reaction (P3), and would make reaction (P2b) metastable; these shifts would result in a completely different sequence of mineral assemblage zones from that observed.

In this regard, it is notable that some of the reactions at Ballachulish involve cordierite, whilst others do not [(e.g., reactions (P3) and (P5)]. The fact that the sequence and spacing of zones involving *both* kinds of reactions correspond so well with Fig. 16.1 suggests that the experimentally determined equilibria are applicable in nature. It therefore appears that the natural cordierite in the Ballachulish aureole may have nucleated in an ordered form (see Maresch et al., Chap. 14, this Vol., for a more detailed treatment of order and disorder in cordierite).

4. The consistency in both location and inferred temperature of isograds in carbonates and pelites (cf. Map 2 and Table 16.3) suggests that *gross* equilibrium was probably attained in both lithologies, at least in the vicinity of the isograds. This is not inconsistent with the observations of Heuss-Aßbichler and Masch (Chap. 11, this Vol.), who argue that, for a given reaction, systematic differences in reaction textures as one approaches the contact indicate increasing amounts of reaction

overstepping. They ascribe this effect to the rate of heating being faster near to the intrusive contact than far from it. However, it seems unlikely that the amount of overstepping in the vicinity of the isograds could exceed 10–15 °C: devolatilization reactions can be reversed to within 10–20 °C over a period of days or weeks in the laboratory (e.g., Newton 1965; Seifert 1976), at temperatures comparable to those experienced in the Ballachulish aureole. In contrast, heating in the aureole occurred over a period of 10^4–10^5 years (Buntebarth, Chap. 18, this Vol.). Thus, the magnitude of reaction overstepping at the isograds probably falls within the uncertainty of the temperature estimates discussed in this chapter.

16.7.1 Andalusite and Sillimanite

Andalusite and sillimanite occur together in the majority of Al_2SiO_5-bearing assemblages in zone V. In most specimens, even though sillimanite is present, andalusite comprises > 90% of the Al_2SiO_5 polymorph, and appears to be in textural equilibrium with the rest of the assemblage (Pattison and Harte, Sect. 8.3.7, this Vol., Pattison 1991). These 'disequilibrium' features involving Al_2SiO_5 minerals contrast with the regular distribution of other pelitic assemblages in the aureole.

For the following reasons, it is likely that sillimanite formed during prograde contact metamorphism in the aureole: (1) it is widespread in a variety of Al_2SiO_5-bearing assemblages in Zone V, even though it is subordinate volumetrically to andalusite; (2) it increases in proportion to, andalusite as the contact is approached; and (3) it increases in coarseness as the contact is approached. With one exception (Pattison and Harte, Sect. 8.8.4, this Vol.), there is no evidence that the growth of sillimanite was associated with late retrogressive fluid movement, such as suggested by Kerrick (1987) for the development of fibrolite in the Ardara aureole, N. Ireland.

The first development of sillimanite at reaction (P3) lies between the experimentally determined And = Sil boundaries of Holdaway (1971) and Richardson et al. (1969); (Fig. 16.1). Neither boundary on its own is satisfactory in explaining the observed distribution of the two polymorph at Ballachulish: Holdaway's boundary suggests that andalusite should not be stable in the aureole, whilst Richardson et al.'s suggests that sillimanite should not be stable. Based on analyses of coexisting andalusite and sillimanite, Pattison (1989; 1991) argue that, because the two polymorphs have negligible minor element content, the location of the boundary at Ballachulish is not the result of a shift in reaction stability due to reduced activity of one or both polymorphs, as suggested by Kerrick and Speer (1988).

Because of the implied absence of sillimanite at the P-T conditions in the aureole, one must discount the applicability of the And = Sil boundary of Richardson et al. (1969). One might argue that the Holdaway (1971) boundary is valid, and ascribe the delay in growth of sillimanite in the aureole to sluggish kinetics of the andalusite-sillimanite inversion. However, this begs the question as to how the equilibrium could be experimentally reversed to within 50 °C over a period of weeks, at temperatures lower than those in the Ballachulish aureole, if sillimanite could not grow in ~10^5

years in the aureole without substantial (ca. 80 °C) overstepping of the equilibrium boundary. Furthermore, the fact that large modal volumes of andalusite nucleated and grew at the reaction (P3) isograd indicates that andalusite did not merely persist from lower grade. Andalusite *stability* must have extended up to this isograd, and therefore well above the Holdaway (1971) andalusite stability field (Pattison 1991).

Salje (1986) has suggested that there is a family of stable And = Sil curves, each particular to the grain size, degree of disorder, non-stoichiometry and minor element content of the two polymorphs, in particular sillimanite. Salje's results appear to reconcile the different experimental results of Richardson et al. (1969) and Holdaway (1971). It is therefore possible that the And = Sil boundary at Ballachulish is particular to the specific grain size, minor element content etc. properties of the two polymorphs in the aureole, in addition to the rate of heating in the aureole. However, as argued by Pattison (1991), the intermediate And = Sil boundary inferred here satisfies all of the petrological arguments used to support either of the Richardson et al. (1969) and Holdaway (1971) determinations, and may therefore be of general applicability.

The common occurrence of andalusite well upgrade of the first occurrence of sillimanite is ascribed to the sluggish kinetics of the And = Sil polymorphic inversion reaction. The very presence of sillimanite shows that it was able to nucleate and grow, which suggests that the rate determining steps in the inversion reaction must have been either the rate of dissolution of andalusite or presence (or lack) of an effective transport medium. Availability of fluid was likely of crucial importance in this regard.

16.8 Summary

Calibration of the pelitic reactions mapped around the Ballachulish Igneous Complex allows an assessment of the pressure of emplacement of the igneous complex and variation in temperature in the thermal aureole. Additional constraints are provided from geothermometry and geobarometry of high grade pelitic assemblages, calibrated carbonate reactions, calcite-dolomite geothermometry, and the microcline-sanidine transformation.

Four independent methods of calculating pressure range between 2.8 and 3.2 kbar, resulting in a best estimate of 3.0 ± 0.5 kbar. Selected maximum temperatures in the aureole include: 370–480 °C for the growth of talc by reaction (C1); 470–490 °C for the transformation of microcline to sanidine; 550 and 560 °C for the first appearance of cordierite [reaction (P1)] in graphitic and non-graphitic pelites, respectively; 620 °C for isobaric invariant assemblage CIII (Fo + Tr + Di + Cc + Dol); 625 and 640 °C for Ms + Qtz = Al_2SiO_5 + Kfs + H_2O [reaction (P3)] in graphitic and non-graphitic pelites, respectively: 670 °C for the onset of migmatisation in non-graphitic, plagioclase-bearing pelites; 750 °C for the formation of garnet in high grade pelites; and 750–800 °C for the formation of hypersthene in high grade pelites and periclase in dolomitic marbles. From these estimates, the maximum contact temperature is estimated to have been 750–800 °C. Four thermal profiles were constructed across different parts of the thermal aureole, including the narrowest and widest parts; these are used to constrain the thermal models of Buntebarth (Chap. 18, this Vol.).

In graphitic pelites such as the Ballachullish Slate, the interaction of graphite with the predominantly hydrous fluid resulted in the addition to the fluid of C-bearing species such as CH_4 and CO_2. The presence of C-bearing fluid species shifted a(H_2O) to lower values, resulting in the onset of pelitic dehydration reactions at lower temperatures and vapour-consuming melting reactions at higher temperatures in graphitic pelites compared to adjacent non-graphitic pelites. Lowered a(H_2O) expanded the stability field of andalusite-bearing assemblage IVa [reaction (P2a)], thereby explaining the restriction of this assemblage to the Ballachulish Slate. The shift in stability of reaction (P2a) also accounts for the anomalously magnesian compositions of cordierite and biotite in assemblage IVa.

With the onset of partial melting, all available free water was consumed. Further melting proceeded at reduced a(H_2O), with the melting reactions internally buffering a(H_2O) to lower values as melting proceeded. The highest grade pelitic assemblage, Grt + Hy + Crd + Bt, records a(H_2O) of about 0.2 at 760 °C. Prograde paths of the most common pelitic and semipelitic assemblages along a(H_2O)-buffering melting reactions are illustrated in a semi-quantitative T-a(H_2O) diagram.

The regular distribution of pelitic assemblages in the aureole, and the consistency between predicted and measured mineral compositions going through a sequence of continuous reactions, suggest that gross equilibrium prevailed during metamorphism of the pelites. There is no evidence for overstepping of experimentally-determined reactions involving cordierite.

Based on the first occurrence of sillimanite at the reaction (P3) isograd, the position of the And = Sil curve in P-T space is constrained to lie between the And = Sil curves of Richardson et al. (1969) and Holdaway (1971). Persistence of andalusite well upgrade of the first occurrence of sillimanite is ascribed to the sluggish kinetics of the polymorphic inversion reaction, which may have been controlled by the availability of fluid.

17 Stable Isotope Geochemistry on the Intrusive Complex and Its Metamorphic Aureole

S. Hoernes, S. MacLeod-Kinsel, R.S. Harmon, D. Pattison and D.F. Strong

17.1 Introduction

Variations in stable isotope ratios can be particularly helpful in studying the origins of igneous rocks, as well as the fluid interactions that may accompany magma generation or which may occur during and after magma emplacement and solidification (see e.g., the recent review in Valley et al. 1986). In certain instances, and to certain extents, granitic magmas may inherit and preserve the isotopic composition of their source region (O'Neil and Chappel 1977; O'Neil et al. 1977; Chivas et al. 1982; Vidal et al. 1984; Hill et al. 1986). More commonly, mafic partial melts may interact with isotopically evolved crust during magma ascent through AFC-type processes which produce correlated stable-radiogenic isotope covariations within genetically related plutonic suites (Michard-Vitrac et al. 1980; Halliday et al. 1980; Clayburn et al. 1983; Fleck and Criss 1985). Finally, interaction with externally derived fluids at either magmatic or subsolidus stages can be an important process (Friedman et al. 1964; Forester and Taylor 1976, 1977; Hildreth et al. 1984; Fourcade and Javoy 1985; Wickham and Taylor 1987; Bickle et al. 1988).

Despite the fact that there are many unaltered granites with primary $\delta^{18}O$ values in excess of +10‰$_{SMOW}$, particularly within the Hercynian of western Europe (Sheppard 1977; Albarede et al. 1980; Michard-Vitrac et al. 1980; Bernard-Griffiths et al. 1985; Wickham and Taylor 1985), most granitoids worldwide have $\delta^{18}O$ values of +6 to +10‰$_{SMOW}$ and δD values of –90 to 40‰$_{SMOW}$ (Taylor 1968, 1974, 1977; O'Neil and Chappell 1977; O'Neil et al. 1977; Harmon and Halliday 1980; Solomon and Taylor 1981; Masi et al. 1981; Blattner et al. 1983; Xu et al. 1984). Generally, lower O- and H-isotope ratios, or internal mineral ^{18}O disequilibrium, are considered to be a reflection of interaction with an external fluid reservoir of different isotopic composition (e.g., magmatic or subsolidus exchange with meteoric hydrothermal fluids or hydrothermally altered country rocks). However, low δD values can also be produced by magma degassing during cooling and crystallization (Nabelek et al. 1983; Taylor et al. 1983).

This work examines the $^{18}O/^{16}O$ and D/H systematics of rocks and minerals of the Caledonian-age Ballachulish Igneous Complex and thermal metamorphic aureole in southwestern Scotland. Previous stable isotope work on Caledonian plutonic rocks of the Scottish Highlands is quite general in scope. On a regional scale, Harmon and Halliday (1980) and Harmon et al. (1984) observed a wide range of O-isotope variation ($\delta^{18}O$ = +7.2 to +11.0‰$_{SMOW}$), but small range of H-isotope variation (δD

G. Voll, J. Töpel, D.R.M. Pattison, F. Seifert
Equilibrium and Kinetics
in Contact Metamorphism

= −85 to −47‰$_{SMOW}$), and considered these values, in general, representative of primary isotopic compositions, because mineral ^{18}O fractionations are as expected for magmatic conditions and petrographic evidence for secondary, hydrothermal alteration generally is lacking. Combined stable-radiogenic isotope relationships were used by these authors to conclude that most of the granitoids of the Scottish Caledonian Orogen were derived from partial melting of deep-seated, igneous sources, and subsequently had been subjected to variable degrees of contamination in the upper crust prior to crystallization. Two quartz diorite samples from the northwest margin of Ballachulish pluton with $\delta^{18}O$ values of < +7.8‰ and $^{87}Sr/^{86}Sr$ ratios of <0.704 were among the least ^{18}O- and ^{87}Sr-enriched rocks analyzed by Harmon and Halliday (1980). This view of Caledonian granitoid petrogenesis was supported by the detailed study conducted by Clayburn et al. (1983) at the Etive igneous complex, a high-level intrusive center located on the opposite side of the Glencoe volcanic center from the Ballachulish Complex that is similar to it, only larger. Here, multiple crustal contaminants to primitive mafic parental melts were isotopically identified.

The Scottish Caledonian plutons were generated in a convergent plate margin tectonic setting after the Caledonian Orogeny. In striking contrast to the intrusive centers of the Scottish Hebrides, which were emplaced at high structural levels during Tertiary continental dilation and rifting, the effects of large-scale meteoric hydrothermal fluids so common in Hebridean plutons (Forester and Taylor 1976, 1977; Ferry 1985) are lacking in the Caledonian granitoids, except at a very few high-level, subvolcanic centers (Harmon and Halliday 1980; Clayburn et al. 1983).

17.2 Geological Relationships

The Ballachulish igneous complex (Fig. 17.1) is typical of the post-tectonic granitoids of the Scottish Caledonian Orogen classified by Read (1961) as 'Newer Granites'. A detailed geological description is presented by Harte and Voll (Chap. 1, this Vol.) and by Pattison and Voll (Chap. 2, this Vol.). Here we only review important details as they pertain to the samples analyzed and interpretation of our analytical data.

The Ballachulish pluton is a composite, calc-alkaline granitoid ranging in composition from diorites with 52 wt% SiO_2 to granite with 74 wt% SiO_2, emplaced into metasediments of lower Dalradian age late in the Caledonian Orogeny (400 ± 20 Ma; P.E. Brown et al. 1968). The stock was emplaced in two distinct stages (Troll and Weiss, Chap. 3; Weiss and Troll, Chap. 4, this Vol.). Initially a relatively dry dioritic melt at 1050 °C was emplaced into the upper crust at a depth of 10 km. After cooling to a temperature of about 800 °C, this dioritic pluton was intruded by a fluid-rich granitic magma that reached its solidus temperature between 700–650 °C. A small body of microgranite occurs within the central granite, and associated with it is poorly developed porphyry copper-style mineralization (Evans et al. 1980). Contacts between the intrusive complex and the host metasedimentary rocks are discordant and the pluton lacks a tectonic fabric. A transcurrent fault of regional scale post-dates the igneous complex and

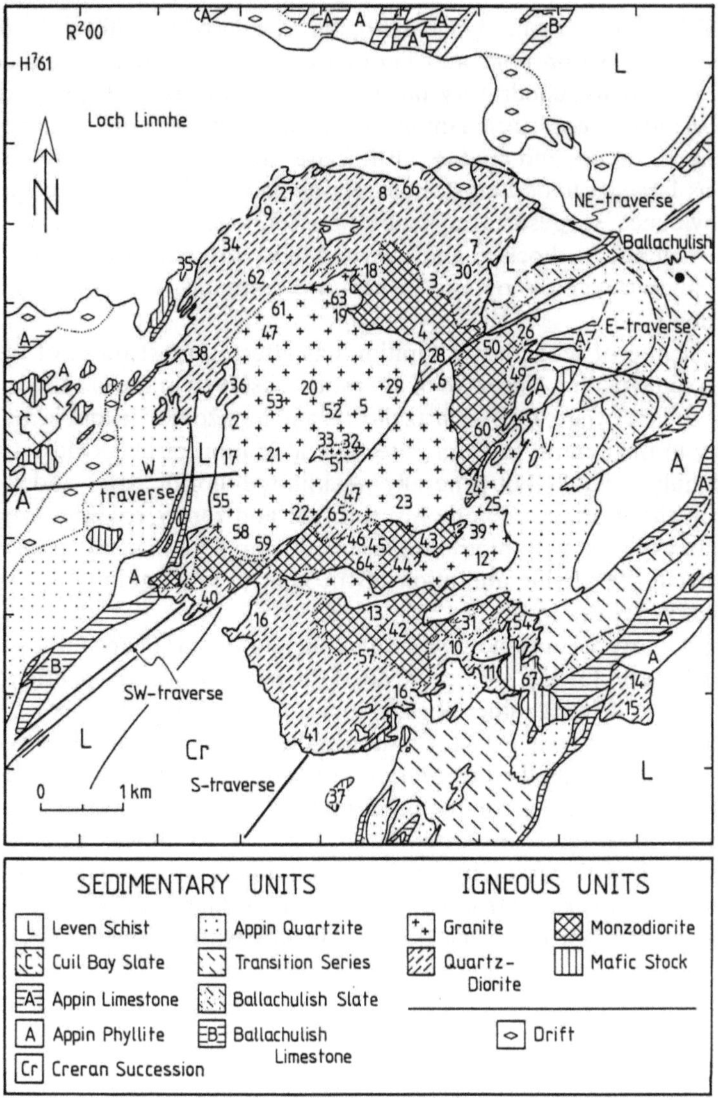

Fig. 17.1. Geological map of the Ballachulish igneous complex and its country rocks, indicating the locations of the igneous samples and of the country rock traverses discussed in this study. *Numbers* refer to samples listed in Table 17.1. SW traverse-D: samples DP83-30, DP358 and DP356; SW traverse-B samples with numbers 13898-13967 listed in Table 17.3

cuts completely through it from NE to SW. Country-rock xenoliths are common in the marginal phases of the intrusive complex.

The mineralogy of the Ballachulish igneous rocks is dominated by plagioclase (25–50 vol%) and alkali feldspar is estimated by Weiss and Troll (Chap. 4, this Vol.) to have constituted 10–20 vol% of the granitic melt at the time of emplacement.

Biotite (5–25 vol%), quartz (5–30 vol%), and K-feldspar (2–40 vol%) occur in all lithologies in variable proportion. Clinopyroxene, orthopyroxene, and hornblende (which occurs both as euhedral crystals and as a reaction product after pyroxene) are present in all but the most silicic samples. Minor and accessory phases include opaque oxides, sphene, zircon, and apatite. Orthoclase first appears as rims on plagioclase and later as large, rounded crystals, rarely as patches interstitial to mafic grains. Textural evidence suggests a crystallization sequence of plagioclase, pyroxene, biotite, hornblende, orthoclase and quartz. Mafic clots consisting of hornblende + biotite + clinopyroxene + opaque oxides are common in rocks of intermediate composition.

Some degree of secondary alteration has affected most rocks of the intrusive complex to some extent. The style of alteration observed in the granite is essentially equivalent to that occurring in quartz diorite and monzodiorite, but is more intense, particularly in the central microgranite. The alteration of plagioclase to sericite ± calcite is highly variable, but generally weak in the dioritic rocks and moderate to extensive in the granites. Commonly chlorite, and rarely epidote, are alteration products after biotite, with biotite replacement complete in some granitic samples. Accompanying the phyllic alteration of the central microgranite are weakly Cu-mineralized quartz veinlets.

The metamorphic aureole surrounding the pluton has been studied in detail by Pattison and Harte (1985), who subdivided the pelitic and semi-pelitic rocks of the aureole into five distinct mineralogical zones. Grade I assemblages of muscovite-quartz-chlorite occur away from the pluton at the very margin or outside the aureole. Zone II is marked by the first appearance of cordierite and Zone III begins where chlorite has reacted away completely. Zone IV is defined by the appearance of andalusite or alkali feldspar. Grade V, cordierite + Al_2SiO_5 + alkali feldspar assemblages, rarely with garnet, hypersthene and spinel, occur adjacent to and within the intrusive complex. Retrograde alteration of high-grade porphyroblasts to micaceous alteration products has occurred locally, within Zone V.

17.3 Analytical Details

Unweathered samples were collected from representative portions of the intrusive complex and along selected traverses into its metamorphic aureole. The analytical work was performed at both Southern Methodist University (silicate $^{18}O/^{16}O$, D/H, and H_2O^+ analyses) and at Bonn University (silicate $^{18}O/^{16}O$, carbonate $^{18}O/^{16}O$ and $^{13}C/^{12}C$ analyses). Whole-rock powders of silicate samples from the intrusive complex were reacted with dilute HCl in order to remove any carbonate present prior to O-isotope analysis. O-isotope ratios were determined for 67 samples from the intrusive complex, for 43 mineral separates and for 96 metamorphic country rocks. H-isotope ratios and contents of H_2O^+ were measured for 51 igneous, 26 mineral separate and 28 metamorphic samples. C-isotope ratios were determined for 23 metamorphic samples.

Methods of silicate oxygen extraction and conversion to CO_2 employed were similar to that described by Clayton and Mayeda (1963). At Dallas, ClF_3 was used as a reagent, whereas F_2 was the reagent used in Bonn. The method of water liberation and conversion to H_2 followed the technique of Bigeleisen et al. (1962). Both whole-rock powders and mineral separates were pre-heated overnight under vacuum at 120 °C to remove observed atmospheric moisture and non-structural water. Total hydrogen yields were measured manometrically to determine precise water contents. Carbonate from metamorphosed limestones and calc-silicate rocks was converted to CO_2 by reaction with anhydrous orthophosphoric acid at 25 °C, following the procedure of McCrea (1950). At Dallas, $^{18}O/^{16}O$ and D/H ratio measurements were made on a Finnigan-251 mass spectrometer; a VG Isogas SIRA-9 was used for $^{18}O/^{16}O$ and $^{13}C/^{12}C$ determinations at Bonn. Major element data utilized in the discussions which follow were obtained by atomic absorbtion and X-ray fluorescence techniques at Memorial University. Isotope ratios are reported here in the usual "δ"-notation as per mill (‰) variations relative to the SMOW standard for O- and H-isotope analyses and relative to PDB standard for C-isotope analyses. Analytical precision is better than ±0.2‰ for oxygen, ±2‰ for hydrogen, ±0.1‰ for carbon isotope ratio measurements, and ±5% of the measured value for water content determinations. In our laboratories, reference samples NBS-28 and NBS-30 yield respective $\delta^{18}O$ and δD values of +9.66‰ and –60‰; NBS-18 has a $\delta^{13}C$ value of –5.0‰ $_{PDB}$.

17.4 Results

Whole-rock O- and H-isotope data for the igneous rocks is given in Table 17.1. Overall within the Ballachulish Igneous Complex, $\delta^{18}O$ variation is +7.5 to +10.9‰ and δD variation is –73 to –48‰ (Figs. 17.2 and 17.3), spanning almost the entire $\delta^{18}O$ and δD ranges previously measured for Caledonian granitoids in the Scottish Highlands (Harmon and Halliday 1980; Harmon et al. 1984). The highest $\delta^{18}O$ value measured of +11.5‰ is for the granite phase of a small stock that is situated some 2 km southeast of the main complex (Fig. 17.1). The quartz diorite phase of this stock has a $\delta^{18}O$ value of +9.9‰, which is also very high relative to most other samples of intermediate composition. The normal mineral ^{18}O fractionations observed for the mineral separates (Tables 17.2 and 17.3) indicate that both $\delta^{18}O$ and δD values for the intrusive complex are essentially primary. From the histogram plots in Fig. 17.2, it is apparent that the range of $\delta^{18}O$ and δD values for the two intrusive phases of the igneous complex are essentially equivalent (I: $\delta^{18}O$ = 7.5 to 10.1‰ and δ = –73 to –52‰; II: $\delta^{18}O$ = 7.5 to 10.9‰; δD = –71 to –48‰), although the phase II rocks, as a group, tend to be enriched in ^{18}O relative to those of phase I. Significantly, the phase I intrusion tends to be more ^{18}O-rich at its margins, where it is in contact with the pelitic and semi-pelitic country rocks of the metamorphic aureole. It may be fortuitous, but rocks of the intrusive complex with the lowest $\delta^{18}O$ values are located in the zones of weakest sericitization, and rocks with the highest $\delta^{18}O$ values are located in the areas where retrograde sericitization is most intense (see Troll and Weiss, Fig.

Table 17.1. Compositional information and whole-rock $\delta^{18}O$, δD, and H_2O^+ values for the igneous complex[a]

Sample	Rock-type	intr.	SiO_2 (wt%)	Larsen index	$\delta^{18}O$ (‰SMOW)	δD (‰SMOW)	H_2O^+ (wt%)
1 Dp 48	Gd	I	—	—	+ 8.8	- 65	0.9
2 DP215	Gr	II	—	—	+ 8.9	- 62	0.6
3 DP217	Gd	I	55.0	6.1	+10.1	- 54	2.0
4 DP218	Gd	I	55.7	2.8	+ 7.9	- 63	1.1
5 DP219	Gr	II	67.0	20.3	+ 9.8	- 48	1.1
6 DP220	Gr	II	—	—	+ 8.3	- 63	1.2
7 DP221	Gd	I	—	—	+ 8.3	- 63	1.1
8 DP227	Gd	I	64.2	14.7	+ 9.1	- 55	1.1
9 DP228	Gd	I	61.1	9.8	+ 7.8	- 68	0.9
10 DP401	Gd	I	—	—	+ 8.9	- 61	1.1
11 DP402	Qd	I	—	—	+ 8.9	- 53	1.6
12 DP405	Gr	II	63.5	16.0	+ 8.6	- 60	0.8
13 DP405	Qmd	I	55.3	5.9	+ 7.5	- 56	1.1
14 DP410	Gr	I	66.2	17.9	+11.5	- 56	1.3
15 DP412	Gd	I	61.1	13.6	+ 9.9	- 52	1.8
16 DP449	Gd	I	61.2	10.5	+ 8.2	- 61	0.9
17 DP454	Gr	II	—	—	+ 8.4	- 59	0.6
18 DP456	Qmd	I	—	—	+ 8.3	- 63	1.0
19 DP459	Gr	II	68.4	21.6	+ 8.9	- 63	0.4
20 DP460	Gr	II	66.8	19.7	+ 7.5	- 68	0.6
21 DP461	Gr	II	—	—	+ 8.9	- 60	0.5
22 DP462	Gr	II	70.4	23.5	+ 9.8	- 49	0.8
23 DP463	Gr	II	—	—	+ 8.7	- 62	0.4
24 DP464	Qmd	I	—	—	+ 8.2	- 61	1.1
25 DP465	Gr	II	64.8	17.1	+ 9.6	- 60	0.7
26 DP524	Gr	II	71.8	24.7	+10.3	- 58	0.5
27 SK29-1	Gd	I	61.9	12.7	+ 8.1	- 63	0.9
28 SK01-1	Gd	I	61.4	16.7	+ 8.8	- 59	0.6
29 SK01-2	Gr	II	67.4	21.3	+ 9.3	- 49	0.6
30 Sk01-3	Qmd	I	61.8	12.0	+ 8.4	- 63	0.8
31 SK01-4	Qmd	I	56.4	3.4	+ 8.0	- 62	0.6
32 B93	Mg	II	—	—	+10.7	- 52	0.8
33 XDR77	Mg	II	—	—	+10.9	- 50	1.2
34 SH3	Gd	I	—	—	+ 8.2	- 67	0.8
35 SH6	GD	I	—	—	+ 9.2	- 71	0.9
36 SH13	Gr	II	—	—	+ 9.4	—	—
37 SH19	A-D	I	—	—	+ 8.9	—	—
38 SH42	Gd	I	—	—	+ 8.2	- 67	1.3
39 WT3	Gd	II	66.7	21.3	+ 9.7	—	—
40 WT32	Qmd	I	57.0	6.5	+ 8.2	- 58	1.2
41 WT83	Qmd	I	58.0	9.4	+ 9.5	- 57	1.4
42 WT86	D	I	52.2	-5.7	+ 8.2	- 57	1.0
43 WT89	Md	I	56.3	3.3	+ 8.3	- 63	0.4
44 WT96	Md	I	53.8	3.8	+ 7.7	—	—
45 WT99	Md	I	56.5	9.3	+ 8.0	—	—
46 WT100	Md	I	56.4	0.1	+ 7.9	- 64	0.2
47 WT102	Gd	II	65.9	19.0	+ 8.7	- 62	0.6
48 WT106	Qmd	II	65.2	20.6	+ 8.2	—	—
49 WT111	Qd	I	54.7	-6.0	+ 8.3	- 65	1.2
50 WT114	D	I	51.9	-0.06	+ 8.0	- 52	2.1
51 WT119	Mg	II	72.3	25.9	+10.4	- 52	0.6
52 WT122	Gd	II	68.9	23.2	+ 9.4	—	—
53 WT124	Gd	II	66.9	21.3	+ 9.0	—	—
54 WT127	Qd	I	—	—	+ 8.3	- 55	1.4

Table 17.1. (continued)

sample	Rock-type	intr.	SiO_2 (wt%)	Larsen index	$\delta^{18}O$ (‰SMOW)	δD (‰SMOW)	H_2O^+ (wt%)
55 WT136	Gd	II	67.5	21.6	+ 9.0	—	—
56 WT156	Qd	I	53.5	7.3	+ 8.4	- 59	1.4
57 WT157	Qd	I	—	—	+ 8.2	- 64	1.3
58 WT162	Qmd	II	65.4	20.5	+ 8.8	—	—
59 WT163	Gd	II	—	—	+ 8.9	—	—
60 WT167	Md	I	55.1	6.8	+ 7.8	- 68	0.6
61 WT203	Mg	II	63.9	19.3	+ 8.9	- 64	
62 WT204	Qmd	I	59.6	14.0	+ 8.7	- 73	1.0
63 WT206	Qmd	II	66.1	20.9	+ 9.1	—	—
64 WT223	Md	I	56.2	7.7	+ 7.9	—	—
65 WT244	Qmd	I	60.9	17.7	+ 8.0	- 63	
66 WT267	Gd	I	61.1	15.7	+ 8.1	—	—
67 WT350	A-D	I	—	—	+ 9.2	—	—

[a]Abbreviations: A-D = Appinitic diorite; D = diorite; Gr = granite; Md = monzodiorite; Mg = monzogranite, Qmd = quartzmonzodiorite; Gd = granodiorite. Chemical data are presented in detail in: Weiss (1986), MacLeod-Kinsel (1987), Weiss and Troll (Chap. 4, this Vol.).

3.10, this Vol.). No pattern of systematic ^{18}O variation is observed for the phase II granite, except that the highest $\delta^{18}O$ values of the main intrusive complex occur in the phyllically-altered and weakly-mineralized microgranite (Fig. 17.3a). H-isotope ratios for both intrusions appear to be randomly distributed, except for the concentration of high δD values in the microgranite and along the fault within the granite (Fig. 17.3b), both of which are portions of the intrusive complex most affected by secondary sericitization and chloritization.

From Fig. 17.2, it is also seen that there is a compositional dependence for the $^{18}O/^{16}O$ ratios that is lacking for D/H ratios. Values of $\delta^{18}O$ tend to exhibit a systematic increase from diorites and monzodiorites with only minor quartz ($x_{13} = +8.1 \pm 0.4‰$), through quartz diorites ($x_{30} = +8.6 \pm 0.6‰$), to granites ($x_{21} = +9.4 \pm 1.0‰$). Also, the evolved rocks tend to be more variable in their stable isotope composition than the more mafic ones. These features are illustrated more clearly in Fig. 17.4, where O-isotope ratios for the intrusive complex are plotted as a function of SiO_2 contents (the same picture would result if plotted against Larsen differentiation index, given in Table 17.1.

Significantly, the two Ballachulish intrusive phases, and the small stock adjacent to the complex in the southeast, exhibit distinct positive trends of $\delta^{18}O$ vs SiO_2. This feature suggests that the original isotopic character of the pluton has been preserved and that there was no isotopic communication between the different intrusive phases via an aqueous fluid during emplacement and cooling of the Ballachulish magmas. The three samples that lie away from the main arrays in Fig. 17.4 (DP217, WT83, and DP460) will be addressed in the discussion which follows.

Mineral separate O-isotope data for 40 samples from 14 rocks is presented in Table 17.2, and H-isotope and water content data for 24 samples from 12 rocks is given in Table 17.3. Observed ranges for $\delta^{18}O$ values are: magnetite = +0.9 to +2.7‰;

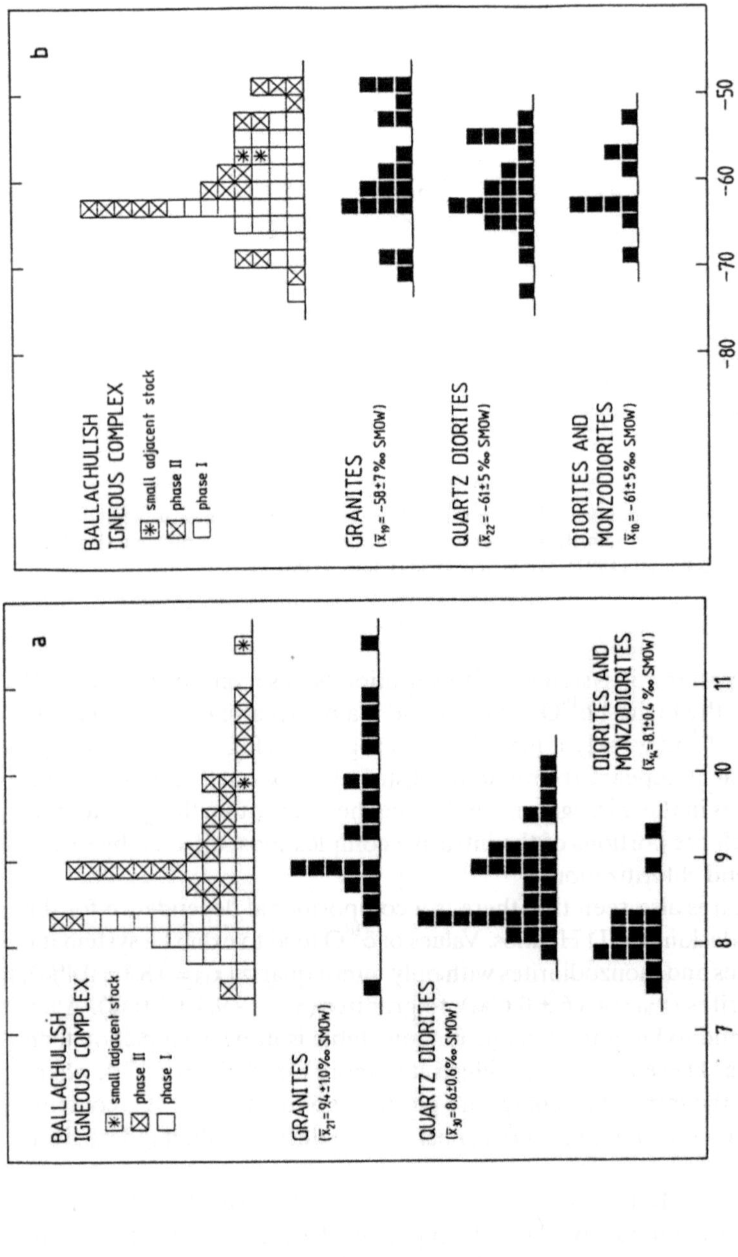

Fig. 17.2. Histograms of whole-rock $\delta^{18}O$ (**a**) and δD (**b**) distributions within the igneous complex. The large range of $\delta^{18}O$ variation (4‰) contrasts sharply with the small range of δD variation (25‰). Although there is no essential difference in D/H between rocks of different composition or intrusive phase, there are systematic variations in $^{18}O/^{16}O$ ratios. In general, there is a tendency for $^{18}O/^{16}O$ ratios to increase with the degree of chemical evolution and for rocks of the first intrusive phase to be less ^{18}O-rich than those of the second phase. A small quartz diorite-granite stock located about 1 km southeast of the main intrusive body is characterized by the highest $\delta^{18}O$ value measured in this study

chlorite = +2.7 to +3.9‰; sphene = +5.2 to +5.5‰; biotite = +4.3 to +6.3‰; hornblende = +5.6 to +6.6‰; and quartz = +9.5 to +10.7‰. Feldspar separates were not analyzed because of the problem of secondary sericitization. For δD values ranges are: hornblende = –79 to –63‰; biotite = –74 to –61‰ sericite (in feldspar) = –57 to –48‰; and chlorite = –43 to –42‰. These orderings of ^{18}O and D enrichments are expected for granitoids that have not been affected by late-stage, sub-solidus ex-

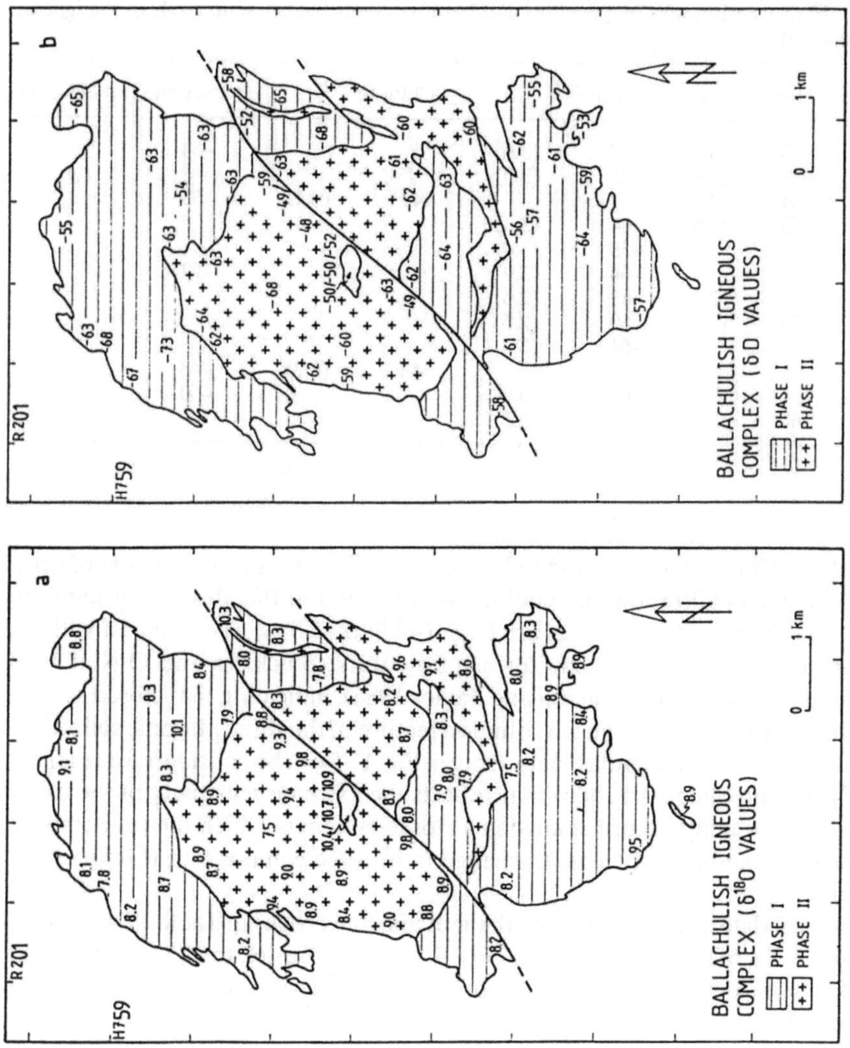

Fig. 17.3. Outline maps of the igneous complex illustrating the geographic distribution of whole-rock $\delta^{18}O$ values (**a**) and δD values (**b**). Note that there are no well-defined spatial patterns of isotopic variation

change with low-D, low-^{18}O meteoric fluids and, therefore, are considered to be magmatic in origin. As discussed in Section 17.5.2, local meteoric waters at the time of emplacement of the melts may not have been extremely depleted in D or ^{18}O. Nevertheless, the combined trends of O- and H-isotope shifts of the secondary mineral phases argue against a significant meteoric component to the involved fluids.

Similarly, $^{18}O/^{16}O$ and D/H fractionations between minerals are typical of sub-solidus, magmatic conditions. This is illustrated in Fig. 17.5, where quartz-biotite ^{18}O fractionations ($^{18}\Delta_{qtz\text{-}bt}$ for the Ballachulish pluton are plotted versus biotite $\delta^{18}O$ values. Shown for reference are the fields documented for the Southern California Batholith by Turi and Taylor (1971) and for the Berridale and New England Batholiths

Table 17.2. Hydrogen isotope compositions and H_2O^+ contents (wt%) of minerals of the igneous complex (in brackets)

Sample No.	Hornblende		Biotite		Chlorite		Sericite (in feldspar)
DP48	–69	(3.0)	–73	(3.6)			–54
DP215	–79	(1.9)	–72	(3.0)			
DP219	–64	(2.6)			–43	(8.4)	–53
DP220	–63	(1.6)	–64	(3.3)			
DP406			–53	(3.4)			
DP454			–64	(3.9)			–57
DP459	–72	(2.2)	–69	(2.6)			
DP460	–79	(2.0)	–69	(2.6)			
DP461			–61	(3.3)			
DP462					–42	(9.2)	–48
DP464			–64	(3.9)			–50
DP465	–69	(1.7)	–74	(2.3)			

by O'Neil and Chappell (1977) and O'Neil et al. (1977). As is characteristic of other plutons where crystallization and cooling has occurred in the absence of meteoric hydrothermal fluids, $^{18}\Delta_{qtz\text{-}bt}$ values are large (+3.9 to +5.8‰) as compared with the restricted range of +1.3 to +3.2‰ for ash-flow tuffs (Stuckless and O'Neil 1973; Lipman and Friedman 1975; Larson and Taylor 1986), indicating that closed system, sub-solidus O-isotope exchange is a common feature of the three plutonic suites. The negative slope of the Ballachulish data array in Fig. 17.5 can be considered a reservoir effect (Taylor and Sheppard 1986) that reflects the extreme sensitivity of the minor phase (biotite) to temperature change in a closed system in which the predominant phase (quartz) exhibits only a small change in isotopic composition (Gregory and Criss 1986).

Analytical data are presented for the metamorphic rocks in Table 17.4. Respective overall $\delta^{18}O$, δD, and $\delta^{13}C$ ranges are +9.3 to +22.3‰; –62 to –22‰; and –1.4 to +4.4‰. The semi-pelitic and pelitic country rocks, in general, are more ^{18}O-rich than the igneous rocks of the intrusive complex, whereas δD values for the metasediments are equivalent or slightly higher than those of the pluton. For the metasediments, $\delta^{18}O$ values increase in the order schists (+9.3–+14.3‰), quartzites (+10.7 to +14.0‰), slates/phyllites (+12.0 to +18.2‰), calc-silicates (+13.3 to +19.4‰) and carbonates (+17.6 to +22.3‰). In general, regional metamorphic country rocks outside the Ballachulish thermal metamorphic aureole tend to be more ^{18}O-rich than inside, although there is no systematic trend of $\delta^{18}O$ with increasing metamorphic grade within the aureole for any particular rock type. For example, country rock slates situated well beyond the aureole zone have $\delta^{18}O$ values of +15.2 to +18.2‰, whereas those phyllites within the contact metamorphic halo have $\delta^{18}O$ values of +12.1 to +15.2‰. Similarly, δD values for the slates (–39 to –21‰) are slightly higher than for the aureole zone phyllites (–62 to –39‰).

Table 17.3. Stable isotope compositions of metasedimentary rocks of the aureole[a]

Area/ traverse	Sample No.	Rock type	Aureole zone	Contact dist. (m)	$\delta^{18}O$ (‰SMOW)	δD	$\delta^{13}C$ (‰PDB)	H_2O^+ (wt%)
SW	DP641	Leven schist	IVb	550	+12.0	−45		2.5
SW	DP358	Leven schist	III	900	+11.0	−40		3.8
SW	DP356	Leven schist	I	1300	+11.1	−46		3.9
S	13898	Xenolith	V		+ 9.7			
S	13911	Leven schist	V	0.3	+11.0			
S	13913	Leven schist	V	10	+11.4			
S	13921	Leven schist	V	35	+11.5			
S	13924	Leven schist	V	100	+11.4			
S	13933	Leven schist	IV	220	+12.4			
S	13936	Leven schist	IV	235	+12.8			
S	13940	Leven schist	IV	240	+12.5			
S	13943	Leven schist	IV	290	+12.5			
S	13946	Leven schist	III	430	+12.0			
S	13951	Leven schist	II	510	+12.8			
S	13954	Leven schist	I	563	+11.2			
S	13954	Leven schist	I	610	+13.7			
S	13955	Leven schist	I	627	+12.3			
S	13963	Leven schist	I	680	+12.4			
S	13967	Leven schist	I	800	+14.3			
S	13970	Leven schist	I	894	+11.1			
S	13975	Leven schist	I	1010	+12.5			
S	13978	Leven schist	I	1240	+12.7			
S	DP626	Leven schist	V	100	+11.6	−52		1.9
S	DP627	Leven schist	IVb	200	+11.9	−50		1.7
S	DP377	Leven schist	III	320	+11.3	−51		2.8
S	DP629	Leven schist	I	670	+11.4	−52		3.9
S	V-LC	Leven schist		13000	+11.7			
W	DP5	Leven schist	V	50	+ 9.3	−55		4.0
W	DP5	Leven schist	V	30	+10.9	−53		2.8
W	A67-1	Lsh-leucosome	V		+11.5			
W	A67-11	Lsh-paleosome	V		+12.2			
W	A93-1	Lsh-leucosome	V		+11.6			
W	A93-2	Lsh-paleosome	V		+12.2			
W	SK29-13	Appin quartzite	IVb		+10.7			
W	SK29-11	Appin phyllite	III		+12.0	−43		2.8
W	SK29-10	Appin quartzite	II		+12.0			
W	DP16	Appin phyllite	I	3200	+14.0	−39		3.3
W	B5	Appin phyllite			+10.6			
W	B20	Appin quartzite		3000	+12.5			
W	B22	Appin quartzite		2800	+11.5			
W	B9	Appin quartzite		150	+11.0			
W	B41	Appin quartzite		100	+12.0			
W	B45	Appin quartzite		30	+12.2			
W	B44	Appin quartzite		25	+12.3			
W	B43	Appin quartzite		5	+12.3			
NE	DP49	Leven schist	V	20	+10.0	−52		2.4
NE	DP50	Leven schist	IVb	150	+ 9.3	−47		4.5
NE	DP482	Leven schist	I	700	+11.3	−53		2.7
NE	DP157	Leven schist	I	850	+11.8	−56		2.7
E	DP601	Appin phyllite	Vb	40	+12.7	−38		4.1
E	DP42	Appin phyllite	IVb	280	+12.9	−56		1.4

Table 17.3. (continued)

Area/ traverse	Sample No.	Rock type	Aureole zone	Contact dist. (m)	$\delta^{18}O$ (‰SMOW)	δD	$\delta^{13}C$ (‰PDB)	H_2O^+ (wt%)
E	DP22	Appin phyllite	IVb	500	+12.1	–55		1.7
E	DP505	Appin phyllite	IVb	550	+13.0	–51		2.6
E	SK01-5	Appin quartzite			+12.2			
E	SK01-6	Appin quartzite			+12.3			
E	SK01-7	Appin quartzite			+13.6			
E	SK01-8	Appin quartzite			+14.0			
E	DP125	Appin phyllite	III		+12.4	–62		2.4
E	DP134A	Appin phyllite	II		+13.4	–56		3.1
E	DP140	Appin phyllite	I		+15.2	–55		2.1
E	DP104	Appin phyllite	I		+14.1	–60		3.3
E	SH150	Dol. marble		700	+19.6		+2.7	
E	SH230	Dol. marble		620	+18.7		+3.8	
E	SH426	Dol. marble		600	+17.6		+3.2	
E	SH435	Dol. marble		430	+22.3			
E	SH307	Dol. marble		410	+19.3			
E	SH309	Dol. marble		350	+19.4		+2.3	
E	SH312	Dol. marble		250	+21.1		+4.4	
E	SH429	Dol. marble		140	+21.0		+3.2	
E	SH314	Dol. marble		100	+21.5		+3.0	
E	SH315	Dol. marble		70	+22.0		+3.8	
E	SH316	Dol. marble		40	+19.4			
E	SH317	Dol. marble		25	+21.4		+3.2	
E	SH421	Calc-silicate rock		790	+14.2		+2.3	
E	SH221	Calc-silicate rock		740	+13.3		–1.0	
E	SH223	Calc-silicate rock		715	+16.8		+3.0	
E	SH224	Calc-silicate rock		640	+16.8		+1.9	
E	SH225	Calc-silicate rock		535	+19.4		+2.7	
E	SH310	Calc-silicate rock		450	+16.0		+2.5	
E	SH311	Calc-silicate rock		390	+16.3		+3.7	
E	SH226	Calc-silicate rock		385	+15.6		+3.3	
E	SH227	Calc-silicate rock		235	+16.4		–0.8	
E	SH228	Calc-silicate rock		230	+16.6		+1.8	
E	SH428	Calc-silicate rock		170	+15.2		–1.4	
E	SH229	Calc-silicate rock		75	+16.7		+1.9	
E	W305	Psammopelite		430	+13.4			
E	W304	Psammopelite		340	+13.5			
E	SH379a	Pelite		330	+14.7			
E	SH379b	Psammopelite		330	+14.3			
E	W293	Psammopelite		265	+12.5			
E	W294	Psammopelite		245	+13.7			
E	W295a	Pelite		210	+15.2			
E	W295b	Psammopelite		210	+14.3			
E	W297a	Pelite		70	+16.1			
E	W297b	Psammopelite		70	+14.3			
E	W298	Pelite		10	+14.8			

[a]Metamorphic zones labelled after Pattison and Harte (1985).
Abbreviations: dol = dolomitic; Lsh = Leven schist; LC = Leven schist from Crearan Inn, Loch Crearan.

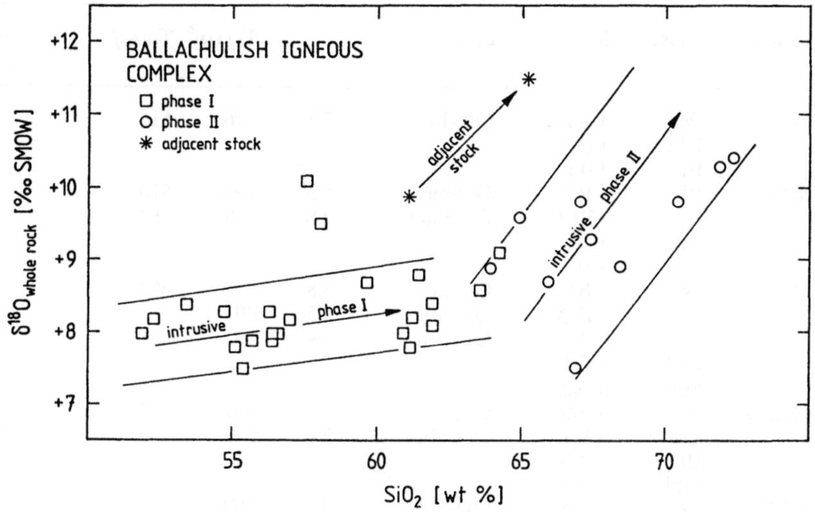

Fig. 17.4. Plot of whole-rock $\delta^{18}O$ values as a function of bulk composition as indicated by SiO_2 content. With three exceptions (DP217, DP460, and WT83), $\delta^{18}O$ values for the samples of each intrusive phase tend to increase systematically with changing bulk composition. The different *slopes* for each sample group suggests a lack of interaction or isotopic communication during emplacement and cooling. The steeper trends for the phase II rocks implies either a greater degree of contamination by a ^{18}O-rich crustal reservoir or assimilation and exchange with a different and more ^{18}O-rich contaminant

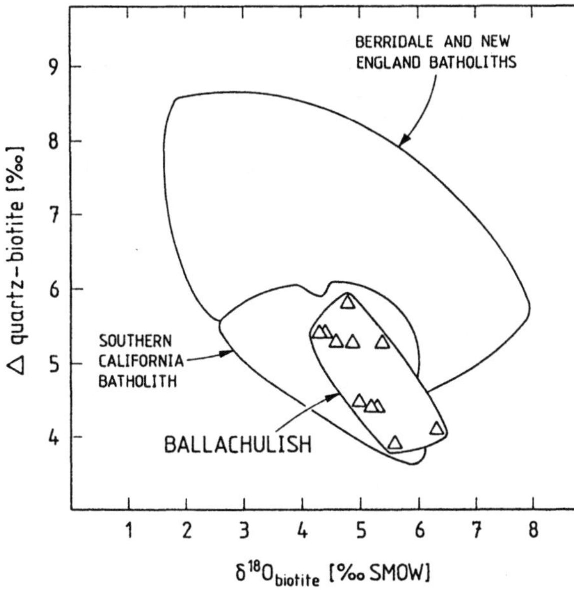

Fig. 17.5. Plot of $\Delta^{18}O_{quartz\text{-}biotite}$ vs $\delta^{18}O_{biotite}$, comparing the Ballachulish igneous complex (denoted by *triangles*) with plutonic granitoids in southeastern Australia (O'Neil and Chappell 1977; O'Neil et al. 1977) and western North America (Turi and Taylor 1971). These normal quartz-biotite fractionations are indicative for subsolidus isotopic exchange during cooling in the absence of meteoric hydrothermal fluids

Table 17.4. Mineral O-isotope compositions (‰ SMOW), fractionations (‰0 and calculated temperatures

Sample	Phase	δ O	$\Delta_{(A-B)}$		$T_{(R\&H)}$[b]	$T_{(B\&J)}$[c]	T_c	Mineral radius (mm)
SH3	Wr[a]	+8.2	Qtz-bio	5.3	496	516		
	Qtz	+10.2						
	Bt	+4.9						
SH6	Wr	+9.2	Qtz-bio	5.3	496	516		
	Qtz	+10.7	Qtz-mag	9.8	555	482		
	Bt	+5.4						
	Mg	+0.9						
DP48	Wr	+8.8	Qtz-bt	5.8	462	494		
	Qtz	+10.6	Qtz-hbl	4.5	493			
	Bt	+4.8	Qtz-sph	5.4	432			
	Hbl	+6.1						
	Sph	+5.2						
DP215	Wr	+8.9	Qtz-Bt	5.4	487	512		
	Qtz	+9.7	Qtz-hbl	3.6	573			
	Bt	+4.3						
	Hbl	+6.1						
DP219	Wr	+9.8	Qtz-hbl	4.3	507			
	Qtz	+9.9	Qtz-chl	7.2	377			
	Hbl	+5.6						
	Chl	+2.7						
D220	Wr	+8.4	Qtz-bt	4.4	559	589		
	Qtz	+9.7	$Qtz\text{-}hbl_{Al\text{-}Fe}$	3.2	617	675	613	0.15
	Bio	+5.3	$Qtz\text{-}hbl_{parg}$	3.2	661		783	0.1
	Hbl	+6.5					788	0.08
DP406	Wr	+7.5						
	Bt	+4.9						
DP454	Wr	+8.4	Qtz-bt	5.3	496	514		
	Qtz	+9.9	Qtz-sph	4.3	501			
	Bt	+4.6						
	Sph	+5.5						
DP459	Wr	+8.9	Qtz-bt	4.4	559	586		
	Qtz	+9.6	$Qtz\text{-}hbl_{Al\text{-}Fe}$	3.8	553	614		
	Bt	+5.2						
	Hbl	+5.8						
DP460	Wr	+7.5	Qtz-bt	5.4	487	511		
	Qtz	+9.8	$Qtz\text{-}hbl_{Al\text{-}Fe}$	3.8	553	597	595	0.1
	Bt	+4.4					724	0.07
	hbl	+6.0					743	0.05
DP461	Wr	+8.9	Qtz-bt	4.5	551	578		
	Qtz	+9.5						
	Bt	+5.0						
DP462	Wr	+9.8	Qtz-chl	6.1	429	564		
	Qtz	+10.0						
	Chl	+3.9						
DP464	Wr	+8.2	Qtz-bt	3.9	604			
	Qtz	+9.5	Qtz-cpx	2.9	646	604		
	Bt	+5.6	Qtz-mg	6.8	720			
	Cpx	+6.6						
	Mg	+2.7						
DP465	Wr	+9.6	Qtz-bt	4.1	585	600		
	Qtz	+10.4	$Qtz\text{-}hbl_{Al\text{-}Fe}$	3.8	553			
	Bt	+6.3						
	Hbl	+6.6						

[a]Abbreviations: wr = whole rock; qtz = quartz; bt = biotite; hbl = hornblende; chl = chlorite; cpx = clinopyroxene; sph = sphene.
[b]R&H = Richter and Hoernes (1988).
[c]B&J = Bottinga and Javoy (1975).
Tc = closure temperatures (see text for explanation).

17.5 Discussion

17.5.1 Magma Origin

In a recent review article discussing stable isotope fractionation and isotope systematics during magmatic processes, Taylor and Sheppard (1986) formulated what they termed the 'central dogma' of the O-isotope geochemistry of igneous rocks, i.e., that all relatively ^{18}O-rich or ^{18}O-depleted silicate melts on earth (e.g., magmas with $\delta^{18}O < 4.5$ or $\delta^{18}O > 7.5$‰) must have been derived at least in part from, or have exchange with, a precursor material that once upon a time resided on or near the earth's surface. It has also been recognized by several workers (e.g., Matsuhisa 1979; Kalamarides 1984; Sheppard and Harris 1985; Woodhead et al. 1987) that closed-system fractional crystallization of a basaltic magma to an evolved differentiate is unlikely to produce an ^{18}O enrichment of more than 1.0 to 1.5‰.

At Ballachulish, the lowest $\delta^{18}O$ values observed are +7.5‰ for monzodiorite DP-406 and granite DP-460, both of which exhibit normal, magmatic mineral ^{18}O fractionations (Table 17.4). An additional 13 samples from the phase I intrusion have $\delta^{18}O$ values between +7.8 and +8.2‰. Of the 16 remaining samples with $\delta^{18}O$ values of >8.3‰, 8 are situated along the eastern and southern margins of the complex where the position of the contact between the pluton and the metasedimentary country rocks is clear. An exception is sample DP-217, which also lies off the main $\delta^{18}O$ -SiO_2 (or LDI) trends in Fig. 17.4. In DP-217, calcite appears as an alteration product in plagioclase and also as a matrix mineral surrounding and replacing hornblende. This feature, which is not observed in other samples from this part of the pluton (e.g., DP-218 and SK01-03), and the presence of limestone xenoliths in the northern margin area of the intrusion (Muir 1953a,b) that presumably were derived from the Appin or Ballachulish limestones which outcrop north of Loch Leven, suggests an incorporation and assimilation of limestone by the magma at this locality prior to crystallization. A similar explanation may apply to sample SH-83, which also lies off the main trend in Fig. 17.4, with the contaminant in this case having been an ^{18}O-rich phyllite rather than limestone.

We infer, therefore, that $\delta^{18}O$ values for parental Ballachulish diorite magma, prior to its emplacement in an upper crustal magma chamber, was in the restricted range of +7.5 to +8.2‰. This is consistent with a derivation from an igneous, rather than metasedimentary type protolith. It is difficult to identify this source region precisely, but it is unlikely that the Ballachulish parent magmas were primary melts derived from the melting of upper mantle peridotite, because such melts would be expected to have $\delta^{18}O < +6.5$‰. Rather, we envisage that the parental mafic magma has an origin deep in the lower crust, or at the crust-mantle boundary, and represents a crustally contaminated and subsequently homogenized primary basaltic melt. Such a melting, assimilation, storage and homogenization (MASH) process has been described recently by Hildreth and Moorbath (1988), and may provide a general explanation for the origin of the post-tectonic ('Newer') granitoids of the Scottish Caledonian Orogen. The higher $\delta^{18}O$ values seen in the diorites of the phase I intrusion at Ballachulish, as well as the compositional dependencies of $\delta^{18}O$ values illustrated in Fig. 17.4, therefore can be viewed as the products of combined fractional crystallization-assimilation processes that occurred during magma transit through and during residence in high-^{18}O upper crust prior to crystallization and cooling.

The higher $\delta^{18}O$ values of the granodiorite and granite of the later phase II intrusion, and the different compositional dependency of $\delta^{18}O$ values for this unit (Fig. 17.4), requires a more complex origin. Either the phase II magmas were derived from a different source region that was much more metasedimentary in character, or contamination was more extensive and/or involved a different and more ^{18}O-rich crustal reservoir. We favor the latter explanation. Weiss and Troll (Chap. 4, this Vol.) consider the phase I mafic/intermediate and phase II intermediate/felsic intrusions to be related by hybrid magma mixing. The overlap of the data arrays for the two intrusions in Fig. 17.4 supports this conclusion. Turi and Taylor (1971) have pointed out that the large volumes of assimilated country rock, which are necessary to produce a shift in O-isotope ratios of 2 to 4‰, should necessarily be correlated to changes in mineralogy if the contaminant is an ^{18}O-rich pelite, as might be the case in Ballachulish. Muscovite and lesser amounts of sillimanite would be expected, and hornblende would be consumed by the reaction: hbl + sill + musc = plag + bt + qtz. Muscovite was not observed in any sample from the main intrusive complex, but is present in samples DP-410 and DP-412, from the small stock to the southwest of the pluton, which are 1–2‰ enriched in ^{18}O at a given SiO_2 content or LDI value compared to rocks from the main intrusive complex (Fig. 17.2). Hornblende is not present in these two samples, although it is common in samples of similar bulk composition from the main complex. Thus, it is likely that assimilation of ^{18}O-rich metasedimentary crust was an important process in the petrogenetic history of the small stock. The similar ^{18}O trends in Fig. 17.4 for the small stock and the Ballachulish phase II rocks suggests that the high $\delta^{18}O$ values for the latter are also a product of variable degrees of contamination of a parental intermediate composition melt by ^{18}O-rich pelitic, metasedimentary upper crust that is similar to that in which the Ballachulish intrusion presently resides.

17.5.2 Fluid Processes

One of the important contributions of stable isotopes to the study of magmatic processes has been the recognition of the role of fluids, both in the magmatic and hydrothermal environments (see e.g. H.P. Taylor 1974; B.E. Taylor et al. 1983; Nabelek et al. 1983; Sheppard and Harris 1985; Wickham and Taylor 1985). The gain, be it direct or indirect, or loss of a volatile fluid phase by a magma should produce a recognizable shift in isotopic composition. Because the amount of oxygen in a silicate melt is extremely high, H-isotope compositions are more responsive to fluid processes than O-isotope compositions.

We have noted previously that the igneous rocks of the Ballachulish Igneous Complex exhibit normal, magmatic mineral $^{18}O/^{16}O$ fractionations and demonstrate that its stable isotope composition was not modified significantly by sub-solidus exchange with a meteoric, hydrothermal fluid. Nabelek et al. (1983) have recognized the isotopic effects on a magma of vapor exsolution during crystallization from their detailed study of $^{18}O/^{16}O$ and D/H variations in the Notch Peak granitoid stock in Utah, USA. In the section which follows, we make direct comparison of the Ballachulish and Notch Peak intrusions, both of which are composite plutons emplaced into metasedimentary country rocks at comparable depths in the crust, to illustrate the different geological processes that may operate in the plutonic environment.

Figure 17.6a is a plot of whole-rock δD values vs whole-rock $\delta^{18}O$ values. In contrast to the Notch Peak situation, where δD variations are large relative to $\delta^{18}O$ variations, at Ballachulish $^{18}O/^{16}O$ ratios are substantially more variable than D/H ratios and no clear differences in isotopic composition are observed for the two phases of intrusion. However, the Ballachulish can be divided into three groups, on the basis of combined δD-$\delta^{18}O$ systematics and position in the intrusive complex. The first group (group A in Fig. 17.6a) consists of granites with δD = –52 to –48‰ and $\delta^{18}O$ = +9.3 to +10.9‰. These six D-rich granites are those that have been subject to extensive secondary alteration that produced chlorite after biotite, sericitization of plagioclase, and oxidation of orthoclase. Two of these samples also contain calcite veinlets (DP-462 and DP-219) and two other samples contain mineralized quartz

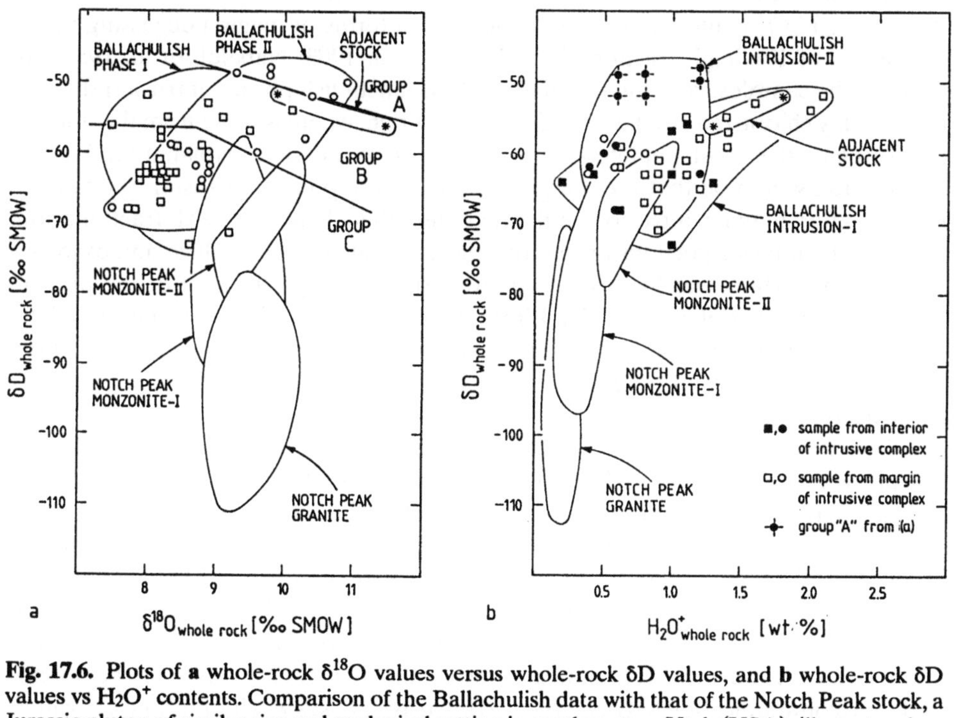

Fig. 17.6. Plots of **a** whole-rock $\delta^{18}O$ values versus whole-rock δD values, and **b** whole-rock δD values vs H_2O^+ contents. Comparison of the Ballachulish data with that of the Notch Peak stock, a Jurassic pluton of similar size and geological setting in southwestern Utah (USA), illustrates clear differences in isotopic composition and water content. *Squares* denote rocks of the intrusive phase I and *circles* denote the granitic rocks of phase II. *Asterisks* label samples of the separate stock SE of the main Complex. See text for discussion

veins (XDR-77 and B-93). The six Group-A samples lie in a narrow zone in the centre of the intrusive complex, parallel and adjacent to the northwest side of the fault that crosscuts the pluton (Fig. 17.1). The second group (Group B in Fig. 17.6a) includes samples from the marginal zone of the pluton that have relatively high H_2O^+ contents (>1 wt%) and high δD and $\delta^{18}O$ values, but are not strongly chloritized. The third group (Group C in Fig. 17.6a) includes samples with low H_2O^+ contents from the interior of the pluton that petrographically are most pristine.

Additional insight into the role of volatiles during crystallization and cooling of the Ballachulish pluton is provided by Fig. 17.6b, in which the whole-rock δD values are plotted as a function of sample water content. Although the δD-H_2O^+ fields for the two phases of the Ballachulish intrusion overlap completely, except for the six altered samples from the center of the complex, there is evidence to suggest that fluid, and therefore isotopic communication between them was limited. The samples of the phase I dioritic intrusion exhibit a larger spread of D/H variation in Fig. 17.6b, with δD values ranging from –73 to –52‰ and H_2O^+ contents ranging from 0.2 to 2.2 wt%, and define a distinct trend of decreasing δD with decreasing water content. Similar, but more pronounced trends of this type are exhibited by the three Notch Peak intrusions (Nabelek et al. 1983). The degassing trend observed by Taylor et al. (1983) for obsidian clasts from tephra and flow obsidians from a subaeric volcanic eruption at the Inyo Domes in California, USA, is described by the composite trend for the Ballachulish-Notch Peak intrusions. The shift to lower D/H values with decreasing H_2O^+ contents has been explained in terms of magma degassing under conditions of Rayleigh fractionation (Taylor et al. 1983; Nabelek et al. 1983). In general, the samples from the margins of the Ballachulish phase I intrusion that are only weakly chloritized are the most D- and H_2O-rich, whereas those from the interior portions are the most depleted in D and H_2O. A possible explanation for this observation would be that the marginal zones of the phase I intrusion crystallized rapidly, after magma intrusion and prevented fluid loss and D/H fractionation, whereas the interior portions of the magma took longer to crystallize and exsolved D-rich vapor which was lost to the country rocks through the roof of the magma chamber. An influx of a D-rich fluid from the country rocks into the margins of the cooling intrusion would be expected to have induced some hydrothermal alteration, but, marginal samples of the phase I intrusion, as previously noted, are not strongly chloritized or otherwise altered.

The situation for the phase II granite at Ballachulish appears to contrast quite sharply with that for the earlier dioritic intrusion. With the exception of the six strongly sericitized and chloritized granites that define Group A in Fig. 17.6a, the phase II granite tends to be very homogeneous in terms of its δD-H_2O^+ characteristics. Values for δD range from only –71 to –59‰ and water contents fall between 0.4 and 1.2 wt%, with no obvious dependence of δD values on H_2O^+ contents. The fact that δD differences on the order of 10‰ are preserved across the contact between the phase I diorites and phase II granite at some localities (Fig. 17.3b), is further evidence that fluid communication between the two intrusive phases was limited and δD values in the earlier intrusion were not affected significantly during crystallization and cooling of the later intrusion. Rather, the distribution of δD values in Fig. 17.3b and the intense alteration of the granite samples situated in the central parts of the intrusion along the northwest side of the fault and in the mineralized microgranite, suggests that the phase I intrusion had essentially crystallized and cooled to some considerable extent before the intrusion of the phase II granite magma. The dioritic shell, therefore, would have provided an impermeable barrier to fluid loss outward and caused volatiles to be concentrated toward the centre of the crystallizing granite magma into the area where secondary alteration is observed to be most intense.

Several important details implicit in the discussion above the clarified in Figs. 17.7 and 17.8, where δD values of the various separated phases from individual samples are plotted vs whole-rock and mineral H_2O^+ contents, respectively. For the phase II intrusion, whole-rock δD values are largely independent of H_2O^+ contents as would be expected if the secondary alteration of this unit was produced internally, as discussed above. For the phase I intrusion, there is a hint of a positive correlation of whole-rock δD values and H_2O^+ contents based upon the three samples analyzed (Fig. 17.7). From Fig. 17.8 it is seen that biotite is more variable in D/H ratio than coexisting hornblende, and that biotite δD values increase linearly with increasing H_2O^+ contents. This increase in δD value for biotite with increasing H_2O^+ content should reflect the amount of D-rich, secondary chlorite and, to a lesser degree, sericite present in a sample. Typically, rocks containing only hornblende or visibly unaltered biotite have δD values of < –60‰, whereas those in which the chloritization of biotite or sericitization of plagioclase has been particularly intense have δD values of < 50‰.

Suzuoki and Epstein (1976) also observed from isotope exchange experiments on hydrous mineral-H_2O systems that at high temperatures (>500 °C), the relative tendency for hydrous minerals in equilibrium with the same water to concentrate D is: Fe-biotite = Fe-hornblende < Mg-biotite = Mg-hornblende < chlorite < muscovite. For the Ballachulish intrusion, hornblende and biotite are close in H-isotope composition and always have lower δD values than their corresponding whole-rock

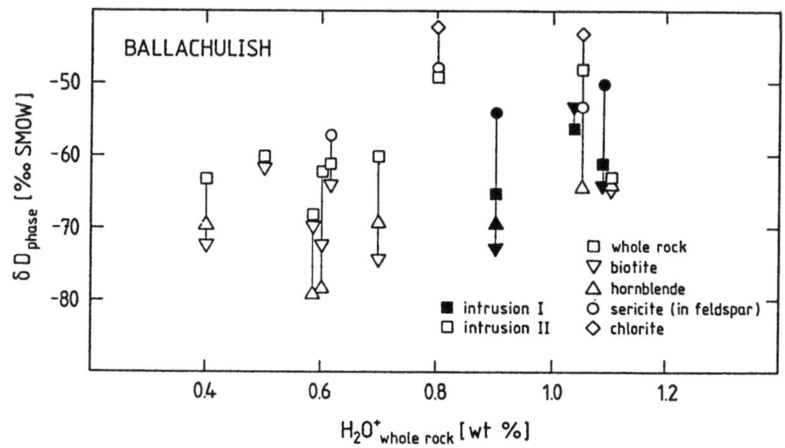

Fig. 17.7. Plot of mineral and whole-rock δD values versus whole-rock total water content. Note that (1) whole-rock δD values tend to increase with increasing water content; (2) sericite and chlorite δD values are significantly higher than biotite or hornblende δD values; and (3) the biotite-hornblende D/H fractionation is quite variable and reversed in two of six cases

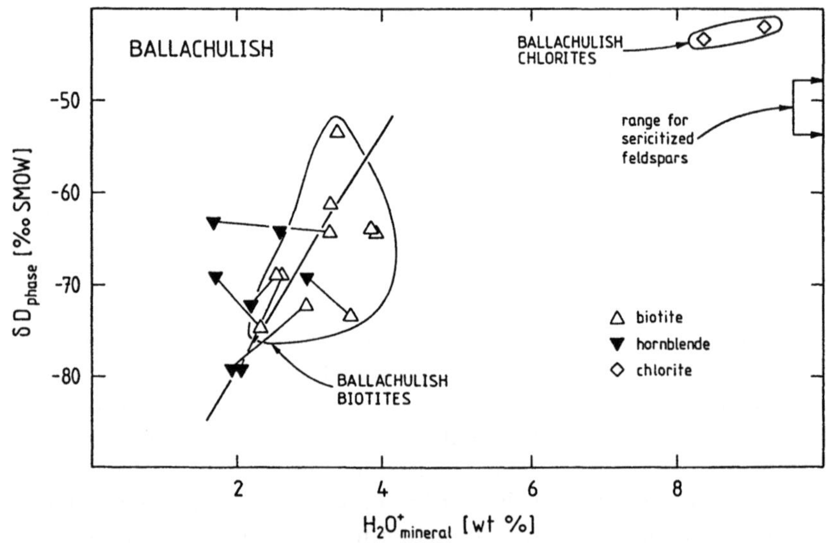

Fig. 17.8. Plot of δ values versus structural water content for hydrous mineral separates. The increase in the δD value in biotite with increasing H_2O^+ content (indicated by the *bold line*) indicates that the δD value of biotite is a function of the extent of replacement by ^{18}O-rich chlorite

samples. By contrast, δD values for chlorite and sericite are higher than for their corresponding whole-rock samples, but the expected D/H fractionation is reversed such that chlorite is enriched significantly in D relative to sericite. This disequilibrium may reflect the fact that sericite and chlorite crystallized at different times in equilibrium with fluids of continually evolving isotopic composition, or that both alterations occurred at a relatively low temperature (≤300 °C) where there is a crossover in the chlorite-muscovite D/H fractionation (Graham et al. 1987). Most importantly though, both secondary chloritization and sericitization at Ballachulish led to an increase in whole rock δD values, a feature which is opposite to that expected when secondary alteration occurs through interaction with a low-D, hydrothermal-meteoric fluid (Taylor 1977).

Despite the intrinsic difficulties in the identification of 'juvenile' waters, as recently discussed by H.P. Taylor and Sheppard (1986) and B.E. Taylor (1986), mineral D/H ratios often can be used to discriminate deep-seated fluid reservoirs (be they either magmatic' or metamorphic' in nature) from surface-derived fluids ('meteoric' waters). Because the stable isotope composition of surface waters depends on temperature and, therefore, on both latitude and altitude (Friedman 1953; Craig 1961; Dansgaard 1964). Ideally, these parameters should be known for the Ballachulish area at the time of intrusion some 400 million years ago, before a meteoric component to the fluids responsible for the secondary, hydrothermal alteration and mineralization at Ballachulish is excluded entirely.

Based on the experimentally determined mineral-water fractionation curves of Suzuoki and Epstein (1976) for biotite, Graham et al. (1984) for hornblende, and

Graham et al. (1987) for chlorite, the hydrogen isotope compositions of coexisting fluids can be calculated from the δD mineral data in Table 17.3, if the temperature of the equilibration is known. Similarly, fluid $\delta^{18}O$ values can be calculated from the mineral data in Table 17.2 using the relationships compiled in Friedman and O'Neil (1977) or more recently in O'Neil (1986), or using calculated fractionations as described by Richter and Hoernes (1988). The calculated isotopic compositions of fluids in equilibrium with hornblende and biotite at 500 °C, chlorite at 400 °C, and mineralized vein quartz fluid inclusion waters at 350–450 °C are plotted in Fig. 17.9 (see also Table 17.5). Shown for comparison are the fields generally assigned to 'primary magmatic waters' (PMW), 'normal' igneous hornblendes and biotites, and 'metamorphic waters' (Sheppard et al. 1969; Sheppard 1986), as well as the 'meteoric water line' of Craig (1961). Importantly, fluids at Ballachulish in equilibrium with hornblende fall within the field for magmatic fluids, whereas those for biotite and chlorite are displaced to significantly higher δD values. A fluid component derived from the local D-rich metasedimentary country rocks by exchange during the alteration of biotite and the formation of chlorite cannot be excluded. However, it is more likely that the high δD values of the late-stage alteration phases reflect D enrichment in residual magmatic fluids, particularly in intrusion II, resulting from both the crystallization of D-depleted hornblende and biotite (Fig. 17.7) and the progressive migration of D-rich exsolved fluids toward the centre of the magmatic complex as

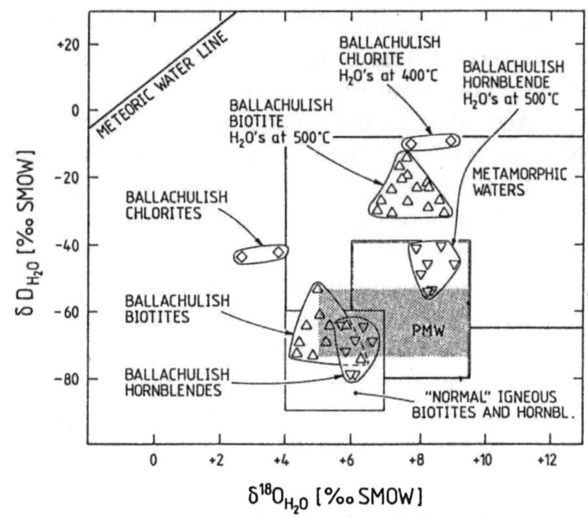

Fig. 17.9. Plot of δD and $\delta^{18}O$ values calculated for fluids in equilibrium with biotite, hornblende, and chlorite. Symbols used are as follows: ∇, Hornblende; ∇, H^2O in equilibrium with hornblende at 500 °C; Δ, Biotite; Δ, H^2O in equilibrium with biotite at 500 °C; ◊, Chlorite; ◊, H^2O in equilibrium with chlorite at 400 °C. The designations *MWL* and *PMW* refer to the "meteoric water line" (Craig 1961) and the field of "primary magmatic water" (Sheppard et al. 1969), respectively. The *shaded area* indicates the possible compositions of mineralizing fluids from fluid inclusion H-isotope and vein-mineral O-isotope analysis (Table 17.5). No evidence is observed for extensive involvement of meteoric hydrothermal fluids

Table 17.5. O- and H-isotope compositions of vein minerals and fluid inclusions from the altered and mineralized portions of the igneous complex

Sample	$\delta^{18}O$ calcite	δ^{18} O quartz	δD_{H2O} meas.	$\delta^{18}O_{H2O}$ calc.[a]
		‰ SMOW		
B93	—	+ 9.7	–73	–0.9 (300 °C) +6.5 (500 °C)
XDR77	—	+13.3	–53	+6.4 (350 °C) +9.2 (450 °C)
DP219	+11.71	—	—	+6.1 (300 °C) +8.4 (400 °C)
DP462	+12.54	—	—	+9.5 (400 °C)

[a]Quartz-H_2O fractionations from Bottinga and Javoy (1975) and calcite-H_2O fractionations from O'Neil et al. (1969).

crystallization and cooling progressed inward. The alteration of biotite to chlorite, and plagioclase to sericite, are attributed to such a situation, where residual D-rich vapor that has been concentrated in the center of a cooling intrusion migrates slowly upward and outward, inducing subsolidus alteration of susceptible phases with which it comes into contact. That chlorite and sericite H-isotope fractionations are only equivalent at ≤300 °C (Suzuoki and Epstein 1976; Graham et al. 1987), indicates that either this was the temperature of the late stage alteration of biotite and plagioclase or that the alteration of these two phases was decoupled, and either occurring in the presence of a fluid of different isotopic composition (i.e., at different times) or at different temperatures.

According to the paleocontinent reconstructions of Briden et al. (1988) and Morel and Irving (1978), the Scottish Caledonides were within 20–30° latitude of the equator during early Ordovician time. By reference to modern meteoric waters (Craig 1961), it would be expected that paleo-meteoric water δD and δ^{18}O values would have been less or equal (–30 to –20‰ and –5 to –4‰, respectively), if elevations were low. However, mountainous areas near the coast in regions of low latitude can have very low δD values, as demonstrated by Ford and Green (1977) for the Ok Tedi porphyry copper deposit in New Guinea. Given that the Ballachulish-Glen-Coe-Etive arc was a Convergent plate margin volcanic region, and the fact that the meteoric hydrothermal alteration at the margin of the Etive Complex is not characterized by D enrichments (Harmon, unpublished data), we do not consider the shift toward higher δD values observed for the most altered Ballachulish samples to reflect the involvement of meteoric-hydrothermal fluids. If that were the case, the secondary chlorite and sericite should exhibit low δ^{18}O values and disequilibrium ^{18}O fractionations, which they do not.

17.6 The Metamorphic Aureole

Variations in D/H, $^{18}O/^{16}O$ and $^{13}C/^{12}C$ ratios were determined for different areas of the Ballachulish metamorphic aureole, and along selected traverses outward from the pluton into the metasedimentary country rocks to examine the nature and extent of isotopic exchange between the pluton and its surrounding environment. As noted above, respective $\delta^{18}O$ and δD values for the regionally metamorphosed, upper-greenschist grade pelitic country rocks are in the range of +15 to +18‰ and –39 to –21‰, whereas quartzites and carbonates have $\delta^{18}O$ values of about +12‰ and > +20‰, respectively.

The overall picture is one of variability, rather than consistency, with isotope ratios exhibiting no consistent pattern of variation with grade in the contact aureole. In some cases, isotopic ratios behave in a coherent manner, decreasing sympathetically toward the pluton, whereas for other traverses changes of the opposite sense or decoupling of the parameters is observed. The various contact zone areas and traverses are discussed in detail in the sections which follow.

17.6.1 The Southwestern and Southern Traverses

In this area, the Leven Schists strike approximately perpendicular to the pluton contact, a situation which provides the maximum potential for the migration of fluids away from the intrusion. In general, the outer dioritic ring lacks abundant hydrous phases, but the southwestern and southern part of the pluton contain the greatest modal proportion of hydrous phases like hornblende and biotite (>25 vol%, see Fig. 4.10b in Weiss and Troll, Chap. 4, this Vol.), suggesting that water-saturated conditions may have been present locally in this area during magma crystallization. Thus, a release of water after magmatic crystallization, and subsequent transport into the surrounding country rocks, is a possibility. The presence of some low-$\delta^{18}O$ Leven Schists suggests that there was a fluid interaction in this part of the aureole on a local scale. Close inspection of the data in Table 17.3 for this part of the aureole, which is governed by pelitic Leven Schists (in this study, for simplicity not specified into Creran Succession or Leven Schist after Litherland 1980), shows that the $\delta^{18}O$ variation is independent of distance from the pluton contact. Given that some Leven Schists as far as 13 km from Ballachulish are low in ^{18}O as well, the $\delta^{18}O$ values around +11‰ must have been acquired by the rock prior to contact metamorphism, probably by interaction with an external fluid during regional metamorphism (Wickman and Taylor 1985; Bickle et al. 1988). This will be the subject of further investigations. The decreasing H_2O contents of the metapelites with decreasing contact distance in this part of the aureole indicate that all retrograde effects in these rocks (like the pinitization of cordierite) is more likely due to internally trapped water.

17.6.2 The Western Area

None of the analyzed rock groups from this part of the aureole (Leven Schists, Appin Phyllites or Appin Quartzites) exhibit systematic isotopic trends as a function of distance from the pluton contact. Not even the majority of samples from the 'Chaotic Zone' of migmatization as described by Pattison and Harte (1988), where Leven Schists form the lower metamorphic protolith are shifted to values lower than 11‰. Pattison and Harte (1988) have demonstrated that country rock melting is only possible through introduction of externally derived fluids into the schists at high temperature (650–680 °C). The quartz diorite of the intrusive complex lies shallowly beneath this zone, and exsolution of magmatic fluid during crystallization ($T^{solidus}$ = 750–680 °C) is thought to be the process responsible for melting and migmatization. Like other high grade rocks in this area, the migmatites have been partially altered to chlorite-muscovite assemblages during a retrograde event, but none of these processes had a significant large-scale imprint on stable isotope composition.

It seems likely, however, that some low $\delta^{18}O$ values (<10‰), such as those of sample 13898, a xenolith of Leven Schists within the pluton (S-traverse), or DP5, a Leven Schist within the 'chaotic zone of migmatization', were influenced by isotopic exchange with intruding melt or with fluids released from the crystallizing pluton.

17.6.3 The Eastern and Northeastern Traverses

In contrast to the southern sector of the aureole, the strike of the metasediments in the eastern sector is approximately parallel to the pluton contact. This situation has reduced the possibilities for fluid migration, and is reflected by the isotopic compositions (Table 17.3). The presence of dolomitic marbles and calc-silicate rocks is an other difference encountered in the eastern sector. The dolomitic marbles (Appin Limestone) are uniform in isotopic composition (x_{12} $\delta^{18}O$ = +20.3 ± 1.5‰ and x_{10} $\delta^{13}C$ = +3.3 ± 0.6‰). The calc-silicate rocks display lower $\delta^{18}O$ values (+13 to +19.4‰), most likely due to internal exchange with the primary silicates present in these rocks (Valley and O'Neil 1984). Pelitic rocks (Appin Phyllites) along the eastern traverse exhibit a trend toward higher $\delta^{18}O$ values approaching the pluton contact, but this feature is probably a compositional artifact. The samples labelled 'A' and 'B' in Table 17.3 are the pelitic and semi-pelitic components of samples, separated in space by only a few centimeters. The preservation of ^{18}O gradients as large as 2‰ on the centimetre scale indicates clearly the absence of fluid flow even as close as 100 m to the contact during contact metamorphism.

The Leven Schists in the NE-traverse again are very low in ^{18}O, compared to normal pelites; the samples closer to the contact from metamorphic zones IVb and V are even lower than those from Zone I. It is an open question if this is due to contact heating and concomittant exchange. It is interesting to note, yet not completely understood, that the Appin Phyllites do not show the precontactmetamorphic isotopic shift that obviously had affected the Leven Schists.

Summarizing the results obtained from whole rock analyses from four country rocks, we have to conclude that no large-scale hydrothermal convection was set in motion by the intrusion, and that there was not a significant flow of fluid from the crystallizing pluton into the laterally adjacent country rocks. The water under-saturated magmas of phase I, forming the outer rim of the cross section as exposed today, and the low permeability of the country rocks at approximately 10 km depth are most likely the reasons for this lack of exchange of matter between pluton and country rocks. The fluids released during and after crystallization of phase II granitoids thus were forced to leave the site of generation in only one direction: upwards. This chanallized flow of fluids is indicated by zones of increased retrograde alteration as observed by Weiss and Troll (Chap. 4, this Vol.).

17.7 Oxygen Isotope Fractionations

The O-isotope fractionations between coexisting minerals of igneous rocks show characteristic differences between plutonic and volcanic rocks. Depending on the cooling rate, minerals become closed systems for oxygen isotope exchange at different temperatures (cf. Taylor and Sheppard 1986). The question is, however, if the minerals exchange the oxygen isotopes during cooling by diffusion only, by surface reaction or recrystallization in the presence of a fluid or by a combination of these mechanisms. A comprehensive treatment on these mechanisms has been presented by Cole and Ohmoto (1986). Based on the equation of Dodson (1973), Giletti (1986) calculated closing temperatures (T_c) taking into account the known diffusion coefficients, some mineral parameters and an assumed cooling rate. Theoretical mineral compositions then are calculated and may be compared with the analyzed $\delta^{18}O$ values. Fitting calculated and measured values the actual cooling rate can be estimated. The mineral data and the calculated temperatures are compiled in Table 17.4. The fractionation factors of Richter and Hoernes (1988) form the basis of the temperature calculation. The increment method used by these authors has the advantage that it takes into consideration the actual mineral composition, which is important if these minerals are members of complicated solid-solution series (e.g., hornblende, chlorite). In addition, fractionation curves can be calculated for minerals where an experimental calibration is lacking. For a comparison, temperatures also are calculated based on the fractionation data of Bottinga and Javoy (1975). The temperature concordance is acceptable for most fractionations (Table 17.4).

17.7.1 Discussion

According to Giletti (1986) oxygen isotope data of slowly cooled igneous and metamorphic rocks may yield some information on the cooling rate, provided several conditions are fulfilled. These conditions are that: (1) the minerals change their isotopic composition during cooling by volume diffusion; (2) the fluid phase, if present, should be stagnant; and (3) the diffusion constants of the minerals, the mode of the rock, the grain size, the whole-rock $\delta^{18}O$ value, and the mineral $\delta^{18}O$ values are known. If the first condition is fulfilled and cooling of the rock is sufficiently slow, then the magmatic O-isotope equilibrium fractionations that existed between mineral phases at the time they crystallized will not be preserved, because each mineral will become a closed system at a different temperature as cooling progresses.

However, many investigations have shown that conditions of O-isotope equilibrium do exist, even in slowly cooled plutonic rocks (O'Neil 1986). Recrystallization and mineral formation have been demonstrated to be the most important mechanisms for oxygen isotope equilibration in metamorphic rocks by Hoernes and Friedrichsen (1978) and Hoernes and Hoffer (1979). The Ballachulish intrusive complex may serve as a model to test the Giletti (1986) hypothesis and as well may contain informations on the mechanisms of post-magmatic O-isotope exchange in plutonic rocks as one important condition, the cooling rate, unknown in most cases, is known for the Ballachulish intrusion.

According to Buntebarth (Chap. 18, this Vol.) the centre of the pluton cooled from about 1050 to 500 °C in 700 000 years. The cooling at the outer portions of the pluton was even faster; the rocks there cooled about 500 °C in only 600 000 years. Using these cooling rates, and the diffusion data given by Giletti (1986), closure temperatures have been calculated for quartz, hornblende, and biotite for some of the Ballachulish samples analyzed (Table 17.4).

In all cases, closure temperatures are higher than the temperature values calculated from O-isotope equilibrium fractionation curves. This, as well as the fact that minerals equilibrated at much lower temperatures than the calculated T_C-values, as can be seen from nearly concordant temperatures in several samples (Table 17.4), suggests strongly that volume diffusion was not the rate determining mechanism for O-isotope exchange. These data, and the isotopic evolution of the fluid phase as reconstructed from oxygen and hydrogen isotope data, suggests that the minerals equilibrated with a fluid phase by surface reaction or recrystallization by a solution-redeposition process. Because only a little information exists on the kinetics of such mechanisms (for a recent compilation of data, see Cole and Ohmoto 1986), we can not verify the cooling rate for the Ballachulish complex as deduced by Buntebarth (Chap. 18, this Vol.) based on the analyzed O-isotope data.

Nevertheless, we observe that a monzodioritic rock of the first intrusion (DP 464), which belongs to the 'chilled margin'-variety of Troll and Weiss (Chap. 3, this Vol.), and exhibits only traces of retrograde phase transformations, displays the highest O-isotope temperatures. The characteristic temperature differences of the individual mineral fractionations of this particular sample (qtz-mg = 720 °C; qtz-cpx = 646 °C; qtz-bt = 604 °C) might be explained by the closure temperature model of Gilletti (1986); unfortunately, several parameters necessary for this calculation are not available.

17.8 Conclusions

The stable isotope data presented in this study have demonstrated that a granitoid magma, intruding at upper crustal levels, does not necessarily promote large hydrothermal convection cells which transport heat from the cooling intrusive body into the country rocks. This is most likely due to: (1) the low permeability and low H_2O-content of the host regional greenschist-amphibolite metamorphic country rocks; and (2) to the two-stage intrusion of the Ballachulish intrusive complex, where the first, relatively dry melts sealed the contact between the second stage, wet granitic magma and the country rocks. Thus, as a whole, the heat transfer was mainly carried out by heat conduction. Any substantial large-scale transport of matter via fluids from the pluton into the laterally adjacent sedimentary country rocks appears unlikely. The stable isotope data allow constraints on the petrogenesis of the magma: the parental mafic magma probably originated from the deep lower crust or the crust-mantle boundary area where it interacted with higher ^{18}O crust through a MASH-type process. Further modification occurred by combined fractional crystallization-assimilation processes during magma transit through and during residence in high-^{18}O upper crust prior to crystallization and cooling. This process was most likely responsible for the large variation in O-isotope composition of the phase II intrusion.

Some of the mineral phases which crystallized from the melt subsequently re-equilibrated during cooling at subsolidus conditions with a fluid phase exsolved from the crystallizing magma. The near equilibrium fractionations at lower temperatures than expected from closure temperatures of the particular minerals indicate a solution-redeposition process as a mechanism for O-isotope exchange. Only in rocks, which cooled in absence of an aqueous fluid phase, as indicated by the lack of retrograde phase transformations, can be observed a fractionation pattern typical for diffusive exchange.

18 Thermal Models of Cooling

G. Buntebarth

18.1 Introduction

The Ballachulish intrusion consists of concentric rings of quartz diorite, monzodiorite and granite in the centre. This suite of rocks intruded Lower Dalradian metasediments, consisting mainly of pelite with minor quartzite. Petrological investigations (Pattison, Chap. 16; Troll and Weiss, Chap. 3, this Vol.) estimate an emplacement depth of the intrusion of 10 km, and a pre-intrusion temperature of the metasediments as approximately 250 but not more than 300 °C (Pattison and Voll, Chap. 2, this Vol.).

In order to model the cooling of the igneous complex and thermal aureole, not only the geometry of the intrusion must be known, but also the thermal properties of the rocks. The thermal conductivity and the thermal diffusivity of the common rock types, along with their temperature dependencies, have therefore been determined. Other factors which are thermally relevant, i.e., dehydration reactions in pelites, and movement of magmatic fluids through the wall rocks, are taken into account.

18.2 Modelling of Rock Types in the Igneous Complex

The intrusive complex consists of three different rock types: granite, quartz-diorite and diorite. As exposed at the surface, they show an approximate concentric structure. Geological data support a very steep or a nearly vertical contact zone of the magma body to the wall rock, except for the western and southeastern portion of the intrusion (Troll and Weiss, Chap. 3, this Vol.). The core of the intrusive body consists of granite, having a diameter of 3400 m at the surface. The core is surrounded by a monzodioritic-dioritic ring of an average thickness of 550 m, enclosed by an outer ring of quartz diorite of an average thickness of 650 m.

Limits for the total vertical extension of the intrusive complex may be given as ≥ 0.5–1.0 km above sea level (Weiss 1986) and ≥ 4 km below sea level (Rabbel and Meissner, Chap. 7, this Vol.). Exposed metasediments within the thermal aureole zone reflect a pressure of $p = 0.3$ GPa (Pattison, Chap. 16, this Vol.), corresponding to a depth of about 10 km. If the cylindrical complex extends about 2 km above and about 4 km below the present surface, it may be considered to be an infinite cylinder, since the distances within the thermal aureole from the contact do not exceed 2 km.

G. Voll, J. Töpel, D.R.M. Pattison, F. Seifert
Equilibrium and Kinetics
in Contact Metamorphism

There are two possible scenarios for the intrusion sequence. (1) In the first case, a central block of country rock subsided within the granitic magma. However, it is neither exposed at the surface nor detected by geophysical methods (Rabble and Meissner, Chap. 7, this Vol.), but it may be inferred from comparison with other intrusive complexes in Scotland (e.g., Ben Nevis). (2) In the second case, just before the fractionated central portion of the cylindrical dioritic magma solidified, the granitic magma was emplaced within the non-solidified core, probably involving the extrusion of a residual rhyodacitic magma and the subsidence of the surrounding, already consolidated diorites (Troll and Weiss, Chap. 3, this Vol.).

In the first case, where a core of country rock exists beneath the inner granite intrusion, there are only two intrusions which affected the wall rocks. The first ring intrusion of a thickness of 550 m was a dioritic magma (assuming no distinction between monzodiorite and diorite). During partial solidification of these intermediate rocks, a quartz diorite intruded between the first ring and the wall rock, forming a second ring of 650 m thickness. For simplification, the thickness of each ring is rounded off to 600 m. This concept yields the model of Fig. 18.1a. The intrusion temperature of the granitic intrusion uprising later was much lower (~800 °C, Weiss and Troll, Chap. 4, this Vol.) and had no effect upon the maximum temperatures reached in the aureole.

The second model considers a cylindrical dioritic intrusion having a radius of 2900 m. When solidification proceeded towards the centre, the inner core which was still viscous was replaced by ascending granitic magma. The solidified dioritic ring at the time of granite intrusion would then have been 1200 m thick. The resulting model is shown in Fig. 18.1b.

The third model which must be taken into consideration is based on the fact that the dip of the contact between intrusion and wall rock is not well known. If the angle was somewhat less than 90 °C, a conical intrusion is necessitated to calculate the cooling history (Fig. 18.1c).

Whereas the models of Figs. 18.1a and 18.1b are two-dimensional, considering an azimuthal heterogeneity in the thermal properties of the rocks, the conical model is three-dimensional. Inhomogeneities in the depth-distance plane can thus be incorporated.

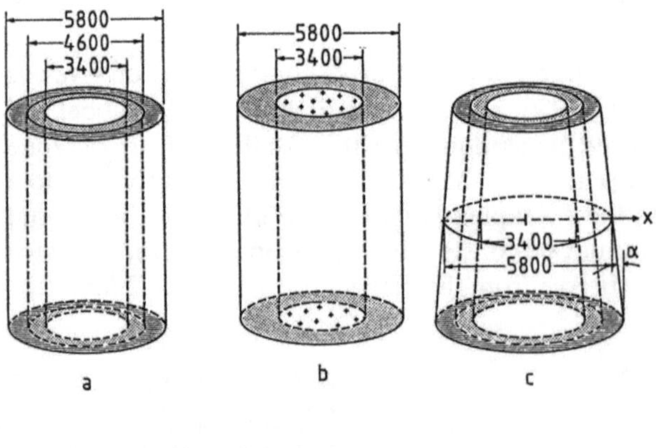

Fig. 18.1. Model bodies of the Ballachulish intrusion. **a** Ring intrusion and **b** cylindrical intrusion of infinite length; **c** conical intrusion

18.3 Thermal Properties of Rocks from the Igneous Complex and Aureole

The thermal properties of rocks were measured using a divided bar apparatus. This heat-flow meter method is described by Francl and Kingery (1954). The divided cylindrical bar is heated by a guard heater, i.e., around the heat source, the heat sink, the sample and the reference. This heating system is insulated by a fibrous ceramic material. The guard heater system minimizes radial heat flow. The diameter of the divided bar is 50 mm and sample thickness may vary between 5 and 10 mm, the thickness commonly used is 8 mm.

The mean grain size of the measured samples usually does not exceed 1 mm, which means that the heat flux penetrates about eight to ten grains of the rocks. The temperature at the surface boundaries between heat source and sample, between sample and reference probe and finally between reference probe and heat sink are measured using three thermocouples in each boundary. The divided bar device is put under a hydraulic press to apply uniaxial pressure of up to 60 MPa.

The thermal diffusivity, as well as the conductivity, can be determined during each run. During a linear increase of the temperature at one side of the divided bar, the temperature increase at the boundary between sample and reference probe results from the diffusivity of the sample and reference probe. After the heat source has reached the maximum temperature and a stationary state is reached in the divided bar, the thermal conductivity can be determined. More details about this method are given by Buntebarth (1987, 1991 in prep.). The temperatures which can be applied to determine the variation of the thermal properties with temperature range between 40 and 550 °C.

The relative accuracy of the apparatus used is somewhat less than 2%. The thermal diffusivity, however, cannot be determined with such precision. In general, the results can be reproduced within 10% of the value.

18.3.1 Thermal Properties of Intrusive Rocks

There are two main rock types in the intrusive complex: (1) granite; and (2) diorite. Both occur as fine-grained rocks at the chilled border zones and also as coarser-grained samples from the more slowly cooled centre of the complex. A systematic variation of the thermal properties with grain size was not deduced, because not only does the grain size vary but also the proportion of the main mineral constituents.

The thermal conductivity (K) of granite is somewhat higher than that of quartz diorite, but the difference is small enough not to be distinguished in the model calculations. A mean value (Fig. 18.2a) for both rock types, including temperature dependency, is given by the expression:

$$K[\mathrm{W/m°C}] = (0.44 + 2.4 \cdot 10^{-4} T[°\mathrm{C}])^{-1}.$$

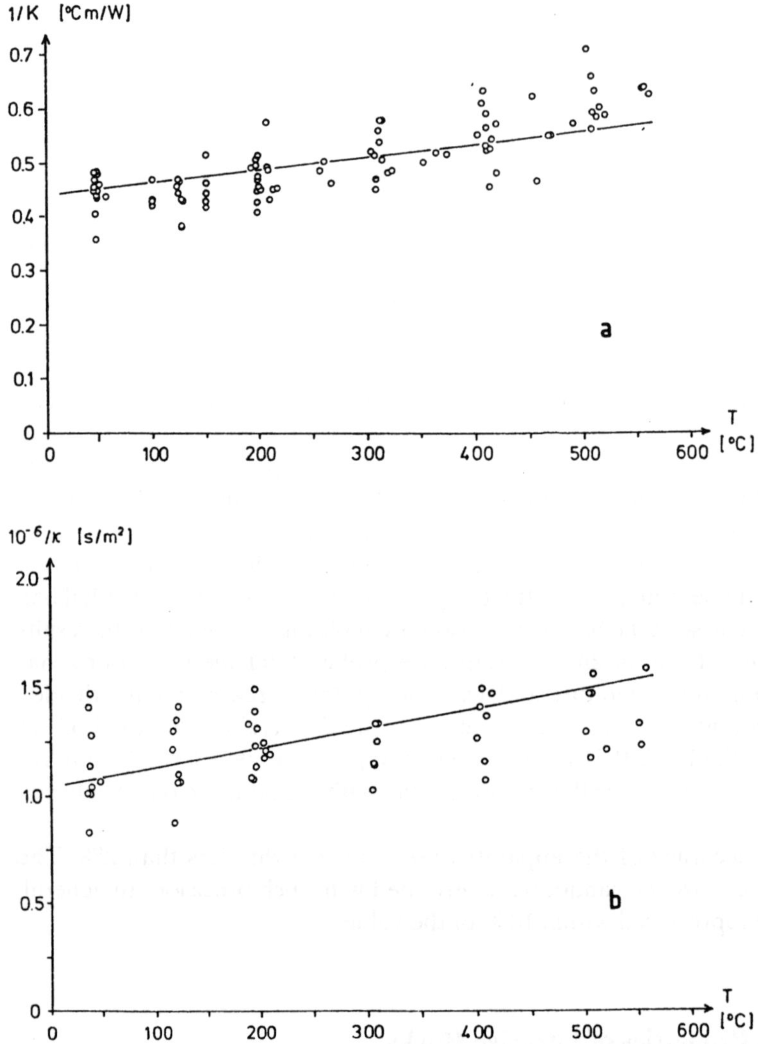

Fig. 18.2 a. Thermal resistivity of granite, quartz diorite, and appinite at uniaxial pressure of 10 MPa; **b** reciprocal thermal diffusivity of granite, quartz diorite, and appinite at pressure of 10 MPa

This expression was determined for a contact pressure of 10 MPa. For measurements above 450 °C, the temperature sensitivity of the conductivity increases in many samples due to thermal cracking, which produces additional thermal resistivities. Thermal cracking can be kept low by applying more elevated contact pressure. If the pressure is increased to 50 MPa, thermal cracking does not influence the matrix properties up to 500 °C.

The thermal diffusivity of granite and quartz diorite is more sensitive to temperature variations than thermal conductivity. As shown in Fig. 18.2b, the thermal diffusivity of both rock types cannot be distinguished. The temperature dependent diffusivity of the two rock types is given by the expression:

$$\kappa[10^{-6}\mathrm{m}^2/\mathrm{s}] = (1.04 + 9.2 \cdot 10^{-4}T[°\mathrm{C}])^{-1}.$$

This expression was determined at a contact pressure of 10 MPa.

18.3.2 Thermal Properties of Wall Rocks

The most common wall rocks at the intrusive complex are Appin Quartzite, Leven Schist, pelitic Creran Succession, Appin Phyllite and Ballachulish Slate. Other wall rocks of much less importance for the cooling history are Appinite, which is a gabbroic rock rich in hornblende, and Ballachulish Limestone. Whereas quartzite is a good heat conductor, in general, pelitic rocks, i.e., Leven Schist, Creran Succession and Ballachulish Slate, conduct heat more poorly.

The pure Appin Quartzite, a medium-grained quartzite with mineral constituents other than quartz less than 10%, has a heat conductivity of 7 W m^{-1} $°C^{-1}$ at a temperature of 40 °C. This value decreases to 3.5 W m^{-1} $°C^{-1}$ at 300 °C, the upper limit of country rock temperature prior to the intrusion. The mean value of several Appin Quartzite samples is given by the expression:

$$K[\mathrm{W/m°C}] = (0.17 + 5.39 \cdot 10^{-4}T[°\mathrm{C}])^{-1}.$$

This gives values about 14% below the maximum value (Fig. 18.3a). An anisotropy in the thermal conductivity cannot be observed.

The thermal diffusivity of Appin Quartzite (Fig. 18.3b) is very temperature sensitive. The mean value, for all quartzite samples is given by the expression:

$$\kappa[10^{-6}\mathrm{m}^2/\mathrm{s}] = (0.306 + 16.9 \cdot 10^{-4}T[°\mathrm{C}])^{-1}.$$

The thermal diffusivity also does not show any significant anisotropy.

In contrast to Appin Quartzite, the pelitic Ballachulish Slate and the pelitic Leven Schist and Creran Succession demonstrate a remarkable anisotropy in their thermal properties. The anisotropy is due to the parallel arrangement and stretching of mineral grains. Figure 18.4 shows the orientation of mica and quartz grains in a sample of Ballachulish Slate, with the following modal mineralogy: 22% muscovite, 36% quartz, 15% graphite and 27% calcite. Whereas all mica grains are strongly arranged parallel to the schistosity, the quartz-grains do not exhibit any preferred orientation. It is possible that the anisotropy in thermal properties of Ballachulish Slate, with its excellent fissility, is due to the oriented muscovite component.

Parallel to the schistosity, the conductivity is given by the temperature-dependent expression:

$$K_{\parallel}[\mathrm{W/m°C}] = (0.20 + 2.54 \cdot 10^{-4}T[°\mathrm{C}])^{-1}.$$

The thermal conductivity perpendicular to the schistosity is given by the expression:

$$K_{\perp}[\mathrm{W/m°C}] = (0.44 + 4.75 \cdot 10^{-4}T[°\mathrm{C}])^{-1}.$$

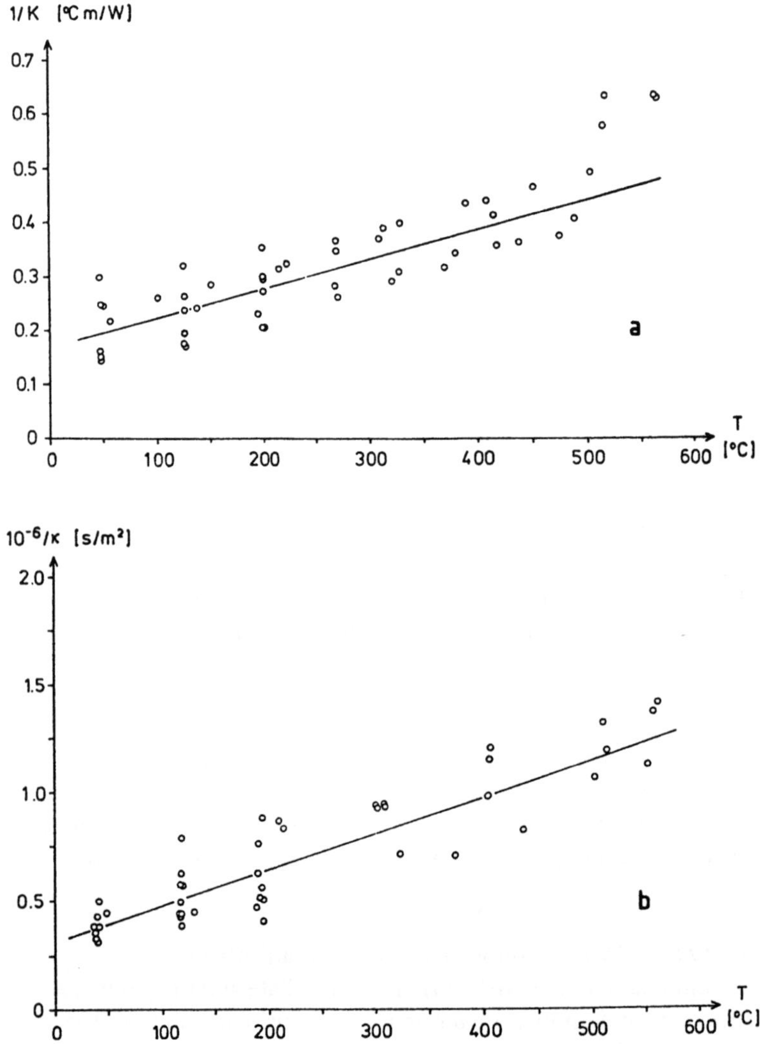

Fig. 18.3 a. Thermal resistivity of Appin quartzite at uniaxial pressure of 10 MPa; **b** reciprocal thermal diffusivity of Appin Quartzite at uniaxial pressure of 10 MPa

The scatter in measured conductivity of individual samples is much less parallel to the schistosity than perpendicular to the schistosity (Fig. 18.5). The thermal anisotropy of individual samples reaches a maximum value of $K_{\parallel}/K_{\perp} = 2.7$ for Ballachulish Slate.

The thermal anisotropy varies with temperature because the temperature sensitivity of thermal conductivity parallel and perpendicular to foliation is somewhat different. The factor of anisotropy at 50 °C is $K_{\parallel}/K_{\perp} = 2.2$; at 500 °C, it has decreased to 2.1.

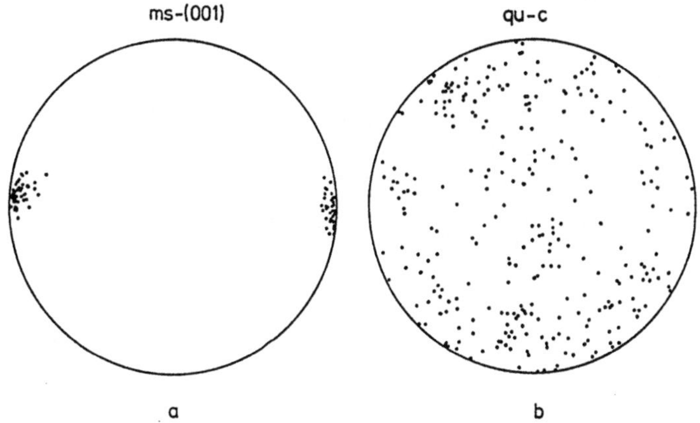

Fig. 18.4 a. Direction of mica (001) in Ballachulish Slate, cut parallel to the schistosity; **b** direction of the optical axis of quartz grains

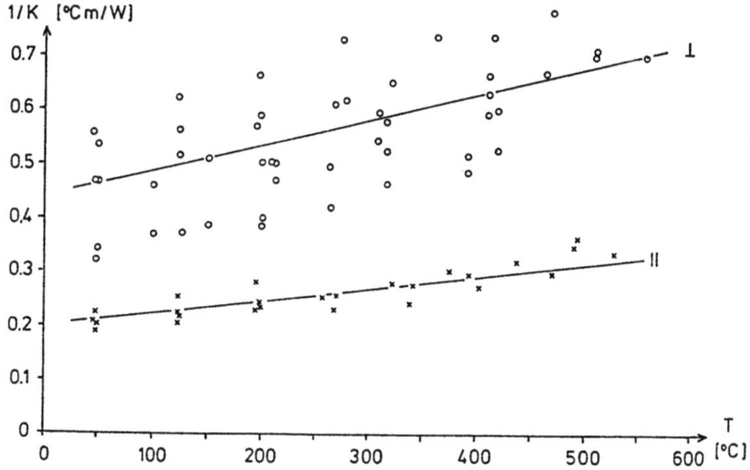

Fig. 18.5. Thermal resistivity of Ballachulish Slate, cut parallel (||) and perpendicular (⊥) to the schistosity (at p = 10 MPa)

The thermal diffusivity of Ballachulish Slate behaves similarly to that of the conductivity. The highest diffusivity (parallel to foliation) is twice the value of the lowest diffusivity (perpendicular to foliation). The mean values for all samples of the two diffusivities are given by the expressions:

$$\kappa_{\parallel}[10^{-6}\mathrm{m^2/s}] = (0.48 + 6.3 \cdot 10^{-4}T[°\mathrm{C}])^{-1},$$
$$\kappa_{\perp}[10^{-6}\mathrm{m^2/s}] = (1.25 + 9 \cdot 10^{-4}T[°\mathrm{C}])^{-1}.$$

Although the Leven Schist and Creran Succession pelites vary more in composition, their thermal properties vary less than that of the Ballachulish Slate. The conductivity of peliterich schists is similar to the conductivity of schist rich in sand. Nevertheless,

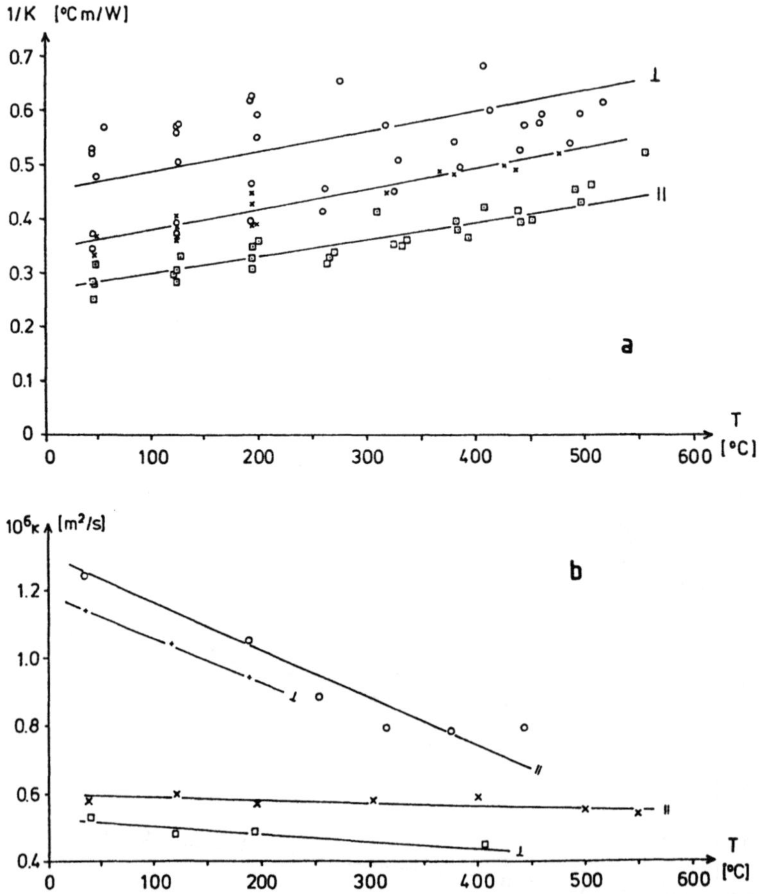

Fig. 18.6 a. Thermal resistivity of Leven Schist, cut parallel (||) and perpendicular (⊥) to the schistosity; *crosses* give values for structurally homogenized pelites at the pluton contact (at p = 10 MPa); **b** thermal diffusivity of a few samples of Leven Schist (p = 10 MPa)

the thermal properties are substantially anisotropic. The thermal conductivity parallel to the foliation is not as high as that of Ballachulish Slate. The mean value for all samples (Fig. 18.6a) is given by the expression:

$$K_{\parallel}[W/m°C] = (0.27 + 3.16 \cdot 10^{-4} T[°C])^{-1}.$$

The mean thermal conductivity perpendicular to the foliation is similar to that of Ballachulish Slate:

$$K_{\perp}[W/m°C] = (0.44 + 3.9 \cdot 10^{-4} T[°C])^{-1}.$$

From these values, a factor of anisotropy of 1.6 results, which is constant up to 500 °C.

Samples taken from near the contact with the pluton have been recrystallized and have therefore lost their anisotropy. They have a conductivity which equals the arithmetic mean of K|| and K⊥. The thermal conductivity of the thermally metamorphosed Leven Schist is given by:

$$K[W/m°C] = (0.34 + 3.8 \cdot 10^{-4} T[°C])^{-1}.$$

The thermal diffusivity of Leven Schist varies over a wide range, especially the diffusivity perpendicular to foliation. A few examples are given in Fig. 18.6b to demonstrate the variety. As mean values for all samples, the following diffusivities are determined:

$$\kappa_{\parallel}[10^{-6} m^2/s] = (0.69 + 14 \cdot 10^{-4} T[°C])^{-1},$$
$$\kappa_{\perp}[10^{-6} m^2/s] = (1.4 + 14 \cdot 10^{-4} T[°C])^{-1}.$$

The diffusivity of samples from near the contact is given by:

$$\kappa[10^{-6} m^2/s] = (0.96 + 16 \cdot 10^{-4} T[°C])^{-1}.$$

Other rock types which are less abundant within the aureole, and therefore less important during cooling of the intrusive complex, are the hornblende-rich gabbroic Appinite and the Ballachulish Limestone. The Appinite has a thermal conductivity nearly equivalent to that of the granite and diorite, although the diffusivity is about 20% higher. The Ballachulish Limestone conducts the heat as well as granite, but the thermal diffusivity is also some percentage higher (Buntebarth 1989).

18.3.3 The Effect of Dehydration Reactions in Pelites on the Cooling History

Pelites release H_2O and CO_2 due to devolatilization reactions during metamorphism. These devolatilization reactions absorb energy, typically about 83.7 kJ per mol of released H_2O or CO_2 (Walther and Orville 1982), but ranging between 59 and 109 kJ according to the type of devolatilization reaction. Considering only typical pelitic dehydration reactions, a kilogram of "average pelite" produces 1.44 mol of H_2O in a temperature interval between 400 and 600 °C (Walther and Orville 1982). This means that E = 120 kJ is needed for dehydration of each kilogram of average pelite. This amount of heat is supplied from the magma body during the time when the adjacent pelites undergo the dehydration reactions.

According to Pattison (Chap. 16, this Vol.), the dehydration of chlorite occurs mainly within the temperature range of 550–570 °C. The main reaction to release water from chlorite involves the mineral reaction (Pattison and Harte, Chap. 8, this Vol.):

chlorite + muscovite + quartz = cordierite + biotite + H_2O

Since the reaction consumes heat energy, it acts as a heat sink during the cooling of the magma body. The strength of the heat sink A is

$$A = E\,\rho/t,$$

with E = energy for dehydration of each kg rock, ρ = rock density, and t = reaction time. The rock density is about $2.7 \cdot 10^3$ kg/m^3. If the reaction is assumed to be a diffusion-controlled reaction according to Fisher (1978), the reaction time may be in the range of 10^4-10^5 years. The total time at which the pelites are exposed to a temperature higher than 550 °C within the aureole covers this range, although the time interval is strongly dependent upon the distance to the contact.

A mean value of $t = 6.3 \times 10^4$ a is applied here which yields a strength of the heat sink of

$$A = 163\ \mu W/m^3.$$

Some uncertainty is due to the fact that the real reaction time is unknown. If the reaction is already completed after a shorter time, the value of A increases correspondingly. However, the heat sink acts during a shorter time.

18.4 Water Migration from the Intrusion into the Wall Rock

The water dissolved in the original magma is gradually enriched in the residual magma during solidification. After oversaturation of the melt, excess water may be expelled into the adjacent rock and could therefore contribute to the heat transport. When the magma has been solidified, the magmatic water may dissipate into the wall rock.

According to Weiss and Troll (Chap. 4, this Vol.), the inner monzodioritic ring of the Ballachulish intrusive complex was relatively dry, with an average initial water content of 0.5–1.5 wt%. An increase of the water concentration towards the centre of the complex can be deduced petrographically from the sequence of crystallization. The outer ring of quartz diorite is supposed to have a slightly higher water content of 2–3 wt% and the central granite to have the highest fraction of 2.5–3.5 wt%.

The ring intrusion and the conical intrusion model do not supply as much water as the cylindrical intrusion model, where the granite also contributes to the thermal state of the country rock. The first two models neglect the heat of the central granite because it intruded later than the diorite magma and, furthermore, at lower temperature, so that the former temperature maximum in the wall rock was not reached again. The convective cooling within the magma body, which was investigated by Norton and Knight (1977), was not considered here because questionable assumptions would be applied along with poorly known boundary conditions.

A certain amount of water migrated perhaps vertically into higher levels of the crust. However, its amount is hard to estimate. The radial flow is restrained in some wall rocks. The permeability of quartzite is so low that a notable fluid flow cannot be assumed. Pelites let the water pass parallel to the layering, but only if their temperature is so low that the water is not consumed in melting reactions. Pattison and Harte (1988) argue that the temperatures around the igneous complex near the contacts were in all cases above the vapour-saturated solidus for pelites, meaning that all water released from the pluton would have been consumed in the anatectic zone, and would not therefore have been available to flow radially into the lower grade pelites farther out in the aureole.

Beside these general considerations, quantitative data are lacking about permeability, porosity and velocity of radial water flow. If in the unlikely event there was any water migration, model calculations using a parameterized migration rate yield that a maximum influence of fluid flow in the country rock is calculated within 1000 m to the contact, if the filtration velocity is about 2.5 cm/a at 3% porosity, with 3 vol% of water passing through the pelites. These values are considered in the profile at the SW contact.

The effect of groundwater motion on cooling as reported by Cathles (1977) is neglected, since extensive water motion can be excluded. Physical properties of the wall rocks and mineralogical (Pattison, Chap. 16, this Vol.) as well as geochemical arguments (Hoernes et al., Chap. 17, this Vol.) exclude a general water convection.

18.5 Numerical Treatment of the Heat Conduction Equation

18.5.1 The Finite-Difference Method to Solve the Heat Conduction Equation

The parabolic partial differential equation for unsteady heat conduction in a heterogeneous, but isotropic medium is

$$2\,c\rho\frac{\partial T}{\partial t} = -\ 2A + \Delta(K\,T) + K\Delta T - T\Delta K, \qquad (1)$$

where c = specific heat, ρ = density, K = thermal conductivity, A = heat sink, T = temperature, t = time and r = radial distance.

This partial differential equation is approximated by using constant thermal conductivities in concentric rings according to Fig. 18.1a,b, and by applying them to one space coordinate only. Simplification reduces Eq. (1) to

$$c\rho\frac{\partial T}{\partial t} = -A + K\left[\frac{\partial^2 T}{\partial r^2} + \frac{1}{r}\frac{\partial T}{\partial r}\right], \qquad (2)$$

which describes the heat conduction in a homogeneous cylindrical system.

Care must be taken at the boundaries between rocks of different thermal properties. Finite difference approximations at the interface between two different media are given, for example, by Carnahan et al. (1969). The derivatives are calculated using Taylor's expansion and the fact that the heat flow through the interface is continuous.

The finite difference equation which approximates Eq. (2) is

$$c\rho\frac{T_{i,n+1}-T_{i,n}}{\Delta t} = -A_i + K\left[\frac{T_{i-1,n}+T_{i+1,n}-2T_{i,n}}{(\Delta r)^2} + \frac{1}{i\,\Delta r}\,\frac{T_{i+1,n}-T_{i-1,n}}{2\Delta r}\right], \qquad (3)$$

where the subscripts i denote the i-th grid point from the centre with $r = i\Delta r$ ($i = 0,1,...,N$) and n denotes the n-th time level with $t = n\Delta t$ ($n = 0,1,...,M$). The space element Δr as well as the time interval Δt are constant.

The applied boundary conditions are (1) vanishing temperature gradients at a distance of r = 8000 m from the centre of the intrusion; and (2) within the first distance element Δr where a mean temperature is applied at each time step Δt. The initial condition is a constant averaged emplacement temperature for the dioritic magma, quickly intruding the country rocks which were at a temperature of 250–300 °C.

The solution of the finite difference equation is found by the explicit method which is stable for $K\Delta t/c\rho\,[\Delta r]^2 \leq 0.5$. The spatial grid element is $\Delta r = 100$ m and the time step $\Delta t = 1$ year.

At the time level n + 1, the temperature at the i-th point is explicitly calculated from the known temperatures at the grid points i + 1, i, and i - 1 at the previous time level n.

The conical model which is shown in Fig. 18.1c adds one more space coordinate to the heat transfer problem. Applying cylindrical coordinates and an inhomogeneous but isotropic medium, the partial differential Eq. (1) becomes

$$2\,c\rho\,\frac{\partial T}{\partial t} = -2A + \frac{\partial^2 (KT)}{\partial r^2} + \frac{1}{r}\frac{\partial (KT)}{\partial r} + \frac{\partial^2 (KT)}{\partial z^2} + K\left(\frac{\partial^2 T}{\partial r^2} + \frac{1}{r}\frac{\partial T}{\partial r} + \frac{\partial^2 T}{\partial z^2}\right)$$
$$-T\left(\frac{\partial^2 K}{\partial r^2} + \frac{1}{r}\frac{\partial K}{\partial r} + \frac{\partial^2 K}{\partial z^2}\right), \tag{4}$$

where K is the thermal conductivity and z is the depth. The first- and second-order space derivatives are replaced by finite difference approximations in the same manner as above. However, the solution method is different. The implicit alternating-direction method is applied, which is discussed by Peaceman and Rachford (1955) and Douglas (1955). One time step Δt is subdivided by two steps. The first step is done in r-direction and using the result of this half time-step, the second step proceeds in z-direction. The symbolic finite difference equations are:

1. $\frac{\Delta t}{2}$ - step: $T_{i,k}^{(n+1/2)} - T_{i,k}^{(n)} = \frac{\Delta t}{2}\left(\delta_r^2 T_{i,k}^{(n+1/2)} + \delta_z^2 T_{i,k}^{(n)}\right) - \bar{A}\Delta t/2$;
2. $\frac{\Delta t}{2}$ - step: $T_{i,k}^{(n+1)} - T_{i,k}^{(n+1/2)} = \frac{\Delta t}{2}\left(\delta_r^2 T_{i,k}^{(n+1/2)} + \delta_z^2 T_{i,k}^{(n+1)}\right) - \bar{A}\Delta t/2$

with $\bar{A} = A/c$.

Each equation leads to a system of equations having the unknown temperature at the grid points i + 1, i, and i - 1 at the time step n + 1/2, n + 1, respectively. This system of linear equations must be solved at every time step $\Delta t/2$ by a Gaussian elimination method.

18.5.2 The Treatment of Latent Heat in the Models Under Consideration

A very large amount of latent heat is released by the magma during crystallization. The heat of fusion really amounts to about one-quarter or more of the heat content of the magma (e.g., Buntebarth 1984). This proportion decreases if the temperature of the magma is high, and increases if the magma intrudes at a temperature only a little above the temperature of solidification.

Two approaches are applied here to take the latent heat of fusion (L) into account. For the ring model, the phase boundary liquid/solid gradually moves from both contact zones of the ring towards the centre, and at the interface between liquid and solid, the heat of fusion is given off. This moving boundary problem is discussed by Crank (1957). The first- and second-order derivatives are approximated using the Lagrange interpolation formula which finally leads to a progress of the phase boundary (Fig. 18.7) of

$$p_{n+1} = p_n + \frac{1}{L\rho}\frac{\Delta t}{(\Delta r)^2}\left[K_2\left\{\frac{2p_n-5}{(p_n-2)(p_n-3)}T_B + \frac{p_n-3}{p_n-2}T^{(n)}_{r+1} + \frac{p_n-2}{3-p_n}T_{r+2}\right\}\right.$$
$$\left. - K_1\left\{\frac{p_n}{p_n+1}T^{(n)}_{r-2} - \frac{p_n+1}{p_n}T^{(n)}_{r-1} + \frac{2p_n+1}{p_n(p_n+1)}T_B\right\}\right], \tag{5}$$

where T_B is the temperature of the phase boundary, K_1 the thermal conductivity of the magma, K_2 that of the solid rock, and p the position of the moving boundary within two grid points dependent on the time step n. Applying the parameter p, the temperature at the grid points r + 1 and r - 1 at the time step n + 1 are

$$T^{(n+1)}_{r-1} = T^{(n)}_{r-1} + 2\frac{\Delta t}{(\Delta r)^2}\kappa_1\left[\frac{T^{(n)}_{r-2}}{p_n+1} - \frac{T^{(n)}_{r-1}}{p_n} + \frac{T_B}{p_n(p_n+1)}\right], \tag{6a}$$

$$T^{(n+1)}_{r+1} = T^{(n)}_{r+1} + 2\frac{\Delta t}{(\Delta r)^2}\kappa_2\left[\frac{T_B}{(p_n-2)(p_n-3)} - \frac{T^{(n)}_{r+1}}{2-p_n} + \frac{T^{(n)}_{r+2}}{3-p_n}\right]. \tag{6b}$$

where κ_1 and κ_2 are the thermal diffusivity of the magma and the solid rock, respectively.

The symmetrical solidification model from both sides of the ring is suitable neither for the cylindrical nor for the conical model. An alternative approach is to

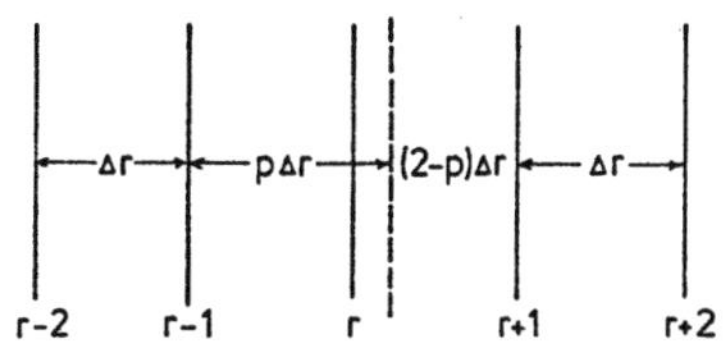

Fig. 18.7. Position of the moving phase boundary (- - -) at the time step n dependent on the parameter p (see text)

assume that during crystallization, additional heat is generated due to the latent heat of fusion. The heat production A is

$$A = \frac{L\rho}{T_L - T_S}\frac{\partial T}{\partial t}, \tag{7}$$

with L = latent heat of fusion, ρ = density, T_L = the liquidus temperature of the magma, and T_S = the solidus temperature.

The heat production A means that the latent heat is given off equally within the temperature range T_L and T_S and uniformly within the whole magma body. This introduces some simplification because solidification proceeds gradually from the contact zone to the centre.

The time-varying part in Eq. (1) becomes

$$c\rho + \frac{\partial T}{\partial t} + A = \left(c\rho + \frac{L\rho}{T_L - T_S}\right)\frac{\partial T}{\partial t}. \tag{8}$$

The numerical values which are applied are L = 310 J/g, c = 0.96 J g^{-1} °C^{-1}, ρ = 2.65 × 10^3 kg/m^3, T_L = 1050 °C for the dioritic magma and 850 °C for the granitic magma, T_S = 750 °C for diorite and 680 °C for granite.

The granite magma, however, was emplaced at a temperature near the 'dry' solidus of the diorites. At this temperature, the granite magma probably contained up to 40–50 vol% crystals, so that approximately one-half of the latent heat was already released during the ascent of this magma. The heat generation term A refers to an interval between the time of intrusion and the time at which the temperature falls below the solidus temperature.

18.5.3 Numerical Treatment of Dehydration Reactions

The heat which is consumed by the pelites per unit volume and time is estimated with A = 163 μW/m^3. This heat sink becomes effective when the temperature at a given grid point exceeds 550 °C and it operates until the temperature falls below 550 °C. It is recognized that the reactions probably would have gone to completion in a shorter period of time near the contact than in a greater distance. Additionally, the energy is not consumed equally at temperatures above 550 °C and furthermore, during heating the reaction temperature of 550 °C is passed most quickly at distances near the contact and more slowly at greater distances. However, this procedure is a first order approximation and it is the same for all models. Because the value of A is rather high, it predominates over the fluid flow from the magma body through the country rock.

18.5.4 Numerical Treatment of Water Migration

The heat carried away with the water migrating from the intrusion into the adjacent rock causes a faster cooling of the intrusive body. It is assumed that the water is released when the temperature of the centre falls below the solidus temperature. The water then migrates as a ring through the rock. Due to the cylindrical coordinates [Eq. (2)] the ring becomes smaller with increasing distance. Its thickness decreases with 1/r. When the migration starts, i.e., when the intrusion releases water, the water ring has a thickness of 1200 m. This value equals the thickness of the ring intrusion, of the conical intrusion and the thickness of the dioritic ring of the cylindrical intrusion. The water migrates only in direction of increasing distance r. The migrating water ring occupies 3 vol% of the pelites, which yield a velocity (v) within the pore space of v = 0.83 m/a. This calculation is made primarily for comparative purposes, since it is unlikely that fluid will have passed through the pelites out into the aureole, due to vapour-consuming anatectic reactions near the contact (see above and Pattison and Harte 1988).

Since the time step Δt is one year, the fluid front moves forward 0.83 m after each time step. At this point, the water equilibrates its temperature (T_w) with that of the rock (T_r) by a simple "mixture" where the equilibrium temperature (T_E) is

$$T_E = \frac{T_w c_w \phi + (1-\phi) c_r T_r}{c_w \phi + (1-\phi) c_r}; \tag{9}$$

c is the specific heat of the fluid, respective of the rock, and ϕ is the porosity which equals the fluid fraction of the system. The rock temperature is interpolated linearly between the grid point i and i + 1. At all points from the front to the back of the flow, the equilibrium temperature must be computed again after each time step. Additionally, the thickness (d) of the concentric fluid flow is reduced gradually with increasing distance (r) from the contact

$$d = r - \sqrt{r^2 - 1200^2}\,. \tag{10}$$

The temperature according to Eq. (1) is corrected at each grid point i involved in the fluid flow. The calculation starts again with the next time step using Eq. (1), and then the correction due to the fluid flow is computed until the fluid has cooled off.

18.6 Cooling of the Intrusion and its Influence on the Aureole

Because the thermal properties of the country rocks are very different, different cooling histories must be considered for different rock types. The models in which Appin Quartzite is involved are simpler than those with pelites, where the additional effects of dehydration reactions and possibly local fluid migration occur.

Since the initial temperature of the magma is not uniform across the pluton, and since neither the pre-intrusion temperature of the country rock nor the time during which the dehydration of pelites operates are exactly known, the effect of changing these parameters has been studied.

18.6.1 Temperature Distribution in Appin Quartzite

The cooling history is modelled most simply with Appin Quartzite, because only the latent heat of fusion influences the conductive cooling of the magma. Time-dependent temperature distributions are shown in Fig. 18.8 for different models at constant distances from the contact. The intrusion temperature is $T_i = 1050$ °C and the country rock temperature is $T_c = 250$ °C for the cylindrical model. A higher country rock temperature of $T_c = 300$ °C is used for the ring model, because the value of 250 °C would yield maximum temperatures in the aureole which are so low that in no case the observed reaction temperature would agree with the model temperature. Additional data are given in Table 18.1.

According to Table 18.1, the maximum temperature is already reached after about 20 000 years at a distance of 300 m and after about 50 000 years at a distance of 500 m from the contact, if the ring model is considered. For the cylindrical model, the maximum temperature is reached at the same distances after 84 000 years and 124 000 years, respectively.

Calcite-dolomite geothermometry from the aureole is thought to indicate minimum temperatures (Masch and Heuss-Aßbichler, Chap. 10, this Vol.). Therefore, these temperature profiles may be somewhat below the expected maximum tempera-

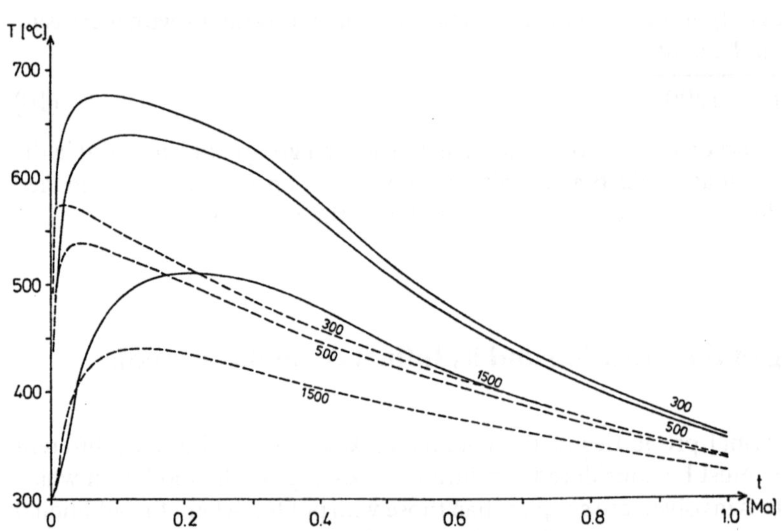

Fig. 18.8. Distribution of temperature vs time of the Appin Quartzite at a distance of 300, 500 and 1500 m from the contact, for the ring (- - -) with $T_i = 1050$ °C and $T_c = 300$ °C and for the cylindrical model (—) with $T_i = 1050$ °C and $T_c = 250$ °C

Table 18.1. Temperature distribution T(°C) in Appin Quartzite near the maximum temperature and heating rates H(°C/1000 a) at different distances r from the contact of the ring model (T_i = 1050 °C, T_c = 300 °C)

Time after intrusion (a)	r = 300 m		Time after intrusion (a)	r = 500 m		Time after intrusion (a)	r = 700 m	
	T	H		T	H		T	H
7000	565.6	7.60	15 000	524.5	3.75	35 000	507.4	0.94
8000	571.0	5.40	20 000	531.9	1.48	40 000	510.8	0.68
9000	574.9	3.90	25 000	536.9	1.00	45 000	513.3	0.50
10 000	577.7	2.80	30 000	540.5	0.72	50 000	515.1	0.36
11 000	579.6	1.90	35 000	542.7	0.44	55 000	516.3	0.24
12 000	580.5	0.90	40 000	544.0	0.26	60 000	517.1	0.16
13 000	581.0	0.50	45 000	544.8	0.16	65 000	517.6	0.10
14 000	581.4	0.40	49 714	545.1	0.06	70 000	517.8	0.04
15 000	581.7	0.30	50 000	545.1	0.00	74 423	517.8	0.00
16 000	581.8	0.10	55 000	545.0	−0.02	75 000	517.8	0.00
17 000	581.9	0.10	60 000	544.5	−0.10	80 000	517.8	0.00
18 000	581.9	0.00	65 000	543.9	−0.12	85 000	517.5	−0.06
19 000	581.9	0.00	70 000	543.2	−0.14	90 000	517.1	−0.08
20 000	581.9	0.00	75 000	542.4	−0.16	95 000	516.6	−0.10
22 500	581.8	−0.04	80 000	541.5	−0.18	100 000	516.0	−0.12
25 000	581.7	−0.04	85 000	540.5	−0.20	105 000	515.4	−0.12
27 500	581.6	−0.04	90 000	539.5	−0.20	110 000	514.7	−0.14
30 000	581.4	−0.08	95 000	538.3	−0.24	115 000	513.9	−0.16
35 000	580.8	−0.12	100 000	537.2	−0.22	120 000	513.1	−0.16
40 000	579.8	−0.20	105 000	536.1	−0.22	125 000	512.3	−0.16
45 000	578.6	−0.24	110 000	534.9	−0.24	130 000	511.4	−0.18
50 000	577.0	−0.32	115 000	533.7	−0.24	135 000	510.5	−0.18
55 000	575.4	−0.32	120 000	532.5	−0.24	140 000	509.6	−0.18
60 000	573.6	−0.32	125 000	531.3	−0.24	145 000	508.7	−0.18
65 000	571.8	−0.36	130 000	530.1	−0.24	150 000	507.7	−0.20
70 000	570.0	−0.36	135 000	528.8	−0.26	155 000	506.7	−0.20
75 000	568.3	−0.36	140 000	527.6	−0.24	160 000	505.6	−0.22
80 000	566.6	−0.34	145 000	526.4	−0.24	165 000	504.5	−0.22
85 000	564.7	−0.36	150 000	525.2	−0.24	170 000	503.4	−0.22
90 000	563.0	−0.34				175 000	502.3	−0.22
95 000	561.2	−0.36				180 000	501.2	−0.22
100 000	559.5	−0.34				185 000	500.0	−0.24
						190 000	498.9	−0.22

ture (see Pattison, Chap. 16, this Vol.). This discrepancy is more accented near the contact zone where the temperature-time peak is small, than at far distances where the maximum temperature is nearly constant for a longer period. For the calculation of reaction kinetics, the heating time near the temperature maximum has to be obtained. Therefore, Fig. 18.9 shows heating rates between the maximum temperature and 10 °C below the maximum. The heating rates for the ring model are tabulated (Table 18.1).

The maximum temperatures reached in the wall rock are shown for the ring model in Fig. 18.10. The temperature decreases almost exponentially with increasing distance from the contact. The good thermal conductivity of Appin Quartzite yields high maximum temperatures in the adjacent rocks. Comparing the ring model with the cylindrical model (Fig. 18.11), the contact temperature has nearly the same value; the

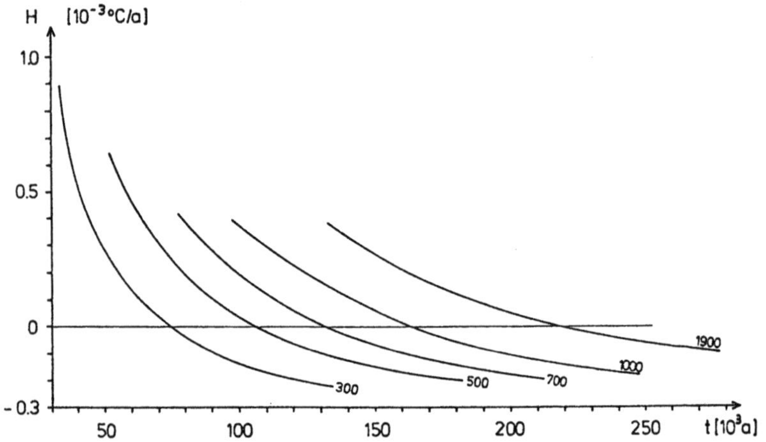

Fig. 18.9. Heating rates for the cylindrical model near the maximum temperature for various distances (m) from the contact. The heating rates begin and end 10 °C below the maximum temperature. At the maximum temperature H = 0

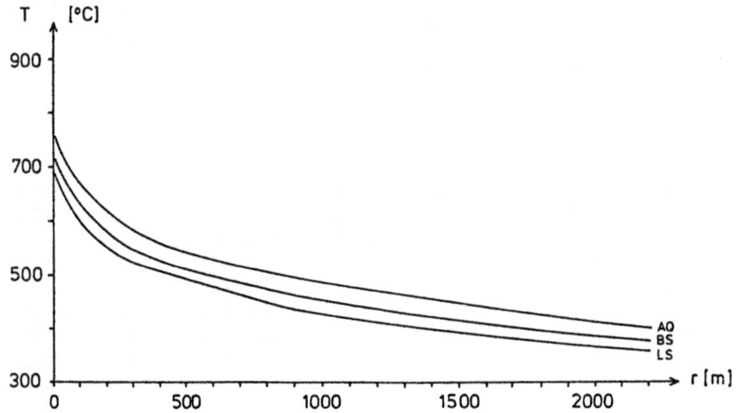

Fig. 18.10. Maximum temperature in the wall rock vs the distances (*r*) from the contact of a ring intrusion. Mean conductivities (average of $K_{\parallel}$ and $K_{\perp}$) are applied for Leven Schist (*LS*) Ballachulish Slate (*BS*) and Appin Quartzite (*AQ*)

temperature at distances far from the intrusion is not more than 50 °C below the values for the cylindrical model. At distances between 200 and 700 m, however, this difference reaches 100 °C. The assumed conditions are as follows: T_i = 1050 °C; T_c = 300 °C for the ring model; and T_c = 250 °C for the cylindrical model. Water flow is not considered. The much larger amount of heat which can be supplied by a cylindrical intrusion is due to this significant difference.

Because the dip of the contact zone is variable, the conical model applies various dip angles α. Even if the angle α of Fig. 18.1c is varied between 0 and 45°, its

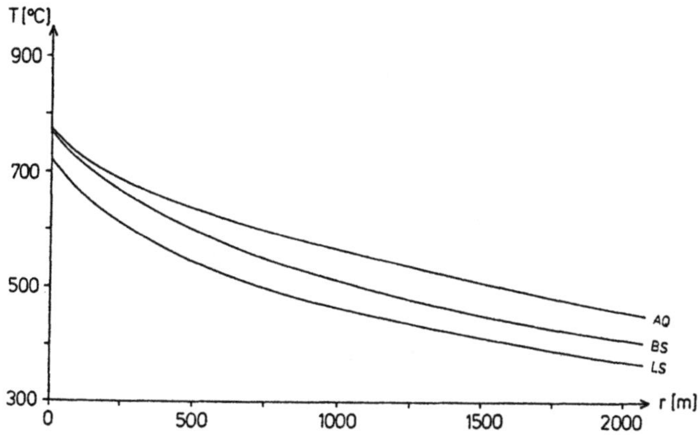

Fig. 18.11. Maximum temperature in the wall rock vs the distance (r) from the contact of a cylindrical intrusion. Mean conductivities (average of $K_{\parallel}$ and $K_{\perp}$) are applied for Leven Schist (*LS*), Ballachulish Slate (*BS*) and Appin Quartzite (*AQ*)

influence on the maximum temperature of the adjacent Appin Quartzite remains small. Angles below $\alpha = 22.5°$ cause a maximum temperature not more than 10 °C above the maximum temperature with $\alpha = 0°$, i.e., the ring model. Only with $\alpha = 45°$ does the temperature significantly exceed the ring model values, yet by not more than 30 °C. Moreover, this higher temperature becomes effective only at distances from the contact greater than about 500 m.

18.6.2 Temperature Distribution in Pelites

As described previously, pelites underwent dehydration reactions and may have been exposed to magmatic fluids. Whereas the first effect consumes heat during the conductive cooling, the second supplies heat. The model calculations demonstrate that the thermal effect of the dehydration reactions predominates over that of the fluid flow. Because of the high thermal anisotropy of Leven Schist and Ballachulish Slate, the temperature distribution can be very different even with the same rock type, depending on the orientation of foliation. Figure 18.12 shows a series of four thermal profiles corresponding to those in Fig. 16.3 of Pattison (Chap. 16, this Vol.). If heat flow from a cylindrical intrusion is parallel to the schistosity, the maximum temperature in the SW profile is about 30 °C above the value of the mean conductivity.

The difference may even reach 100 °C for Ballachulish Slate. The temperature difference is much smaller if a ring intrusion is assumed, causing at most a difference of 40 °C for Ballachulish Slate, and less for Leven Schist. The total heat content in the ring intrusion is two-thirds of the heat from a cylindrical intrusion. Whereas the heat of the ring model dissipates through two contact zones, the heat of the cylindrical model flows in one direction only.

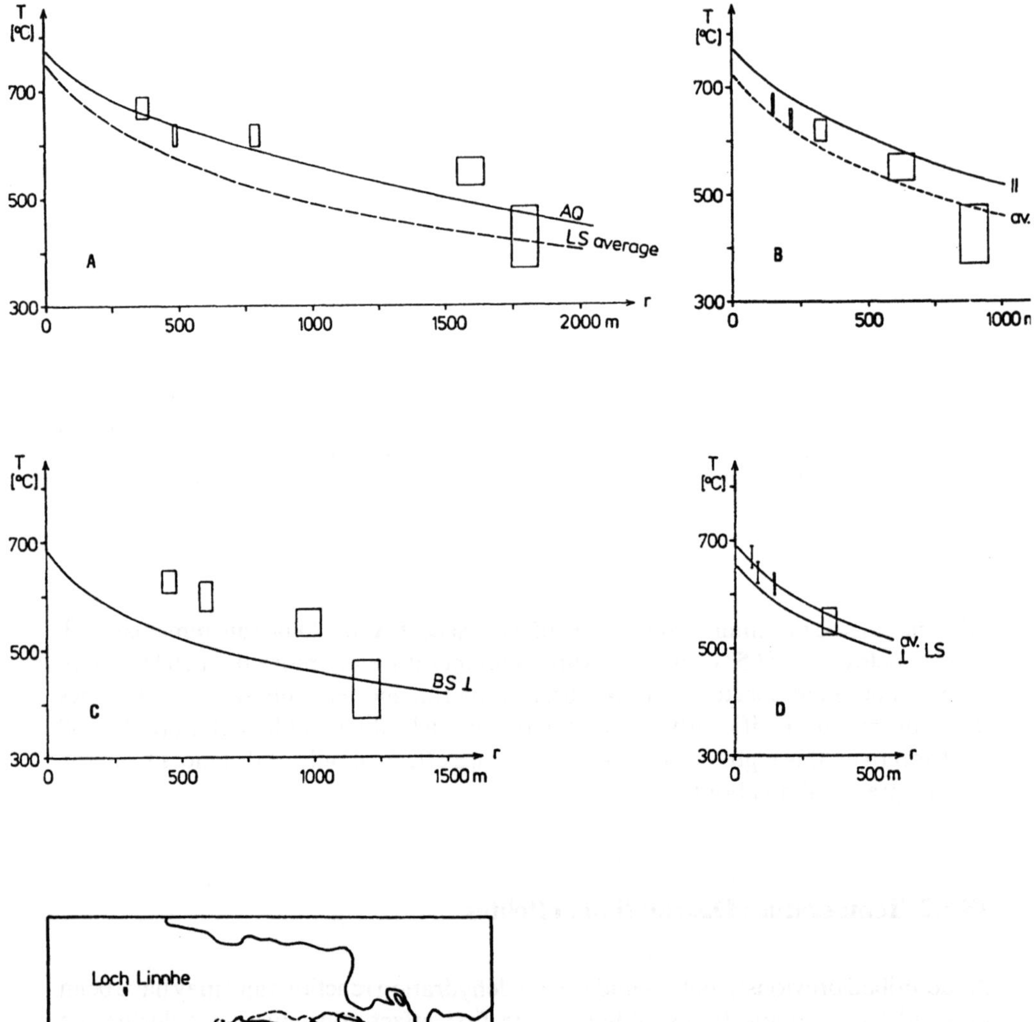

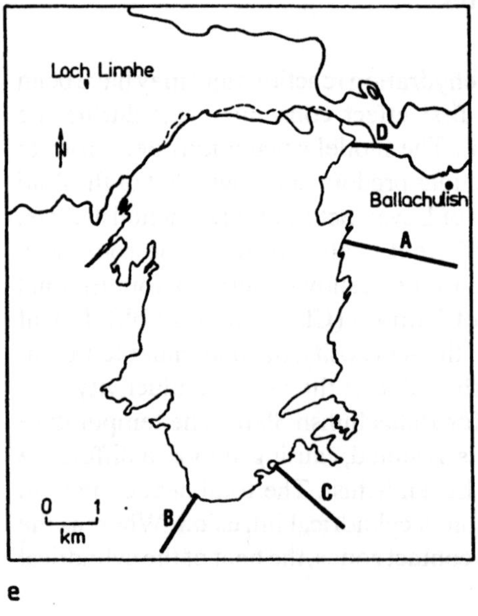

e

Fig. 18.12a-d. Maximum temperatures in the Ballachulish aureole for four sectors (*A* to *D*) along with mineral reaction geothermometer data of high priority (Pattison, Chap. 16; Masch and Heuss-Aβbichler, Chap. 10, this Vol.); pre-intrusion temperature: $T_c = 250$ °C, intrusion temperature: $T_i = 1050$ °C; in profile D is $T_i = 1000$ °C; *AQ* Appin Quartzite; *LS* Leven Schist; *BS* Ballachulish Slate. **e** Location of the four thermal profiles *A* to *D*. The profiles correspond to those in Fig. 16.3

The maximum temperature reached at a distance of 200 m from the contact is 625 °C for a cylindrical intrusion and 550 ° for a ring intrusion, if a mean conductivity for Leven Schist is considered. If the higher conductivity parallel to the schistosity is applied, the maximum temperatures are 685 and 560 °C, respectively. At distances of 2000 m, the difference decreases to 10 °C in both cases. The higher anisotropy of Ballachulish Slate causes differences that are somewhat higher.

18.6.3 Effect of Changing Initial Conditions

Petrological investigations support an initial magma temperature (T_i) being variable in different sections of the pluton (Weiss and Troll, Chap. 4, this Vol.). The coldest part (T_i = 1000 °C) is assumed at the NE sector. Other sectors of the pluton are considered to have T_i = 1050 °C. The effect of changing the initial temperature is demonstrated for Appin Quartzite in Fig. 18.13. Applying magma temperatures of T_i = 1050 °C and T_i = 1000 °C, the temperatures at the contact are 774 °C and 744 °C, respectively. The difference of 30 °C is nearly independent of the distance from the contact. At 2000 m from the contact, the respective temperatures are 461 and 428 °C, when the country rock temperature of T_c = 250 °C is assumed.

The most likely pre-intrusion temperature of the country rock may be T_c = 250 °C, but a maximum value of T_c = 300 °C is possible, too. For both cases along with T_i = 1050 °C, temperatures of 774 and 803 °C are obtained at the contact to Appin Quartzite; at a distance of 2000 m, the temperatures are 461 and 481 °C. It can be seen that the temperature difference decreases slowly with distance from the contact (see Fig. 18.14).

A larger uncertainty is due to the reaction time of dehydration reactions in the pelites (see above). The applied time of t = 63 000 a is an estimated mean value for

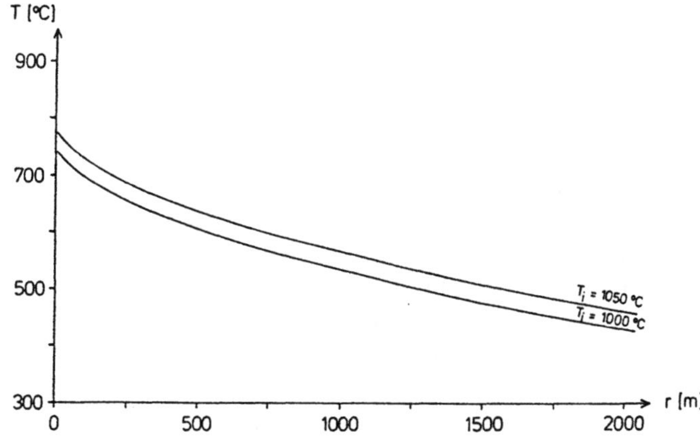

Fig. 18.13. The effect of a variable intrusion temperature (T_i) on the maximum temperature of the country rock (Appin Quartzite). The pre-intrusion temperature is T_c = 250 °C

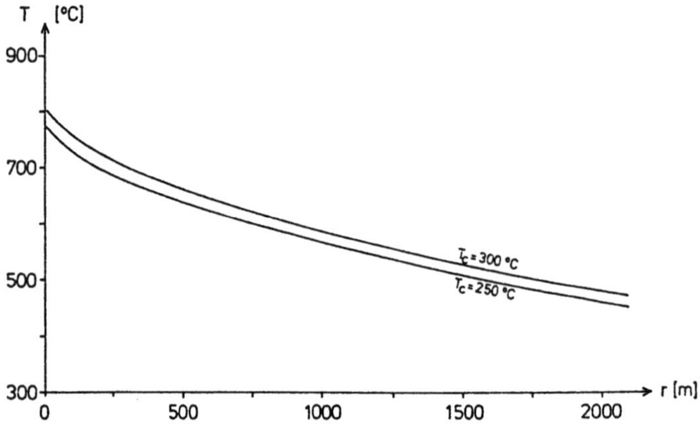

Fig. 18.14. The effect of a variable pre-intrusion temperature of the country rock (T_c) on the maximum temperature for Appin Quartzite. The intrusion temperature is T_i = 1050 °C

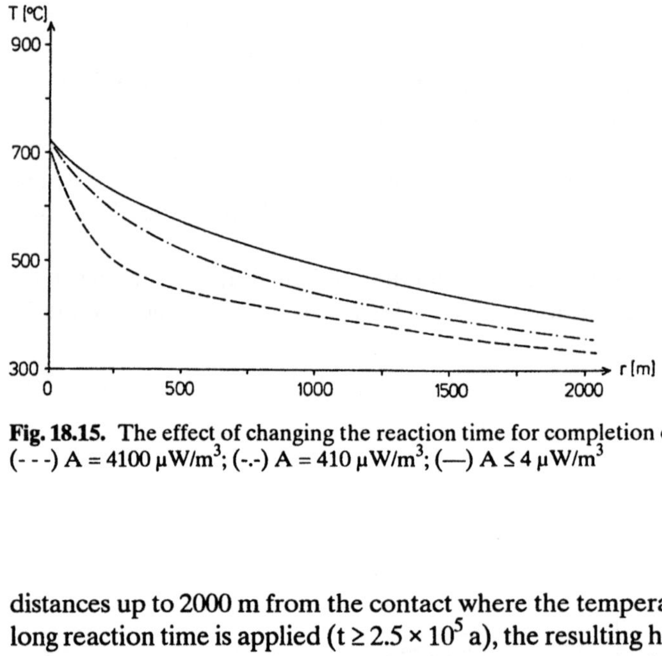

Fig. 18.15. The effect of changing the reaction time for completion of the dehydration of pelites: (- - -) A = 4100 $\mu W/m^3$; (-.-) A = 410 $\mu W/m^3$; (—) A $\leq$ 4 $\mu W/m^3$

distances up to 2000 m from the contact where the temperature exceeds 550 °C. If a long reaction time is applied ($t \geq 2.5 \times 10^5$ a), the resulting heat sink of A = 40 $\mu W/m^3$ has a small effect. It reduces the maximum temperature by 10 °C. Longer reaction times have no effect. Shorter times, however, have a large effect (Fig. 18.15). If A = 400 $\mu W/m^3$ is applied, which equals a reaction time of t = 25 000 a, the resulting temperature decreases by 50 °C at the distance of 500 m, and A = 4 000 $\mu W/m^3$ corresponding to t = 2500 a reduces the maximum temperature which would be reached without heat sink by 130 °C at 500 m. It should be noted that the heat sink becomes effective for the whole time period when the country rock exceeds 550 °C.

18.7 The Models Fitting Best to the Geothermometer Data

The calculations of the cooling history are based on many facts which are not easy to quantify, e.g., the geometry of the intrusive body, magma convection after the setting of the intrusion, the amount of released water and its migration rate, the permeability of the rocks, the homogeneity of the rocks, and others. A conservative estimate of the accuracy of the calculated maximum temperature yields an uncertainty of about ±70 °C within a few hundred meters distance from the contact. The model is substantially improved if mineral reaction temperatures are used to constrain the calculations. Pattison (Chap. 16, this Vol.) provides a detailed discussion of temperature variation in the aureole based on a wide variety of mineral equilibria.

Various mineral geothermometers from the eastern profile (Masch and Heuss-Aβbichler, Chap. 10; Kroll et al. Chap. 13, this Vol.) are compared with the maximum temperature in Appin Quartzite considering a ring model and a cylindrical model in Fig. 18.16. The temperatures calculated from the ring model appear too low, in general. The pre-intrusion country rock temperature would have to be about 350 °C in order to match the constraints from the mineral equilibria. Furthermore, the decrease of maximum temperature on the temperature-distance plot is too steep to fit the mineral equilibria data, particularly within a distance of 500 m from the contact.

Whereas the ring model does not seem to supply enough heat, the temperatures of the cylindrical model exceed slightly the given reaction temperatures, even when a pre-intrusion temperature of the country of T_c = 250 °C is applied. Since Appin Phyllite comprises about one-third of the aureole on the eastern flank, a profile is included in Fig. 18.12 for Leven Schist, similar lithologically to Appin Phyllite. This gives a profile similar to that for the Appin Quartzite ring intrusion model.

The mineral equilibria data lie more or less between these curves, so that the cylindrical model along with a country rock composed of Appin Quartzite and Leven Schist would seem to give the best fit. In the SW profile, Creran Succession pelites

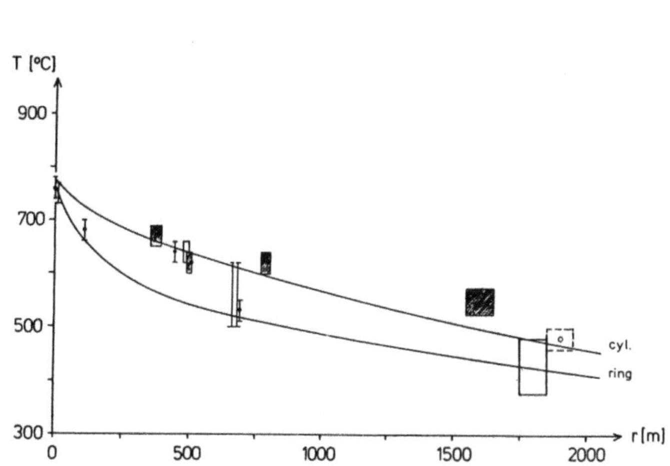

Fig. 18.16. Geothermometer data (⬚ Kroll et al., Chap. 13; I Masch and Heuss-Aβbichler, Chap. 10; □, I Pattison, Chap. 16, this Vol.); compared with the maximum temperature in Appin Quartzite for a ring and a cylindrical intrusion model. ▨ Data of high priority

form the contact. The modelled profiles agree very well with mineral equilibria data when a mean value of conductivity is used. In the narrow NE profile, the calculated temperatures are somewhat lower than inferred from mineral equilibria, especially at the contact. The difference may be due to either magma convection which yields a higher contact temperature, or to a dip of the contact which raises the temperatures near the contact. A shorter time for the dehydration of Leven Schist yields a steeper decrease of the temperature near the contact. However, to fit the constraints from mineral equilibria, this model would require a higher magma temperature than the assumed temperature of $T_i = 1000$ °C. This adjusting time would also improve the fit of the southern profile.

The discrepancy between the modelled profiles and the mineral equilibria data is largest in the SE profile. However, a substantial improvement would result from using mean thermal properties of Ballachulish Slate instead of the applied properties perpendicular to the schistosity. Examining Fig. 18.12, it is apparent that the aureole bulges outward in this area to encompass the small stock in the southeast. This may indicate that the quartz diorite extends beneath the SE profile, which would explain the higher mineral equilibria temperatures compared to the modelled temperatures for a subvertical contact.

18.8 Summary

A cooling history of the Ballachulish intrusive complex is modelled by using three different intrusive model bodies, because of the imprecise knowledge of the intrusion geometry.

The thermal conductivity and the thermal diffusivity of the common rock types involved in the cooling process are determined dependent on the temperature. Further thermally relevant properties are considered, i.e., the dehydration of pelites, and very locally, a movement of magmatic fluids into the wall rocks.

The heat conduction equation is solved numerically using finite difference methods in a two-dimensional space for a ring intrusion and a cylindrical intrusion model. A three-dimensional model is applied with a conical intrusion. The model results are compared with mineral equilibria data from several mineralogical reactions. Heat transfer from a vertical cylindrical intrusive body agrees best with the mineral equilibria temperatures on the investigated four profiles.

Part V
Concluding Discussion

19 Evidence of Fluid Phase Behaviour and Controls in the Intrusive Complex and Its Aureole

B. Harte, D.R.M. Pattison, S. Heuss-Aßbichler, S. Hoernes, L. Masch and S. Weiss

19.1 Introduction

This chapter brings together evidence pertaining to the presence, composition, influence and mobility of fluids rich in volatile species (principally H_2O and CO_2) in the development of the Ballachulish intrusive complex and its aureole. We examine fluid-mineral-rock interactions over the complete range of scales from that of the small-scale systems represented by individual rocks, to that of the large-scale system represented by the combined intrusive complex and its aureole. This will allow comparison with models of fluid behaviour identified in other low- to high-grade metamorphic situations (see reviews in Walther and Wood 1986; and Valley 1986), such as the hydrothermal circulation system of the Hebridean intrusive complexes (Taylor and Forester 1971; Forester and Taylor 1977), and the high fluid-pressure model of regional metamorphism in orogenic belts (Etheridge et al. 1983, 1984).

We shall begin by considering data concerning fluid presence and behaviour on the small scale within individual rocks and lithologies, thereby gathering information for the subsequent consideration of the overall pattern of fluid behaviour in the Ballachulish igneous-metamorphic Complex as a whole. The data we shall summarize is given in detail and illustrated in other chapters in this volume, to which we shall make frequent reference.

For simplicity in this chapter the use of the word fluid will be restricted to volatile-rich (H-C-O) fluids, and will not include melts (very silica-rich fluids) which will be designated separately. The pressure of the fluid phase will be symbolized by P_F, and the pressure on the solid phases, which is expected to be essentially a function of depth of burial or load pressure, will be symbolized by P_L.

Pattison (Chap. 16, this Vol.) shows that the best estimates from the fluid-absent equilibrium garnet + plagioclase + orthopyroxene + quartz are 0.30 to 0.32 GPa, and these estimates are supported by other data from the high-grade pelitic assemblages. A solid pressure (P_L) of 0.3 GPa indicates that at the time of contact metamorphism, the rocks had a depth of burial of approximately 10 km, and we shall use these pressure/depth estimates in the following discussion of fluid behaviour.

G. Voll, J. Töpel, D.R.M. Pattison, F. Seifert
Equilibrium and Kinetics
in Contact Metamorphism

19.2 Fluid Presence and Properties Within Particular Lithologies

The assemblages of minerals and their compositions in particular rocks or rock types provide information concerning any fluid phase which may have been present at the time of assemblage formation. This section is concerned with assembling such evidence on fluid presence and properties for each of the various lithologies of the intrusive igneous complex and its aureole. Where evidence of fluid presence is found, information is sought concerning whether the fluid appears to be solely *derived internally* from within the rock concerned (closed-system behaviour), or whether the fluid appears to be *derived externally* to the rock (open-system behaviour).

At the outset of these considerations of fluid phase behaviour, we should emphasize that fluid conditions vary with time, and that the mineral assemblage data obtained from different lithologies and places will not all refer to the same temporal period. In the case of igneous rocks, petrographic and mineral chemical features allow some reconstruction of fluid conditions over the time range of melt crystallization and subsolidus modification. In the metamorphic rocks of the aureole, a history of mineral assemblage characteristics is rarely preserved and the data are generally believed to refer to the time of peak metamorphism (highest temperature conditions), which will have varied with distance from the intrusive complex (Buntebarth, Chap. 18, this Vol.). Some indications of fluid behaviour with time in the metamorphic rocks may be given by retrograde assemblages.

The presence (or absence) of fluid may influence the mineral assemblages of rocks both directly and indirectly:

1. *Direct influence* — in which the presence of the fluid as a distinct phase with particular chemical components, modifies the equilibrium phase assemblages and compositions within the rocks
2. *Indirect influence* — in which the presence of the fluid modifies the kinetics of reaction within the rocks and the ease of attainment of equilibrium or the susceptibility to deformation; for example by changing diffusion rates without necessarily modifying the actual equilibrium mineral assemblage of the rocks.

Clearly, the evidence of these two types of fluid influence in any given rock type should be complementary. The evidence of direct fluid involvement in mineral formation in each lithology will be considered first, then that of indirect involvement.

19.2.1 Evidence of Direct Fluid Interaction with Mineral Assemblages

19.2.1.1 The Igneous Complex

Amongst the three major igneous rock groups distinguished by Troll and Weiss (Chap. 3, this Vol.; see Fig. 19.1 and Maps 1 and 2 for distribution of rock types), the monzodiorites stand apart from the quartz-diorites and the granites. The preservation of both hypersthene and augite, and the limited development of biotite and lack of magmatic amphibole, indicate the limited availability of H_2O during the magmatic

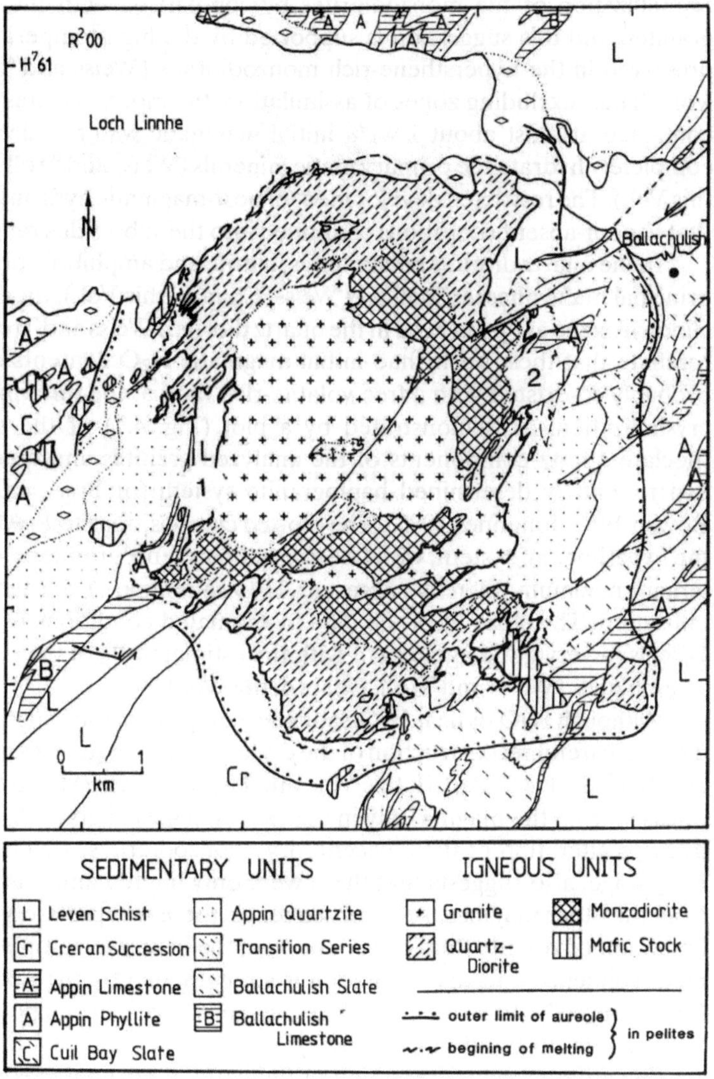

Fig. 19.1. Map showing distribution of principal rock types in and around the Ballachulish intrusive complex. The approximate outer limit of the aureole and the approximate limit of partial melting in pelitic and semi-pelitic rocks are marked. Note that the granite, which is the most hydrous of the igneous intrusions is largely surrounded by diorites, of which the monzodiorites show particularly 'dry' mineral assemblages (see text). Locations *1* and *2* show respectively the position of the 'Chaotic Zone' (where partial melting is particularly conspicuous); and the Allt Guibhsachain area (where metamorphic zones are best seen in the calcareous assemblages)

crystallization of the monzodiorites by comparison with the quartz diorites and granites; and this suggestion is supported by the high-temperature phase relations preserved in the hypersthene-rich monzodiorites (Weiss and Troll, Sect. 4.3.1, this Vol.). Thus, excluding zones of assimilation, the monzodiorites are thought to have contained at most about 1 wt% initial magmatic water, which was insufficient to completely hydrate the primary mafic minerals (Weiss and Troll 1989; and Sect. 4.3.1, this Vol.). The restricted development of post-magmatic hydrous phases also suggests that vapour-absent conditions continued into the subsolidus range in monzodiorites.

In the quartz diorites and granites, biotite and amphibole or biotite alone are the principal mafic phases (Troll and Weiss, Chap. 3, this Vol.), though small amounts of clinopyroxene are preserved in the quartz diorites. Weiss and Troll (Chap. 4, this Vol.) estimate that these rocks had initial magmatic H_2O contents of 2.0–3.0 wt%, and probably co-existed with a free volatile-rich fluid at the terminal stage of magmatic crystallization, as demonstrated by a plot (Fig. 4.7) of the normative albite-orthoclase-quartz components of the analyzed granites and quartz-diorites into the experimentally determined haplogranite system for P_{H_2O} = 0.3 GPa (Tuttle and Bowen 1958; Johannes 1985). The plotted compositions follow the thermal valley of the experimental system, with late-crystallizing leucogranites plotting on or near the ternary minimum, thereby suggesting H_2O saturation at the time of magmatic crystallization. The close fit of the data to estimated conditions with P_{H_2O} of 0.25–0.30 GPa, is in very good agreement with the estimated P_L of 0.3 GPa from the aureole assemblages (Weiss and Troll, Chap. 4, this Vol.).

Although H_2O or near H_2O-saturated compositions are suggested for granites and diorites, careful reconstruction of the crystallization sequences and conditions (Weiss and Troll, Sect. 4.3, Part II, this Vol.; and Fig. 4.12) suggests that hydrous fluids were released from the magma only in the closing stages of crystallization. The restricted hydrous alteration of the adjacent 'dry' monzodiorites, which were intruded at an early stage, also suggests that there were only limited amounts of vapours released from the other magmas. As a consequence, there was probably limited potential for liberation of volatile fluid or vapour from the intrusive complex to the surrounding metasediments, and this likelihood is strengthened by the occurrence of the most H_2O-saturated intrusives, the granites, within a partial envelope of monzodiorites and diorites (Fig. 19.1 and Maps 1 and 2).

In addition, the preservation of unhydrated magmatic assemblages shows that there could not have been pervasive flow of hydrous fluid of any origin through the crystallized igneous rocks at high temperatures.

19.2.1.2 Pelites and Semi-Pelites in the Contact Aureole

Metamorphic Zones up to the Onset of Partial Melting

For the most part, at temperatures below those of anatexis, the presence of a fluid phase rich in H_2O is anticipated in pelites and semi-pelites during prograde contact metamorphism because their reaction sequence in the aureole is dominated by dehydration reactions, in which phyllosilicates (chlorite, muscovite, biotite) are replaced by less hydrous or anhydrous minerals (cordierite, K-feldspar and Al_2SiO_5 minerals). This conjecture is supported by other evidence given by Pattison and Harte (Chap. 8, this Vol.) and Pattison (Chap. 16, this Vol.), which may be summarized as follows:

1. Within any given pelitic lithology in the aureole, the metamorphic zone boundaries, which correspond to dehydration mineral reactions, are regular and well defined, suggesting relatively uniform conditions of a_{H_2O} in each lithology. Comparisons of zonal sequences and boundaries across different lithologies suggest close similarities for all litho-stratigraphic units except the Ballachulish Slate, which unlike other pelites is graphitic (see items 4. and 5. in this list for further discussion).
2. The common zonal sequence of pelitic mineral assemblages and reactions may be modelled by experimentally-based petrogenetic grids for which $P_{H_2O} = P_F = P_L$.
3. Temperature estimates from pelite petrogenetic grids and reactions in which it is assumed that $P_{H_2O} = P_L = 0.32$ GPa, are in good agreement with other geothermometric constraints. Also pressure estimates based on the petrogenetic grids, and assuming that $P_{H_2O} = P_L$, are 0.31 GPa (Pattison, Chap. 16, this Vol.); in excellent agreement with the average estimate of 0.32 GPa based on equilibria which do not involve fluid.
4. The widespread conditions of $P_{H_2O} = P_F = P_L$ suggested by items 2. and 3. in this list will satisfy the general conditions of constant a_{H_2O} noted in 1. above, because the presence of a pure H_2O fluid phase will dictate constant $a_{H_2O} = 1$ at the given $P_F = P_L$ conditions.
5. In the Ballachulish slate lithology, the presence of graphite in equilibrium with hydrous vapour will generate C-bearing fluid species and lower a_{H_2O} (Fig. 19.2). Pattison (Chap. 16, this Vol.) demonstrates that the temperatures of zone reactions and their modified sequence in the Ballachulish Slate, may be explained by these graphite-H_2O equilibria. In particular, it is shown that the Zone IV andalusite-bearing assemblage of the Ballachulish Slate, is favoured against the K-feldspar-bearing assemblage of other lithologies, by lowering of a_{H_2O} where C-bearing species are present in the fluid. This shift in relative stability also explains otherwise anomalous Mg/(Mg + Fe) values of the ferromagnesian phases in the andalusite-bearing assemblage.

The above features provide strong evidence in favour of the presence of a fluid phase in the pelites at the time of development of the mineral zone assemblages (i.e., time of peak metamorphism for each zone); and this fluid was usually close to pure H_2O in composition with $P_F = P_L$.

It may be debated whether evidence of $P_{H_2O} = P_L$ *only* applies to the times of major reaction dehydration in each rock. At the onset of contact metamorphism, the rocks had already undergone dehydration reactions during regional metamorphism at a considerably earlier date (Harte and Voll, Chap. 1 and Pattison and Voll, Chap. 2, this Vol.), and subsequent to that event had undergone a net uplift prior to contact metamorphism of approximately 10 km (since regional metamorphic pressures are estimated at 0.6 GPa; see Pattison and Voll, Chap. 2, this Vol.). H_2O generated during regional metamorphism clearly leaves the rocks so that pore space is kept to very

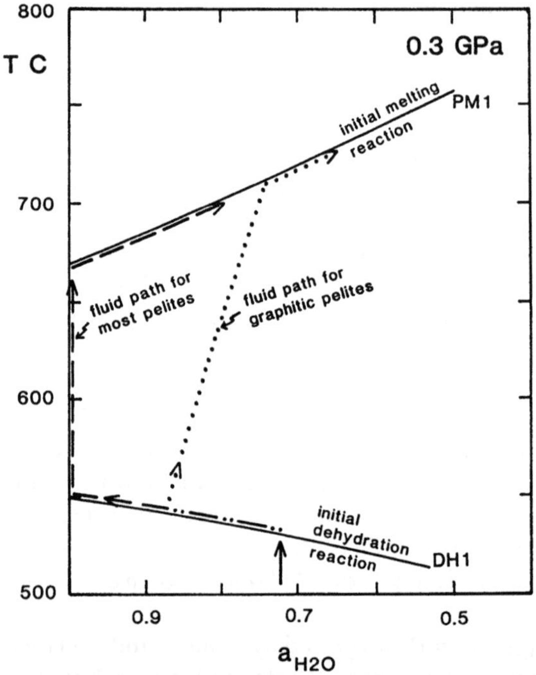

Fig. 19.2. Schematic evolution of fluid conditions (represented by a_{H_2O}) with increasing temperatures in pelitic and semi-pelitic rocks. *Dashed line* refers to most lithologies, the *dotted line* to the Ballachulish Slate lithology. Assuming fluid absence at the start of contact metamorphism a_{H_2O} is 1 (the value adapted is arbitrary) and remains so until the first dehydration reaction (DH^1) is encountered. Buffering along this reaction curve eventually gives rise to a virtually pure H_2O fluid phase ($a_{H_2O} = 1$) in most pelites and semi-pelites (*dashed line*). This fluid phase is presumed to remain present through a sequence of dehydration reactions (not shown) until the first reaction generating partial melt (*PM1*) is encountered, whereupon consumption of H_2O by the melt leads to a_{H_2O} being buffered along the melting curve to progressively lower values. In the Ballachulish lithology (*dotted line*), the general pattern of evolution is similar except that the presence of graphite leads to the generation of C-bearing fluid species; thus a virtually pure H_2O phase never forms and a_{H_2O} is 1 even when fluid is present in the middle temperature range (see Pattison, Chap. 16, this Vol., for further details)

small volumes (Walther and Orville 1982). Any water retained in the rocks at the end of peak regional metamorphism, may have been consumed by subsequent retrograde reactions, or have escaped along fractures during uplift. Thus it is probably unlikely that before contact metamorphism, the rocks were saturated with fluid, as might be expected for sediments undergoing burial (Fyfe et al. 1978; Etheridge et al. 1983).

If we envisage that at the commencement of contact metamorphism, fluid-absent assemblages began reaction at $a_{H_2O} < 1$, the release of H_2O by reaction would then give rise to a fluid pressure which would come to equal P_L as reaction progressed and thus give $a_{H_2O} = 1$ (Fig. 19.2). The amounts of reaction would be small until a_{H_2O} reached 1.0 (Pattison, Chap. 16, this Vol.), so that the bulk of the reaction would occur at $a_{H_2O} = 1.0$, and this will be the major condition of formation of mineral assemblages delineating the metamorphic zones (up to the conditions of partial melting).

To what extent the conditions of $P_F = P_L$ persisted in individual rocks throughout the whole time cycle of contact metamorphism is difficult to assess. Thompson (1983) has argued, on the basis of the general loss of water during metamorphism, that assemblages may readily become fluid absent at times other than those of major dehydration reactions. However, the loss of fluid by hydrofracturing (Walther and Orville 1982) or hydraulic fracturing (Etheridge et al. 1983, 1984) does not imply that all fluid is removed, but merely the excess of water consistent with maintaining very low porosity. Surface tension will also effect the fluid distribution.

The distribution of retrograde metamorphism provides a little evidence in this matter. Pattison and Harte (Chap. 8, this Vol.) note that retrograde reactions, particularly the extensive conversion of cordierite to pinite, are common at all grades within the aureole and often are the reverse of the prograde reactions. In general, isotope data (Sect. 19.3.2) do not suggest that the fluid for retrogressive reaction was of external rather than internal origin. Thus there is a possibility that the rocks retained a proportion of the fluid liberated during prograde metamorphism.

Conditions in the Subzone of Partial Melting

The presence of melts in the pelites and semi-pelites at the highest grades of metamorphism probably generated an important exception to the widespread occurrence of H_2O-rich fluid in the pelites during contact metamorphism. The high levels of H_2O needed to effect saturation of the acidic melts concerned, means that they will have absorbed H_2O. Petrogenetic analysis (Pattison and Harte, Chap. 8, this Vol.) of the mineral assemblages associated with melt conditions works well on the assumption that where melt was present, there was no separate volatile-rich fluid phase. Thus where melting occurred, we assume that it led to the loss of a separate H_2O-rich volatile phase and therefore to conditions where $a_{H_2O} < 1$. The precise behaviour would depend upon the exact reaction associated with melting, but as melt increased in abundance the a_{H_2O} would generally decrease and would be buffered by the melt reaction. This is illustrated schematically in Fig. 19.2 and discussed in detail in Pattison (Chap. 16, this Vol.), who suggests buffering of a_{H_2O} to values as small as 0.2. The extent of these effects in removing the H_2O phase in pelites and quartzo-feldspathic rocks throughout the aureole is difficult to assess, because it is difficult to quantify the amounts of melt produced (Harte et al., Chap. 9, this Vol.) and even more difficult to know the amount of H_2O available at the beginning of melting. As noted above some H_2O may have been mechanically expelled from the rocks during the occurrence of subsolidus dehydration reactions. However, in these highest grade aureole rocks, the development of peak metamorphic temperatures will have occurred quite rapidly following intrusion (less than 0.2 Ma, Buntebarth, Chap. 18, this Vol.), and possibly with considerable overstepping of reactions. It is possible that the rocks may have jumped to high temperature assemblages without progressively going through the whole reaction sequence now seen as a function of grade in the aureole.

During cooling the water held in silicate melt will have been released on crystallization, and there is a notable tendency to extensive retrograde alteration in the pelitic to psammitic rocks which have undergone partial melting (Pattison and Harte (Sect. 8.8, this Vol.). One area where the effect of melting in absorbing H_2O is likely to have been considerable is in the 'Chaotic Zone' of Leven Schists on the west flank of the aureole (Fig. 19.1), where melting was clearly more extensive than elsewhere (Harte et al., Chap. 9, this Vol.). In this area there is generally a greater extent of retrograde alteration, affecting K-feldspar and biotite in addition to cordierite, than elsewhere in the aureole (Pattison and Harte, Sect. 8.8, this Vol.).

19.2.1.3 Carbonate and Calc-Silicate Rocks

These rock types, because of their extensive reactions and the dependence of those reactions on fluid composition, provide very useful monitors of the presence, pressure, composition and behaviour of fluids during metamorphism. In the present case, the evidence is largely limited to the area of outcrop of marbles and calc-silicate rocks in the northeast of the aureole, between the intrusive contact and the Allt Guibhsachain (Fig. 19.1).

Evidence Concerning P_F

Masch and Heuss-Aßbichler (Chap. 10, this Vol.) provide temperature estimates for the carbonate rocks at various distances from the intrusive contact, using the Mg content of calcite in equilibrium with dolomite, and show that the best estimates are in accord with temperature estimates from the pelitic rocks (Pattison, Chap. 16, Fig. 16.4, this Vol.). In addition an estimate of total pressure in the marbles is made possible by the occurrence of the assemblage tremolite + forsterite + diopside + calcite + dolomite, which (with six phases, including fluid) is univariant (or isobarically invariant) in the system CaO-MgO-SiO_2-CO_2-H_2O. The total pressure estimate obtained is 0.28 ± 0.04 GPa, which is in excellent agreement with the estimate of 0.30 to 0.32 GPa obtained from pelites (Pattison, Sect. 16.3, this Vol.). This estimate therefore provides direct evidence of $P_F = P_L$; a conclusion which is further supported by the fact that the temperature estimates for all reactions fit a T–X_{CO_2} calibrated petrogenetic grid for P_F = O.3 GPa (Fig. 10.2).

Textural and structural features also support the presence of a fluid phase with pressures similar to lithostatic pressures, during the metamorphism. Masch and Heuss-Aßbichler (Chap. 10, Part III, this Vol.) note that fluid migration and channelling in the Ballachulish marbles is shown by the occurrence of small fluid-induced strings of reaction products at reaction sites, as well as calcite veinlets and veins. A hydraulic fracturing mechanism is thus suggested, which indicates fluid pressures similar to solid stresses in order to reduce effective stresses (Hubbert and Rubey 1959; Etheridge et al. 1983, 1984).

Evidence Concerning Fluid Composition and Mobility

Masch and Heuss-Aßbichler (Chap. 10, this Vol.) show that variations in fluid composition, X_{CO_2}, in siliceous dolomites and impure limestones may be inferred. The evidence comes from the character and distribution of mineral assemblages, and by making quantitative estimates of X_{CO_2} using temperature estimates (from calcite-dolomite geothermometry) and a calibrated T-X_{CO_2} grid of the relevant mineral reactions (Fig. 10.12, reproduced here as Fig. 19.3). The data are suggestive of different types of behaviour at different grades of metamorphism, with more open-system behaviour at low and high grades of metamorphism, and more closed-system behaviour at middle grades.

Considering the middle grades of metamorphism, as best displayed in the northeastern (Allt Guibhsachain) area of the aureole, Masch and Heuss-Aßbichler (Chap. 40, this Vol.) report the occurrence of mineral assemblages corresponding to the following reactions in the system CaO-MgO-SiO_2-H_2O-CO_2:

$$1\,Tr + 11\,Dol \Leftrightarrow 8\,Fo + 13\,Cal + 9\,CO_2 + 1\,H_2O \quad (C9)$$
$$1\,Tr + 5\,Cal \Leftrightarrow 11\,Di + 2\,Fo + 5\,CO_2 + 1\,H_2O \quad (C10)$$
$$1\,Tr + 3\,Cal \Leftrightarrow 4\,Di + 1\,Dol + 1\,CO_2 + 1\,H_2O \quad (C7)$$
$$1\,Di + 3\,Dol \Leftrightarrow 2\,Fo + 4\,Cal + 2\,CO_2 \quad (C11)$$

(abbreviations: Tr — tremolite; Dol — dolomite; Fo — forsterite; Cal — calcite; numbers correspond to reaction labels in Fig. 19.3 and Chap. 10, this Vol.). The rocks demonstrating these assemblages and reactions are also associated with the occur-

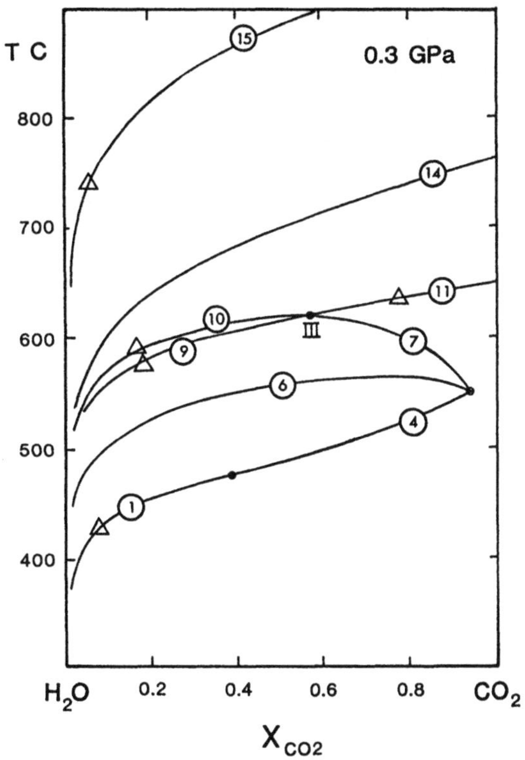

Fig. 19.3. T-X_{CO_2} section for siliceous dolomite and impure limestone (after Masch and Heuss-Aßbichler, Chap. 10, this Vol.). The reactions shown are: *1* 3 dolomite + 4 quartz + 1 H_2O = 1 talc + 3 calcite + 3 CO_2 (C1); *4* 8 quartz + 5 dolomite + 1 H_2O = 1 tremolite + 3 calcite + 7 CO_2 (C4); *6* 2 quartz + 1 tremolite + 3 calcite = 5 diopside + 3 CO_2 + 1 H_2O (C6); *7* 1 tremolite + 3 calcite = 4 diopside + 1 dolomite + 1 CO_2 + 1 H_2O (C7); *9* 1 tremolite + 11 dolomite = 8 forsterite + 13 calcite + 9 CO_2 + 1 H_2O (C9); *10* 1 tremolite + 5 calcite = 11 diopside + 2 forsterite + 5 CO_2 + 3 H_2O (C10); *11* 1 diopside + 3 dolomite = 2 forsterite + 4 calcite + 2 CO_2 (C11); *14* 1 quartz + 1 calcite = 1 wollastonite + 1 CO_2 (C14); *15* 1 dolomite = 1 periclase + 1 calcite + 1 CO_2 (C15)

rence in some cases of the isobarically invariant assemblage: Tr + Dol + Fo + Cal + Di (+ fluid), which occurs at invariant point III in Fig. 19.3, with an estimated X_{CO_2} of 0.55. The lowest grade isobarically univariant assemblage of reaction (C9) has an estimated temperature of 580 °C giving an X_{CO_2} value of ca. 0.2, whilst the higher grade (640 °C) assemblage of reaction (C11) suggests X_{CO_2} of 0.8 (Masch and Heuss-Aßbichler, Chap. 10, this Vol.).

The above relationships are strongly indicative of local buffering of X_{CO_2} by reactions like (C9) which have high product CO_2/H_2O ratios (Greenwood 1975; Rice and Ferry 1982). Thus the fluid phase composition appears to have been driven along the univariant reaction curve to high X_{CO_2} and invariant point III where other reactions arise. Clearly, in such situations the fluid produced during reaction appears to be retained in the rocks and closed-system behaviour is approximated.

The behaviour noted above for middle grades, contrasts with the widespread occurrence of isobarically divariant calcareous assemblages at other grades in the aureole, and with determinations of X_{CO_2} for the following low- and high-grade (Fig. 19.3) assemblages:

$$3\ Dol + 4\ Qtz + 1\ H_2O = 1\ Tlc + 3\ Cal + 3\ CO_2 \quad (C1)$$
$$1\ Dol = 1\ Per + 1\ Cal + 1\ CO_2 \quad (C15)$$

Respectively these isobarically univariant reaction assemblages yield temperature estimates of 420 and 760 °C, and corresponding X_{CO_2} values of ca. 0.1 or less. Here there is little evidence of buffering of fluid composition by internally generated fluid. In another case, the isobarically univariant assemblage wollastonite + calcite + quartz [reaction (C14), Fig. 19.3] is abundant locally. Superficially, the occurrence of this univariant assemblage suggests internal X_{CO_2} buffering, but the large amount of wollastonite found in these rocks must indicate dilution of the CO_2 fluid product of the reaction by other fluid (e.g., see Ferry 1983).

Overall, the calcareous assemblages provide evidence, therefore, of both virtually-closed and more-open system behaviour. The cause of the open-system behaviour is uncertain. The infiltration of externally-derived fluid into the calcareous rocks from the magmas of the intrusive complex, or from pelitic rocks, is not supported by stable isotope studies (Hoernes et al., Chap. 17, this Vol.). Masch and Heuss-Aßbichler (Chap. 10, this Vol.) emphasize the difference in reaction textures between middle-grade 'closed' systems with localized reaction at intergrowths and coronas, and high-grade, more open, systems with granoblastic textures, extensive calcite veins and widely distributed reaction products. They suggest that the more open behaviour does not necessarily indicate infiltration of H_2O-rich fluid from outside the calcareous rocks, but may be a result of more extensive exchange and mixing of pore fluids within the calcareous rocks themselves. They also make the important point that the mineral assemblages and textures do not support the concept of progressive metamorphism, whereby each high-grade rock has evolved through the sequence of low- and middle-grade assemblages seen in other parts of the aureole. Thus they refute the suggestion that the difference between middle- and high-grade assemblages, implies a history of closed-, then open-system behaviour for the high-grade rocks.

19.2.1.4 The Appin Quartzite

The quartz and feldspar dominated mineral assemblages in this lithology do not suggest that significant fluid would be liberated from them during prograde metamorphism, nor do they allow a basis for assessing fluid phase involvement as in the case of the carbonate rocks. Some negative evidence for the presence of fluid in these rocks is given by the lack of identified melting features. Given that K-feldspar and plagioclase are minor phases in the quartzites and often show some evidence of detrital origin, a low-temperature melting assemblage must have been available if H_2O-rich fluid had been present. A detailed consideration of oxygen isotope compositions also provides indirect kinetic evidence for the limited availability of fluid in the Appin Quartzite (see Sect. 19.2.2).

19.2.2 Indirect Fluid Control of Rock-Phase Assemblages

Evidence of fluid presence or absence may be obtained from the purely kinetic viewpoint of the importance of hydrous fluids in promoting mineral reaction and equilibrium. Such evidence is indicated in the instances itemized below.

19.2.2.1 Alkali Feldspar Transformations in the Magmatic Rocks

Troll and Weiss (Chap. 3, this Vol.) and Weiss and Troll (1989) find a correlation of Al, Si-ordering phenomena and exsolution development in the alkali feldspars with the hydration state of different magmatic rocks. The monzodiorites showing the 'dryest' mineral assemblages (with more pyroxene and less biotite and amphibole), show the least amount of Al,Si ordering by comparison with the more hydrous diorites and granites. Similarly, the 'dryest' monzodiorites show cryptoperthites, fine-lamellar film perthites and complexly shaped symplectites; whilst the more hydrous monzodiorites, diorites and granites show patch- to flame-perthites and in some cases coarse patch- to vein-perthites. The extent of Al,Si ordering and the coarsening of exsolution structures in the alkali feldspars have both been shown to be kinetically favoured by greater fluid activity (Parsons 1978; Yund 1984), and the above features support the previous arguments that vapour-absent conditions persisted under subsolidus conditions in the 'dry' monzodiorites.

19.2.2.2 Isotope Distribution, Feldspar Mineralogy and Quartz-Grain Growth in the Appin Quartzite

Despite its restricted mineralogy, the Appin Quartzite shows many interesting features associated with the presence of relict clastic pebbles of quartz and feldspar. Detailed oxygen-isotope measurements on single grains within a given specimen show that there is a large spread in both quartz and K-feldspar oxygen-isotope compositions even up to 40 m from the contact with the intrusive complex (Hoernes and Voll, Chap. 15, this Vol.). For K-feldspar, consideration of diffusion coefficients suggests that equilibrium would have been attained by this phase if a hydrous fluid had been present, because of the higher diffusion rates in the presence of H_2O. Thus the lack of equilibration suggests the absence of an H_2O-rich fluid phase.

In contrast, aspects of ordering, inversion and unmixing in the clastic alkali feldspar grains (Kroll et al., Chap. 13; Kroll, Chap. 20, this Vol.) indicate that some H_2O must have been present. This is also supported by Buntebarth and Voll's (Chap. 12, this Vol.) study showing a low activation energy for quartz grain growth, which indicates H_2O presence.

These conflicting observations may be partly explained by localized availability of H_2O, as for example in fluid released from fluid inclusions in clastic quartz grains (Buntebarth and Voll, Chap. 12, this Vol.). Taken in conjunction with the previous evidence of the lack of melting in the Appin Quartzite, it seems unlikely that a pervasive hydrous fluid phase with $P_F = P_L$ was present for any length of time during the metamorphism of this lithology.

19.3 Overall Pattern of Fluid Behaviour in the Intrusive Complex and Aureole

19.3.1 General Considerations

In considering the extent of migration of hydrothermal fluids in the intrusive complex and its aureole, and the part such migration may have played in modifying phase distributions and compositions, much useful evidence comes from the stable isotope data presented in Chapters 15 and 17 of this Volume. However, it is pertinent to summarize the evidence outlined above for the presence of a fluid phase in the various lithologies of the intrusive complex and its aureole. We have seen evidence for a fluid phase in the diorites and granites of the intrusive complex, and the pelites and carbonate rocks of the thermal aureole. Definite evidence for the lack of a fluid phase comes from the monzodiorites of the intrusive complex, whilst in the Appin Quartzite there seems to have been restricted fluid availability. In the case of pelitic rocks an exception to the presence of fluid occurs in the case of the highest grade rocks containing melt, where a free hydrous phase is expected to be absorbed by the melt.

The lack of evidence for a fluid phase in some lithologies at any time during the igneous-metamorphic history immediately suggests that lack of a pervasive hydrothermal circulation system operating across the intrusive complex and its aureole. Furthermore, even where there is evidence for a fluid phase, the rocks concerned provide little evidence for extensive fluid migration between lithologies, such as the development of distinctive mineral assemblages or alteration zones adjacent to lithological contacts. In the calcareous rocks, widely variable X_{CO_2} and the presence of assemblages of different variance clearly refute fully open-system behaviour where H_2O or CO_2 appear to behave as perfectly mobile components (e.g., Zen 1963). The fluid phases in the diorites, granites, pelites and carbonates may all have sources within their own lithology, either as products of magma crystallization or as products of metamorphic devolatilization reactions.

In the pelitic and calcareous rocks, in which devolatilization proceeded, the loss of fluids is evident on present day examination. However, the time scale of this loss is extremely uncertain. Also it is evident that histories of mineral reaction and fluid behaviour in individual rocks cannot be reliably interpreted by reference to the grade sequence of mineral zones in the aureole (see Sect. 19.2.1.3).

When retrogressive changes in mineralogy are considered, the broader pattern of alteration shows restricted behaviour which reflects the primary phase relationships. Thus, in the intrusive complex, sericitization of feldspar occurs least in the monzodiorites which appear to have lacked a fluid phase during primary magmatic or immediate subsolidus crystallization (Weiss and Troll 1989). In the pelitic lithologies, retrogression often reverses the prograde sequence. Furthermore, it is most intense in the 'Chaotic Zone' (Fig. 19.1; Pattison and Harte, Chap. 8, this Vol.), where there is the highest degree of anatectic melting, and thereby the greatest amount of H_2O to be liberated during crystallization of that melt. Thus, the observed retrogression supports the evidence of the rocks having a strong internal control on their fluid phase relationships and tending to retain their own fluid, rather than being subject to externally imposed fluid conditions.

19.3.2 Stable Isotope Data

In some high-level intrusions (e.g., the Scottish Hebridean province), stable isotopes have provided evidence of extensive convective fluid circulation through the rocks (of both igneous and metasedimentary origin), which was driven by the large thermal gradients caused by the igneous intrusions. In the case of Ballachulish (Hoernes et al., Chap. 17, this Vol.), stable isotopes do not provide evidence of the operation of any large scale circulation system. This is shown by the following:

1. For the intrusive complex, oxygen and hydrogen isotope data show absolute concentration values and fractionation factors between minerals, which are largely typical primary magmatic values for a situation where a crustal component to the initial magmas has been suggested. Thus the $\delta^{18}O$ and δD igneous whole-rock values range largely from 8–10‰ and –50–65‰, respectively. There is some re-equilibration of O-isotopes at subsolidus conditions in rocks of the younger granitic intrusive phase, but it indicates the presence of an internally generated magmatic fluid, exsolved during crystallization of the magma, in accord with the evidence given previously.

2. Oxygen isotope ratios in all rocks are in general accord with those expected for the different lithologies involved. More particularly, there are no clear differences between rocks outside the aureole and the low-grade or high-grade contact metamorphic rocks within the aureole. For marbles in the northeastern sector of the aureole, $\delta^{18}O$ and $\delta^{13}C$ are respectively 20 ± 2.5 and 3 ± 1‰, irrespective of distance from the intrusive contact. The Appin Quartzite shows no systematic variation with grade

either on east or west flanks of the intrusive complex. Out of 60 $\delta^{18}O$ determinations for pelitic and semi-pelitic rocks, 51 fall within the range 11–15‰, and the variations for particular lithologies along narrow transects across the metamorphic zones are much smaller than this range (Hoernes et al., Chap. 17, Table 17.3, this Vol.). A very small number of data from the Chaotic Zone (showing most extensive melting of pelites-psammites) indicate slightly lower $\delta^{18}O$, and some contribution of fluid from the underlying quartz diorite (Pattison and Harte, Chap. 8; Harte et al., Chap. 9, this Vol.) may be speculated in this limited case.

3. There are sharp changes in isotope concentrations (which cannot purely be functions of mineralogy and temperature) across contacts between different members of the igneous complex, and between rock types in the metamorphic aureole.

The stable isotope data clearly rule out any extensive hydrothermal circulation of meteoric waters like that seen in the Hebridean province. Furthermore, they demonstrate the lack of circulation of fluid between different members of the intrusive complex, and the general lack of any fluid circulation between the intrusive complex and the metasediments of the aureole. Finally, the isotopes provide no evidence of fluid movement between different lithologies of the aureole.

These observations are generally in harmony with those expected from detailed consideration of the internal phase relationships preserved within individual rocks (as presented in the preceding section). The lack of circulation of fluid between the aureole and intrusive rocks may be partly explained by the hygroscopic character of the granitic melts generated within the aureole at the highest grades.

19.4 Discussion and Comparison with Other Areas

The evidence presented above clearly indicates that a fluid phase with $P_F = P_L$ was present in some lithologies during cooling of the igneous complex and metamorphism of the aureole rocks. However, a pervasive fluid phase was absent from other lithologies, and the local internal buffering of X_{CO_2} in some carbonate rocks indicates restricted circulation of externally derived fluid; a conclusion that is also supported by evidence of retrogressive effects. Extensive stable isotope data also show that there was little communication of fluid between the different lithologies, with no imposition of particular isotope signatures which would suggest the widespread circulation or passage of fluid through all lithologies. It is important to note that the thermal evolution of the Ballachulish intrusive complex and aureole (Buntebarth, Chap.18, this Vol.) also fits with this evidence of fluid behaviour. Thus the thermal conditions in the aureole may be modelled essentially by conduction, and without any convective movement of heat which fluid circulation would cause (cf. Norton and Knight 1977).

These features contrast sharply with the situation seen about some high-level (often <2-km deep) intrusive complexes, as most particularly demonstrated for example by intrusive complexes in the Tertiary igneous province of the Scottish

Hebrides, where depletion in ^{18}O shows the formation of extensive hydrothermal circulations involving meteoric ground waters (Taylor and Forester 1971; Forester and Taylor 1977). The situation at Ballachulish also contrasts with that of the Notch Peak Pluton, Utah (Nabelek et al. 1984), with estimated pressures of metamorphism of 0.15 GPa (ca. 4 km); here stable isotope data show that fluids from the crystallizing pluton penetrated previously unmetamorphosed, calcareous argillites but not limestones. At Notch Peak, there is no evidence of a hydrothermal circulation system; the differences between the argillites and the carbonates are attributed to permeability differences. These are themselves attributed to permeability enhancement in the calcareous argillites as a consequence of high reactivity with devolatilization reactions (see, for example, Fyfe et al. 1978; Valley 1986); whilst the virtually pure carbonate rocks show little reaction and fluid generation.

At Ballachulish, all the aureole lithologies, except the Appin Quartzite, show a succession of devolatilization reactions, which would have provided some scope for potential permeability enhancement. However, some reasons, involving depth of emplacement and fluid availability, may readily be put forward to explain the different behaviour at Ballachulish compared to the shallower level contact aureoles. The depth of intrusion and metamorphism now exposed at Ballachulish is estimated as 10 km. At pressures above 0.1–0.2 GPa (3–6 km), dry cracks in common rock types will become closed (Wang and Simmons 1978). Thus at the depths of formation of the Ballachulish rocks there would be no readily available crack system to aid permeability and provide access for meteoric waters.

If one considers a system of fluid filled cracks open to the earth's surface and suitable for convective cycling of meteoric waters, the fluid pressure will be hydrostatic (equivalent to the load of head of the fluid). Evidence from fluid pressures in sedimentary basins (e.g., Hubbert and Rubey 1959; Wood and Walther 1986) and considerations of the strength of rocks (e.g., Fyfe et al. 1978), consistently indicate that such an open crack system and associated hydrostatic pressures can only be present to *maximum* depths of about 6 km. At greater depths, rock material will collapse around the fluid filled cracks, (which will thereby cease to form a system open to the surface) and cause fluid pressures to equal lithostatic pressures ($P_F = P_L$). According to these criteria, with a depth of metamorphism of 10 km in the Ballachulish aureole, any fluid present would be expected to be at lithostatic pressures, and abundant direct evidence of this has been documented above. With $P_F = P_L$ at Ballachulish the convective cycling of meteoric or other fluid at hydrostatic pressures was impossible, and marked distinctions to the Hebridean Tertiary and other shallow intrusive complexes are readily explained.

The Ballachulish evidence fits in with Valley and O'Neil's (1982) observation that isotopic evidence of ground water circulation around plutons is limited to depths less than 7–10 km. Wickham and Taylor's (1985) advocacy of convective circulation in a regional metamorphic setting at up to a 12-km depth in the Pyrenees is at variance with other data.

The question of whether a convective circulation of any volatile-rich fluid can occur with P_F approximately equal to P_L leads to divided opinions: Etheridge et al. (1983) suggest the occurrence of large scale convection in cases of regional metamorphism; Ferry (1986) contemplates the possibility; Wood and Walther (1986) discount it; and Valley (1986) emphasizes the unlikelihood of there being adequate bouyancy forces or permeabilities for convection under the proposed conditions. Clearly, the strong preservation of original isotopic characteristics in many lithologies throughout the Ballachulish complex and aureole provides no support for the occurrence of convective circulation under the circumstances applying there. However, other features of the Ballachulish situation, such as limited fluid availability and the presence of fluid 'sinks' (see below) may have an important bearing on this as well as the fundamental restrictions to convection.

In addition to the limitations on the permeability of the Ballachulish rocks as a function of the depth of metamorphism, the lithological characteristics of the rocks themselves were probably resistant to extensive fluid infiltration. Even in the case of shallow intrusions, the extent of isotopic modification by percolating fluids varies with rock type. Thus at Mull, ^{18}O depletion is less extensive in shales than fractured basalts (Forester and Taylor 1976). At Ballachulish, the rocks had already undergone regional metamorphism at upper greenschist facies conditions, and therefore extensive recrystallization and plastic deformation must have severely reduced permeability, with rock porosity probably having become well below 1% (Walther and Orville 1982), though this does not exclude connectivity of the pore fluid.

On top of these limited permeabilities, we may further note that without an external source of fluid, there were limitations on fluid availability at Ballachulish. Some lithologies (e.g., the monzodiorites) are water deficient and at no time had H_2O to yield to the fluid budget; whilst at the time of peak metamorphism, the highest grade aureole rocks were probably consuming water during melting. These potential fluid 'sinks' also formed a barrier between more hydrous igneous rocks and the lower grade aureole rocks.

The evidence of very restricted fluid movement at Ballachulish also contrasts markedly with the conditions shown in many situations of regional metamorphism, where there is evidence of widespread high P_F approximately equal to or greater than P_L (Etheridge et al. 1983), combined with extensive fluid migration as indicated by: mass transport (Etheridge et al. 1983, 1984; Ferry 1986); high fluid/rock ratios controlling fluid composition during reactions in calcareous rocks (Ferry 1986, and references therein); stable isotope data (Rumble et al. 1982; Graham et al. 1983; Wickham and Taylor 1985).

By comparison with the regional metamorphic terrains concerned, Ballachulish shows similar features of $P_F = P_L$, and extensive mineral reaction in pelitic and calcareous rocks, so that permeability enhancement within individual rocks is unlikely to have been a distinguishing factor. However, other features provide clear contrasts. In the first place, there are the previously noted facts that the Ballachulish aureole rocks had already undergone regional metamorphism of upper greenschist facies conditions and been uplifted from depths of approximately 20 km to 10 km before the contact metamorphism occurred. The rocks had already undergone considerable devolatilisation therefore, and there were no effects of compaction under increasing load to increase fluid pressures (see also Sect. 19.2.2.2). Thus the amount of potential fluid for release and upward movement would be considerably reduced in this contact metamorphic setting by comparison with the regional one (cf. Fyfe et al. 1978; Wood and Walther 1986).

Secondly, little deformation affected the aureole rocks during contact metamorphism. The tectonic foliations and folds seen in the aureole rocks are clearly relics of the earlier regional metamorphic event. Thus, in contrast to typical low-middle-grade regional metamorphism, there would be little contribution to the metamorphic porosity and permeability of the rocks as a consequence of transient microcracking in response to deformation (Etheridge et al. 1983, 1984).

19.5 Conclusions

Considering the evolution of the Ballachulish complex and aureole, we may note the following features with respect to volatile-rich fluids:

1. Fluid developed during late-stages of magmatic crystallization of the quartz diorites and granites with P_F approximately equal to P_L (0.3 GPa).
2. Fluid was generated in the pelites, semi-pelites and calcareous rocks as a consequence of metamorphism leading to devolatilization. Independent evidence from both pelitic and calcareous rocks suggests that P_F was approximately equal to P_L (0.3 GPa).
3. Some parts of the intrusive complex and aureole probably did not contain a fluid phase during their crystallization and metamorphism.
4. The extent of fluid mobility in the calcareous rocks was limited, and internal rock buffering of fluid to high X_{CO_2} values is indicated locally.
5. Partial melting of pelites and semi-pelites in the innermost aureole probably absorbed any locally available H_2O during prograde metamorphism and acted as a barrier to fluid transport between the intrusive complex and the aureole rocks.
6. Retrogressive development of hydrous minerals suggests that the fluid for its occurrence was internally derived (i.e., from within the retrograde rocks themselves), because it is most extensive in igneous rocks and country rocks where melts are expected to have released H_2O on crystallization.
7. Stable isotope data show no evidence of widespread fluid flow between the major lithologies of either the intrusive complex or the aureole or both. No system of hydrothermal fluid circulation was established.
8. There is no evidence of heat transport by fluid flow. Thermal modelling adequately accounts for the T distribution in the aureole by conduction.

The distinctive features of the Ballachulish situation, by comparison with very high-level intrusions showing hydrothermal circulation, and many models of regional metamorphism showing extensive fluid transport, may be accounted for by:

1. A depth of burial of 10 km, which is greater than that possible for maintenance of a system of cracks open to the surface. Thus, $P_F = P_L$, and the circulation of meteoric water was impossible.
2. The intrusive magmas released little volatile-rich fluid during crystallization. Furthermore, some diorites did not develop fully hydrated assemblages in the subsolidus and could act as consumers of available volatiles.

3. The aureole rocks had been metamorphosed and significantly dehydrated during previous regional metamorphism. Thus these rocks had both limited fluid production and restricted permeabilities. Partial melts formed at the peak of metamorphism in the inner aureole would have consumed any available H_2O.
4. The lack of deformation during contact metamorphism led to diminished transient microcracking and therefore permeability.

20 Intracrystalline Processes

H. Kroll

20.1 Introduction

The thermal history of minerals may be recorded by examining intracrystalline processes, notably, ordering and exsolution. We will deal in this paper with both these processes, using K-feldspar, cordierite, Mg-calcite and pyroxenes as examples. The discussion is based on the contributions by Kroll et al. (Chap. 13, this Vol.), Maresch et al. (Chap. 14, this Vol.), Masch and Heuss-Aβbichler (Chap. 10, this Vol.) and Feuer et al. (Chap. 6, this Vol.). H. Feuer, Frankfurt, provided additional personal communication on pyroxenes.

If the cooling behaviour of a solid solution is treated in an elementary thermodynamic way by pairwise nearest neighbour interactions, ordering and exsolution appear to be mutually exclusive processes. Defining W_{AA}, W_{BB}, W_{AB} as bond interaction energies between AA, BB, AB pairs in an AB solid solution, three different types of behaviour can be expected depending on the relative sizes of the interaction energies. If $W_{AB} = \frac{1}{2} (W_{AA} + W_{BB})$, the solid solution should behave as an ideal mixture (no tendency to order or exsolve); if $W_{AB} < \frac{1}{2} (W_{AA} + W_{BB})$, bonds between unlike neighbours are favoured, i.e., there exists a tendency to order, and finally, if $W_{AB} > \frac{1}{2} (W_{AA} + W_{BB})$, the solid solution tends to unmix.

However, when considering the ordering process in an AB compound which is off stoichiometry towards B-rich compositions, $A_{1-x}B_{1+x}$, it is clear that some of the second nearest neighbours of A which ideally should be only A atoms are B atoms. In order to maximize the number of A atoms in the second coordination sphere, B atoms must segregate. Thus ordering and unmixing in a refined model of solid solution behaviour are not mutually exclusive, but may occur at the same time in a given crystal. Such models are discussed in the metallurgical and the recent mineralogical literature (e.g., Carpenter 1985, on order/disorder transformations).

Examples of order dependent unmixing in mineral systems given by Carpenter (1985) are peristerites, dolomites, and omphacites. If Na-feldspar is off stoichiometry towards Ca-feldspar or dolomite towards calcite or (ideal) omphacite towards diopside or jadeite, the ordering of (Al,Si), (Ca,Mg) and (Ca,Na; Mg,Al), respectively, is inhibited, so that the mixed crystal is forced to purify itself via exsolution, which is known as conditional unmixing. On the contrary, Na-feldspar does not get off stoichiometry with respect to Al:Si, when K-feldspar is substituted, and thus Al,Si ordering in the alkali feldspar system does not drive unmixing, but only modifies the miscibility gap.

G. Voll, J. Töpel, D.R.M. Pattison, F. Seifert
Equilibrium and Kinetics
in Contact Metamorphism

We will start our discussion with the disordering and re-ordering behaviour of K-rich feldspars in the Ballachulish aureole, we then consider cordierite, and finally discuss the unmixing processes in Mg-calcite and pyroxenes.

20.2 K-Feldspars

20.2.1 Changes on Heating

In the Appin Quartzite, clastic K-feldspars occur which during the Caledonian regional metamorphism had all been converted to the low microcline structural state. These low microclines provide the rare event in which a K-feldspar population is homogeneous with respect to structural state, thereby providing an excellent reference state.

When approaching the igneous contact, a transition zone extending from 1900 m to ≈1100 m is crossed in which orthoclase is found together with relic low microcline. The amount of orthoclase increases at the expense of low microcline, which finally disappears at a distance of ≈1100 m (Kroll et al., Chap. 13, this Vol.; Figs. 13.1, 13.3). Orthoclase is a re-ordering product of sanidine into which low microcline had been converted.

Transition zones have also been observed by Steiger and Hart (1967) in the seven traverses they investigated around the Eldora quartz-monzonite stock near Boulder, Colorado. The Eldora stock has a diameter of ≈3 km, whereas the Ballachulish intrusion measures ≈4 km • 7 km. Steiger and Hart (1967) could relate the widths of the transition zones and their distances from the contact to the temperature profiles in the various traverses: the less steep the temperature decreases, the wider and more distant the transition zone. The Ballachulish transition zone fits into this scheme, being ≈800 m wide, compared to ≈100 – ≈300 m at the Eldora stock.

Bambauer and Bernotat (1982) investigated K-feldspars along a traverse in regional metamorphic rocks of the Central Swiss Alps. Their results can also be interpreted in terms of a transition zone. When considering their Fig. 7 of X_{Al}^{T1o}–X_{Al}^{T1m} vs distance it is seen that in a northern zone where the Alpine metamorphic temperatures exceeded ≈300 °C, but did not exceed ≈480 °C, low microcline is by far the predominant phase. This zone corresponds to the low microcline zone of the Ballachulish aureole (Kroll et al., Chap. 13, this Vol.; Figs. 13.1, 13.2, 13.3). To the south, at a fairly sharp boundary, namely, the microcline-sanidine transition isograd, high microcline and orthoclase set in, but over a distance of several kilometres low microcline is still observed. This zone would correspond to the Eldora and Ballachulish transition zones.

The existence of the transition zone is discussed by Kroll et al. (Chap. 13, this Vol.; Fig. 13.25) from two different points of view: (1) if the microcline-sanidine transformation temperature is strongly dependent on composition, the transition zone is created by variable re-solution of perthitic Na-feldspar by microcline, as is actually observed; and (2) if the transformation temperature is virtually independent of composition, the transition zone must result from overheating.

The character of the transformation is taken to be first order, but close to tricritical, in analogy to the high-to-low albite inversion when freed from the modifying displacive effect.

The dominant composition of low microcline in the zone of no contact metamorphism is around Or96. In the transition zone, some re-solution of perthitic Na-feldspar is observed: at first, film-perthite, and at higher temperatures spindle-perthite also disappears, whereas patchy perthite survives. As schematically drawn in Fig. 13.25a, various degrees of resolution create a range of Na-containing microclines which all invert at different temperatures. The lower and upper transformation temperatures were taken from Fig. 13.26 at the outer and inner transition zone boundaries. There is, however, one argument against such a large range of transformation temperatures (≈70 °C: the displacive-free, high-to-low albite inversion would occur at ≈600 °C (Salje et al. 1985), just ≈120 °C above the lower temperature of the microcline-sanidine transition. This observation does not favour the assumption that between ≈Or80 and Or100 the transformation temperature should increase by ≈70 °C.

The alternative possibility assumes that the transformation temperature is virtually independent of composition. To create a transition zone under this assumption, overheating must have occurred. Experimental results on the Al,Si-exchange kinetics in sanidine (Kroll et al., Chap. 13, this Vol.) have shown that at least some hundred bars of partial water vapour pressure must have been present in the aureole to convert sanidine to orthoclase or microcline during cooling. Under these pressure conditions, ordered sanidine becomes disordered within $t \leq 25\,000$ years even at temperatures as low as 550 °C, which is close to the maximum temperature at the inner boundary of the transition zone (555 °C, Fig. 13.26). The heating and cooling cycle from 480 to 555 °C and back lasted about 20 times as long (Fig. 13.28). It is highly unlikely that low microcline should have survived this long time overheating without a specific reason. One such reason could be that dry conditions dominated during heating, but 'wet' conditions governed the late stage of cooling. This appears to be a somewhat artificial assumption, especially in view of the regional metamorphic conditions in the Central Swiss Alps, where transition zones are also observed. The other possibility assumes that non-hydrostatic stresses exerted on the microcline grains by surrounding quartz grains hindered the inversion from triclinic to monoclinic geometry. However, at the moment this is merely a speculation.

It appears that the occurrence of transition zones in both contact and regional metamorphic terrains still awaits a satisfactory explanation.

20.2.2 Changes on Cooling

When sanidine coexisting with low microcline in the transition zone was cooled down, it transformed to orthoclase. The Or content of orthoclase does not follow the metastable extension of the albite-sanidine solvus, such that there would be a composition gap between orthoclase and low microcline, but we found that in fact their compositions are practically the same. This is not surprising because orthoclase consists of triclinic, highly short-range ordered domains, and thus its solution behaviour is similar to low microcline. Therefore, it can not be expected that a sanidine-microcline transition loop as drawn in Fig. 13.25a is reflected in the Or-content of the re-ordering products orthoclase and low microcline.

The zone next to the contact is called the sanidine zone, where all low microcline had been converted into sanidine. According to X-ray data, in this zone sanidine turned during the cooling process into orthoclase and intermediate microcline, whereas in the transition zone, no intermediate microcline is found, so that the outer boundary of the occurrence of intermediate microcline coincides with the boundary between the sanidine and transition zones. When the feldspars are looked at by TEM, low microcline is detected in the sanidine zone and intermediate microcline in the transition zone, both about 100 m to either side of the boundary. Their amounts, however, must be so small that they cannot be seen in the X-ray diagram. In any case, the question arises, why at some distance off the intrusion contact sanidine does not convert into intermediate microcline, but only into orthoclase. Kroll et al. (Chap. 13, this Vol.) ascribe this, at least in part, to the influence of inhomogeneous external and internal stresses. When at the transition temperature sanidine is under non-hydrostatic stress, it is conceivable that it will preferentially transform into microcline rather than orthoclase. The stress accumulated must be larger the higher the temperature was from which sanidine cooled down to the transition temperature. As the stress necessary to transform sanidine to microcline must exceed a threshold value, this process must cease at a certain temperature above the transition temperature. In the field, this corresponds to the boundary between the sanidine zone and the transition zone.

External stress is built up by the quartz grains surrounding K-feldspars; internal stress is built up due to deformation textures inside the K-feldspars which they acquired during the regional metamorphism. This deformation texture may be responsible for the highly irregular, chaotic domain structure seen using the electron microscope. It thus constitutes a memory effect. When deformed grains are recrystallized, they form regular film-perthite lamellae on cooling and a much more regular inversion twinning.

20.2.3 TTT-Behaviour

In the whole aureole, no sanidine was found that did not re-order to either orthoclase or intermediate microcline. Kroll et al. (Chap. 13, this Vol.) examine the conditions under which sanidine may transform. From the work of Knitter (1985), Papageorgiou and Kroll (1986) and Papageorgiou (pers. communication) two temperature-time-transformation (TTT)-diagrams have been calculated for Al,Si ordering to demonstrate the influence of fluid pressure (Figs. 20.1, 20.2).

From a TTT-diagram, the extent of isothermal transformation can be read off as a function of time. For example, if sanidine is quenched from an equilibrium temperature of 773 °C where the Al content of the T_1 site is $X_{Al}^{T1} = 0.293$ to, say, 550 °C, and is then annealed isothermally, X_{Al}^{T1} increases with time and approaches the

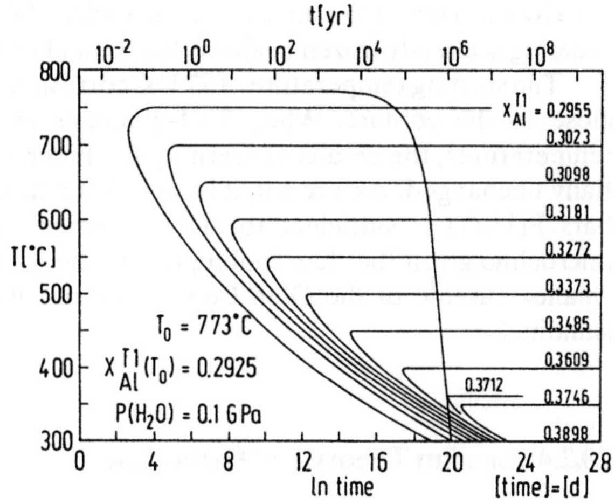

Fig. 20.1. Temperature-time-transformation (TTT) diagram for Al,Si order in sanidine at a pressure of $P(H_2O) = 0.1$ GPa. The cooling curve applies to the country rock at the contact. It was calculated from the results of Buntebarth (Chap. 18, this Vol.; and Fig. 13.22) using the relation $1/T = 1/T_o + \eta \cdot t$, where η is a characteristic cooling rate constant

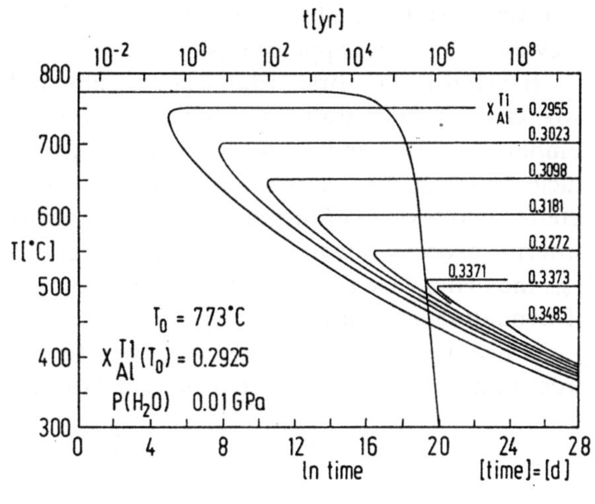

Fig. 20.2. TTT diagram for Al,Si order in sanidine at a pressure of 0.01 GPa (cf. Fig. 20.1)

equilibrium value 0.327. The horizontal parts of the iso-X_{Al}^{T1} curves are not truly horizontal, but only appear to be so within the accuracy of drawing.

From the results of Buntebarth (Chap. 18, this Vol. and Fig. 13.22) cooling curves for the Appin Quartzite at the contact have been drawn into Figs. 20.1 and 20.2 using the relation $1/T = 1/T_o + \eta \cdot t$, where η is a cooling rate constant. Strictly speaking, this procedure is not correct because the effect of continuous cooling is mixed with isothermal annealing, but for the present purpose this is a good approximation (for another type of diagram see Figs. 13.23, 13.24). During cooling, sanidine passes through various states of order until the ordering process is effectively frozen in. This happens at ≈ 365 °C, if a fluid pressure of 0.1 GPa is assumed (Fig. 20.1), but at ≈505 °C at a fluid pressure of 0.01 GPa (Fig. 20.2). The lowest inversion temperature was taken as 480 °C (Fig. 13.25). Thus, if sanidine is cooled under $P(H_2O) \geq 0.1$ GPa, it has a potential for further ordering below 480 °C and therefore can invert to orthoclase or microcline. When cooled at $P(H_2O) \leq 0.01$ GPa, ordering is already frozen in above 480 °C, and inversion is no longer possible.

The starting temperature of 773 °C corresponds to the maximum temperature close to the contact. When TTT-diagrams are calculated for lower starting temperatures, the results concerning the kinetic cut-off temperatures are essentially unchanged. As a result, it is seen that a fluid pressure of just some hundred bars $P(H_2O)$ is sufficient to allow sanidine to transform into orthoclase or microcline given the slow cooling rate in the Ballachulish aureole. In the nearby smaller aureole of the Glen Coe intrusion, some sanidine is preserved as low sanidine.

20.2.4 Landau Theory and Orthoclase

Landau theory has been successfully applied to displacive and order/disorder phase transitions in feldspars by Salje (1985) and Salje et al. (1985). Carpenter (1988) gives a review and further applications.

If a transition is first order, as we assume to be true for K-feldspars, there are two 'critical' temperatures associated with it: the equilibrium transformation temperature T_{diff} and a second temperature $T_c < T_{diff}$, at which the barrier to nucleation vanishes, so that continuous ordering becomes possible. If the modulated tweed texture of orthoclase can be interpreted as the result of a continuous ordering process, it would be of interest to get an idea at what temperature below T_{diff} the orthoclase texture will develop. Some speculation within the framework of Landau theory may be allowed.

If just one order parameter (Q_{od}) is relevant, the Landau potential for a first-order transition can be written [Carpenter 1988, (Eq. 1)]:

$$\Delta G = \frac{1}{2} a(T - T_c) Q_{od}^2 + \frac{1}{4} bQ_{od}^4 + \frac{1}{6} cQ_{od}^6 , \qquad (1)$$

where $\Delta G = G_{ord}-G_{dis}$ is the free energy difference between a (partly) ordered and a disordered state ($Q_{od} \neq O$ as opposed to $Q_{od} = O$), a,b,c are parameters which are *assumed* to be independent of temperature.

The order parameter varies with temperature according to Carpenter [1988, Eq. (17)]:

$$Q_{od}^2 = \frac{2}{3} Q_o^2 \left\{ 1 + \left[1 - \frac{3}{4} \left(\frac{T - T_c}{T_{diff} - T_c} \right) \right]^{1/2} \right\} , \qquad (2)$$

where Q_o is the jump in Q_{od} at T_{diff}, i.e. $Q_{od} = O$ goes to $Q_{od} = Q_o$. When solved for T_c, we have

$$T_c = \frac{T - T_{diff} \cdot A}{1 - A} , \qquad (3)$$

where

$$A = \frac{4}{3}\left[1 - \left(\frac{3Q_{od}^2}{2Q_o^2} - 1 \right)^2 \right] . \qquad (4)$$

We will evaluate Eq. (3) at T_{diff} = 480 °C, which is the lower temperature of the transition loop (Fig. 13.25). A reasonable choice for Q_{od} is 0.95 at T = 350 °C. Arbitrary values chosen for Q_o are 0.4, 0.5, and 0.6. The results are listed in the following table:

T_{diff} = 480 °C

Q_o	T_c/°C	T_{diff}-T_c/°C
0.4	478	2
0.5	475	5
0.6	467	13

It is seen that T_c only slightly depends on the choice of Q_o, such that the orthoclase tweed texture may develop just a few degrees below the equilibrium transition temperature T_{diff}.

We emphasize that these calculations are highly speculative. It should also be noted that the treatment of microcline ordering by Carpenter (1988) is different from ours in two respects. (1) Carpenter (1988) assumes tricritical behaviour, which implies that there is no jump in Q_o; (2) based on the experiments by Openshaw et al. (1979), he suggests that a second (displacive) order parameter is operative. However, when repeating Openshaw's experiments, we could not reproduce any of his results and, in particular, could not find evidence for a 'smeared' transition of low microcline near room temperature. When we calculate Q_{od} at T = OK from Eq. (2) using the data in the table, $Q_{od(0K)}$ is much greater than 1 (ca. 1.4). Possibly the assumption that the parameters a, b, c in Eq. (1) are independent of temperature is not valid. It is clear that some new experimental data are needed.

20.3 Cordierite

20.3.1 The Transformation Process

Pure Mg-cordierite, $Mg_2Al_4Si_5O_{18}$, with no Al,Si long-range order is hexagonal (P6/mcc). At $T_c \approx$ 1450 °C, it transforms via an order/disorder transition into its orthorhombic modification (Cccm) (Schreyer and Schairer 1961). For symmetry reasons the transition must be first order (McConnell 1985).

New experimental results on the transition have recently been reported by Putnis et al. (1987) and are discussed by Maresch et al. (Chap. 14, this Vol.) with regard to observations on natural cordierites, especially those of the Ballachulish aureole.

Putnis et al. (1987) performed kinetic experiments to follow the changes in Al,Si order, lattice parameters and submicroscopic texture that occur in the hexagonal to orthorhombic transformation process. They annealed glass of cordierite composition at various temperatures below T_c. First, hexagonal cordierite crystallized, which after some time developed a modulated structure, and finally the orthorhombic modification nucleated within the modulated crystals. At this point in the annealing process the lattice parameters abruptly changed from hexagonal to orthorhombic. The orthorhombic phase appeared with a high degree of long range order as seen in the distortion index

$$\Delta = 2\theta_{131} - (2\theta_{511}\text{-}2\theta_{421})/2,$$

which measures the splitting of the hexagonal (211) peak into three orthorhombic peaks. Δ may vary between 0 in a hexagonal Mg-cordierite to 0.25 in a fully ordered orthorhombic crystal. Putnis et al. (1987) report Δ ≈ 0.22 at the modulated to orthorhombic transformation, i.e., Δ does not change continuously with time as reported earlier (Putnis 1980a). By contrast, the Al,Si short range order determined by MAS-NMR measurements is a continuous function of ln (time). It is very large already in the 'disordered' hexagonal cordierite ($Q_{od} \approx 0.8$).

Putnis et al. (1987) monitored the changes in the powder patter using high resolution Synchroton X-ray powder diffractometry. The earlier observation that Δ increases continuously with time was explained by the insufficient resolution of conventional powder diffractometry used by the authors at that time. This then leaves the problem of interpreting all the intermediate Δ values reported on natural cordierites. Maresch et al. (Chap. 14, this Vol.) address this problem and conclude that the Δ index, although it may not measure long-range order, measures at least the progress of the hexagonal-to-orthorhombic reaction. In order to solve the resolution problem, it would be highly desirable to remeasure synthetic and natural samples that gave intermediate Δ indices using a Synchroton powder diffractometer.

The chemical composition of natural cordierites may deviate significantly from the ideal formula $Mg_2Al_4Si_5O_{18}$. As Maresch et al. (Chap. 14, this Vol.) suggest, this deviation may reduce the jump of the Δ parameter when the hexagonal crystal transforms to the orthorhombic form, thereby increasing the range of allowed intermediate values. However, the jump cannot diminish completely as the observation of very small Δ values may suggest, because the transformation is first order.

20.3.2 Natural Occurrence

Carpenter and Putnis (1985) summarized observations on the relation between growth kinetics and ordering kinetics in minerals that undergo order/disorder transformations. There is ample evidence that crystals may grow disordered in the stability field of the ordered modification because nucleation and growth kinetics in a disordered phase are faster than in an ordered one. Among the examples listed are adularia, which have grown in alpine veins below ≈480 °C within the low-microcline stability field, but nevertheless exhibit a large variety of metastable structural states. Omphacite contains antiphase domain boundaries, which prove that it grew disordered, subsequently ordering. Plagioclases grown at low temperatures also show evidence for post-crystallization ordering.

Due to its high transformation temperature, natural cordierite invariably grows in the stability field of the ordered phase. In fact, in natural cordierite, sector trilling and complex trilling are observed which are difficult to explain as a result of growth twinning, but are consistent with transformation twinning (Schreyer and Yoder 1964; Zeck 1972; Putnis and Holland 1986). Zeck (1972) has shown that in twinned crystals, optic axial planes are related by rotations of 120.0(1)°. For metrical reasons, this is not possible for individuals that grew and twinned with orthorhombic symmetry. Contrary to expectation, Maresch et al. (Chap. 14, this Vol.), did not find evidence for twinning, apart from two exceptions, in the whole traverse which they investigated through the Ballachulish aureole. They explain this finding with the slow rate of heating that precludes overstepping of cordierite-forming reactions. There are two such reactions in the aureole, occurring at (1) 560 ± 20 °C and (2) 620 ± 20 °C (Pattison, Chap. 16, this Vol.):

muscovite + chlorite + quartz ⇔ cordierite + biotite + H_2O (P1)
muscovite + biotite + quartz ⇔ cordierite + K-feldspar + H_2O (P2b)

The given temperatures are valid only when *ordered* cordierite grew in the reactions. If hexagonal cordierite rather than orthorhombic cordierite grew, reactions (P1) and (P2b) would have been delayed by 80–85 °C, according to Putnis and Holland (1986). However, this would completely destroy the regular sequence of metamorphic zones established by Pattison and Harte (Chap. 8, this Vol.) from cordierite-bearing and cordierite-absent reactions around the Ballachulish igneous complex.

Maresch et al. (Chap. 14, this Vol.) give a TTT-diagram for the Al,Si ordering process in cordierite (Fig. 14.6). Two series of curves representing increasing Δ values were determined by Putnis and Bish (1983) for anhydrous Mg-cordierite at 1 atm, and by Langer and Stürzebecher (1983) for hydrous Mg-cordierite at $P(H_2O) = 0.5$ GPa, respectively. Notwithstanding the open interpretation of intermediate Δ values, the experiments suggest that all cordierites have zero Δ values at very short run times. However, a TTT-diagram only reflects the experimental conditions by which it was determined and may not be valid for different conditions. In the dry and the hydrothermal experiments, cordierite probably crystallized instantaneously in the disordered state with subsequent ordering. In the Ballachulish aureole, ordered cordierite can crystallize by reaction (P1) at 560 °C at a distance of about 600 m to the contact (Figs. 14.2, 14.3, 14.4). The heating/cooling path at this distance, calculated by Buntebarth (Chap. 18, this Vol.) for the cylinder model, reaches a maximum temperature of ≈580 °C, which is well below 560 + 80 °C required for disordered cordierite to nucleate. If due to slow nucleation kinetics, ordered cordierite should not have crystallized at 560 °C, there would be more than 50 000 years available for crystallization during the temperature path from 560 to 580 °C. A TTT-diagram determined at these conditions could look completely different from Fig. 14.6, showing only ordered cordierite.

At a distance of 300 m, the maximum temperature was ≈670 °C, i.e., cordierite could have nucleated from reaction (P1) in the disordered form. If it did so, its temperature-time path would follow the '300 m'-curve in Fig. 14.6. As Maresch et al. (Chap. 14, this Vol.) point out, crystals following this curve would now be fully ordered. Consequently, they should be twinned in contrast to those grown at ≈600 m. However, such a zonation of the aureole caused by twinned and untwinned crystals does not exist. Furthermore, it is unreasonable to expect that disordered cordierite would crystallize at 300 m, when ordered cordierite crystallized at 600 m.

In summary, all evidence points to the fact that cordierite has nucleated from reactions (P1) and (P2b) in the ordered low-temperature phase.

20.4 Mg-Calcites

20.4.1 Subsolidus Behaviour

The calcite limb of the calcite-dolomite solvus exhibits a clear dependence on temperature, and therefore the Mg-content of calcites which coexist with dolomites serves as a potential geothermometer. The dolomite limb is too steep to be useful. Calcite has space group $R\bar{3}c$; in the ordered dolomite structure it is reduced to $R\bar{3}$. The boundary between the ordered and disordered phase fields intersects the solvus at its crest. The size difference of Ca and Mg and their strong ordering tendency in dolomite act as driving forces for unmixing. When looking at the subsolidus behaviour in the calcite-magnesite system, it is seen that dolomite tolerates as little excess Ca as Mg, whereas calcite tolerates much more Mg than magnesite tolerates Ca (Carpenter 1985). This is a common observation: small cations (like Mg) can substitute large ones (like Ca) more easily than large cations can substitute small ones. This is reflected in the volume behaviour. The *partial* molar volume of the small volume end member at infinite dilution is usually much larger than its *molar* volume, resulting in positive excess volumes (Newton and Wood 1980).

20.4.2 External vs Internal Buffering in Decarbonation Reactions

Masch and Heuss-Aβbichler (Chap. 10, this Vol.) have investigated decarbonation reactions in siliceous dolomites and have used the Mg-content of product calcites coexisting with dolomites to estimate reaction temperatures.

In decarbonation reactions two extreme conditions may be considered: complete external buffering of the fluid phase composition and internal buffering (Powell 1978). Consider a T-X_{CO_2} diagram with a schematically drawn reaction boundary at which the assemblage

$$A + B\ (+H_2O) \leftrightarrow C + D + CO_2$$

is at equilibrium (Fig. 20.3). In case of external buffering, the composition of the fluid phase is controlled by an outer reservoir. On heating, A + B react at the boundary to give C + D + CO_2, assuming no overheating to occur. Because of external buffering, the fluid phase does not change composition, and the reaction goes to completion at T_1. In the field, the isobaric univariant four-phase assemblage A + B + C + D would constitute an isograd, i.e., an ideally sharp boundary separating dolomites with A + B from those containing C + D.

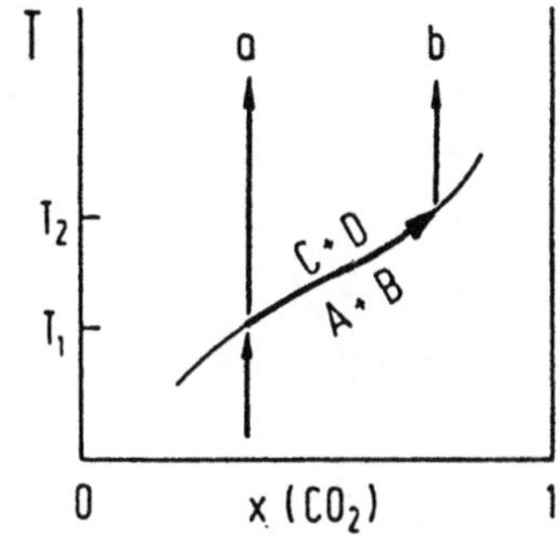

Fig. 20.3. Schematic drawing of a reaction boundary in a T-x(CO_2) diagram. During prograde metamorphism A + B (+H_2O) reacts at T_1 to give C + D + CO_2. In case of external buffering the reaction goes to completion at T_1 with no change of the fluid phase composition (*path a*). If the fluid phase is internally buffered, the reaction proceeds along the reaction boundary until one of the reactants is used up (*path b*) or another reaction boundary is crossed

In the case of internal buffering, the composition of the fluid phase is controlled by the reaction assemblage itself. Therefore, the reaction starts at T_1, but does not go to completion at the same temperature. The fluid will change its composition towards CO_2-rich compositions, and the reaction will proceed along the reaction line to higher temperatures until one of the reactants is used up (at T_2) or an isobaric invariant point is attained. On moving along the reaction boundary in Fig. 20.3, small portions of the reactants will consecutively be transferred into products. As a result of internal buffering, the reaction assemblage will occur in the field over a distance rather than at a line. However, it might be difficult to detect the reaction products in thin sections when they are generated in too small amounts (Greenwood 1975).

Masch and Heuss-Aßbichler (Chap. 10, this Vol.) observe isobaric univariant reaction assemblages near to the isograds C1, C9, C14 (Figs. 10.1 and 10.2). They are missed in the zones of reaction (C1) and (C9), but are observed in the zone of reaction (C14). Likewise, five-phase assemblages Cal-Dol-Tr-Di-Fo representing the isobaric invariant point (CIII) are found.

Ideal external buffering can be excluded in the Ballachulish aureole (see Harte et al., Chap. 19, this Vol.). Masch and Heuss-Aßbichler (Chap. 10, this Vol.) suggest an interplay of reaction control and internal dilution of fluid phase compositions. Decarbonation reactions liberate CO_2-rich fluids, whereas a fluid transport system interconnecting portions of carbonate layers and possibly adjacent country rocks may change the compositions to H_2O-rich values. Consequently, reactions will be driven to completion and isobaric univariant assemblages within zones can be transferred into isobaric divariant assemblages.

20.4.3 Mg-Calcite Thermometry

Several calcite samples from a prograde sequence of thermometamorphosed siliceous dolomites were analyzed by the electron microprobe. Typically the calcite grains consist of a core of fairly constant composition and a rim with Mg decreasing in concentration towards the boundary. The zonation results from retrogressive resetting and may be understood from a consideration of the calcite-dolomite solvus (Fig.

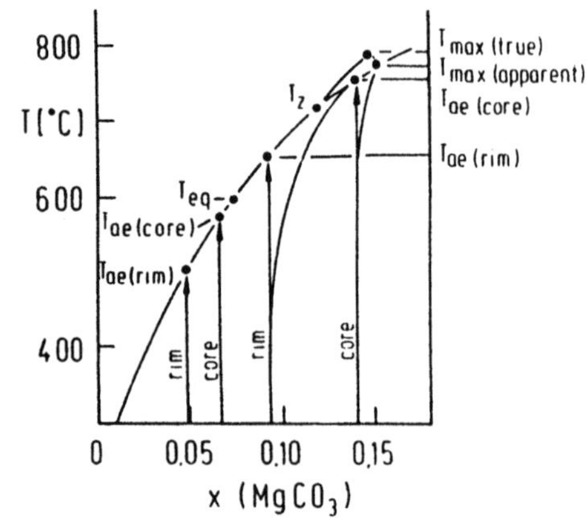

Fig. 20.4. The calcite limb of the calcite-dolomite solvus with cooling paths for two Mg-calcites schematically drawn. (1) At T_{eq}, Mg-calcite is assumed to crystallize with its equilibrium composition during a decarbonation reaction at the maximum temperature reached. On cooling from T_{eq}, the Mg content decreases in the core and the rim of the grain at different rates and finally freezes in at some lower temperature. (2) At T_z, Mg-calcite is assumed to equilibrate within a zone with its composition plotting on the solvus limb. During subsequent increase of temperature, the Mg content is supposed not to follow the solvus limb, but to increase along a loop that intersects the solvus at a temperature $T_{max(apparent)} < T_{max(true)}$. Below this temperature the change of the Mg content is similar to case (1)

20.4). Assume that a calcite grain which has grown in a decarbonation reaction is in equilibrium with a dolomite grain across a grain boundary at temperature T_{eq}. With the onset of cooling the calcite grain will change its composition. If the rate of equilibration is faster than the cooling rate, the calcite composition follows the solvus limb. At lower temperatures, the rate of equilibration will slow down due to the exponential decrease of the Ca,Mg-interdiffusion coefficient, so that the actual calcite composition falls inside the solvus, until below a certain temperature (the kinetic cut-off temperature T_{co}) the composition is frozen in. This composition corresponds to a temperature $T_{ae} > T_{co}$ on the solvus, which is called the 'apparent equilibrium' temperature. Because the grains have a finite size, each part of the grain will follow its own time-temperature-composition path. The Ca,Mg exchange in the interior of the grains will freeze in before the outer parts will do, resulting in a chemical zonation. If the core of a grain shows a chemically homogeneous plateau, this may be taken to approximately represent the original composition acquired at T_{eq}.

The Mg-content of the calcite grains was converted by Masch and Heuss-Aβbichler (Chap. 10, this Vol.) into temperatures. The maximum temperatures, usually seen in the cores of the grains, are plotted in Fig. 20.5. Toward the rim, the Mg-content decreases due to retrogressive reequilibration. Sample B317 is an exception; it seems that retrogressive resetting not only affected its rim, but the core, too, and lasted to lower temperatures. In this sample retrogressive alteration of silicates is far more advanced than in the other samples.

Comparing the position of carbonate samples within the thermal profile of calculated temperatures (Fig. 20.5) and that of pelite samples (Fig. 16.4), it is

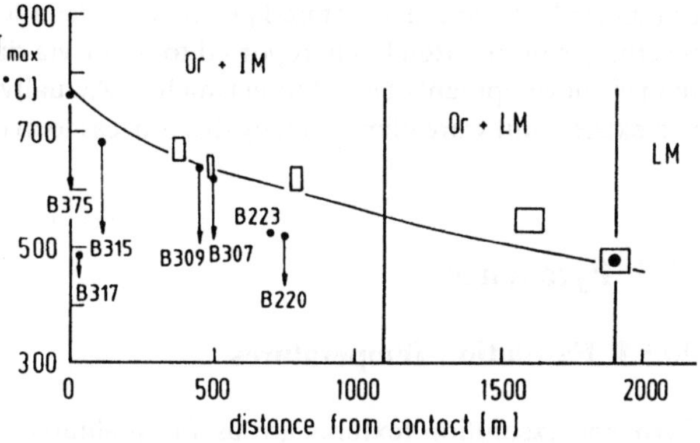

Fig. 20.5. Variation of the maximum temperatures in Appin Quartzite with distance from contact. The curve was calculated by Bunterbarth (Chap. 18, this Vol.) for a cylinder-type intrusion assuming 1050 °C as the intrusion temperature and 250 °C as the temperature of the country rocks. *Open rectangles* indicate isogrades in pelites (Pattison, Chap. 16, this Vol.). *Dots* with *arrows* give temperatures derived from the Mg-calcite geothermometer, where *dots* represent 'core temperatures' and *arrowheads* give 'rim temperatures' (Masch and Heuss-Aßbichler, Chap. 10, this Vol.). The *square* with a *dot* indicates the microcline-sanidine transition isograd

conspicuous that aside from more or less coinciding estimates of the contact temperature, the best match is for the calcites of samples B307 and B309. Both samples are from isograds, whereas the other carbonate samples are located within zones. The small number of samples does not exclude accidental differences in retrogressive resetting. However, Masch and Heuss-Aßbichler (Chap. 10, this Vol.) point at the partly different processes of equilibration of calcites at isograds and in zones. At the mapped isograds, decarbonation reactions occur at the maximum temperature attained at this distance. In the zones, the temperature increased further after completion of reactions. The reactions produce calcite by a precipitation process; these reaction calcites are supposed to have the composition corresponding to the equilibration temperature. The mismatch of the samples B220, B223, B317 with the thermal profile may be the result of sluggish intercrystalline equilibration of carbonates after the decarbonation reactions have come to an end. These grains may have followed some kind of a loop, as indicated in Fig. 20.4. Their Mg contents correspond to an apparent equilibration temperature and do not reflect the maximum temperature.

The difference between temperatures deduced from the Mg-content in the rim and the core, respectively, seems to decrease with increasing distance from the contact (Fig. 20.5). This follows from the fact that the temperature range over which unmixing is effective is the larger, the higher the temperature was at which cooling and unmixing began, as indicated in Fig. 20.4.

Masch and Heuss-Aßbichler (Chap. 10, this Vol.) report that "with the exception of one sample, the grains do not show microscopic exsolution lamellae, nor do microprobe analysis indicate submicroscopic exsolution". It might be interesting to see whether exsolution lamellae exist in the core of a grain on a scale below microprobe resolution (5–10 μm). If they ever existed in the rim, they have been transported outward in a chemical potential gradient towards dolomite. Retrogressive resetting is more often being reported to occur via diffusion of Mg out of the grain than via development of exsolution lamellae. Actually, exsolution lamellae may have pre-existed, but were eliminated by this process to avoid strain and surface energies.

20.5 Pyroxenes

20.5.1 Exsolution Temperatures

Pyroxene exsolution textures are useful qualitative and quantitative indicators of cooling history. Feuer et al. (Chap. 6, this Vol.) investigated the exsolution microstructures of pyroxenes from two different varieties of the Ballachulish monzodiorite, the 'normal' coarse- to medium-grained facies and the fine-grained or 'chilled' marginal facies, the cooling rates of which differ by a factor of 10 to 50. Weiss and Troll (Chap. 4, this Vol.) suggest that in the normal facies, diorite orthopyroxene (hypersthene) started crystallizing at ≈1100–1050 °C with augite (and pigeonite?) co-crystallizing. During cooling, resetting due to Ca-(Mg,Fe) exchange occurred, so that the two-pyroxene thermometer of Lindsley (1983) now gives apparent equilibration temperatures around 800 °C. For the pyroxenes of the marginal facies, Weiss and Troll (Chap. 4, this Vol.) assume metastable crystallization from a considerably supercooled melt. However, Feuer et al. (Chap. 6, this Vol.) estimated exsolution temperatures of 950 ± 50 °C, suggesting that augite from the 'chilled margin' crystallized at T > 950 ± 50 °C.

During cooling, the ortho- and clinopyroxenes develop fine-scale lamellar exsolution, investigated by Feuer et al. (Chap. 6, this Vol.) using electron microscopy. The orthopyroxenes exsolve augite lamellae on (100) which is the only orientation possible. It minimizes dimensional strain, and structurally (100) is the plane least dissimilar in the two structures. The augite grains on the other hand develop orthopyroxene plus pigeonite lamellae or sometimes just pigeonite, and often the orthopyroxene lamellae in augite are rimmed by pigeonite. Orthopyroxene lamellae are ||(100) of the host augite for the reasons given, but pigeonite occurs as '100' and '001' lamellae in orientations close to (100) and (001).

The '100' and '001' planes are irrational planes which provide an exact dimensional fit between augite and pigeonite having different a, c and β, but essentially identical b dimensions (Fig. 20.6). Thus either one or both planes are used for exsolution. Their positions are a function of composition and temperature. To calculate them, it is not only necessary to determine the temperature dependence of the lattice parameters of both augite and pigeonite, but also to make a guess at the chemical changes as a function of temperature. Feuer et al. found that at 1070 °C the chemical changes due to cooling could essentially be reversed, whereas no reversal was observed after prolonged heating at 925 °C. From the lattice parameters of the reversed material at 1070 °C and of the original composition at T ≤ 925 °C, they could calculate the orientations of the '100' and '001' lamellar interfaces to compare with those actually measured.

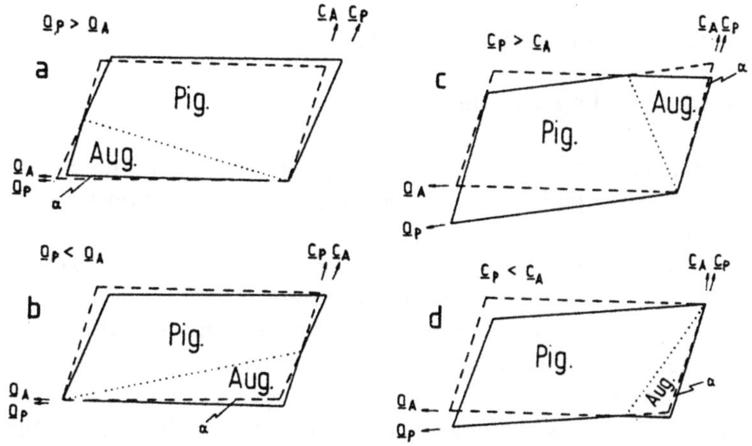

Fig. 20.6. Relative orientations of augite (*Aug*) and pigeonite (*Pig*) unit cells providing exact phase boundaries of the '001' (**a,b**) and '100' (**c,d**) type. The relative difference between cell dimensions has been exaggerated for clarity. The angle α indicates the 'exact phase boundary lattice rotation'. '001' boundaries result when the (001) lattice planes are common to augite and pigeonite. The position of the boundary is controlled by the relative sizes of the *a* dimensions. For '100' boundaries the common plane is (100) and *c* is controlling the position of the boundary (after Robinson et al. 1977)

For the '100' lamellae of pigeonite found in augites from the 'normal' facies, they report two different positions, corresponding to two exsolution events at 1000 ± 30 °C and 920 ± 50 °C, respectively, and one position for the '001' lamellae, corresponding to exsolution occurring at 935 ± 30 °C.

In augites from the chilled margin '001' pigeonite lamellae occupy a position that points to an unmixing event at 950 ± 50 °C, whereas for the '100' lamellae the situation is less clear.

Inherent in these estimates is the assumption that the composition of the pigeonite lamellae did not further change below ≈925 °C. This may be too high a closure temperature, especially when compared with the apparent equilibration temperatures determined from the coexisting ortho- and clinopyroxene compositions using Lindsley's (1983) geothermometer (cf. Fig. 4.7). Therefore, the estimated exsolution temperatures provide an upper limit.

A lower limit is given by the onset of amphibole crystallization. The upper thermal stability limit of amphiboles from monzodiorites covers a range of ≈720–650 °C, as reported by Weiss and Troll (Chap. 4, this Vol.). This temperature postdates the development of '100' pigeonite and (100) orthopyroxene exsolution lamellae, because (010) amphibole lamellae terminate against them (Feuer et al., Sect. 6.2.3, this Vol.).

20.5.2 Exsolution Assemblages

20.5.2.1 Orthopyroxene

As stated above, the orthopyroxene crystals contain unmixed augite lamellae oriented ||(100). The augite c-axis dimension is adapted to the host, i.e., c is strained due to coherent intergrowth. The b dimensions are virtually identical.

Lateral growth is more difficult for a coherent lamella that for an incoherent one, because atoms have to cross the coherent interface via an interstitial mechanism. Feuer et al. observed ledges at the (100) interfaces of those lamellae which were broader than 20 nm. When material is diffusing to the ledge, the coherent interface layer can grow by moving its incoherent edge. Lamellae smaller than 20 nm did not show ledges. Possibly they could not become broader because they did not develop ledges.

The augite lamellae are found in both (100) twin positions at equal frequency, that is, augite originally did not nucleate within pigeonite, which later on transformed into orthopyroxene, but orthopyroxene was the host at the time of exsolution. Otherwise the balanced distribution of twin positions could not be understood.

In plutonic and high-grade metamorphic rocks, another type of exsolution products may be observed: Guinier-Preston (G.P.) zones, of which Champness and Lorimer (1976) give beautiful examples. G.P. zones occur when during cooling, a stranded Ca-diffusion profile is built up between previously exsolved augite lamellae. Such G.P. zones are not found in the Ballachulish orthopyroxenes. The cooling rate was probably too fast to allow further unmixing once the augite lamellae had formed.

20.5.2.2 Clinopyroxene

Figure 20.7 is a pseudobinary phase diagram corresponding to a section through the pyroxene quadrilateral at about the $En_{70}Fs_{30}$-$Di_{75}Hd_{25}$ join. It is derived from Lindsley (1983, Fig. 9). The diagram is not a true binary because the tie lines move in and out of the diagram with changing temperature. The temperature range over which three pyroxenes (orthopyroxene, augite and pigeonite) coexist is estimated to extend over 20 °C.

The position of the coherent solvus was calculated by Tullis and Yund (1979) for iron-free clinopyroxenes. Using high temperature lattice parameters, they found that the coherent and the strain-free solvi almost coincide for lamellar orientations of zero dimensional strain, '100' and '001'. I have assumed the same to be valid for the section chosen.

Feuer et al. (Chap. 6, this Vol.) determined the temperature of the three-pyroxene field. They heated an orthopyroxene that contained augite lamellae at 1070 °C where pigeonite started to develop, indicating that they had reached the three phase field. Weiss and Troll (Chap. 4, this Vol.) estimated the pyroxene crystallization temperature of the 'normal' monzodiorite as ≈1100–1050 °C, whereas they assume from textural evidence that the crystallization temperature in the marginal facies was lower by up to 200 °C. From the similarity of exsolution textures in pyroxenes from both

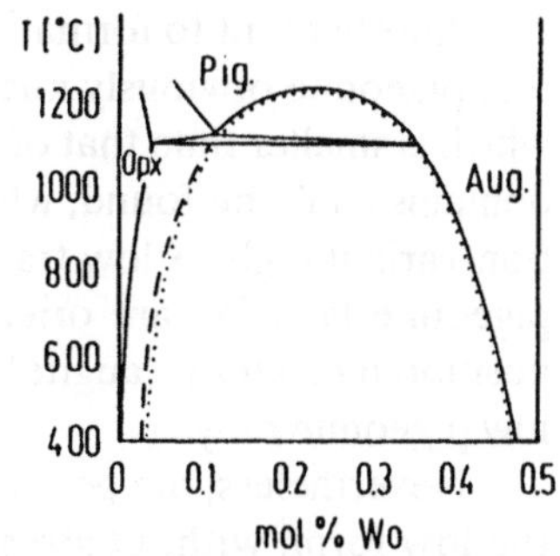

Fig. 20.7. Pseudobinary section through the pyroxene quadrilateral at the $En_{70}Fs_{30}$-$Di_{75}Hd_{25}$ join for P = 0.1 MPa (after Lindsley 1983, Fig. 1). *Pig*: Pigeonite; *Aug*: augite. *Dashed curve*: metastable extension of the pigeonite limb of the incoherent pigeonite-augite solvus. *Dotted curve*: assumed position of the coherent pigeonite-augite solvus. When calculated from high temperature lattice parameters for the '001' and '100' orientations of zero compositional strain, its position virtually coincides with the coherent solvus (Tullis and Yund 1979). The pigeonite C2/c-$P2_1/c$ transition is not shown because its location is not well constrained

facies, it can be assumed that crystallization in each case occurred below the temperature of the three-pyroxene field.

On cooling, only one exsolution product, orthopyroxene, should have developed in augite. However, very frequently three-phase assemblages are observed, and sometimes augite is found solely with exsolved pigeonite. Pigeonite has probably exsolved metastably: its structure is similar to augite and the compositional change to pigeonite is smaller than to orthopyroxene which facilitates nucleation. Pigeonite is also found as a rim along orthopyroxene lamellae. In this case, it may have nucleated at a lower temperature at sites of enhanced Ca concentration in augite next to orthopyroxene out of which Ca may have migrated during cooling, thereby creating an area of Ca supersaturation in a stranded diffusion profile. The supersaturation has then initiated a late stage nucleation event producing metastable pigeonite.

The pigeonite lamellae deserve some more discussion. Feuer et al. annealed a three-phase assemblage at 1070 °C. They found that after 2 weeks the pigeonite lattice parameters had changed, likely because the composition had equilibrated at 1070 °C. The '001' lamellae had adapted their a (and b) parameters to the augite host, and the '100' lamellae had adapted their c (and b) parameters, indicating coherency. The volume had increased due to the uptake of Ca. If the lattice parameters indicate coherency after heating at 1070 °C, one would expect the lamellae were coherent before heating, too. However, the lattice parameters of the unheated pigeonite are not adapted to the augite. They are in fact smaller than estimated for an unstrained crystal of the same composition. It appears that the lattice parameters are strained by the surrounding augite to smaller values than in a normal 'free' crystal, although thermal contraction is larger in pigeonite than in augite! A strong indication for this compression is the fact that on heating to 1150 °C, pigeonite does not invert from the low ($P2_1/c$) to the high temperature modification (C2/c).

Another hint to formation of pigeonite lamellae under stress is the observation that pigeonite obviously nucleated in the low-temperature form, the cell volume of which is smaller than that of the high-temperature modification. First, no anti-phase domains could be found, which would result from the class-equivalent, but translation-variant high → low transformation. Second, their Fig. 6.4 shows that the '001' pigeonite lamellae are oriented in the acute angle β [c Λ '001' > c Λ (001)]. This position indicates a (augite) > a (pigeonite) (Robinson et al. 1977). This is true for low pigeonite only.

Nevertheless, the possibility should not be ruled out that pigeonite nucleated in the low form, without stress being involved simply because exsolution occurred below the transformation temperature. The C2/c → $P2_1/c$ transition varies from 500 °C for Fe-rich compositions to 1000 °C for Mg-rich pigeonites (Prewitt et al. 1971). When considering the exsolution temperatures estimated from 'exact phase boundaries', crystallization in the low form seems to be unlikely. However, as noted, the estimated temperatures are upper limits. The behaviour of the pigeonite lamellae in the augite host remains puzzling.

20.5.3 Comparison of Pyroxenes from the 'Normal' and Marginal Facies

The pyroxene microstructures of the slowly cooled 'normal' and the quickly cooled marginal monzodiorite facies are not significantly different. Orthopyroxene crystals of both facies are similar with respect to widths and distribution of exsolved augite lamellae. Thin lamellae (1.8 nm wide) were found only in samples of the marginal facies.

Similarly, in augite crystals the width of exsolution lamellae is not specific for either one of the facies. However, the quantity of exsolution products is different. In crystals from the marginal facies orthopyroxene and pigeonite are always found, whereas they are less abundant in the 'normal' facies.

The orthopyroxene lamellae themselves contain thin lamellae (mostly 2.7 nm wide) of clinopyroxene in both twin positions. They are much more abundant in samples of the marginal facies. They are considered as growth faults that formed by the propagation of 0.9-nm ledges during lateral growth of orthopyroxene lamellae.

Pigeonites that have nucleated as a late-stage event at the (100) interfaces between orthopyroxene lamellae and augite host are observed in samples of both facies. However, they are common to the marginal facies, but are rare in the normal facies.

In total, the frequent occurrence of non-equilibrium features in samples from the marginal facies is consistent with the assumption of Weiss and Troll (Chap. 4, this Vol.) that it cooled much faster than the 'normal' facies.

20.6 Summary

Investigation of intracrystalline processes and associated (micro-)structural features of various minerals of the Ballachulish pluton and aureole may be briefly summarized as follows:

1. Low microcline transforms into sanidine over a range of temperatures: from ≈480 to ≈550 °C, corresponding to a range of distances extending from 1900 to 1100 m in Appin Quartzite east of the contact (Fig. 13.3). This transition zone can be explained in two ways: (a) If the microcline-sanidine transformation temperature is strongly dependent on composition, the transition zone is created by variable re-solution of perthitic Na-feldspar by microcline, (b) if the transformation temperature is virtually independent of composition, overheating is necessary to produce the transition zone. Both interpretations, however, face either thermodynamic or kinetic problems. On cooling, sanidine transforms into either ('tweed') orthoclase or irregular microcline. The chaotic microtexture of intermediate microcline is thought to be controlled by the internal deformation texture acquired by the low microcline crystals during the previous regional Caledonian metamorphism. For triclinic twin-like domains to develop in sanidine after cooling below the transformation temperature, a certain threshold of internal and/or external inhomogeneous stress is necessary. This assumption explains the distribution of irregular microcline and orthoclase in the aureole, with irregular microcline being restricted to the inner aureole (Fig. 13.2).

2. When cordierite is produced in metapelites from reacting Ms + Chl + Qtz [reaction (P1)] or Ms + Bt + Qtz [reaction (P2b)] it may grow either as an initially Al,Si disordered hexagonal phase or in the ordered orthorhombic modification, depending on the heating rate. For the disordered phase to nucleate, the reaction isograd involving low cordierite has to be overstepped by ≈80 °C. In case of disordered nucleation, the high-to-low transition results in crystals showing sector trilling. Apart from rare exceptions, this is not found in cordierites of the Ballachulish aureole, indicating that cordierite nucleated in the low form and that overstepping of the reaction isograd was small.

Intermediate distortion indices are difficult to interpret in the light of recent experimental results by Putnis et al. (1987). High-resolution Synchroton powder diffractometry and electron microscopy may have to be combined to attack the problem.

3. Mg-calcite occurs as a product of decarbonation reactions in siliceous dolomites, with the Mg-content originally equilibrated at the temperature of reaction. If Mg-calcite grows at an isograd, the Mg-content reflects the maximum temperature reached at the isograd. If it occurs within a divariant field, the Mg-content reflects the temperature of growth at the isograd, being lower than the maximum temperatures reached during subsequent heating. This has to be accounted for when using the Mg-content as a geothermometer. On cooling, retrogressive resetting of the thermometer occurs. Mg concentrates in exsolving dolomite, which can be reintegrated during a microprobe analysis, or it migrates out of the grain. Both features were observed in Ballachulish Mg-calcites. In the latter case, a stranded Mg diffusion profile was generated with a plateau in the center of the grain. Probably the plateau represents the original Mg-content and is thus useful for geothermometry.

4. The pyroxenes of the Ballachulish monzodiorite grew below the temperature of the three-pyroxene field as co-crystallizing orthopyroxene and augite. During cooling, they developed equilibrium and non-equilibrium exsolution assemblages. Augites in particular commonly show a three-pyroxene assemblage with orthopyroxene, '100' and '001' pigeonite lamellae within the augite host. The pigeonite lamellae represent three different exsolution events at temperatures below 1000 °C, estimated from the position of the '100' and '001' lamellae relative to the crystallographic (100) and (001) planes of the augite host. Pigeonite nucleated in the low temperature ($P2_1/c$) form, either due to stress imposed by the surrounding augite or because the nucleation temperature was lower than the inversion temperature. Exsolution and inversion temperatures are not known with sufficient accuracy to decide this point. On heating, the pigeonite lamellae show a thermal expansion similar to augite and, in addition, they do not transform to the high form at $T \leq 1150$ °C, because they are strained by the surrounding augite. Comparison of microstructures of pyroxenes from slowly and quickly cooled areas of the monzodiorite reveals that non-equilibrium microtextures are more abundant in the quickly cooled samples than in the slowly cooled ones.

21 Summary and Outlook

G. Voll

21.1 Summary

The multidisciplinary study of the Ballachulish Igneous Complex and thermal aureole represents the most comprehensive investigation of contact metamorphism yet published. The aim of this study was to investigate equilibrium and kinetic aspects of mineral- and rock forming processes in a well-constrained natural setting. The Ballachulish Igneous Complex and its aureole were chosen because of the simple shape of the intrusion and the lack of any later episodes of metamorphism, deformation or retrogression that postdates emplacement and cooling of the complex. A great variety of metasedimentary rocks run into the thermal aureole: feldspathic quartzites, psammopelites with and without graphite and pyrite, impure limestones and dolomites. All rock types experienced regional upper greenschist-lower amphibolite metamorphism and acquired fine-grained equilibrium mineral assemblages prior to emplacement of the complex. Regional structures are simple and uniform. The psammopelites show a penetrative slaty cleavage and heat was transferred either parallel or normal to it in different parts of the contact; thus, the effects of different heat conductivities and anisotropy effects could be assessed.

In the first part of the chapter, the results of these studies are summarized. References to the individual contributions in this book are given by chapter numbers (in parentheses). The second part of the chapter briefly outlines areas for further study.

The input of heat into the aureole and, therefore, its thermal history, is controlled by the Ballachulish Igneous Complex, and knowledge of its size, shape, intrusion temperature, intrusion sequence and crystallization history had to be established at the outset. Therefore, a detailed map of the intrusion and its rock types was produced (Chaps. 3 and 4). From whole-rock chemical analyses of different phases of the complex and from the compositions of coexisting plutonic minerals, the intrusion sequence and temperatures of intrusion and crystallization of the magmas were derived. Although the Ballachulish Igneous Complex is polyphase in nature, the thermal history of the complex itself and of its aureole was mainly controlled by the intrusion of the hot and 'dry' diorite pluton. More detailed information on intrusion- and crystallization temperatures of the diorites was obtained from the nucleation and growth history of pyroxenes (Chap. 5) and from TEM and X-ray studies of the exsolution textures in pyroxenes (Chap. 6).

From surface- and aeromagnetic data as well as from gravity measurements it can be inferred that the present mean depth of the intrusion is at least 4 km (Chap. 7).

G. Voll, J. Töpel, D.R.M. Pattison, F. Seifert
Equilibrium and Kinetics
in Contact Metamorphism
©Springer-Verlag Berlin Heidelberg 1991

The contact is steep in the south and east, but lies shallowly beneath the surface up to 1.5 km from the W and SE contacts. The relative amount of dioritic rocks increases abruptly beneath the surface.

Water given off from the melt is not regarded to play an important role. The monzodiorite melts intruded with very little dissolved water and crystallized without developing a fluid phase (Chaps. 4 and 5). On the other hand, the quartz-diorite and the granite melts were more hydrous and developed a fluid phase during cooling. Stable isotope data (Chap. 17) also indicate that no fluid convection was set up by the intrusion, and that influx of volatiles into the aureole occurred only locally.

In the aureole, the mineral reactions in metapelitic (Chaps. 8, 9 and 16) and metacarbonate (Chaps. 10 and 11) rocks, correlated fabric changes and derived temperatures were the central object of several studies. These established a detailed subdivision of the aureole. In pelite rocks, zones with different mineral assemblages separated by reaction isograds were mapped. Partial melting in pelitic rocks was found to be common near the contact.

The occurrences of quartzites having high heat conductivity correlate with an outward shift of the zone boundaries. This and the anisotropy of heat conduction in pelitic rocks leads to a width-variation of the aureole (defined by the first appearance of cordierite) from 400 to 1700 m. Original slaty cleavage is found to be well preserved in the outer part of the aureole. It disappears abruptly and gives way to massive hornfelses in the inner part of the aureole, where K-feldspar and cordierite are abundant. Whole-rock and mineral compositions are given for the metapelitic aureole rocks and Schreinemakers' analysis is applied. From the KFMASH model system, petrogenetic grids are derived separately for the outer part of the aureole without partial melt, and for the inner part where assemblages coexisted with partial melts. All mineral assemblages found in pelitic rocks of the aureole could thus be explained by a continuous petrogenetic grid.

The sequence of reactions in quartz-bearing and quartz-free pelitic rocks, the changes of Mg-Fe ratios in minerals involved, the observed effect of F in biotite, the exchange of (Fe,Mg)Si for 2Al and the a_{H_2O}-lowering effect of graphite all follow expectations based on experimental and thermodynamic equilibrium data. It is therefore concluded that to a first approximation, the reactions established equilibrium at each step and that no obvious overstepping due to fast heating occurred. On the other hand, adjacent to a smaller intrusion SE of the complex, which had its own independent time-temperature-history, signs of overstepping are observed (Chaps. 8 and 16). It is therefore indicated that as far as metapelite reactions are concerned, a spread of disequilibrium courses is anticipated between the Ballachulish Aureole and smaller intrusions with shorter times of heating and cooling.

In a narrow zone close to the contact, widespread partial melting in pelites is observed (Chap. 9). Partial melting resulted in the separation of patches and veins of leucosome with variable composition, and led to hydraulic fracturing and mobility of the quartzo-feldspatic melts. These effects are strongest within Leven Schists west of the contact, which are part of an anticline and underlain by quartz-diorites. The quartz-diorites gave off water which fluxed the production of the melts. Grain coarsening also reaches a maximum at this site.

In the siliceous dolomites and impure limestones a series of contact metamorphic reactions was also established (Chap. 10). The reaction sequence is incomplete, since the carbonate rocks do not form as large and homogeneous bodies as the pelites. Seven isograds were mapped based on devolatilization reactions, and temperatures were determined from a calibrated petrogenetic grid and from the calcite-dolomite thermometer. Newly formed minerals range from talc at the low temperature end to periclase found near the contact and in xenoliths. Mg contents of calcite coexisting with dolomite reflect the temperature of the decarbonating reactions, and do not appear to be changed further by continued heating. Zoned carbonates have higher Mg contents in the core, which is interpreted to reflect incomplete reequilibration as temperatures fall. A fluid pressure equal to lithostatic pressure during these reactions is derived (Chap. 10). Temperatures obtained from carbonate reactions are in very good agreement with temperatures derived from hydration reactions in the pelites. X_{CO_2} in the fluid is regarded as internally controlled by the decarbonating reactions. However, the amount of fluid is controlled by diffusion of fluid along grain boundaries. As the fluid moves away from the reaction site, it changes composition by dilution. Marked differences in X_{CO_2} are found for near-contact, middle and outer parts of the aureole. Textural evidence and the observed separate X_{CO_2} evolution paths in each zone point towards different metamorphic paths (Chap. 11), and the present day *spatial* sequence of isograds does not represent the *temporal* sequence of events during progressive metamorphism at a given location.

Combining the results from dehydrating (Pattison 1989) and decarbonating reactions (Chap. 10) and the microcline/sanidine isograd (Chap. 12) the P-T-a_{H_2O} conditions for different contact distances within the aureole round the Ballachulish Igneous Complex are derived (Chap. 16). A pressure of 3.0 ± 0.5 kbar is obtained for the country rocks and the intrusion during formation of the aureole. Temperatures of contact assemblages range from near 400 °C at the outer margin to near 800 °C at the contact. Four different temperature-distance profiles, including the narrowest and widest parts of the aureole, are constructed (Chap. 16) which serve as constraints for the temperature-time modelling (Chap. 18).

In general, the evolution of the fluid phase is controlled on a local scale by the host rocks. For the Ballachulish Slates, rich in graphite, it is shown that a_{H_2O} was decreased (Chaps. 16 and 19) most probably due to the dilution of water by C-bearing fluid species. This leads to lower temperature appearance of first cordierite, and the development of andalusite rather than the K-feldspar development in non-graphitic pelites. The onset of melting is shown to occur at about 670 °C, and resulting melt buffers a_{H_2O} to water-undersaturation. The partial melts attract and dissolve water set free by reactions or given off by the underlying quartz-diorite, such as at the western contact (Chaps. 9, 16 and 19). In the remaining aureole, fluid exchange between the intrusion and the aureole was very limited, as shown by the limited extent of partial melting.

Those parts of the aureole in which dehydrating and decarbonating reactions dominate are considered to be fluid-saturated (Chaps. 8, 10, and 19). In contrast, in the Appin Quartzite, heterogeneity of oxygen isotopes indicate that there could not have been a pervasive fluid because this would have led to isotopic equilibration and homogenization (Chap. 15). On the other hand, clastic alkali feldspar crystals from the same quartzite show features of sanidinization and mixing on heating, and features of ordering, inversion and unmixing during cooling which apparently cannot be explained without the presence of at least some water (Chaps. 13 and 20). In addition, the low activation energy for quartz grain growth, the enrichment of alkali feldspar in Na and the homogenization between clastic feldspar grains with respect to Na content all point to the existence of at least some water in the contact quartzites. It is suggested that grain coarsening of quartz liberates water from fluid inclusions in original quartz grains, which may have been sufficient to flux these transitions without causing homogenization and equilibration of oxygen isotopes.

In the metapelites the rarity of retrogressive hydrating reactions during cooling indicates that water set free by reactions on heating was not stored but driven out. Where, however, water was enriched in melts, it was set free on cooling and caused more pronounced retrograde reactions (Chap. 9). Similarly, rehydration was found to be more common within the granitic intrusives which crystallized from a magma richer in water (Chaps. 4 and 5). Such water migration, however, was not part of a circulating convective system involving the intrusion and its aureole.

Data on the shape, composition and magma temperature of the igneous complex formed the basis for modelling the heat transport and time-temperature curves for different distances from the contact (Chap. 18). The temperature-dependant thermal properties of the magmatic and various country rocks were also determined and taken into account in the calculations of the temperature-time-contact distance relationships. Two models have been developed: (1) one with a subsided block of country rocks surrounded by magmatic rings; and (2) another with a solid cylindrical block of magmatic rock. The assumed shape of the second model has been varied from purely cylindrical to conical. The possible effects of water released by the intrusion, the heat-consuming effect of dehydrating reactions in the aureole, and variations in heat transport parallel or normal to the pervasive cleavage have been taken into account. The best fit with temperatures derived from phase petrological equilibrium data (Chaps. 8, 10, and 13) is achieved with the cylindrical model.

Information on the kinetics and mechanisms of reactions has been derived from studies of microtextures, grain coarsening, oxygen isotope exchange, as well as order-disorder relationships and exsolution processes in individual minerals.

From the investigation of microtextures in the heated carbonate rocks, an attempt was made to establish relationships between reaction mechanisms and resulting parageneses and textures for single decarbonation reactions (Chap. 11). Support is drawn from a contact aureole round the Alpine Monzoni Intrusion where the same reactions cause clearer development of such textural differences. For a single reaction drastically different grain morphologies and grain-neighbourhood relations result between reactants and products. The observed phenomena are classified into four different texture types which are thought to be caused by differences between open- and closed-system behaviour. Reactions were controlled by kinetics of transport or dissolution or growth at interfaces, and super- or undersaturation with respect to dissolving phases. From the start of a reaction, these parameters change with time and temperature and lead to development of different textures, related to changing reaction mechanisms. The sequence of textures is thought to reflect systematic changes of the controlling parameters during progress of the same reaction. The heating and cooling rates seem to be an important factor in modifying the sequence of reaction textures, although this was not quantified.

Grain coarsening of quartz in the Appin Quartzite, driven by interfacial free energy, represents a less complicated case of fabric change set up by static annealing within the aureole (Chap. 12). Coarsening starts at about 620 °C and increases steadily towards the contact. Subgrains of deformed quartzes, produced during regional metamorphism, served as seeds for the growing grains. Coarsening affects quartzites both with and without sharp preferred c-axes-orientation, and has little effect on the preferred orientation. A low activation energy of 8.5 kcal/mol is ascribed to the effect of water which may be liberated from deformed quartz grains while they recrystallize. Heated quartz inverts to high-quartz and reinverts on cooling. The associated volume changes cause strain-induced boundary migration and weak subgrain-formation. No significant equilibration, redistribution or change of oxygen isotopes was found in separated clastic quartz and alkali-feldspar pebbles from outside and from various parts inside of the aureole, nor in contact-coarsened quartzes derived from such pebbles (Chap. 15).

The same alkali-feldspar-pebbles from the Appin Quartzite were studied (Chap. 13), to assess phase changes, disordering and reordering, and mixing and unmixing during heating and cooling. Preceding regional metamorphism left behind maximum microclines. Contact heating causes: (1) gradual replacement of microcline by sanidine; (2) disordering and remixing of unmixed albite, and an increase in average Na-content; (3) homogenization of grain composition; and (4) in a narrow zone near the contact, coarsening. Cooling causes: (1) reordering, (2) formation of intermediate microcline and orthoclase in the inner and outer parts of the aureole, respectively, and (3) unmixing. Deformational faults acquired during regional metamorphism survive all these changes and cause highly irregular features of reordering, unmixing and inversion twinning or tweed structure.

Cordierite apparently grew as low cordierite in the metapelites throughout the aureole, and its Δ index increases slightly and develops more scatter towards the contact (Chap. 14). The reasons for differences in the Δ index are still unknown, but do not appear to be caused by metastable growth of a hexagonal phase and its subsequent ordering.

The features of unmixing observed in pyroxenes from the monzodiorites suggest that the pyroxenes did not equilibrate (Chap. 6). A marked change from (001)- to (100)-unmixing of pigeonite lamellae in augites is attributed to lower strain energy for the (100) lamellae.

Kinetics and mechanisms of intracrystalline processes in alkali feldspars, cordierite, (Ca,Mg) carbonate, and pyroxene are summarized in Chap. 20, and the effect of fluids on reaction kinetics, as well as their mobility, is discussed in Chap. 19. Although fluid pressure was approximately equal to lithostatic pressure where fluids were present, there was little mobility of fluids between lithological units.

21.2 Suggestions for Future Work

The Ballachulish Igneous Complex and aureole is probably the most thoroughly studied model case for contact metamorphism. This has been achieved due to a very instructive pluton and aureole and the applications of diverse lines of approach. Many intrusions and their aureoles are of similar size, consolidated at similar depth and affected similar country rocks, so that the results presented here may have broad applications.

Nevertheless, our understanding of plutonism and contact metamorphic processes and their time dependence could be further deepened, both by further studies on the Ballachulish rocks themselves and, in particular, by comparison to other complexes with different internal and external conditions. In the following, a few examples of such projects and their goals are listed:

1. Ballachulish Igneous Complex
   - Studies of the geochemistry, in particular trace elements and Sr isotopes: origin of the melts, contribution of the mantle, the lower and middle crust.
   - Geophysical studies of the deeper levels of the intrusion: shape, homogeneity, extent of rock types, relation to contemporaneous intrusions and dykeswarms.
   - Experimental studies on the melts by programmed cooling: equilibrium-disequilibrium relationships for the crystallization sequence.
2. Aureole rocks
   - Study of trace elements, specifically REE: information about metamorphic migration on different scales, formation of migmatites, mass balance constraints.
   - Grain growth of calcite, dolomite, quartz, and micas and its dependence on inhibiting effects of the presence of other phases.
   - Fluid inclusion studies: survival of regional metamorphic fluids in contact metamorphism; information on history of fluid movement during heating and cooling.
   - Detailed examination of pelite/quartzite contacts at different grades in the aureole: information on extent and nature of fluid migration and metasomatic mass transfer between layers.
   - Contact metamorphic reactions in the Appinite Suite bodies within the thermal aureole: mineral reactions in metabasic assemblages at well constrained P-T-t conditions could be elucidated.
3. Mineralogy of the aureole rocks
   - Oxide minerals: kinetics on heating and especially cooling from mixing and unmixing behaviour of magnetite, ilmenite and hematite; variation of oxygen fugacity with time and temperature.
   - Pyrrhotite-pyrite isograd in the Ballachulish Slate (Neumann 1950): information on relationships between sulphides and silicates in Ballachulish Slate, C-O-H-S fluid evolution.
   - Coexisting (?) low microcline and orthoclases in the Appin Quartzite: Is the sanidinization of low-microcline in the transition zone ruled by a first order transition loop or not?

- Recrystallization of KNa-feldspars in the Appin Quartzite: in the aureole studied, it starts at ca. 650 °C, clearly overstepping the threshold temperature of ca. 300 °C typical of regional metamorphism.
- Defects (*Realbau*) in sheet silicates: the type and frequency of stacking faults or polymorphs should be related to coarsening, heating and compositional change.
- Homogenization of clastic zircons: resetting of ages for single grains and the population, cut-off temperature for self-diffusion processes in zircons.
- Further examination of calc-silicate assemblages (e.g., grossular + wollastonite + idocrase): information on fluid composition in these layers, external vs internal buffering.

4. Other intrusions and their aureoles

   The Ballachulish Igneous Complex and its aureole have unique time-temperature histories. Studies of the kinetics of mineral and rock-forming processes should be extended to other intrusions and aureoles with different internal and external conditions. Such systems might include:
   - Larger and smaller intrusions with longer or shorter heating durations.
   - Intrusions with higher or lower starting temperatures.
   - Aureoles showing differing fluid phase composition and mobilities, including convective systems set up by heating.
   - Aureoles around dry, hot melts attracting water from the aureole, others around melts giving off water to the aureole.
   - Very high level intrusions adjoining porous rocks or rocks with intricate joint systems.
   - Aureoles with obvious metasomatism.
   - Aureoles from deeper levels, causing contact effects at higher pressures.
   - Aureoles where contact heating interacts with deformation.

Historically, regional metamorphism has attracted much more attention than contact metamorphism. Contact metamorphism, however, provides a better model for detailed equilibrium and kinetic studies because boundary conditions can be assessed more precisely, and the influence of single factors can be separated more clearly. Especially for kinetics of processes, contact metamorphism promises more insight. On the scale of the Ballachulish intrusion, it seems that many reactions occurred without significant overstepping, proceeding at essentially equilibrium conditions and producing results consistent with equilibrium predictions. Most likely, further insight into the infinitely varied disequilibrium processes is gained best from studying smaller intrusions and aureoles with shorter times of heating and cooling. Results from these should be gradually connected to those from larger intrusions which have approached equilibrium better, and ultimately to the long-lived processes of orogenies and their regional metamorphism. On this extended journey, we have made a first step.

References

Abbott, R.N.Jr.; Clarke, D.B. (1979): Hypothetical liquidus relations in the subsystem Al_2O_3-FeO-MgO projected from quartz, alkali feldspar and plagioclase for a (H_2O)≤ 1. Can. Mineral. 17: 549-560

Abercrombie, H.J.; Skippen, G.B.; Marshall, D.D. (1987): F-OH substitution in natural tremolite, talc and phlogopite. Contrib. Mineral. Petrol. 97: 305-312

Abraham, K.; Schreyer, W. (1973): Elementverteilung in koexistierenden Festkörperphasen. Forsch. Ber. Landes NRW 2374: 1-101

Akima, H. (1978): A method of bivariate interpolation and smooth surface fitting for irregularly distributed data points. ACM Trans. Math. Software 2: 160-164

Albarede, F.; Dupuis, C.; Taylor, H.P. (1980): ^{18}O:^{16}O evidence for non-cogenetic magmas associated in a 300 Ma old concentric pluton at Ploumanac'h (Brittany, France). J. Geol. Soc. London 137: 641-647

Aleksandrov, K.S.; Ryzhova, T.V.; Belikov, B.P. (1964): The elastic properties of pyroxenes. Sov. Phys. Crystallogr. 8: 589-591

Andersen, D. J.; Lindsley, D.H. (1984): Application of a two-pyroxene thermometer: correlation of apparent temperature with Al_2O_3 in augite. Lunar Planet. Sci. 15: 7-8

Anderson, E.M. (1937): The dynamics of formation of cone-sheets, ring-dykes and cauldron subsidences. Proc. R. Soc. Edinburgh 56: 128-157

Anderton, R. (1985): Sedimentation and tectonics in the Scottish Dalradian. Scott. J. Geol. 21: 407-464

Armbruster, T.; Bloss, F.D. (1981): Mg-cordierite: Si/Al ordering, optical properties, and distortion. Contrib. Mineral. Petrol. 77: 332-336

Ashworth, J.R. (1985): Introduction. In: Ashworth, J.R. (ed.) Migmatites. Blackie, Glasgow, pp 1-35

Asprey, L.B. (1976): The preparation of very pure fluorine gas. J. Fluorine Chem. 7: 359-361

Atherton, M.P. (1968): The variation in garnet, biotite and chlorite composition in medium grade pelitic rocks from the Dalradian of Scotland, with particular reference to the zonation in garnet. Contrib. Mineral. Petrol. 18: 347-371

Atherton, M.P. (1977): The metamorphism of the Dalradian rocks of Scotland. Scott. J. Geol. 13: 331-370

Bachinski, S.W.; Müller, G. (1971): Experimental determinations of the microcline-low albite solvus. J. Petrol. 12: 329-356

Bailey, E.B. (1910): Recumbent folds in the schists of the Scottish Highlands. Q. J. Geol. Soc. London 66: 586-620

Bailey, E.B. (1923): The metamorphism of the South-Western Highlands of Scotland. Geol. Mag. 60: 317-331

Bailey, E.B.; Maufe, H.B. (1960): The geology of Ben Nevis and Glen Coe and the surrounding country: explanation of Sheet 53 - update of 1916 memoir. Mem. Geol. Surv. Scot. H.M.S.O., Edinburgh, 307 pp

Bambauer, H.U.; Bernotat, W.H. (1982): The microcline/sanidine transformation isograd in metamorphic regions. I. Composition and structural state of alkali feldspars from granitoid rocks of two N-S traverses across the Aar Massif and Gotthard Massif, Swiss Alps. Schweiz. Mineral. Petrogr. Mitt. 62: 185-230

Bambauer, H.-U.; Krause, C.; Kroll, H. (1989): TEM-investigation of the sanidine/microcline transition in metamorphic gradients: the K-feldspar varieties. Eur. J. Min. 1: 47-58

Bamford, D. (1979): Seismic constraints on the deep geology of the Caledonides of northern Britain. In: Harris, A.L.; Holland, C.H.; Leake, B.E. (eds.) The Caledonides of the British Isles - reviewed. Spec. Publ. Geol. Soc. London 8: 93-96

Bamford, D.; Nunn, K.; Prodehl, C.; Jacob, B. (1977): LISPB - III Upper crustal structure of northern Britain. J. Geol. Soc. London 133: 481-488

Barrow, G. (1893): On an intrusion of muscovite-biotite gneiss in the south-eastern Highlands of Scotland and its accompanying metamorphism. Q. J. Geol. Soc. London 49: 330-358

Barrow, G. (1912): On the geology of lower Deeside and the southern Highland Border. Proc. Geol. Assoc. 23: 274-290

Beeré, W. (1975): A unifying theory of the stability of penetrating liquid phases and sintering pores. Acta Metal. 23: 131-138

Bell, I.A.; Wilson, C.J.L. (1977): Growth defects in metamorphic biotite. Phys. Chem. Minerals 2: 153-169

Bence, A.E.; Albee, A.L. (1968): Empirical correction factors for the electron microanalysis of silicates and oxides. J. Geol. 76: 382-403

Berman, R.G. (1988): Internally consistent thermodynamic data for minerals in the system Na_2O-K_2O-CaO-MgO-FeO-Fe_2O_3-SiO_2-TiO_2-H_2O-CO_2. J. Petrol. 29: 445-552

Bernard-Griffiths, J.; Peucat, J.J.; Sheppard, S.; Vidal, Ph. (1985): Petrogenesis of Hercynian leucogranites from the southern Armorican massif: contribution of REE and isotopic (Sr, Nd, Pb and O) geochemical data to the study of source rock characteristics and ages. Earth Planet. Sci. Lett. 74: 235-250

Bhattacharyya, B.K. (1964): Magnetic anomalies due to prism-shaped bodies with arbitrary polarisation. Geophysics 29: 517-531

Bickle, M.J.; Powell, R. (1977): Calcite - dolomite geothermetry for iron - bearing carbonates. Contrib. Mineral. Petrol. 59: 281-292

Bickle, M.J.; Wickham, S.M.; Chapman, H.J.; Taylor, H.P. (1988): A strontium, neodymium, and oxygen isotope study of hydrothermal metamorphism and crustal anatexis in the Trois Seigneurs Massif, Pyrenees, France. Contrib. Mineral. Petrol. 100: 399-417

Bigeleisen, J.; Perlman, M.L.; Prosser, H.C. (1962): Conversion of hydrogenic materials to hydrogen for isotopic analysis. Anal. Chem. 24: 1356-1357

Blattner, P.; Dietrich, V.; Gansser, A. (1983): Contrasting ^{18}O enrichment and the origins of High Himalayan and Transhimalayan intrusives. Earth Planet. Sci. Lett. 65: 276-286

Bluck, B.J. (1985): The Scottish paratectonic Caledonides. Scott. J. Geol. 21: 437-464

Boberski, C.; Schreyer, W. (1990): Synthesis and water contents of Fe^{2+}-bearing cordierites. Eur. J. Mineral. 2: 565-584

Bohlen, S.R.; Wall, V.J.; Boettcher, A.L. (1983): Experimental investigation and application of garnet granulite equilibria. Contrib. Mineral. Petrol. 83: 52-61

Bottinga, Y.; Javoy, M. (1975): Oxygen isotope partitioning among minerals in igneous and metamorphic rocks. Rev. Geophys. Space Phys. 13: 401-418

Bowers, T.S.; Jackson, K.J.; Helgeson, H.C. (1984): Equilibrium activity diagrams for coexisting minerals and aqueous solutions at pressures and temperatures to 5 kb and 600 °C. Springer, Berlin, Heidelberg, New York

Bowes, D.R.; Wright, A.E. (1967): The explosion breccia-pipes near Kentallen, Scotland, and their geological setting. Trans. R. Soc. Edinburgh 67: 109-143

Bowman, J.R.; Essene, E.J. (1982): P-T-X$(CO)_2$ conditions of contact metamorphism in the Black Butte aureole, Elkhorn, Montana. Am. J. Sci. 282: 341-357

Bradbury, H.J.; Harris, A.L.; Smith, R.A. (1979): Geometry and emplacement of nappes in the Central Scottish Highlands. In: Harris, A.L.; Holland, C.M.; Leake, B.E.(eds.) The Caledonides of the British Isles - reviewed. Spec. Publ. Geol. Soc. London 8: 213-220

Brantley, S.L.; Crane, S.R.; Crerar, D.A.; Hellman, R.; Stallard, R. (1986): Dissolution at dislocation etch pits in quartz. Geochim. Cosmochim. Acta 50: 2349-2361

Breemen, O. van; Piasecki, M.A.J. (1983): The Glen Kyllachy Granite and its bearing on the nature of the Caledonian orogeny in Scotland. J. Geol. Soc. London 140: 47-62
Breemen, O. van; Aftalion, M.; Pankhurst, R.J.; Richardson, S.W. (1979): Age of the Glen Dessary syenite, Inverness-shire: diachronous Palaeozoic metamorphism across the Great Glen. Scott. J. Geol. 15: 49-62
Briden, J.C.; Kent, D.V.; Lapointe, P.V.; Livermore, R.A.; Roy, G.L.; Sequin, M.K.; Smith, A.G.; van der Voo, R.; Watts, D.R. (1988): Paleomagnetic constraints on the evolution of the Caledonian-Appalachian orogen. In: Harris, A.L.; Fettes, D.J. (eds.) The Caledonian-Appalachian Orogen. Spec. Publ. Geol. Soc. London 38: 35-48
Brook, M.; Brewer, M.S.; Powell, D. (1976): Grenville ages for rocks in the Moine of north-western Scotland. Nature (London) 260: 515-551
Brown, G.C. (1979): Geochemical and geophysical constraints on the origin and evolution of Caledonian granites. In: Harris, A.L.; Holland, C.M.; Leake, B.G. (eds.) Caledonides of the British Isles - reviewed. Spec. Publ. Geol. Soc. London 8: 645-651
Brown, J.F.(1975): The geochronology of northwest Argyllshire. Ph.D.Thesis, Oxford Univ.
Brown, P.E.; Miller, J.A.; Grasty, R.L. (1968): Isotopic ages of late Caledonian granitic intrusions in the British Isles. Proc. Yorksh. Geol. Soc. 36: 251-276
Brown, W.L.; Parsons, I. (1985): Calorimetric and phase-diagram approaches to two-feldspar geothermometry: a critique. Am. Mineral. 70: 356-361
Buddington, A.F.; Lindsley, D.H. (1964): Iron-titanium oxide minerals and synthetic equivalents. J. Petrol. 5: 310-357
Buntebarth, G. (1984): Geothermics - an introduction. Springer, Berlin, Heidelberg, New York
Buntebarth, G. (1987): Laboratory thermal conductivities applied to crustal conditions. In: Hamza, V.M.; Frangipani, A.; Beck A.E.; Ribeiro, F.B. (eds.) Geothermics and geothermal energy. Rev. Bras. Geofis. 5: 103-109
Burnham, C.W. (1979): The importance of volatile constituents. In: Yoder, H.S. (ed.) The evolution of the igneous rocks - 50th Anniv perspectives. Univ. Press, Princeton, pp 439-482
Buseck, P.R.; Nord, G.L.; Veblen, D.R. (1980): Subsolidus phenomena in pyroxenes. In: Prewitt, C.T. (ed.) Reviews in mineralogy, vol. 7. Pyroxenes. Mineral. Soc. Am., Washington, DC, pp 117-211
Butler, R.W.H.; Coward, M.P. (1984): Geological constraints, structural evolution and deep geology of the northwest Scottish Caledonides. Tectonics 3: 347-365
Carnahan, B.; Luther, H.A.; Wilkes, J.O. (1969): Applied numerical methods. John Wiley & Sons, New York
Carpenter, M.A. (1985): Order-disorder transformations in mineral solid solutions. In: Kieffer, S.W.; Navrotsky, A. (eds.) Microscopic to macroscopic: atomic environments to mineral thermodynamics. Reviews in mineralogy 14. Mineral. Soc. Am, pp 187-223
Carpenter, M.A. (1988): Thermochemistry of aluminium/silicon ordering in feldspar minerals. In: Salje, E.K.H. (ed.). Physical properties and thermodynamic behaviour of minerals. Reidel, Dordrecht, pp 265-323
Carpenter, M.A.; Putnis, A. (1985): Cation order and disorder during crystal growth: some implications for natural mineral assemblages. In: Thompson, A.B.; Rubie, D.C. (eds.) Metamorphic reactions. Advances in physical geochemistry, vol. 4. Springer, Berlin, Heidelberg, New York, pp 1-26
Carpenter, M.A.; Putnis, A.; Navrotsky, A.; McConnell, J.D.C. (1983): Enthalpy effects associated with Al-Si ordering in anhydrous Mg-cordierite. Geochim. Cosmochim. Acta 47: 899-906
Cathles, L.M. (1977): The analysis of cooling intrusives by groundwater convection. Econ. Geol. 72: 804-826
Champness, P.E.; Lorimer, G.W. (1973): Precipitation (exsolution) in an orthopyroxene. J. Mater. Sci. 8: 467-474
Champness, P.E.; Lorimer, G.W. (1976): Exsolution in silicates. In: Wenk, H.R. (ed.) Electron microscopy in mineralogy. Springer, Berlin, Heidelberg, New York, pp 74-204
Chapell, B.W.; Stephens, W.E. (1988): Origin of infracrustal (I-type) granite magmas. Trans. R. Soc. Edinburgh Earth Sci. 79: 71-86

Chinner, G.A. (1962): Almandine in thermal aureoles. J. Petrol. 3: 310-340
Chivas, A.R.; Andrew, A.S.; Sinha, A.K.; O'Neil, J.R. (1982): Geochemistry of a Plio-Pleistocene oceanic arc plutonic complex, Guadalcanal. Nature (London) 300: 139-143
Clayburn, J.A.P.; Harmon, R.S.; Pankhurst, R.J.; Brown, J.F. (1983): Sr, O, and Pb isotope evidence for origin and evolution of Etive igneous complex, Scotland. Nature (London) 303: 492-497
Clayton, R.N.; Mayeda, T.K. (1963): The use of bromine pentafluoride in the extraction of oxygen from oxides and silicates for isotopic analysis. Geochim. Cosmochim. Acta 27: 43-52
Clough, C.T. (1897): In: Gunn, W. et al. (eds.) The geology of cowal. Mem. Geol. Surv. Scot.
Cole, D.R.; Ohmoto H. (1986): Kinetics of isotopic exchange at elevated temperatures and pressures. In: Valley, J.W.; Taylor, H.P.Jr.; O'Neil, J.R. (eds.) Reviews in mineralogy, vol. 16. Stable isotopes in high temperature geological processes. Min. Soc. Am., pp 41-90
Cox, K.G.; Bell, J.D.; Pankhurst, R.J. (1979): The interpretation of igneous rocks. Allen & Unwin, London, 450 pp
Craig, G.Y. (ed.) (1991): Geology of Scotland, (3rd edn.). Scottish Academic Press, Edinburgh
Craig, H. (1961): Isotopic variations in meteoric waters. Science 133: 1702-1703
Crank, J. (1957): Two methods for the numerical solution of moving boundary problems in diffusion and heat flow. Q. J. Mech. and Appl. Math. 10: 220-231
Criss, R.E.; Taylor, H.P. (1986): Meteoric-hydrothermal systems. In: Valley, J.W.; Taylor, H.P. Jr.; O'Neil, J.R. (eds.) Reviews in mineralogy, vol. 16. Stable isotopes in high temperature geological processes. Min. Soc. Am., pp 373-424
Czank, M.; Schumacher, J.C. (1986): TEM-Untersuchung zur Polytypie im Biotit. Z. Kristall 174: 38-39
Dachs, E.; Metz, P. (1988): The mechanism of the reaction 1 tremolite + 3 calcite + 2 quartz = 5 diopside + 3 CO_2 + 1 H_2O: results of powder experiments. Contrib. Mineral. Petrol. 100: 542-551
Dansgaard, W. (1964): Stable isotopes in precipitation. Tellus 16: 436-468
Deines, P. (1977): On the oxygen isotope distribution among mineral triplets in igneous and metamorphic rocks. Geochim. Cosmochim. Acta 41: 1709-1730
Del Monte, M.; Paganelli, L.; Simboli, G. (1967): The Monzoni rocks, a modal and chemical study. Mineral. Petrogr. Acta 13: 75-118
Dempster, T.J. (1985): Uplift patterns and orogenic evolution in the Scottish Dalradian. J. Geol. Soc. London 142: 111-128
Dewey, J.F.; Pankhurst, R.J. (1970): The evolution of the Scottish Caledonides in relation to their radiometric age patterns. Trans. R. Soc. Edinburgh 68: 361-389
Didier, J. (1987): Contribution of enclave studies to the understanding of origin and evolution of granitic magmas. Geol. Rundsch. 76: 41-50
Dodson, M.H. (1973): Closure temperature in cooling geochronological and petrological systems. Contrib. Mineral. Petrol. 40: 259-274
Douglas, J., Jr. (1955): On the numerical integration of $\delta^2u/\delta x^2 + \delta^2w/\delta y^2 = \delta u/\delta t$ by implicit methods. J. Soc. Ind. Appl. Math. 3: 42-55
Drake, M.J.; Weill, D.F. (1975): Partition of Sr,Ba,Ca,Y,Eu^{2+},Eu^{3+}, and other REE between plagioclase feldspar and magmatic liquid: an experimental study. Geochim. Cosmochim. Acta 39: 689-712
Droop, G.T.R. (1989): Reaction history of garnet-sapphirine granulites and conditions of Archaean high-pressure granulite-facies metamorphism in the Central Limpopo Mobile Belt, Zimbabwe. J. Metamorphol. Geol. 7: 383-404
Droop, G.T.R.; Charnley, N.R. (1985): Comparative geobarometry of pelitic hornfelses associated with the Newer Gabbros: a preliminary study. J. Geol. Soc. London, 142: 53-62
Droop, G.T.R.; Treloar, P.J. (1981): Pressures of metamorphism in the thermal aureole of the Etive Granite Complex. Scott. J. Geol. 17: 85-102

Drury, S.A. (1980): Lewisian pyroxene gneisses from Barra and the geochemistry of the Archean lower crust. Scott. J. Geol. 16: 199-207
Eggler, D.H. (1972): Water-saturated and undersaturated melting relations in a Paricutin andesite and an estimate of water content in the natural magma. Contrib. Mineral. Petrol. 34: 261-271
Eggler, D.H.; Burnham, C.W. (1973): Crystallization and fractionation trends in the system andesite-H_2O-CO_2-O_2 at pressures to 10 kb. Geol. Soc. Am. Bull. 84: 2517-2532
Elles, G.L.; Tilley, C.E. (1930): Metamorphism in relation to structure in the Scottish Highlands. Trans. R. Soc. Edinburgh 56: 621-646
Elliott, D.; Johnson, M.R.W. (1980): The structural evolution of the northern part of the Moine thrust belt. Trans. R. Soc. Edinburgh 71: 69-96
Elphick, S.C.; Graham, C.M.; Dennis, P.F. (1988): An ion microprobe study of anhydrous oxygen diffusion in anorthite: a comparison with hydrothermal data and some geological implications. Contrib. Mineral. Petrol. 100: 490-495
England, P.; Thompson, A.B. (1984): Pressure-temperature-time paths of regional metamorphism. Part I. Heat transfer during the evolution of regions of thickened continental crust. J. Petrol. 25: 894-928
Entwisle, D.C. (1984): Determination of density, porosity, magnetic susceptibility and sonic velocity data for rocks of the Ballachulish pluton and the surrounding area. Br. Geol. Surv., Geophys. Lab. Rep. 144, Keyworth, Nottinghamshire
Etheridge, M.A.; Wall, V.J.; Vernon, R.H. (1983): The role of the fluid phase during regional metamorphism and deformation. J. Metam. Geol. 1: 205-226
Etheridge, M.A.; Wall, V.J.; Cox, S.F.; Vernon, R.H. (1984): High fluid pressures during regional metamorphism and deformation: implications for mass transport and deformation mechanism. J. Geophys. Res. 89: 4344-4358
Evans, A.L.; Mitchell, J.G. Embleton, B.J.J.; Creer, K.M. (1971): Radiometric ages of the Devonian polar shift relative to Europe. Nature (Phys. Sci.) 229: 50-51
Evans, A.M.; Haslam, H.W.; Shaw, R.P. (1980): Porphyry style copper-molybdenum mineralization in the Ballachulish granite, Agryllshire, Scotland. Trans. Inst. Min. Metall. 86: B152-B153
Ferry, J.M. (1983): Regional metamorphism of the Vassalboro Formation, south-central Maine, USA: a case study of the role of fluid in metamorphic petrogenesis. J. Geol. Soc. London 140: 551-576
Ferry, J.M. (1985): Hydrothermal alteration of Tertiary igneous rocks from the Isle of Skye, northwest Scotland. II Granites. Contrib. Mineral. Petrol. 91: 283-304
Ferry, J.M. (1986): Reaction progress: a monitor of fluid-rock interaction during metamorphic and hydrothermal events. In: Walther, J.V.; Wood, B.J. (eds.) Fluid-rock interactions during metamorphism. Springer, Berlin, Heidelberg, New York, pp 60-88
Ferry, J.M.; Spear, F.S. (1978): Experimental calibration of the partitioning of Fe and Mg between biotite and garnet. Contrib. Mineral. Petrol. 66: 113-118
Fettes, D.J. (1979): A metamorphic map of the British and Irish Caledonides. In: Harris, A.L.; Holland, C.M.; Leake, B.E. (eds.). The Caledonides of the British Isles: reviewed. Spec. Publ. Geol. Soc. London 8: 307-321
Fettes, D.J.; Harris, A.L.; Hall, L.M. (1984): The Caledonian geology of the Scottish Highlands. In: Fettes, D.J.; Harris, A.L. (eds.) Synthesis of the Caledonian rocks of Britain. Reidel, Dordrecht, pp 303-334
Fettes, D.J.; Graham, C.M.; Harte, B.; Plant, J.A. (1986): Lineaments and basement domains: an alternative view of Dalradian evolution. J. Geol. Soc. London 143: 453-464
Fielitz, K. (1971): Untersuchungen von Kompressions- und Scherwellengeschwindigkeiten in Gesteinen unter erhöhtem Druck. Z. Geophys. 37: 943-956
Fisher, G.W. (1970): The application of ionic equilibria to metamorphic differentiation: an example. Contrib. Mineral. Petrol. 29: 91-103
Fisher, G.W. (1973): Nonequilibrium thermodynamics as a model for diffusion-controlled metamorphic processes. Am. J. Sci. 273: 897-924
Fisher, G.W. (1978): Rate laws in metamorphism. Geochim. Cosmochim. Acta 42: 1035-1050

Fisher, G.W.; Elliott, D. (1974): Criteria for quasi-steady diffusion and local equilibrium in metamorphism. In: Hofmann, A.W.; Giletti, B.J.; Yoder, H.S. Jr.; Yund, R.A. (eds.). Geochemical transport and kinetics. Carnegie Inst. Washington Publ. 637: 231-242

Fleck, R.J.; Criss, R.E. (1985): Strontium and oxygen isotopic variations in Mesozoic and Tertiary plutons of central Idaho. Contrib. Mineral. Petrol. 90: 291-308

Flowers, G.C. (1979): Correction of Holloway's (1977) adaptation of the modified Redlich-Kwong equation of state for calculation of the fugacities of molecular species in supercritical fluids of geologic interest. Contrib. Mineral. Petrol. 69: 315-318

Ford, J.H.; Green, D.C. (1977): An oxygen- and hydrogen-isotope study of the Panguna porphyry-copper deposit, Bougainville. J. Geol. Soc. Aust. 24: 63-80

Forester, R.W.; Taylor, H.P. Jr. (1976): ^{18}O depleted igneous rocks from the Tertiary complex of the Isle of Mull, Scotland. Earth Planet. Sci. Lett. 32: 11-17

Forester, R.W.; Taylor, H.P. Jr. (1977): $^{18}O/^{16}O$, D/H and $^{13}C/^{12}C$ studies of the Tertiary igneous complex of Skye, Scotland. Am. J. Sci. 277: 136-177

Fourcade, S.; Javoy, M. (1985): Preliminary investigations of $^{18}O/^{16}O$ and D/H compositions in rhyo-ignimbrites in the In Hihaou (In Zize) magmatic center, central Ahaggar, Algeria. Contrib. Mineral. Petrol. 89: 285-295

Francl, J.; Kingery, W.D. (1954): Thermal conductivity: IV. Apparatus for determining thermal conductivity by a comparative method. J. Am. Ceramic Soc. 37: 80-84

Freeman, B.; Klemperer, S.L.; Hobbs, R. (1988): The deep structure of northern England and the Iapetus Suture zone from BIRPS deep seismic reflection profiles. J. Geol. Soc. London 145: 727-740

Freer, R. (1981): Diffusion in silicate minerals and glasses: a data digest and guide to the literature. Contrib. Mineral. Petrol. 76: 440-454

Freer, R.; Dennis, P.F. (1982): Oxygen diffusion studies. I. A preliminary ion microprobe investigation of oxygen diffusion in some rock-forming minerals. Mineral. Mag. 45: 179-192

Frey, M; Bucher, K.; Frank, E.; Mullis, J. (1980): Alpine metamorphism along the geotraverse Basel - Chiasso. A review. Eclogae Geol. Helv. 73: 527-546

Friedman, I. (1953): Deuterium content of natural waters and other substances. Geochim. Cosmochim. Acta 4: 89-103

Friedman, I.; O'Neil, J.R. (1977): Compilation of stable isotope fractionation factors of geochemical interest. Geol. Surv. Prof. Pap. 440 KK

Friedman, I.; Redfield, A.C.; Schoen, B.; Harris, J. (1964): The variation of the deuterium content of natural waters in the hydrologic cycle. Rev. Geophys. 2: 177-224

Frost, C.D.; O'Nions, R.K. (1985): Caledonian magma genesis and crustal recycling. J. Petrol 26: 514-544

Fuhrman, M.L.; Lindsley, D.H. (1988): Ternary feldspar modeling and thermometry. Am. Mineral. 73: 201-215

Fyfe, W.S.; Price, N.J.; Thompson, A.B. (1978): Fluids in the earth's crust. Elsevier, Amsterdam, 383 pp

Ghiorso, M.S. (1984): Activity/composition relations in the ternary feldspars. Contrib. Mineral. Petrol. 87: 282-296

Gibbs, G.V. (1966): The polymorphism of cordierite I: the crystal structure of low cordierite. Am. Mineral. 51: 1068-1087

Giletti, B. (1986): Diffusion effects on oxygen isotope temperatures of slowly cooled igneous and metamorphic rocks. Earth Planet. Sci. Lett. 77: 218-228

Giletti, B.; Yund, R.A. (1984): Oxygen diffusion in quartz. J. Geophys. Res. 89: 4039-4046

Goldsmith, J.R. (1988): Enhanced Al/Si diffusion in $KAlSi_3O_8$ at high pressures: the effect of hydrogen. J. Geol. 96: 109-123

Goldsmith, J.R.; Jenkins, D.M. (1985): The high-low albite relations revealed by reversal of degree of order at high pressures. Am. Mineral. 70: 911-923

Goldsmith, J.R.; Laves, F. (1954): The microcline-sanidine stability relations. Geochim. Cosmochim. Acta 5: 1-19

Goldsmith, J.R.; Newton, R.C. (1969): P-T-X relations in the system $CaCO_3$-$MgCO_3$ at high temperatures and pressures. Am. J. Sci. 267-A: 160-190

Goldsmith, J.R.; Newton, R.C. (1974): An experimental determination of the alkali feldspar solvus. In: MacKenzie, W.S.; Zussman, J. (eds.) The feldspars Manchester, Univ. Press, Manchester, pp 337-359

Goldsmith, J.R.; Graf, D.L.; Heard, H.C. (1961): Lattice constants of the calcium-magnesium carbonates. Am. Mineral. 46: 453-457

Gottschalk, M. (1984): Konsistente Gleichgewichtsdaten für Reaktionen in metamorphen kieseligen Dolomiten. Diplom-Arbeit Min. Petrol. Inst. Univ. Tübingen

Gottschalk, M.; Metz, P. (1984): Konsistente Gleichgewichtsdaten für Reaktionen in metamorphen kieseligen Dolomiten. Fortschr. Mineral. 62, 1: 78-79

Graf, D.L.; Goldsmith, J.R. (1965): Dolomite - magnesian calcite relations at elevated temperatures and CO_2 pressures. Geochim. Cosmochim. Acta 7: 109-128

Graham, C.M. (1986): The role of the Cruachan Lineament during Dalradian evolution. Scott. J. Geol. 22: 257-270

Graham, C.M.; Greig, K.M.; Sheppard, S.M.F.; Turi, B. (1983): Genesis and mobility of the H_2O-CO_2 fluid phase during regional greenschist and epidote amphibolite facies metamorphism: a petrological and stable isotope study in the Scottish Dalradian. J. Geol. Soc. London 140: 577-599

Graham, C.M.; Harmon, R.S.; Sheppard, S.M.F. (1984): Experimental hydrogen isotope studies: hydrogen isotope exchange between amphibole and water. Am. Mineral. 69: 128-138

Graham, C.M.; Viglino, J.A.; Harmon, R.S. (1987): Experimental study of hydrogen isotope exchange between aluminous chlorite and water and of hydrogen diffusion in chlorite. Am. Mineral. 72: 566-579

Grant, J.A. (1985): Phase equilibria in low-pressure partial melting of pelitic rocks. Am. J. Sci. 285: 409-435

Grant, J.A. (1986): Quartz-phlogopite-liquid equilibria and origins of charnockites. Am. Mineral. 71: 1071-1075

Greenwood, H.J. (1975): Buffering of pore fluids by metamorphic reactions. Am. J. Sci. 275: 573-593

Gregory, R.T.; Criss, R.E. (1986): Isotope exchange in open and closed fluid/rock systems. In: Valley, J.W.; Taylor, H.P., Jr.; O'Neil, J.R. (eds.) Reviews in mineralogy, vol. 16. Stable isotopes in high temperature geological processes. Min. Soc. Am., pp 91-127

Haggerty, S.E. (1976): Opaque mineral oxides in terrestrial igneous rocks. In: Rumble, D., III (ed.) Oxide minerals. Mineral. Soc. Am., Rev. Mineral. 3: HG 141-174

Hall, J.; Brewer, J.A.; Matthews, D.J.; Warner, M.R. (1984): Crustal structure across the Caledonides from the "WINCH" seismic reflection profile: influences on the evolution of the Midland Valley of Scotland. Trans. R. Soc. Edinburgh, Earth Sci. 75: 97-109

Halliday, A.N. (1984): Coupled Sm-Nd and U-Pb (zircon) systematics in the late Caledonian granites and the nature of the basement under northern Britain. Nature (London) 307: 229-233

Halliday, A.N.; Stephens, W.E. (1984): Crustal controls on the genesis of the 400 Ma old Caledonian granites. Phys. Earth Planet. Int. 35: 89-104

Halliday, A.N.; Stephens, W.E.; Harmon, R.S. (1980): Rb-Sr and O isotopic relationships in three zoned Caledonian granitic plutons, Southern Uplands, Scotland: evidence for varied sources and hybridization of magma. J. Geol. Soc. London 137: 329-349

Halliday, A.N.; Graham, C.M.; Aftalion, M.; Dymoke, P. (1989): The depositional age of the Dalradian Supergroup: U-Pb and Sm-Nd isotopic studies of the Tayvallich Volcanics, Scotland. J. Geol. Soc. London 146: 3-6

Hammarstrom, J.M.; Zen, E-an (1986): Aluminium in hornblende: an empirical igneous geobarometer. Am. Mineral. 71: 1297-1313

Harker, A. (1954): Petrology for students - an introduction to the study of rocks under the microscope. Univ. Press, Cambridge, 283 pp

Harker, R.I.; Tuttle, O.F. (1955): Studies in the system CaO-MgO-CO_2, limits of solid solution along the binary join $CaCO_3$-$MgCO_3$. Am. J. Sci. 253: 274-282

Harley, S.L. (1984): An experimental study of the partitioning of Fe and Mg between garnet and orthopyroxene. Contrib. Mineral. Petrol. 86: 359-373

Harley, S.L.; Green, D.H. (1982): Garnet-orthopyroxene barometry for granulites and garnet peridotites. Nature (London) 300: 697-700

Harmon, R.S. (1983): Oxygen and strontium isotopic evidence regarding the role of continental crust in the origin and evolution of British Caledonian granites. In: Atherton, M.P.; Gribble, C.D. (eds.) Migmatites, melting and metamorphism. Shiva, Kent, pp 62-79

Harmon, R.S.; Halliday, A.N. (1980): Oxygen and strontium relationships in the British late Caledonian granites. Nature (London) 283: 21-25

Harmon, R.S.; Halliday, A.N.; Clayburn, J.A.P.; Stephens, W.E. (1984): Chemical and isotope systematics of the Caledonian intrusions of Scotland and northern England: a guide to magma source region and magma-crust interaction. Philos. Trans. R. Soc. London Ser. A 310: 709-742

Harper, C.T. (1967): The geological interpretation of potassium-argon ages of metamorphic rocks from the Scottish Caledonides. Scott. J. Geol. 3: 46-66

Harris, A.L.; Fettes, D.J. (1972): Stratigraphy and structure of Upper Dalradian rocks at the Highland Border. Scott. J. Geol. 8: 253-264

Harris, A.L.; Pitcher, W.S. (1975): The Dalradian Supergroup. In: Harris, A.L.; Shackleton, R.M.; Watson, J.; Downid, C.; Harland, W.B.; Moorbath, S. (eds.) A correlation of the Precambrian rocks in the British Isles. Spec. Rep. Geol. Soc. London 6: 52-75

Harris, A.L.; Bradbury, H.J.; McGonigal, N.H. (1976): The evolution and transport of the Tay Nappe. Scott. J. Geol. 12: 103-113

Harris, A.L.; Baldwin, C.T.; Bradbury, H.J.; Johnson, H.D.; Smith, R.A. (1978): Ensialic basin sedimentation: the Dalradian Supergroup. In: Bowes, D.R.; Leake, B.E. (eds.) Crustal evolution in northwestern Britain and adjacent regions. Spec. Issue Geol. Soc. London 10: 115-138

Hart, S.R. (1964): The petrology and isotopic-mineral age relations of a contact zone in the Front Range, Colorado. J. Geol. 72: 493-525

Harte, B. (1988): Lower Palaeozoic metamorphism in the Moine-Dalradian belt of the British Isles. In: Harris, A.L.; Fettes, D.J. (eds.) The Caledonian-Appalachian Orogen. Spec. Publ. Geol. Soc. London 38: 123-134

Harte, B.; Dempster, T.J. (1987): Regional metamorphic zones: tectonic controls. Philos. Trans. R. Soc. London Ser. A 321: 3-22

Harte, B; Hudson, N.F.C. (1979): Pelite facies series and the temperatures and pressures of Dalradian metamorphism. In: Harris, A.L.; Holland, C.H.; Leake, B.E. (eds.) The Caledonides of the British Isles - reviewed. Spec. Publ. Geol. Soc. London 8: 323-337

Harte, B.; Johnson, M.R.W. (1969): Metamorphic history of Dalradian rocks in Glens Clova, Esk and Lethnot, Angus, Scotland. Scott. J. Geol. 5: 54-80

Harte, B.; Linklater, C.M. (1989): Melt connectivity and the nature of lenkosomes and melanosomes in partially molten rocks from the Ballachulish aureole. Terra Abstracts 1: 304

Harte, B.; Booth, J.E.; Dempster, T.J.; Fettes, D.J.; Mendum, J.R.; Watts, D. (1984): Aspects of the post-depositional evolution of Dalradian and Highland Border Complex rocks in the Southern Highlands of Scotland. Trans. R. Soc. Edinburgh, Earth Sci. 75: 151-163

Harwood, D.S.; Larson, R.R. (1969): Variations in the delta index of cordierite around the Cupsuptic Pluton, West-Central Maine. Am. Mineral. 54: 896-908

Haslam, H.W. (1968): The crystallization of intermediate and acid magmas at Ben Nevis, Scotland. J. Petrol. 9: 84-104

Haslam, H.W.; Kimbell, G.S. (1981): Disseminated copper-molybdenum mineralisation near Ballachulish, Highland Region. Mineral. Reconn. Progr. Rep. 43, Inst. Geol. Sci. London, 45 pp

Heinrich, W.; Metz, P.; Bayh, W. (1986): Experimental investigation of the mechanism of the reaction: 1 tremolite + 11 dolomite = 8 forsterite + 13 calcite + 9 CO_2 + 1 H_2O. A SEM study. Contrib. Mineral. Petrol. 93: 215-221

Helgeson, H.C. (1968): Evaluation of irreversible reactions in geochemical processes involving minerals and aqueous solutions - I. Thermodynamic relations. Geochim. Cosmochim. Acta 32: 853-877

Helgeson, H.C., Delany, J.M., Nesbitt, N.W., Bird, D.K. (1978): Summary and critique of the thermodynamic properties of rock-forming minerals. Am. J. Sci. 278A: 229 pp

Helz, R.T. (1982): Phase relations and compositions of amphiboles produced in studies of the melting behaviour of rocks. In: Veblen, D.R.; Ribbe, P.H. (eds.) Amphiboles: petrology and phase relations. Mineral. Soc. Am., Rev. Mineral. 9B: 279-354

Hensen, B.J.; Harley, S.L. (1990): Graphical analysis of P-T-X relations in granulite facies metapelites. In: Ashworth, J.R.; Brown, M. (eds.) High-temperature metamorphism and crustal anatexis. Unwin & Hyman, London: 19-56

Herman, J.S.; White, W.B. (1985): Dissolution kinetics of dolomite: effects of lithology and fluid flow velocity. Geochim. Cosmochim. Acta 49: 2017-2026

Hess, H.H.; Phillips, A.H. (1938): Orthopyroxenes of the Bushveld type. Am. Mineral. 26: 450-456

Heuss-Aßbichler, S. (1987): Mikrotexturen, Reaktionsmechanismen und Stofftransport bei Dekarbonatisierungsreaktionen in Thermoaureolen. Ph.D. Thesis, Univ. München, 191 pp

Hickman, A.H. (1975): The stratigraphy of late Precambrian metasediments between Glen Roy and Lismore. Scott. J. Geol. 11: 117-142

Hickman, A.H. (1978): Recumbant folds between Glen Roy and Lismore. Scott. J. Geol. 14: 191-212

Hickman, A.H. (1980): Structural interpretation of the Loch Leven area: reply (letter to the editor). Scott. J. Geol. 16: 76-80

Hildreth, W.; Moorbath, S. (1988): Crustal contributions to arc magmatism in the Andes of Central Chile. Contrib. Mineral. Petrol. 98: 455-489

Hildreth, W.; Christiansen, R.L.; O'Neil, J.R. (1984): Catastrophic isotopic modification of rhyolitic magma at times of caldera subsidence, Yellowstone Plateau volcanic field. J. Geophys. Res. 89: 8339-8369

Hill, R.J.; Silver, L.T.; Taylor, H.P. (1986): Coupled Sr-O isotope variations as an indicator of source heterogeneity for the Northern Peninsular Ranges Batholith. Contrib. Mineral. Petrol. 92: 351-361

Hodges, K.V.; Spear, F.S. (1982): Geothermometry, geobarometry and the Al_2SiO_5 triple point at Mt. Moosilauke, New Hampshire. Am. Mineral. 67: 1118-1134

Hoefs, J. (1987): Stable isotope geochemistry, 3rd. edn. Springer, Berlin, Heidelberg, New York

Hoernes, S.; Friedrichsen H. (1978): Oxygen and hydrogen isotope study of the polymetamorphic area of the northern Ötztal-Stubai Alps (Tyrol). Contrib. Mineral. Petrol. 67: 305-315

Hoernes, S.; Hoffer, E. (1979): Equilibrium relations on prograde metamorphic mineral assemblages. A stable isotope study of rocks of the Damara Orogen, Namibia. Contrib. Mineral. Petrol. 68: 377-389

Hoernes, S.; Hoffer, E. (1985): Stable isotope evidence for fluid-present and fluid-absent metamorphism in metapelites from the Damara Orogen, Namibia. Contrib. Mineral. Petrol. 90: 322-330

Hoernes, S.; MacLeod, S.; Harmon, R.S.; Troll, G.; Weiss, S.; Pattison, D. (1985): Stable isotope geochemistry of the Ballachulish igneous complex and contact metamorphic aureole, Southwestern Grampian Highlands, Scotland. Terra Cognita 5, K88 (EUG III)

Hoffer, E. (1983): Compositional variations of minerals in metapelites involved in low to medium grade isograde reactions in the southern Damara Orogen, Namibia/SW-Africa. In: Martin, H.; Eder, F.W. (eds) Intracontinental fold belts. Springer, Berlin, Heidelberg, New York, pp 745-766

Holdaway, M.J. (1971): Stability of andalusite and the aluminum silicate phase diagram. Am. J. Sci. 271: 97-131

Holdaway, M.J.; Lee, S.M. (1977): Fe-Mg cordierite stability in high-grade pelitic rocks based on experimental, theoretical and natural observations. Contrib. Mineral. Petrol. 63: 175-198

Holden, P.; Halliday, A.N.; Stephens, W.E. (1987): Neodymium and strontium isotope content of microdiorite enclaves points to mantle input to granitoid production. Nature (London) 330: 53-56

Holder, M.T. (1983): Discussion on convection and crystallization in the Criffell-Dalbeattie pluton. J. Geol. Soc. London 138: 351-366
Holland, J.G.; Lambert, R.J. (1975): The geochemistry and origin of the Lewisian gneisses of the Scottish mainland: the Scourie and Inver assemblages and sub-crustal accretion. Precamb. Res. 2: 161-188
Hollister, L.S.; Grissom, G.C.; Peters, E.K.; Stowell, H.H.; Sisson, V.B. (1987): Confirmation of the empirical correlation of Al in hornblende with pressure of solidification of calc-alkaline plutons. Am. Mineral. 72: 231-239
Horne, J. (1886): The origin of the andalusite schists of Aberdeenshire. Min. Mag. 6: 98-100
Hubbard, C.A. (1983): New standard reference materials for X-ray powder diffraction. Adv. X-ray Anal. 26: 45-51
Hubbert, M.K.; Rubey, W.W. (1959): Role of fluid pressure in mechanics of overthrust faulting I. Mechanics of fluid-filled porous solids and its applications to overthrust faulting. Geol. Soc. Am. Bull. 709: 115-166
Hunter, R.H. (1987): Textural equilibrium in layered igneous rocks. In: Parsons, I. (ed.) Origins of igneous layering. Reidel, Dordrecht, pp 473-503
Hurford, A.J. (1977): A preliminary fission track dating survey of Caledonian 'Newer and Last Granites' from the Highlands of Scotland. Scott. J. Geol. 13: 271-284
Iijima, S.; Buseck, P.R. (1978): Experimental study of disordered mica structures by high-resolution electron microscopy. Acta Cryst. A34: 709-719
Iijima, S.; Zhu, J. (1982): Electron microscopy of a muscovite-biotite interface. Am. Mineral. 67: 1195-1205
Indares, A.; Martignole, J. (1985): Biotite-garnet geothermometry in the granulite facies: the influence of Ti and Al in biotite. Am. Mineral. 70: 272-278
Inderst, B. (1987): Mischkristalle und fluide Phase (H_2O-CO_2) bei Hochtemperatur-Dekarbonatisierungsreaktionen in zwei Thermoaureolen. Ph. D. Thesis, Univ. München
Institute of Geological Sciences (ed.) (1979): Argyll. Bouguer gravity anomaly map, sheet 56° N-06° W. 1:250000 series, Keyworth, Nottinghamshire
Institute of Geological Sciences (ed.) (1981): Argyll. Aeromagnetic anomaly map, sheet 56° N-06° W. 1:250000 series, Keyworth, Nottinghamshire
Irvine, T.N.; Baragar, W.R.A. (1971): A guide to the chemical classification of the common volcanic rocks. Can. J. Earth Sci. 8: 523-548
Ishii, T. (1975): The relation between temperature and composition of pigeonite in some lavas and their application to geothermometry. Mineral. J. (Jpn.) 8: 48-57
I.U.G.S. Subcommission (ed.) (1973): Classification and nomenclature of plutonic rocks: recommendations. N. Jahrb. Mineral., Monatsh. 4: 149-164
Jaffe, H.W.; Jaffe, E.B. (1973): Bedrock geology of the Monroe quadrangle, New York. N.Y. State Mus. Sci. Serv., Map Chart Ser. 20, Albany, N.Y.
Jaffe, H.W.; Robinson, P.; Tracy, R.J.; Ross, M. (1975): Orientation of pigeonite exsolution lamellae in metamorphic augite: correlation with composition and calculated optimal phase boundaries. Am. Mineral. 60: 9-28
Joesten, R. (1974): Local equilibrium and metasomatic growth of zoned calc-silicate nodules from a contact aureole, Christmas Mts., Big Ben region, Texas. Am. J. Sci. 273: 876-901
Joesten, R. (1977): Evolution of mineral assemblage zoning in diffusion metamorphism. Geochim. Cosmochim. Acta 41: 649-670
Johannes, W. (1969): An experimental investigation of the system MgO-SiO_2-H_2O-CO_2. Am. J. Sci. 267: 1083-1104
Johannes, W. (1983): Metastable melting in granite and related systems. In: Atherton, M.P.; Gribble, C.D. (eds.) Migmatites, melting and metamorphism. Shiva, Kent, pp 27-36
Johannes, W. (1984): Beginning of melting in the granite system Qz-Or-Ab-An-H_2O. Contrib. Mineral. Petrol. 86: 264-273
Johannes, W. (1985): The significance of experimental studies for the formation of migmatites. In: Ashworth, J.R. (ed.) Migmatites. Blackie, Glasgow, pp 36-85

Johnson, M.R.W., Frost, R.T.C. (1977): Fault and lineament patterns in the Southern Highlands of Scotland. Geol. Mijnbouw 56: 287-294

Johnson, M.R.W.; Kelley, S.P.; Oliver, G.J.H.; Winter, D.A. (1985): Thermal effects and timing of thrusting in the Moine Thrust zone. J. Geol. Soc. London 142: 863-873

Jurewicz, S.R.; Watson, E.B. (1985): The distribution of partial melt in a granitic system: the application of liquid phase sintering theory. Geochim. Cosmochim. Acta 49: 1109-1121

Kalamarides, R.I. (1984): Kiglapait geochemistry VI: oxygen isotopes. Geochim. Cosmochim. Acta 48: 1827-1836

Kamineni, D.C. (1974): Variation in the distortion index of cordierite east of the Sparrow Lake granite pluton, District of Mackenzie. Can. Mineral. 12: 419-421

Kennedy, W.Q. (1948): On the significance of thermal structure in the Scottish Highlands. Geol. Mag. 85: 229-234

Kern, H. (1978): The effect of high temperature and high confining pressure on compressional wave velocities in quartz-bearing and quartz-free igneous and metamorphic rocks. Tectonophysics 44: 185-203

Kerrick, D.M. (1974): Review of metamorphic mixed-volatile (H_2O-CO_2) equilibria. Am. Mineral. 59: 729-762

Kerrick, D.M. (1987): Fibrolite in contact aureoles of Donegal, Ireland. Am. Mineral. 72: 240-254

Kerrick, D.M.; Jacobs, C.K. (1981): A modified Redlich-Kwong equation for H_2O, CO_2 mixtures at elevated pressures and temperatures. Am. J. Sci. 281: 735-767

Kerrick, D.M.; Speer, A. (1988): The role of minor element solid solution on the andalusite-sillimanite equilibrium in metapelites and paraluminous granitoids. Am. J. Sci. 288: 152-192

Kirkpatrick, R.J. (1981): Kinetics of crystallization of igneous rocks. In: Lasaga, A.C.; Kirkpatrick, R.J. (eds.) Kinetics of geochemical processes. Mineral. Soc. Am., Rev. Mineral. 8: 321-398

Kitamura, M.; Hiroi, Y. (1982): Indialite from Unazuki pelitic schist, Japan, and its transition texture to cordierite. Contrib. Mineral. Petrol. 80: 110-116

Kitamura, M.; Yamada, H. (1987): Origin of sector trilling in cordierite in Daimonji hornfels, Kyoto, Japan. Contrib. Mineral. Petrol. 97: 1-6

Kitamura, M.; Yasuda, M.; Morimoto, N. (1981): Morphology change of exsolution lamellae of pigeonite in Bushveld augite - an electron microscopic observation. Proc. Jpn. Acad. Ser. B 57: 183-187

Knitter, R. (1985): Kinetik des Al, Si-Ordnungs-/Unordnungsprozesses in Alkali-Feldspäten. Diplomarbeit, Inst. Mineral., Westf. Wilhelms-Univ. Münster

Kohlstedt, D.L.; Vander Sande, J.B. (1976): On the detailed structure of ledges in an augite-enstatite interface. In: Wenk, H.-R. (ed.) Electron microscopy in mineralogy. Springer, Berlin, Heidelberg, New York, pp 234-237

Korzhinskii, D.S. (1959): Physicochemical basis of the analysis of the paragenesis of minerals. Consultants Bur., New York, Chapman & Hall, London

Kouchi, A.; Sunagawa, I. (1985): A model for mixing basaltic and dacitic magmas as deduced from experimental data. Contrib. Mineral. Petrol. 89: 17-23

Krause, C. (1986): Transmissionselektronenmikroskopische Untersuchung von metamorphen K-Feldspäten: die Sanidin-Mikroklin-Umwandlung. Ph.D. Thesis, Westf. Wilhelms-Universität Münster

Kretz, R. (1983): Symbols for rock-forming minerals. Am. Mineral. 68: 277-279

Kroll, H. (1973): Estimation of the Al,Si distribution of feldspars from the lattice translations Tr[110] and Tr[110]. I. Alkali feldspars. Contrib. Mineral. Petrol. 39: 141-156

Kroll, H.; Knitter, R. (1985): Kinetik des Al,Si- Ordnungsvorgangs in Alkali-Feldspäten. Fortschr. Mineral. 63/1: 105

Kroll, H.; Ribbe, P.H. (1983): Lattice parameters, composition and Al,Si order in alkali feldspars. Mineral. Soc. Am., Rev. Mineral. 2, 2nd edn: 57-99

Kroll, H.; Ribbe, P.H. (1987): Determining (Al,Si) distribution and strain in alkali feldspars using lattice parameters and diffraction-peak positions: a review. Am. Mineral. 72: 491-506

Kroll, H.; Bambauer, H.-U.; Schirmer, U. (1980): The high albite-monalbite transitions. Am. Mineral. 65: 1192-1211

Kroll, H.; Schmiemann, I.; von Cölln, G. (1986): Feldspar solid solutions. Am. Mineral. 71: 1-16

Langer, K.; Schreyer, W. (1969): Infrared and powder X-ray diffraction studies on the polymorphism of cordierite $Mg_2(Al_4Si_5)O_{18}$. Am. Mineral. 54: 1442-1459

Langer, K.; Stürzebecher, M. (1983): Experimental study of the influence of water on the high-low-transition of synthetic Mg-cordierite. Fortschr. Mineral. 61, Beih. 1: 126-128

Larson, P.B.; Taylor, H.P. (1986): $^{18}O/^{16}O$ ratios in ash-flow tuffs and lavas erupted from the central Nevada caldera complex and the central San Juan caldera complex, Colorado. Contrib. Mineral. Petrol. 92: 146-156

Lasaga, A.C. (1983): Geospeedometry: an extension of geothermometry. In: Saxena, S.K. (ed.). Adv. Phys. Geochem. 3: 81-114

Lasaga, A.C. (1986): Metamorphic reaction rate laws and development of isograds. Mineral. Mag. 50: 359-373

Laves, F. (1952): Phase relations of the alkali feldspars: I. Introductory remarks. II. The stable and pseudostable phase relations in the alkali feldspar system. J. Geol. 60: 436-450, 549-574

Laves, F. (1960): Al/Si Verteilungen, Phasen-Transformationen und Namen der Alkalifeldspäte. Z. Krist. 113: 265-296

Leake, B.E. (1978): Nomenclature of amphiboles. Miner. Mag. 42, 533-563

Le Breton, N.; Thompson, A.B. (1988): Fluid-absent (dehydration) melting of biotite in metapelites in the early stages of crustal anatexis. Contrib. Mineral. Petrol. 99: 226-237

Lindsay, N.G.; Haselock, P.J.; Harris, A.L. (1989): The extent of Grampian orogenic activity in the Scottish Highlands. J. Geol. Soc. London 146: 733-735

Lindsley, D.H. (1983): Pyroxene thermometry. Am. Mineral. 68: 477-493

Lindsley, D.H.; Andersen, D.J. (1983): A two-pyroxene thermometer. J. Geophys. Res. 88: A887-906

Lipman, P.W.; Friedman, I. (1975): Interaction of meteoric water with magma: an oxygen isotope study of ash-flow sheets from southern Nevada. Geol. Soc. Am. Bull. 86: 695-702

Litherland, M. (1970): The stratigraphy and structure of the Dalradian rocks around Loch Creran, Argyll. Ph.D. Thesis, Univ. Liverpool

Litherland, M. (1980): The stratigraphy of the Dalradian rocks around Loch Creran, Argyllshire. Scott. J. Geol. 16: 105-123

Litherland M. (1982): The structure of the Loch Creran Dalradian and a new model for the S.W. Highlands. Scott. J. Geol. 18: 205-225

Litherland, M. (1983): The structure of the Loch Creran Dalradian and a new model for the S.W. Highlands: reply (letter to the editor). Scott. J. Geol. 19: 268-270

Lofgren, G. (1974): An experimental study of plagioclase crystal morphology: isothermal crystallization. Am. J. Sci. 274: 243-273

Lonker, S.W. (1981): The P-T-X relations of the cordierite-garnet-sillimanite-quartz equilibrium. Am. J. Sci. 281: 1056-1090

Loomis, T.P. (1982): Numerical simulations of crystallization processes of plagioclase in complex melts: the origin of major and oscillatory zoning in plagioclase. Contrib. Mineral. Petrol. 81: 219-229

Maaløe, S. (1978): The origin of rhythmic layering. Mineral. Mag. 42: 337-345

Maaløe, S. (1985): Principles of igneous petrology. Springer, Berlin, Heidelberg, New York, 371 pp

Maaløe, S.; Wyllie, P.J. (1975): Water content of a granite magma deduced from the sequence of crystallization determined experimentally with water-undersaturated conditions. Contrib. Mineral. Petrol. 52: 175-191

MacCulloch, J. (1817): Observations on the mountain Cruachan in Argyllshire, with some remarks on the surrounding country. Trans. Geol. Soc. Great Britain 4: 117-138

MacLeod-Kinsel, S. (1987): An oxygen and hydrogen isotope study of the Ballachulish igneous complex and metamorphic aureole SW Grampian Highlands, Scotland. MS-Thesis, South. Methodist Univ. Dallas

Magaritz, M.; Taylor, H.P. (1976): $^{18}O/^{16}O$ and D/H studies of igneous and sedimentary rocks along a 500 km traverse across the Coast Range Batholith into Central British Columbia at latitudes 54°-55° N. Can. J. Earth Sci. 13: 1514-1530

Maresch, W.V.; Czank, M. (1985): The optical and X-ray properties of $Li_2Mg_2[Si_4O_{11}]$, a new type of chain silicate. N. Jahrb. Mineral. Monatsh. 1985: 289-297

Marquardt, D.W. (1963): An algorithm for least-squares estimation of non-linear parameters. J. Soc. Inst. Appl. Math. 11: 431-441

Marsh, B.D. (1982): On the mechanics of igneous diapirism, stoping, and zone melting. Am. J. Sci. 282: 808-855

Martignole, J.; Sisi, J.C. (1981): Cordierite-garnet-H_2O equilibrium: a geological thermometer, barometer and water fugacity indicator. Contrib. Mineral. Petrol. 77: 38-46

Masch, L. (1977): Die Aureole des Monzoni-Intrusivkomplexes (Norditalien) - ein Modell progressiver Thermometamorphose kieseliger Karbonate. Habil., Univ. München

Masi, U.; O'Neil, J.R.; Kistler, R.W. (1981): Stable isotope systematics in Mesozoic granites of central and northern California and southwestern Oregon. Contrib. Mineral. Petrol. 76: 116-126

Mather, J.D. (1970): The biotite isograd and the lower greenschist facies in the Dalradian rocks of Scotland. J. Petrol. 11: 253-275

Matsuhisa, Y. (1979): Oxygen isotopic compositions of volcanic rocks from the east Japan island arcs and their bearing on petrogenesis. J. Volcanic Geotherm. Res. 5: 271-296

McBirney, A.R.; Murase, T. (1984): Rheological properties of magmas. Annu. Rev. Earth Planet. Sci. 12: 337-357

McCarthy, T.S.; Hasty, R.A. (1976): Trace element distribution patterns and their relationship to the crystallization of granitic melts. Geochim. Cosmochim. Acta 40: 1351-1358

McConnell, J.D.C. (1985): Symmetry aspects of order-disorder and the application of Landau theory. In: Kieffer, S.W., Navrotsky, A. (eds.). Microscopic to macroscopic: atomic environments to mineral thermodynamics. Mineral. Soc. Am., Rev. Mineral. 14: 165-186

McCrea, J.M. (1950): The isotopic chemistry of carbonates and a paleotemperature scale. J. Chem. Phys. 18: 849-857

McKenzie, D.P. (1984): The generation and compaction of partially molten rock. J. Petrol. 25: 713-765

McKerrow, W.S. (1987): The Southern Uplands controversy. J. Geol. Soc. London 144: 735-736

McKerrow, W.S. (1988): The development of the Iapetus Ocean from the Arening to the Wenlock. In: Harris, A.L.; Fettes, D.J. (eds.) The Caledonian-Appalachian Orogen. Spec. Publ. Geol. Soc. London 38: 404-412

McLaren, A.C. (1984): Transmission electron microscope investigation of the microstructures of microclines. In: Brown, W.L. (ed.) Feldspars and feldspathoids. Reidel, Dordrecht, pp 373-410

McLaren, A.C.; Fitz Gerald, J.D. (1987): CBED and ALCHEMI investigation of local symmetry and Al,Si ordering in K-feldspars. Phys. Chem. Minerals 14: 281-292

Meagher, E.P.; Gibbs, G.V. (1977): The polymorphism of cordierite II: the crystal structure of indialite. Can. Mineral. 15: 43-49

Medenbach, O. (1985): A new microrefractometer spindle-stage and its application. Fortschr. Mineral. 63: 111-133

Mehnert, K.R. (1968): Migmatites and the origin of granitic rocks. Elsevier, Amsterdam

Mendum, J.R.; Fettes, D.J. (1985): The Tay nappe and associated folding in the Ben Ledi-Loch Lomond area. Scott. J. Geol. 21: 41-56

Metz, P.; Trommsdorff, V. (1968): On phase equilibria in metamorphosed siliceous dolomites. Contrib. Mineral. Petrol. 18: 305-309

Michard-Vitrac, A.; Albarede, F.; Dupuis, C.; Taylor, H.P. (1980): The genesis of Variscan (Hercynian) plutonic rocks: inferences from Sr, Pb, and O isotope studies on the Maladeta igneous complex, central Pyrenees (Spain). Contrib. Mineral. Petrol. 72: 57-72

Miller, J.A.; Brown, P.E. (1965): Potassium-argon age studies in Scotland. Geol. Mag. 102: 106-134

Mirwald, P.W.; Schreyer, W. (1977): Die stabile und metastabile Abbaureaktion von Mg-Cordierit in Talk, Disthen und Quarz und ihre Abhängigkeit vom Gleichgewichtswassergehalt des Cordierits. Fortschr. Mineral. 55: 95-97

Miyashiro, A. (1957): Cordierite-indialite relations. Am. J. Sci. 255: 43-62

Miyashiro, A; Shido, E. (1984): Tschermak subsitution in low- and middle-grade pelitic schists. J. Petrol. 26: 449-487

Montel, J.-M.; Weber, C.; Pichavant, M. (1986): Biotite-silimanite-spinel assemblages in high grade rocks: occurrences, chemographic analysis and thermobarometric interest. Bull. Mineral. 109: 555-573

Morel, P.; Irving E. (1978): Tentative paleocontinental maps for the early Phanerozoic and Proterozoic. J. Geol. 86: 535-561

Morimoto, N.; Tokonami, M. (1969): Oriented exsolution of augite in pigeonite. Am. Mineral. 54: 1101-1117

Morse, S.A. (1986): Convection in aid of adcumulus growth. J. Petrol. 27: 1183-1214

Mueller, R.F. (1967): Model for order-disorder kinetics in certain quasi-binary crystals of continuously variable composition. J. Phys. Chem. Solids 28: 2239-2243

Mueller, R.F. (1969): Kinetics and thermodynamics of intracrystalline distribution. Mineral. Soc. Am. Spec. Pap. 2: 83-93

Muir, I.D. (1953a): A local potassic modification of the Ballachulish granodiorite. Geol. Mag. 90: 182-193

Muir, I.D. (1953b): Quartzite xenoliths in the Ballachulish granodiorite. Geol. Mag. 90: 409-428

Munoz, J.L. (1984): F-OH and Cl-OH exchange in micas with applications to hydrothermal ore deposits. In: Bailey, S.W. (ed.), Micas. Reviews in mineralogy, vol. 13. Mineral. Soc. Am., pp 469-544

Nabelek, C.R.; Lindsley, D.H. (1985): Tetrahedral Al in amphibole: a potential thermometer for some mafic rocks. Geol. Soc. Am., Abstr. Progr., 68384: 673

Nabelek, P.I.; O'Neil, J.R.; Papike, J.J. (1983): Vapour phase exsolution as a controlling factor in hydrogen isotope variation in granitic rocks: the Notch Peak granitic stock, Utah. Earth Planet. Sci. Lett. 66: 137-150

Nabelek, P.I.; Labotka, T.C., O'Neil, J.R.; Papike, J.J. (1984): Contrasting fluid/rock interaction between Notch Peak granitic intrusion and argillites and limestones in western Utah: evidence from stable isotopes and phase assemblages. Contrib. Mineral. Petrol. 86: 25-34

Nabelek, P.I.; Lindsley, D.H.; Bohlen, S.R. (1987): Experimental examination of two-pyroxene graphical thermometers using natural pyroxenes with application to metaigneous pyroxenes from the Adirondack Mountains, New York. Contrib. Mineral. Petrol. 97: 66-7

Nakajima, Y.; Ribbe, P.H. (1981): Texture and structural interpretation of the alteration of pyroxene to other biopyriboles. Contrib. Mineral. Petrol. 78: 230-239

Naney, M.T. (1983): Phase equilibria of rock-forming ferromagnesian silicates in granitic systems. Am. J. Sci. 283: 993-1033

Naney, M.T.; Swanson, S.E. (1980): The effect of Fe and Mg on crystallization in granitic systems. Am. Mineral. 65: 639-653

Neumann, H. (1950): Pseudomorphs of pyrrhotine after pyrite in the Ballachulish Slates. Mineral. Mag. 29: 234-238

Newton, R.C. (1965): The thermal stability of zoisite. J. Geol. 73: 431-441

Newton, R.C.; Perkins, D. III (1982): Thermodynamic calibration of geobarometers based on the assemblages garnet-plagioclase-orthopyroxene (clinopyroxene)-quartz. Am. Mineral. 67: 203-222

Newton, R.C.; Wood, B.J. (1979): Thermodynamics of water in cordierite and some petrological consequences of cordierite as a hydrous phase. Contrib. Mineral. Petrol. 68: 391-405

Newton, R.C.; Wood, B.J. (1980): Volume behaviour of silicate solid solutions. Am. Mineral. 65: 733-745

Nicolas, A.; Jackson, M. (1982): High temperature dikes in peridotites: origin by hydraulic fracturing. J. Petrol. 23: 568-582

Nobugai, K.; Morimoto, N. (1979): Formation mechanism of pigeonite lamellae in Skaergaard augite. Phys. Chem. Minerals 4: 361-371

Nobugai, K.; Tokonami, M.; Morimoto, N. (1978): A study of subsolidus relations of the Skaergaard pyroxenes by analytical electron microscopy. Contrib. Mineral. Petrol. 67: 111-117

Nockolds, S.R. (1941): The Garabal Hill - Glen Fyne igneous complex. Q. J. Geol. Soc. London 96: 451-511

Norton, D.; Knight, J.E. (1977): Transport phenomena in hydrothermal systems: cooling plutons. Am. J. Sci. 277: 937-981

O'Hara, M.J. (1968): The bearing of phase equilibria studies in synthetic and natural systems on the origin and evolution of basic and ultrabasic rocks. Earth Sci. Rev. 5: 49-133

Ohmoto, H.; Kerrick, D.M. (1977): Devolatilization equilibria in graphitic systems. Am. J. Sci. 277: 1013-1044

O'Neil, J.R. (1986): Theoretical and experimental aspects of isotopic fractionation. In: Valley, J.W.; Taylor, H.P. Jr.; O'Neil, J.R. (eds.) Reviews in mineralogy, vol. 16. Stable isotopes in high temperature geological processes. Mineral. Soc. Am., pp 1-36

O'Neil, J.R.; Chappell, B.W. (1977): Oxygen and hydrogen isotope relations in the Berridale batholith. J. Geol. Soc. London 133: 559-571

O'Neil, J.R.; Clayton, R.N.; Mayeda, T.K. (1969): Oxygen isotope fractionations in divalent metal carbonates. J. Chem. Phys. 51: 5547-5558

O'Neil, J.R.; Shaw, S.E.; Flood, R.H. (1977): Oxygen and hydrogen isotope compositions as indicators of granite genesis in the New England batholith, Australia. Contrib. Mineral. Petrol. 62: 313-328

Openshaw, R.E.; Henderson, C.M.B.; Brown, W.L. (1979): A room-temperature phase transition in maximum microcline. Unit cell parameters and thermal expansion. Phys. Chem. Minerals 5: 95-104

Otten, M.T. (1985): The subsolidus history of the Artfjället gabbro: a TEM study of olivine, augite and orthopyroxene. J. Petrol. 26: 488-514

Pankhurst, R.J. (1970): The geochronology of the basic igneous complexes. Scott. J. Geol. 6: 83-117

Pankhurst, R.J. (1982): Geochronological tables from British igneous rocks. In: Sutherland, D.S. (ed.) Igneous rocks of the British Isles. John Wiley & Sons, New York, pp 575-581

Pankhurst, R.J.; Sutherland, D.S.; Brown, G.C.; Pitcher, W.S. (1982): Caledonian granites and diorites of Scotland and Ireland. In: Sutherland, D.S. (ed.) Igneous rocks of the British Isles. John Wiley & Sons, New York, pp 149-190

Papageorgiou, D.; Kroll, H. (1986): Al, Si-Austauschkinetik in Alkali-Feldspäten. Fortschr. Mineral. 64 Beih. 1: 141.

Parker, R.L. (1977): Understanding inverse theory. Annu. Rev. Earth Planet. Sci. 5: 35-64

Parsons, I. (1978): Feldspars and fluids in cooling plutons. Mineral. Mag. 42: 1-17

Parsons, I.; Brown, W.L. (1983): A TEM and microprobe study of a two-perthite alkali gabbro: implications for the ternary feldspar systems. Contrib. Mineral. Petrol. 82: 1-12

Parsons, I.; Brown, W.L.; Jacquemin, H. (1986): Mineral chemistry and crystallization conditions of the Mboutou layered gabbro - syenite - granite complex, North Cameroon. J. Petrol. 27: 1305-1329

Pattison, D.R.M. (1985): Petrogenesis of pelitic rocks in the Ballachulish thermal aureole. Ph.D. Thesis, Univ. Edinburgh, 590 pp

Pattison, D.R.M. (1987): Variations in Mg/(Mg+Fe), F, and (Fe,Mg) = 2Al in pelitic minerals in the Ballachulish thermal aureole, Scotland. Am. Mineral. 72: 255-272

Pattison, D.R.M. (1989): P-T-a(H_2O) conditions and the influence of graphite on pelitic phase relations in the Ballachulish aureole, Scotland. J. Petrol. 30: 1219-1244

Pattison, D.R.M. (1991): Stability of andalusite and sillimanite and the Al_2SiO_5 triple point: Constraints from the Ballachulish aureole, Scotland. J. Geol. (submitted)

Pattison, D.R.M.; Harte B. (1985): A petrogenetic grid for pelites in the Ballachulish and other Scottish thermal aureoles. J. Geol. Soc. London 142: 7-28
Pattison, D.R.M.; Harte, B. (1988): Evolution of structurally contrasting anatectic migmatites in the 3-kbar Ballachulish aureole, Scotland. J. Metam. Geol. 6: 475-494
Peaceman, D.W.; Rachford, H.H., Jr. (1955): The numerical solution of parabolic and elliptic differential equations. J. Soc. Ind. Appl. Math. 3: 28-41
Pearce, J.A.; Harris, N.B.W.; Tindle, A.G. (1984): Trace element discrimination diagrams for the tectonic interpretation of plutonic rocks. J. Petrol. 25: 956-983
Perkins, D. III; Chipera, S.J. (1985): Garnet-orthopyroxene-plagioclase-quartz barometry: refinement and application to the English River subprovince and the Minnesota River Valley. Contrib. Mineral. Petrol. 89: 69-80
Phillips, W.J.; Fuge, R.; Phillips, N. (1981): Convection and crystallization in the Criffell-Dalbeattie pluton. J. Geol. Soc. London 138: 351-366
Piasecki, M.A.J.; Breemen, O. van (1983): Field and isotopic evidence for a c. 750 Ma tectonothermal event in Moine rocks in the Central Highland region of the Scottish Caledonides. Trans. R. Soc. Edinburgh, Earth Sci. 75: 151-163
Pidgeon, R.T.; Aftalion, M. (1978): Cogenetic and inherited zircon U-Pb systems in granites: Palaeozoic granites of Scotland and England. In: Bowes, D.R.; Leake, B.E. (eds.) Crustal evolution in northwestern Britain and adjacent regions. Geol. J. Spec. Issue 10: 183-220
Pitcher, W.S. (1979): The nature, ascent and emplacement of granitic magmas. J. Geol. Soc. London 136: 627-662
Pitcher, W.S. (1982): Granite type and tectonic environment. In: Hsu, K. (ed.) Mountain building processes. Academic Press, New York, London, pp 19-40
Piwinskii, A.J.; Wyllie, P.J. (1968): Experimental studies on igneous rock series: a zoned pluton in the Wallowa batholith, Oregon. J. Geol. 76: 205-234
Plant, J.; Simpson, P.R.; Green, P.M.; Watson, J.V.; Fowler, M.B. (1983): Metalliferous and mineralised Caledonian granites in relation to regional metamorphism and fracture systems in northern Scotland. Trans. Inst. Min. Metall., Ser. B, 92: 33-42
Platten, I.M. (1982): Partial melting of feldspathic quartzite around late Caledonian minor intrusions in Appin, Scotland. Geol. Mag. 119: 413-419
Platten, I.M. (1983): Partial melting of semi-pelite and the development of marginal breccias around a late Caledonian minor intrusion in the Grampian Highlands of Scotland. Geol. Mag. 120: 37-49
Powell, D.; Baird, A.W.; Charnley, N.R.; Jordan, P.J. (1981): The metamorphic environment of the Sgurr Beag Slide; a major crustal displacement zone in Proterozoic Moine rocks of Scotland. J. Geol. Soc. London 128: 661-673
Powell, D.; Andersen, T.B.; Drake, A.A.; Hall, L.; Keppie, J.D. (1988): The age and distribution of basement rocks in the Caledonide orogen of the N. Atlantic. In: Harris, A.L.; Fettes, D.G. (eds.) The Caledonian-Appalachian Orogen. Spec. Publ. Geol. Soc. London 38: 63-74
Powell, R. (1978): Equilibrium thermodynamics in petrology. An introduction. Harper & Row, London, 248 pp
Powell, R. (1983): Processes in granulite facies metamorphism. In: Atherton, M.P.; Gribble, C.D. (eds.), Migmatites, melting and metamorphism. Shiva, Cheshire, Kent, pp 127-139
Powell, R.; Evans, J.A. (1983): A new geobarometer for the assemblage biotite-muscovite-chlorite-quartz. J. Metam. Geol. 1: 331-336
Prewitt, C.T.; Brown, G.E.; Papike, J.J. (1971): Apollo 12 clinopyroxenes: high temperature X-ray diffraction studies. In: Proc. 2nd., Lunar Sci. Conf. Houston, pp 59-68
Puhan, D. (1979): Petrographie und geothermometrische Untersuchungen an silikatführenden Dolomit-Calcit-Marmoren zur Ermittlung der Metamorphosebedingungen im Zentral-Damara-Orogen (SW-Afrika). Habil. Univ. Göttingen
Pupin, J.P. (1980): Zircon and granite petrology. Contrib. Mineral. Petrol. 73: 207-220
Putnis, A. (1980a): The distortion index in anhydrous Mg-cordierite. Contrib. Mineral. Petrol. 74: 135-141

Putnis, A. (1980b): Order-modulated structures and the thermodynamics of cordierite reactions. Nature (London) 287: 128-131

Putnis, A.; Angel, R.J. (1985): Al,Si ordering in cordierite using magic angle spinning NMR II: models of Al,Si order from NMR data. Phys. Chem. Minerals 12: 217-222

Putnis, A.; Bish, D.L. (1983): The mechanism and kinetics of Al,Si ordering in Mg-cordierite. Am. Mineral. 68: 60-65

Putnis, A.; Holland, T.J.B. (1986): Sector trilling in cordierite and equilibrium overstepping in metamorphism. Contrib. Mineral. Petrol. 93: 265-272

Putnis, A.; Salje, E.K.H.; Redfern, S.A.T.; Fyfe, C.A.; Strobl, H. (1987): Structural states of cordierite I: order parameters from synchrotron X-ray and NMR data. Phys. Chem. Minerals 14: 446-454

Puziewicz, J.; Johannes, W. (1988): Phase equilibria and compositions of Fe-Mg-Al minerals and melts in water saturated peraluminous granitic systems. Contrib. Mineral. Petrol. 100: 156-168

Rabbel, W. (1987): Seismische Erkundung oberflächennaher Störzonen: strahlentheoretische Grundlagen und Feldbeispiele. Ph.D. Thesis. Univ. Kiel

Read, H.H. (1961): Aspects of Caledonian magmatism in Britain. Liverpool Manch. Geol. J. 2: 653-683

Redfern, S.A.T.; Salje, E.K.H.; Maresch, W.V.; Schreyer, W. (1989): Powder diffraction and infrared study of the hexagonal-orthorhombic phase transition in K-bearing cordierite. Am. Mineral. 74: 1293-1299

Rice, A. (1985): The mechanism of the Mt. St. Helens eruption and speculations regarding Soret effects in planetary dynamics. Geophys. Surv. 7: 303-384

Rice, J.M.; Ferry, J.M. (1982): Buffering, infiltration, and the control of intensive variables during metamorphism. In: Ferry, J.M. (ed.) Characterisation of metamorphism through mineral equilibria. Mineral. Soc. Am., Rev. Mineral. 10: 263-326

Richardson, S.W.; Powell, R. (1976): Thermal causes of the Dalradian metamorphism in the Central Highlands of Scotland. Scott. J. Geol. 12: 237-268

Richardson, S.W.; Gilbert, M.C.; Bell, P.M. (1969): Experimental determination of kyanite-andalusite and andalusite-sillimanite equilibria: the aluminum silicate triple point. Am. J. Sci. 267: 259-272

Richter, R.; Hoernes, S. (1988): The application of the increment method in comparison with experimentally derived and calculated O-isotope fractionations. Chem. Erde 48: 1-18

Ridley, J.; Thompson, A.B. (1986): The role of mineral kinetics in the development of metamorphic microtextures. In: Walther, J.V.; Wood, B.J. (eds.) Advances in geophysical geochemistry, vol. 5. Fluid-rock interactions during metamorphism. Springer, Berlin, Heidelberg, New York, pp 154-193

Rietmeijer, F.J.M. (1979): Pyroxenes from iron-rich igneous rocks in Rogaland, SW Norway. Ultraiectina 21, Rijksuniv. Utrecht

Rietmeijer, F.J.M.; Champness, P.E. (1982): Exsolution structures in calcic pyroxenes from the Bjerkreim-Sokndal lopolith, SW Norway. Mineral. Mag. 45: 11-24

Roberts, A.M.; Smith, D.I., Harris, A.L. (1984): The structural setting and tectonic significance of the Glen Dessary syenite, Inverness-shire. J. Geol. Soc. London 141: 1033-1042

Roberts, J.L. (1974): The structure of the Dalradian rocks in the SW Highlands of Scotland. J. Geol. Soc. London 130: 93-124

Roberts, J.L. (1976): The structure of the Dalradian rocks in the North Ballachulish district of Scotland. J. Geol. Soc. London 132: 139-154

Roberts, J.L.; Treagus, J.E. (1977): Polyphase generation of nappe structures in the Dalradian rocks of the Southwest Highlands of Scotland. Scott. J. Geol. 13: 185-188

Roberts, J.L.; Treagus, J.E. (1979): Stratigraphical and structural correlation between the Dalradian rocks of the SW and Central Highlands of Scotland. In: Harris, A.L.; Holland, C.H.; Leake, B.E. (eds.) The Caledonides of the British Isles - reviewed. Spec. Publ. Geol. Soc. London 8: 199-204

Roberts, J.L.; Treagus, J.E. (1980): The structural interpretation of the Loch Leven area: discussion. (letter to the editor). Scott. J. Geol. 16: 73-75

Roberts, J.L.; Treagus, J.E. (1983): The structure of the Loch Creran Dalradian and a new model for the S.W. Highlands: discussion (letter to the editor). Scott. J. Geol. 19: 265-267

Robertson, J.K.; Wyllie, P.J. (1971a): Rock-water systems, with special reference to the water-deficient region. Am. J. Sci. 271: 252-277

Robertson, J.K.; Wyllie, P.J. (1971b): Experimental studies on rocks from the Deboullie Stock, northern Maine, including melting relations in the water-deficient environment. J. Geol. 79: 549-571

Robinson, P.; Ross, M.; Nord, G.L. Jr.; Smyth, J.R.; Jaffe, H.W. (1977): Exsolution lamellae in augite and pigeonite: fossil indicators of lattice parameters at high temperature and pressure. Am. Mineral. 62: 857-873

Rogers, G.; Dempster, T.J.; Bluck, B.J.; Tanner, P.W.G. (1989): A high precision U-Pb age for the Ben Vuirich granite: implications for the evolution of the Scottish Dalradian Supergroup. J. Geol. Soc. London 146: 789-798

Rubie, D.C. (1986): The catalysis of mineral reactions by water and restrictions on the presence of aqueous fluid during metamorphism. Mineral. Mag. 50: 399-415

Rubie, D.C.; Thompson, A.B. (1985): Kinetics of metamorphic reactions at elevated temperatures and pressures: an appraisal of available experimental data. In: Thompson, A.B.; Rubie, D.C. (eds): Advances in physical geochemistry, vol. 4. Kinetics, textures and deformation. Springer, Berlin, Heidelberg, New York, pp 27-79

Rumble, D., III; Ferry, J.M.; Hoering, T.C.; Boucot, A.J. (1982): Fluid flow during metamorphism at the Beaver Brook fossil locality. Am. J. Sci. 282: 866-919

Salje, E.K.H. (1985): Thermodynamics of sodium feldspar I: order parameter treatment and strain induced coupling effects. Phys. Chem. Minerals. 12: 93-98

Salje, E.K.H. (1986): Heat capacities and entropies of andalusite and sillimanite: the influence of fibrolitisation on the phase diagram of the Al_2SiO_5 polymorphs. Am. Mineral. 71: 1366-1371

Salje, E.K.H. (1987): Structural states of Mg-cordierite II: Landau theory. Phys. Chem. Minerals. 14: 455-460

Salje, E.K.H.; Kuscholke, B.; Wruck, B.; Kroll, H. (1985): Thermodynamics of sodium feldspar II: experimental results and numerical calculations. Phys. Chem. Minerals. 12: 99-107

Sanford, R.F. (1980): Textures and mechanisms of metamorphic reactions in the Cockeysville marble near Texas, Maryland. Am. Mineral. 65: 654-669

Schreyer, W. (1966): Synthetische und natürliche Cordierite III. Polymorphiebeziehungen. N. Jahrb. Mineral. Abh. 105: 211-244

Schreyer, W. (1986): The mineral cordierite: structure and reactions in the presence of fluid phases. Ber. Bunsenges. Phys. Chem. 90: 748-755

Schreyer, W.; Schairer, J.F. (1961): Compositions and structural states of anhydrous Mg-cordierites: a reinvestigation of the central part of the system $MgO-Al_2O_3-SiO_2$. J. Petrol. 2: 324-406

Schreyer, W.; Yoder, H.S. (1964): The system Mg-cordierite-H_2O and related rocks. N. Jahrb. Mineral 101: 271-342

Schreyer, W.; Gordillo, C.E.; Werding, G. (1979): A new sodian-beryllian cordierite from Soto, Argentina, and the relationship between distortion index, Be content, and state of hydration. Contrib. Mineral. Petrol. 70: 421-428

Seck, H.A. (1971): Koexistierende Alkalifeldspäte und Plagioklase im System $NaAlSi_3O_8$ - $KAlSi_3O_8$ - $CaAl_2Si_2O_8$ - H_2O bei Temperaturen von 650° C bis 900° C. N. Jahrb. Mineral. 115: 315-345

Seifert F. (1976): Stability of the assemblage cordierite + K-feldspar + quartz. Contrib. Mineral. Petrol. 57: 179-185

Seifert, F.; Schumacher, J.C. (1986): Cordierite-spinel-quartz assemblages: a potential geobarometer. Bull. Geol. Soc. Finl. 58: 95-108

Selkregg, K.R.; Bloss, F.D. (1980): Cordierites: compositional controls of Δ , cell parameters, and optical properties. Am. Mineral. 65: 522-533

Senderov, E.E.; Yas'kin, G.M. (1975): Conversion of sanidine to microcline under hydrothermal conditions. Geochem. Int. 12/3: 139-145

Senderov, E.E.; Yas'kin, G.M. (1976): On stability of monoclinic potash feldspars. Geokhimiya 7: 138-154 (in Russian)
Shackleton, R.J. (1958): Downward-facing structures of the Highland Border. Q. J. Geol. Soc. London 113: 361-392
Shackleton, R.M. (1979): The British Caledonides: comments and summary. In: Harris, A.L.; Holland, C.H.; Leake, B.E. (eds.) The Caledonides of the British Isles - reviewed. Spec. Publ. Geol. Soc. London 8: 389-403
Sheppard, S.M.F. (1977): The Cornubian batholith, SW England: D/H and $^{18}O/^{16}O$ studies of kaolinite and other alteration minerals. J. Geol. Soc. London 137: 573-591
Sheppard, S.M.F. (1986): Characterization and isotopic variations in natural waters. In: Valley, J.W.; Taylor, H.P. Jr.; O'Neil, J.R. (eds.) Reviews in mineralogy, vol. 16. Stable isotopes in high temperature geological processes. Mineral. Soc. Am., pp 165-183
Sheppard, S.M.F.; Harris, C. (1985): Hydrogen and oxygen isotope geochemistry of Ascension Island lavas and granites: variation with crystal fractionation and interaction with sea water. Contrib. Mineral. Petrol. 91: 74-81
Sheppard, S.M.F.; Nielsen, R.L.; Taylor, H.P. (1969): Oxygen and hydrogen isotope ratios of clay minerals from porphyry copper deposits. Econ. Geol. 66: 515-542
Skippen, G.B. (1971): Experimental data for reactions in siliceous marbles. J. Geol. 79: 457-481
Skippen, G.B. (1974): An experimental model for low pressure metamorphism of siliceous dolomitic marble. Am. J. Sci. 274: 487-509
Slaughter, J.; Kerrick, D.M.; Wall, V.J. (1975): Experimental and thermodynamic study of equilibrium in the system $CaO-MgO-SiO_2-H_2O-CO_2$. Am. J. Sci. 275: 143-162
Smart, R.M.; Glasser, F.P. (1977): Stable cordierite solid solutions in the $MgO-Al_2O_3-SiO_2$ system: composition, polymorphism and thermal expansion. Sci. Ceram. 9: 256-263
Smith, J.V. (1974): Feldspar minerals. Vol. 1. Crystal structure and physical properties. Springer, Berlin, Heidelberg, New York, 627 pp
Smith, J.V.; Brown, W.L. (1988): Feldspar minerals. Vol. 1. Crystal structures, physical, chemical and microtextural properties, 2nd edn. Springer, Berlin, Heidelberg, New York
Smith, P.P.K. (1977): An electron microscope study of amphibole lamellae in augite. Contrib. Mineral. Petrol. 59: 317-322
Smith, P.; Parsons, I. (1974): The alkali-feldspar solvus at 1 kilobar water-vapour pressure. Mineral. Mag. 39: 747-767
Smyth, J.R. (1974): The high temperature crystal chemistry of clinohypersthene. Am. Mineral. 59: 1069-1082
Snyder, D.B.; Flack, C.A. (1990): A Caledonian age for reflectors within the mantle lithosphere north and west of Scotland. Tectonics 4: 903-922
Solomon, G.C.; Taylor, H.P. (1981): The geographic distribution of ^{18}O values in Mesozoic and early Cenozoic granitic rocks of the southwestern North American cordillera. Geol. Soc. Am. Abstr. 1981: 558
Soper, N.J. (1988): Timing and geometry of collision, terrane accretion and sinistral strike-slip events in the British Caledonides. In: Harris, A.L.; Fettes, D.J. (eds.) The Caledonian-Appalachian Orogen. Spec. Publ. Geol. Soc. London 38: 481-492
Soper, N.J.; Anderton, R. (1984): Did the Dalradian slides originate as extensional faults? Nature (London), 307: 357-360
Sparks, R.S.J.; Huppert, H.E.; Turner, J.S. (1984): The fluid dynamics of evolving magma chambers. Philos. Trans. R. Soc. London Ser. A 310: 511-534
Spicker, G.; Huckenholz, H.G. (1986): Zur Petrographie und Geochemie der Intrusion von Monzoni. Fortschr. Mineral. 64, Beih. 1: 172
Spicker, G.; Huckenholz, H.G. (1988): Fraktionierungsmodellrechnungen am Beispiel des Monzoni-Intrusivkomplexes. Fortschr. Mineral. 66: Beih. 1: 149
Steiger, R.H.; Hart, S.R. (1967): The microcline-orthoclase transition within a contact aureole. Am. Mineral. 52: 87-116
Stephens, W.E. (1988): Granitoid plutonism is the Caledonian orogen of Europe. In: Harris, A.L.; Fettes, D.J. (eds.) The Caledonian-Appalachian Orogen. Spec. Publ. Geol. Soc. London 38: 389-403

Stephens, W.E.; Halliday, A.N. (1984): Geochemical contrasts between late Caledonian granitoid plutons of northern, central and southern Scotland. Trans. R. Soc. Edinburgh Earth Sci. 75: 259-273

Stewart, F.H. (1946): The gabbroic complex of Belhelvie in Aberdeenshire. Q. J. Geol. Soc. London, 102: 465-98

Stuckless, J.S.; O'Neil, J.R. (1973): Petrogenesis of the Superstition-Superior volcanic area as inferred from strontium and oxygen isotope studies. Geol. Soc. Am. Bull. 84: 1987-1997

Su, S.C.; Bloss, F.D.; Ribbe, P.H. (1984): Optic axial angle, a precise measure of Al,Si ordering in T_1 tetrahedral sites of K-rich alkali feldspars. Am. Mineral. 69: 440-448

Su, S.-C.; Ribbe, P.H.; Bloss, F.D. (1986): Alkali feldspar: structural state determined from composition and optic axial angle 2V. Am. Mineral. 71: 1285-1296

Sunagawa, I. (1981): Characteristics of crystal growth in nature as seen from morphology of mineral crystals. Bull. Mineral. 104: 81-87

Sunagawa, I. (1984): Growth of crystals in nature. Bull. Mineral. 104: 81-87

Suzuoki, T.; Epstein, S. (1976): Hydrogen fractionation between OH-bearing minerals and water. Geochim. Cosmochim. Acta 40: 1229-1240

Swanson, S.E. (1977): Relation of nucleation and crystal-growth rate to the development of granitic textures. Am. Mineral. 62: 966-978

Tait, S.R.; Huppert, H.E.; Sparks, R.S.J. (1984): The role of compositional convection in the formation of adcumulus rocks. Lithos 17: 139-146

Talwani, M.; Heirtzler, J.R. (1962): The mathematical expression for the magnetic anomaly over a two-dimensional body of polygonal cross-section. Tech. Rep. 6, Lamont-Doherty Geol. Observ., Columbia Univ., New York

Tanner, P.W.G. (1970): The Sgurr Beag Slide - a major break within the Moinian of the western Highlands of Scotland. Q. J. Geol. Soc. London 126: 435-463

Taylor, B.E. (1986): Magmatic volatiles: isotopic variation of C,H, and S. In: Valley, J.W.; Taylor, H.P. Jr; O'Neil, J.R. (eds.) Reviews in mineralogy, vol. 16. Stable isotopes in high temperature geological processes. Mineral. Soc. Am., pp 185-225

Taylor, B.E.; Eichelberger, J.C.; Westrich, H.R. (1983): Hydrogen isotopic evidence of rhyolithic magma degassing during shallow intrusion and eruption. Nature (London) 306: 541-545

Taylor, H.P. (1968): The oxygen isotope geochemistry of igneous rocks. Contrib. Mineral. Petrol. 19: 1-71

Taylor, H.P. (1974): Oxygen and hydrogen evidence for large scale circulation and interaction between groundwaters and igneous intrusions with particular reference to the San Juan volcanic field, Colorado. In: Hofmann, A.W.; Hofman, A.W; Giletti, B.; Yoden, H.S.; Yund, R. (eds.) Geochemical transport and kinetics. Carnegie Inst., Washington, pp 299-323

Taylor, H.P. (1977): Water rock interactions and the origin of H_2O in granitic batholiths. J. Geol. Soc. London 133: 509-558

Taylor, H.P.; Forester, R.W. (1971): Low-O^{18} igneous rocks from the intrusive complexes of Skye, Mull, and Ardnamurchan, western Scotland. J. Petrol. 12: 465-497

Taylor, H.P.; Sheppard, S.M.F. (1986): Igneous rocks I: processes of isotopic fractionation and isotope systematics. In: Valley, J.W.; Taylor, H.P. Jr.; O'Neil, J.R. (eds.) Reviews in mineralogy, vol. 16. Stable isotopes in high temperature geological processes. Mineral. Soc. Am., pp 227-271

Taylor, H.P. Jr.; Forester, R.W. (1979): An oxygen and hydrogen isotope study of the Skaergaard intrusion and its country rocks: a description of a 55 my. old fossil hydrothermal system. J. Petrol. 20: 355-419

Thirlwell, M.F. (1988a): Wenlock to mid-Devonian volcanism of the Caledonian-Appalachian orogen. In: Harris, A.L.; Fettes, D.J. (eds.) The Caledonian-Appalachian Orogen. Spec. Publ. Geol. Soc. London 38: 415-428

Thirlwell, M.F. (1988b): Geochronology of late Caledonian magmatism in northern Britain. J. Geol. Soc. London 145: 951-967

Thomas, P.R. (1979): New evidence for a Central Highland root zone. In: Harris, A.L.; Holland, C.H.; Leake, B.E. (eds.) The Caledonides of the British Isles - reviewed. Spec. Publ. Geol. Soc. London 8: 205-212

Thompson, A.B. (1976): Mineral reactions in pelitic rocks: I. Prediction of P-T-X (Fe-Mg) phase relations. II. Calculation of some P-T-X (Fe-Mg) phase relations. Am. J. Sci. 276: 401-454

Thompson, A.B. (1982): Dehydration melting of pelitic rocks and the generation of H_2O-undersaturated granitic liquids. Am. J. Sci. 282: 1567-1595

Thompson, A.B. (1983): Fluid-absent metamorphism. J. Geol. Soc. London 140: 533-547

Thompson, A.B.; Tracy, R.J. (1979): Model systems for anatexis of pelitic rocks: II. Facies series and reactions in the system CaO-$KAlO_2$-$NaAlO_2$-Al_2O_3-SiO_2-H_2O. Contrib. Mineral. Petrol. 70: 429-438

Thornton, C.P.; Tuttle, O.F. (1960): Chemistry of igneous rocks: I Differentiation Index. Am. J. Sci. 258: 664-684

Tilley, C.E. (1924): Contact metamorphism in the Comrie area of the Perthshire Highlands. Q. J. Geol. Soc. London, 80: 22-71

Tindle, A.G.; Pearce, J.A. (1981): Petrogenetic modelling of in situ fractional crystallization in the zoned Loch Doon pluton, Scotland. Contrib. Mineral. Petrol. 78: 196-207

Tracy, R.J. (1982): Compositional zoning and inclusions in metamorphic minerals. In: Ferry, J.M. (ed.). Characterization of metamorphism through mineral equilibria. Mineral. Soc. Am., Rev. Mineral. 10: 762-775

Treagus, J.E. (1974): A structural cross-section of the Moine and Dalradian rocks of the Kinlochleven area, Scotland. J. Geol. Soc. London 130: 525-544

Trommsdorff, V.; Skippen, G. (1986): Vapor loss ("boiling") as a mechanism for fluid evolution in metamorphic rocks. Contrib. Mineral. Petrol. 94: 317-322

Tullis, J. (1983): Deformation of feldspars. In: Ribbe, P.H. (ed.) Feldspar mineralogy. Mineral. Soc. Am., Rev. Mineral. 2: 297-324

Tullis, J.; Yund, R.A. (1979): Calculation of coherent solvi for alkali feldspar, iron-free clinopyroxene, nepheline-kalsilite, and hematite-ilmenite. Am. Mineral. 64: 1063-1074

Tullis, J.; Yund, R.A. (1982): Grain growth kinetics of quartz and calcite aggregates. J. Geol. 90: 301-318

Turi, B.; Taylor, H.P. (1971): An oxygen and hydrogen isotope study of a granodiorite pluton from the southern California batholith. Geochim. Cosmochim. Acta 35: 383-406

Turnock, A.C.; Lindsley, D.H.; Grover, J.E. (1973): Synthesis and unit cell parameters of Ca-Mg-Fe pyroxenes. Am. Mineral. 58: 50-59

Tuttle, O.F.; Bowen, N.L. (1958): Origin of granite in the light of experimental studies in the system $NaAlSi_3O_8$ - $KAlSi_3O_8$ - SiO_2 - H_2O. Geol. Soc. Am. Mem. 74: 153 pp

Tyler, I.M.; Ashworth, J.R. (1983): The metamorphic environment of the Foyers Granitic Complex. Scott. J. Geol. 19: 271-285

Valley, J.W. (1986): Stable isotope geochemistry of metamorphic rocks. In: Valley, J.W.; Taylor, H.P.Jr.; O'Neill, J.F. (eds.) Stable isotopes in high temperature geological processes. Mineral. Soc. Am., Rev. Mineral. 16: 446-489

Valley, J.W.; O'Neil, J.R. (1982): Oxygen isotope evidence for shallow emplacement of Adirondack anorthosite. Nature (London) 300: 497-500

Valley, J.W.; O'Neil, J.R. (1984): Fluid heterogeneity during granulite facies metamorphism in the Adirondacks: stable isotope evidence. Contrib. Mineral. Petrol. 85: 158-173

Valley, J.W.; Taylor, H.P.; O'Neil, J.R. (eds.) (1986): Reviews in mineralogy, vol. 16. Stable isotopes in high temperature geological processes. Mineral. Soc. Am., 569 pp

Vance, J.A. (1969): On synneusis. Contrib. Mineral. Petrol. 24: 7-29

Van der Molen, I. (1981): The shift of the α - β transition temperature of quartz associated with the thermal expansion of granite at high pressure. Tectonophysics 73: 323-342

Van der Molen, I.; Paterson, M.S. (1979): Experimental deformation of partially-melted granite. Contrib. Mineral. Petrol. 70: 299-318

Velde, B. (1965): Phengitic micas: synthesis, stability and natural occurrence. Am. J. Sci. 253: 65-103

Venkatesh, V. (1954): Twinning in cordierite. Am. Mineral. 39: 636-646

Vidal, P.; Bernard-Griffiths, J.; Cocherie, A.; Le Fort, P.; Peucat, J.J.; Sheppard, S.M.F. (1984): Geochemical comparison between Himalayan and Hercynian leucogranites. Phys. Earth Planet. Int. 35: 179-190

Vidale, R. (1969): Metasomatism in a chemical gradient and formation of calc-silicate bands. Am. J. Sci. 267: 857-87

Vielzeuf, D. (1983): The spinel and quartz associations in high grade xenoliths from Tallante (S. E. Spain) and their potential use in geothermometry and geobarometry. Contrib. Mineral. Petrol. 82: 301-311

Voll, G. (1960): New work on petrofabrics. Liverpool and Manchester Geol. J., Cent. Vol. 2: 503-567

Voll, G. (1964): Deckenbau und Fazies im Schottischen Dalradian. Geol. Rundsch. 21: 590-612

Voll, G. (1969): Klastische Mineralien aus den Sedimentserien der Schottischen Highlands und ihr Schicksal bei aufsteigender Regional- und Kontaktmetamorphose. Habil., Tech. Univ. Berlin

Voll, G. (1976): Recrystallization of quartz, biotite and feldspars from Erstfeld to the Leventina Nappe, Swiss Alps, and its geological significance. Schweiz. Mineral. Petrogr. Mitt. 56: 641-647

Walker, F. (1924): A marginal modification of the Ballachulish granite. Geol. Mag. 61: 550-554

Walther, J.V. (1983): Description and interpretation of metasomatic phase relations at high pressures and temperatures: 2. Metasomatic reactions between quartz and dolomite at Campolungo, Switzerland. Am. J. Sci. 283-A: 459-485

Walther, J.V.; Helgeson, H.C. (1980): Description and interpretation of metasomatic phase relations at high pressures and temperatures: 1. Equilibrium activities of ionic species in nonideal mixtures of CO_2 and H_2O. Am. J. Sci. 280: 575-606

Walther, J.V.; Orville, P.M. (1982): Volatile production and transport in regional metamorphism. Contrib. Mineral. Petrol. 79: 252-257

Walther, J.V.; Wood, B.J. (1984): Rate and mechanism in prograde metamorphism. Contrib. Mineral. Petrol. 88: 246-259

Walther, J.V.; Wood, B.J. (1986): Fluid-rock interactions during metamorphism. Springer, Berlin, Heidelberg, New York

Wang, H.F.; Simmons, G. (1978): Microcracks in crystalline rocks from 5.3 km depth in the Michigan basin. J. Geophys. Res. 83: 5849-5856

Watson, J. (1984): The ending of the Caledonian orogeny in Scotland. J. Geol. Soc. London 141: 193-214

Watson, J.; Dunning, F.W. (1979): Basement-cover relations in the British Caledonides. In: Harris A.L.; Holland, C.M.; Leake, B.E. (eds.) The Caledonides of the British Isles - reviewed. Spec. Publ. Geol. Soc. London 8: 67-92

Weiss, S. (1986): Petrogenese des Intrusivkomplexes von Ballachulish, Westschottland: Kristallisationsverlauf in einem zonierten Kaledonischen Pluton. Ph.D. Thesis, Univ. München

Weiss, S.; Troll, G. (1989): The Ballachulish Igneous Complex, Scotland: petrography, mineral chemistry and order of crystallization in the monzodiorite - quartz diorite suite and in the granite. J. Petrol. 30: 1069-1115

Wenner, D.B., Taylor, H.P. (1973): Oxygen and hydrogen isotope studies of the serpentinization of ultramafic rocks in oceanic environments and continental ophiolite complexes. Am. J. Sci. 273: 207-239

White, A.J.R.; Chappell, B.W. (1983): Granitoid types and their distribution in the Lachlan Fold Belt, southeastern Australia. Mem. Geol. Soc. Am. 159: 21-34

Wickham, S.M. (1987): The segregation and emplacement of granitic magmas. J. Geol. Soc. London 144: 281-297

Wickham, S.M.; Taylor, H.P.Jr. (1985): Stable isotopic evidence for large-scale seawater infiltration in a regional metamorphic terrane: the Trois Seigneurs Massif, Pyrenees, France. Contrib. Mineral. Petrol. 91: 122-137

Wickham, S.M.; Taylor, H.P. (1987): Stable isotope constraints on the origin and depth of penetration of hydrothermal fluids associated with Hercynian regional metamorphism and crustal anatexis in the Pyrenees. Contrib. Mineral. Petrol. 95: 255-268

Wones, D.R.; Eugster, H.P. (1965): Stability of biotite: experiment, theory and application. Am. Mineral. 50: 1228-1272

Wones, D.R.; Gilbert, M.C. (1982): Amphiboles in the igneous environment. In: Veblen, D.R.; Ribbe, P.H. (eds.) Amphiboles: petrology and phase relations. Mineral. Soc. Am., Rev. Mineral. 9B: 355-389

Wood, B.J.; Walther, J.V. (1986): Fluid flow during metamorphism and its implications for fluid-rock ratios. In: Walther, J.V.; Wood, B.J. (eds.) Fluid-rock interactions during metamorphism. Springer, Berlin, Heidelberg, New York, pp 89-108

Woodhead, J.D.; Harmon, R.S.; Fraser, D.G. (1987): O, S, Sr, and Pb isotope variations in volcanic rocks from the northern Mariana Islands: implications for crustal recycling in intra-oceanic arcs. Earth Planet. Sci. Lett. 83: 39-52

Wright, A.E.; Bowes, D.R. (1979): Geochemistry of the Appinite suite. In: Harris, A.L.; Holland, C.M.; Leake, B.E. (eds.) Caledonides of the British Isles - reviewed. Spec. Publ. Geol. Soc. London 8: 699-704

Wright, T.L. (1967): The microcline-orthoclase transformation in the contact aureole of the Eldora Stock, Colorado. Am. Mineral. 52: 117-136

Xu, K.; Sun, N.; Wang, D.; Hu, S.; Liu, Y.; Ji, S. (1984): Petrogenesis of the granitoids and their metallogenic relations in south China. In: Xu, K.; Tu, G. (eds.) Geology of granites and their metallogenic relations. Science Press, Beijing

Yardley, B.W.D. (1977): The nature and significance of the mechanism of sillimanite growth in the Connemara Schists, Ireland. Contrib. Mineral. Petrol. 65: 53-58

Yoder, H.S. (1976): Generation of basaltic magma. US Nat. Acad., Washington, 265 pp

Yoder, H.S.; Tilley, C.E. (1962): Origin of basaltic magma: an experimental study of natural and synthetic rock systems. J. Petrol. 3: 342-532

Yoder, H.S.; Stewart, D.B.; Smith, J.R. (1957): Ternary feldspars. Carnegie Inst. Washington Yearb. 56: 206-214

Yund, R.A. (1984): Alkali feldspar exsolution: kinetics and dependence of alkali interdiffusion. In: Brown, W.L. (ed.) Feldspars and feldspathoids. Reidel, Dordrecht, pp 281-316

Yund, R.A., Anderson, T.F. (1974): Oxygen isotope exchange between potassium feldspar and KCl solutions. In: Hofmann, A.W.; Giletti, B.; Yoder, H.S.; Yund, R.A. (eds.) Geochemical transport and kinetics. Carnegie Inst. Washington 634: 99-105

Yund, R.A.; Ackermand, D.; Seifert, F. (1980): Microstructures in the alkali feldspars from the granulite complex of Finnish Lapland. N. Jahrb. Mineral., Monatsh., 109-117

Zeck, H.P. (1972): Transformation trillings in cordierite. J. Petrol. 13: 367-380

Zen, E-An (1963): Components, phases and criteria of chemical equilibrium in rocks. Am. J. Sci. 261: 929-942

Zhou, J. (1985): The timing of calc-alkaline magmatism in parts of Alpine-Himalayan collision zone and its relevance to the interpretation of Caledonian magmatism. J. Geol. Soc. London 142: 309-317

Zschau, J. (1979): Auflastgezeiten. Habil., Univ. Kiel

Subject Index

Geological Map of the Ballachulish Igneo

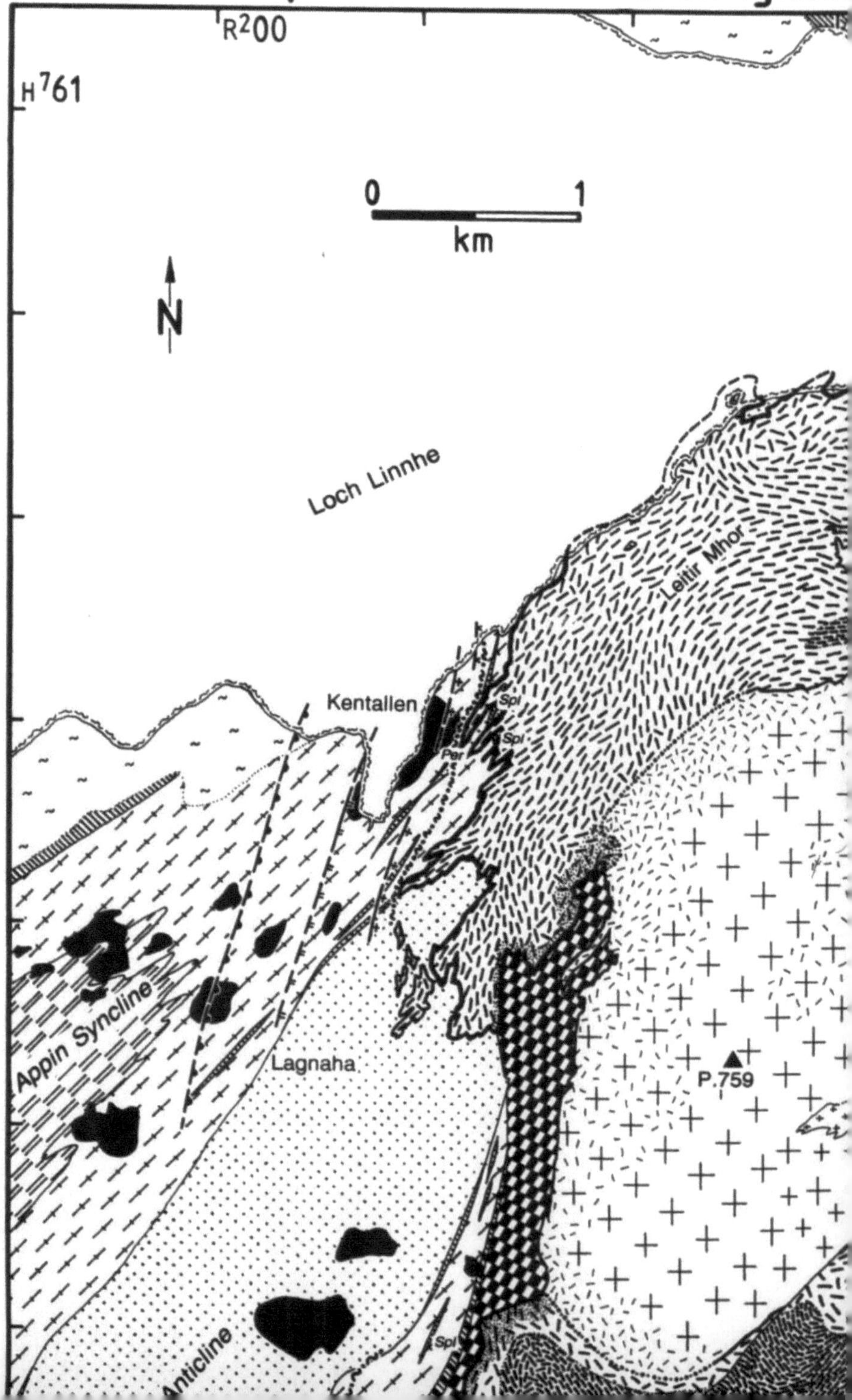

lex and its Thermal Aureole Map 2

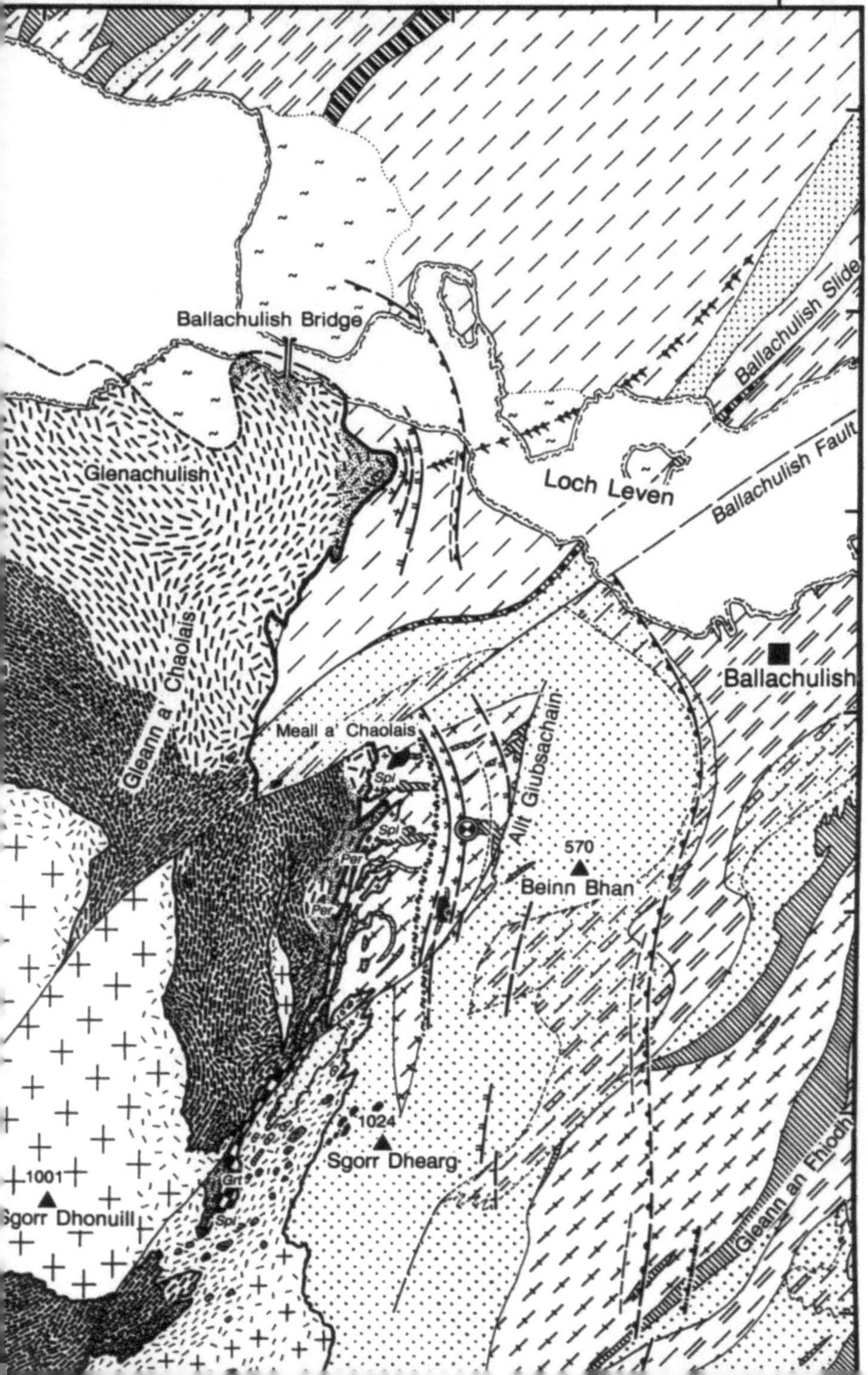

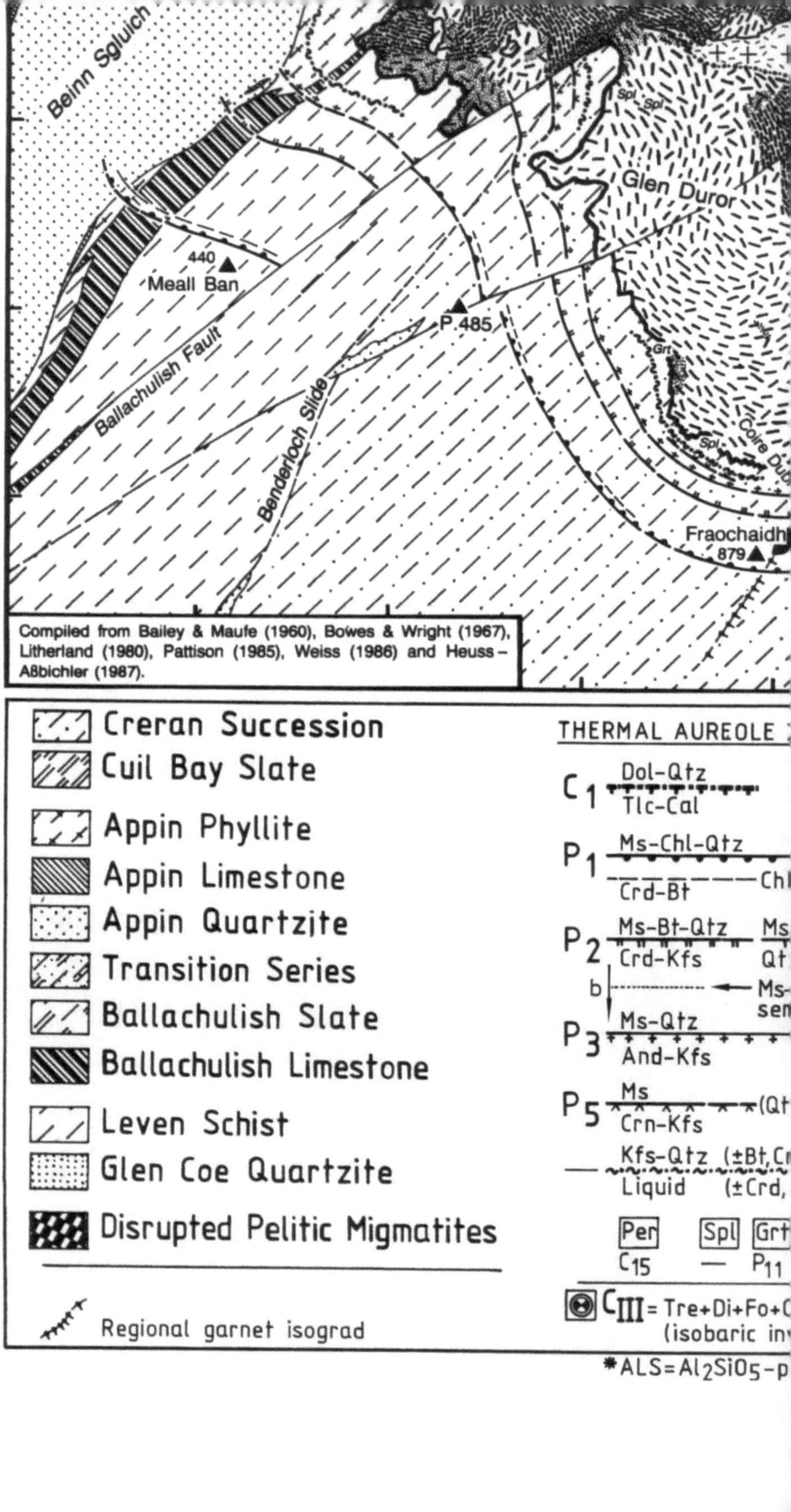

Beinn Sgluich
Glen Duror
440
Meall Ban
P.485
Ballachulish Fault
Benderloch Slide
Coire Dub
Fraochaidh
879
Spl
Grt
Compiled from Bailey & Maufe (1960), Bowes & Wright (1967), Litherland (1980), Pattison (1985), Weiss (1986) and Heuss-Aßbichler (1987).
Creran Succession
Cuil Bay Slate
Appin Phyllite
Appin Limestone
Appin Quartzite
Transition Series
Ballachulish Slate
Ballachulish Limestone
Leven Schist
Glen Coe Quartzite
Disrupted Pelitic Migmatites
Regional garnet isograd
THERMAL AUREOLE
C1 Dol-Qtz / Tlc-Cal
P1 Ms-Chl-Qtz / Crd-Bt
P2 Ms-Bt-Qtz / Crd-Kfs
b
P3 Ms-Qtz / And-Kfs
P5 Ms / Crn-Kfs
Kfs-Qtz / Liquid
Per C15
Spl
Grt P11
CIII = Tre+Di+Fo+C
(isobaric in
*ALS=Al2SiO5-p

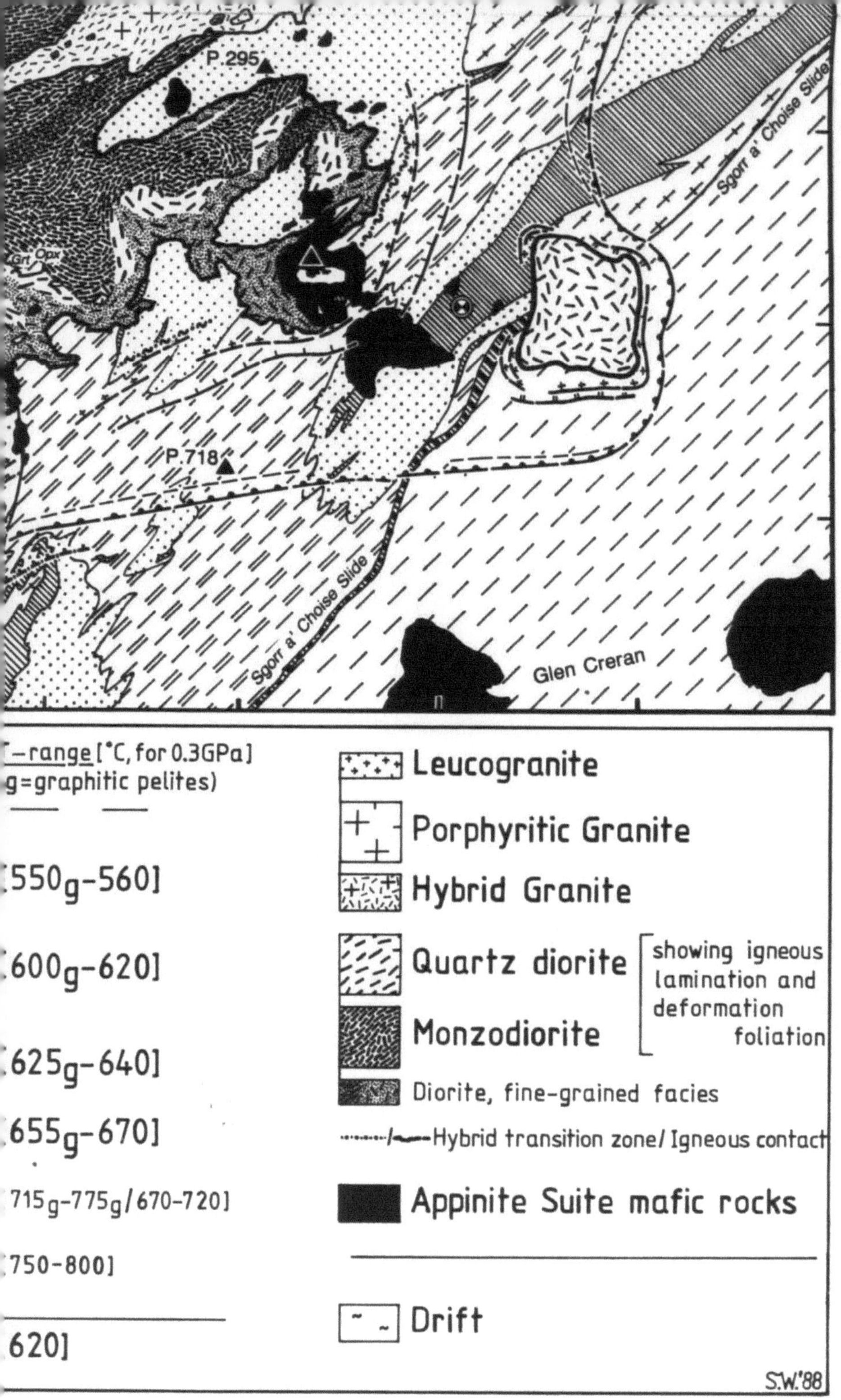
P.295
Grt Opx
P.718
Sgorr a' Choise Slide
Sgorr a' Choise Slide
Glen Creran
T–range [°C, for 0.3GPa]
g=graphitic pelites)
[550g–560]
[600g–620]
[625g–640]
[655g–670]
[715g–775g/670–720]
[750–800]
620]
Leucogranite
Porphyritic Granite
Hybrid Granite
Quartz diorite
Monzodiorite
showing igneous lamination and deformation foliation
Diorite, fine-grained facies
Hybrid transition zone/ Igneous contact
Appinite Suite mafic rocks
Drift
S.W.'88

Geological Map of the

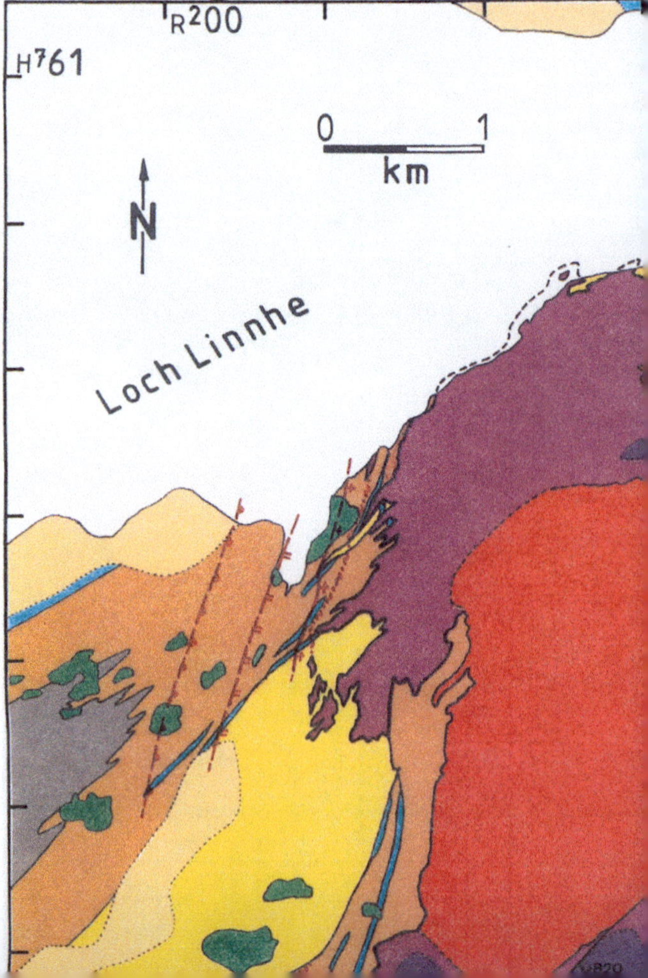

chulish Area Map 1

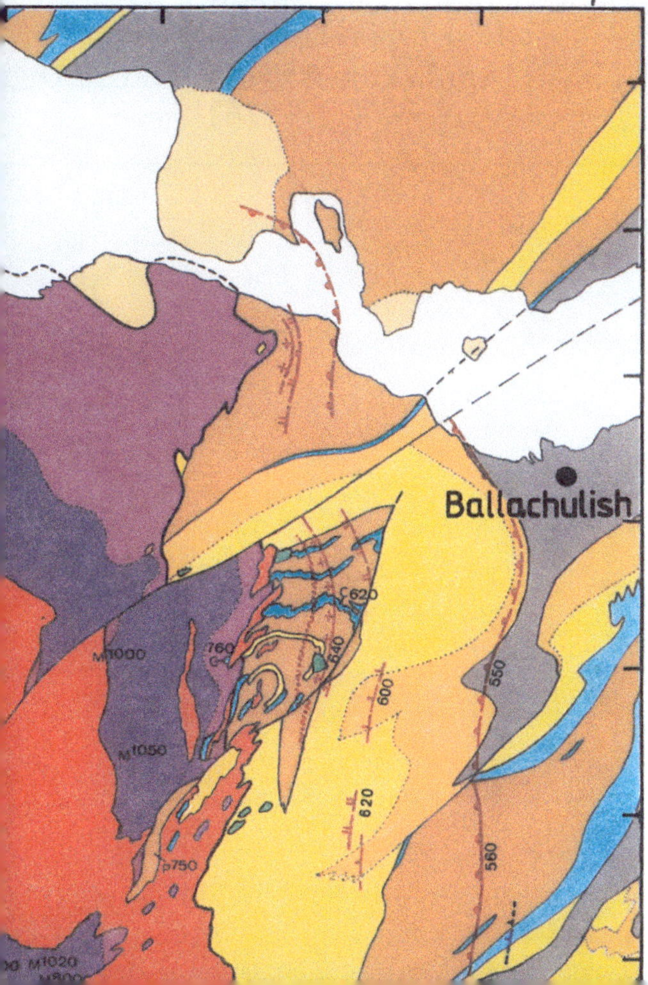

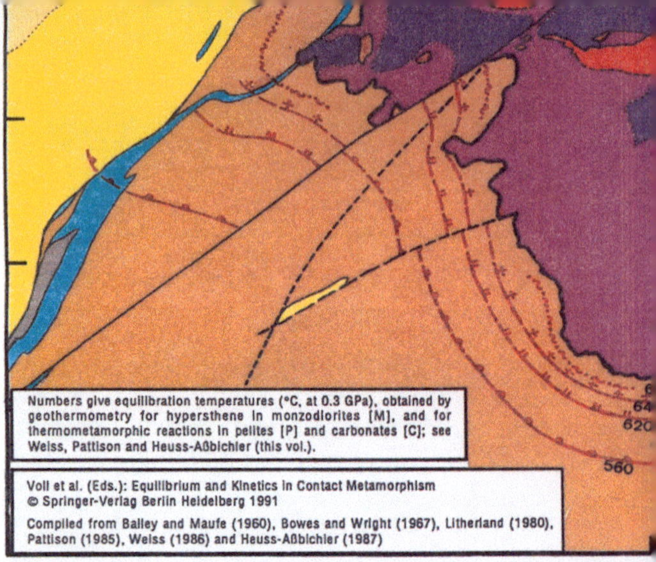

- Cuil Bay Slate; Ballachulish [carbonaceous / graphitic slate]
- Appin Phyllite; Leven & Creran [semipelitic - pelitic phyllite]
- Appin Limestone; Ballachulish L [dolomite, limestone, calcareous schist
- Transition Series [interbedded quartzite / slate]
- Appin Quartzite; Glen Coe Qua [feldspathic quartzite]

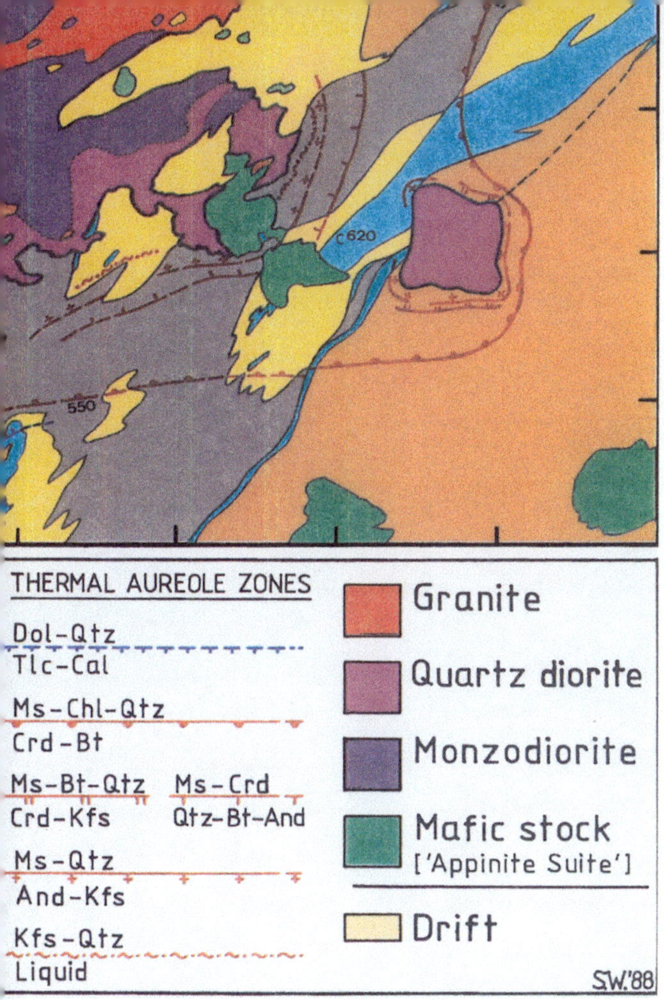

c620
550
THERMAL AUREOLE ZONES
Dol-Qtz
Tlc-Cal
Ms-Chl-Qtz
Crd-Bt
Ms-Bt-Qtz
Crd-Kfs
Ms-Crd
Qtz-Bt-And
Ms-Qtz
And-Kfs
Kfs-Qtz
Liquid
Granite
Quartz diorite
Monzodiorite
Mafic stock
['Appinite Suite']
Drift
S.W.'88

MIX
Papier aus verantwortungsvollen Quellen
Paper from responsible sources
FSC® C105338

If you have any concerns about our products,
you can contact us on
ProductSafety@springernature.com

In case Publisher is established outside the EU,
the EU authorized representative is:
Springer Nature Customer Service Center GmbH
Europaplatz 3, 69115 Heidelberg, Germany

Printed by Libri Plureos GmbH
in Hamburg, Germany